Applied RNAi

From Fundamental Research to Therapeutic Applications

Edited by

Patrick Arbuthnot

Antiviral Gene Therapy Research Unit
Department of Molecular Medicine and Haematology
University of the Witwatersrand
Johannesburg
South Africa

and

Marc S. Weinberg

Department of Molecular and Experimental Medicine
The Scripps Research Institute
La Jolla, CA
USA; and
Antiviral Gene Therapy Research Unit
School of Pathology
University of the Witwatersrand
Johannesburg
South Africa

Caister Academic Press

Caister Academic Press
Norfolk, UK

www.caister.com

British Library Cataloguing-in-Publication Data
A catalogue record for this book is available from the British Library

ISBN: 978-1-908230-43-0 (hardback)
ISBN: 978-1-908230-67-6 (ebook)

Cover design adapted from Figure 2.5

Contents

	Contributors	v
	Preface	ix
1	Overview of Biogenesis and Applications of MicroRNA Patrick Arbuthnot and Marc S. Weinberg	1
2	Non-canonical MicroRNA Biogenesis and Function Thomas C. Roberts and Matthew J.A. Wood	19
3	Non-coding RNAs and the Epigenetic Control of Gene Expression Kevin V. Morris	43
4	From Mice to Men: Towards the Clinical Translation of MicroRNA Technologies for Somatic Cell Reprogramming Elena Senís and Dirk Grimm	51
5	Systems Biology Tools to Understand the Role of Host MicroRNAs in Infection: a Closer Look at HIV Jerolen Naidoo, Robyn Brackin, Rethabile Khutlang, Anca Savulescu and Musa M. Mhlanga	79
6	Synthetic MicroRNA Blocking Agents Peter Järver, Adrian G. Torres and Michael J. Gait	105
7	Exploiting MicroRNAs to Regulate Transgene Expression Virginie Pichard, Dejana Ivacik and Nicolas Ferry	127
8	Use of Artificial MicroRNAs for Gene Silencing Mohube B. Mowa and Abdullah Ely	139
9	Harnessing RNAi for the Treatment of Viral Infections Lorea Blázquez and Puri Fortes	151
10	Roles of MicroRNAs in Cancers Associated with Human Tumour Viruses Xianzhi Lin, Deguang Liang and Ke Lan	181

11 MicroRNAs as Cancer Biomarkers 201
David Otaegui and Charles H. Lawrie

12 MicroRNA Deregulation in Trinucleotide Repeat Expansion Disorders 227
Edyta Koscianska, Emilia Kozlowska, Edyta Jaworska and Wlodzimierz J. Krzyzosiak

Index 247

Contributors

Patrick Arbuthnot
Antiviral Gene Therapy Research Unit
Department of Molecular Medicine and Haematology
University of the Witwatersrand
Johannesburg
South Africa

patrick.arbuthnot@wits.ac.za

Lorea Blázquez
Department of Hepatology and Gene Therapy
Center for Applied Medical Research (CIMA)
University of Navarra
Pamplona
Spain

lblazquez@unav.es

Robyn Brackin
Gene Expression and Biophysics Group
CSIR Emerging Health Technologies Platform
Pretoria
South Africa

robyn@mhlangalab.org

Abdullah Ely
Antiviral Gene Therapy Research Unit
Department of Molecular Medicine and Haematology
University of the Witwatersrand
Johannesburg
South Africa

abdullah.ely@wits.ac.za

Nicolas Ferry
Laboratory of Hepatic Biotherapies
INSERM UMR 948
University Hospital Center
Nantes
France

nicolas.ferry@ansm.sante.fr

Puri Fortes
Department of Hepatology and Gene Therapy
Center for Applied Medical Research (CIMA)
University of Navarra
Pamplona
Spain

pfortes@unav.es

Michael J. Gait
Medical Research Council
Laboratory of Molecular Biology
Cambridge
UK

mgait@mrc-lmb.cam.ac.uk

Dirk Grimm
Department of Infectious Diseases, Virology
Cluster of Excellence CellNetworks
Heidelberg University Hospital
Heidelberg
Germany

dirk.grimm@bioquant.uni-heidelberg.de

Dejana Ivacik
Antiviral Gene Therapy Research Unit
Department of Molecular Medicine and Haematology
University of the Witwatersrand
Johannesburg
South Africa

dejanaivacik@yahoo.co.uk

Peter Järver
Center for Infectious Medicine
Karolinska Institutet
Karolinska University Hospital
Stockholm
Sweden

peter.jarver@ki.se

Edyta Jaworska
Department of Molecular Biomedicine
Institute of Bioorganic Chemistry
Polish Academy of Sciences
Poznan
Poland

edyjaw@ibch.poznan.pl

Rethabile Khutlang
Gene Expression and Biophysics Group
CSIR Emerging Health Technologies Platform
Pretoria
South Africa

rethabile@mhlangalab.org

Edyta Koscianska
Department of Molecular Biomedicine
Institute of Bioorganic Chemistry
Polish Academy of Sciences
Poznan
Poland

edytak@ibch.poznan.pl

Emilia Kozlowska
Department of Molecular Biomedicine
Institute of Bioorganic Chemistry
Polish Academy of Sciences
Poznan
Poland

emiliak@ibch.poznan.pl

Wlodzimierz J. Krzyzosiak
Department of Molecular Biomedicine
Institute of Bioorganic Chemistry
Polish Academy of Sciences
Poznan
Poland

wlodkrzy@ibch.poznan.pl

Ke Lan
Unit of Tumor Virology
Institut Pasteur of Shanghai
Chinese Academy of Sciences
Shanghai
China

lanke@ips.ac.cn

Charles H. Lawrie
Biodonostia Research Institute
San Sebastián
Spain; and
Nuffield Department of Clinical Laboratory Sciences
University of Oxford
Oxford
UK

charles.lawrie@biodonostia.org
charles.lawrie@ndcls.ox.ac.uk

Xianzhi Lin
Unit of Tumor Virology
Institut Pasteur of Shanghai
Chinese Academy of Sciences
Shanghai
China

linxianzhi@sibs.ac.cn

Deguang Liang
Unit of Tumor Virology
Institut Pasteur of Shanghai
Chinese Academy of Sciences
Shanghai
China

dgliang01@sibs.ac.cn

Musa M. Mhlanga
Gene Expression and Biophysics Group
CSIR Emerging Health Technologies Platform
Pretoria
South Africa; and
Gene Expression and Biophysics Unit
Institute for Molecular Medicine
Lisbon
Portugal

musa@mhlangalab.org

Kevin V. Morris
Biotechnology and Biomedical Sciences
The University of New South Wales
Sydney
Australia; and
Department of Molecular and Experimental Medicine
The Scripps Research Institute
La Jolla, CA
USA

kmorris@scripps.edu

Mohube B. Mowa
Antiviral Gene Therapy Research Unit
Department of Molecular Medicine and Haematology
University of the Witwatersrand
Johannesburg
South Africa

betty.mowa@wits.ac.za

Jerolen Naidoo
Gene Expression and Biophysics Group
CSIR Emerging Health Technologies Platform
Pretoria
South Africa

jerolen@mhlangalab.org

David Otaegui
Biodonostia Research Institute
San Sebastián
Spain

david.otaegui@biodonostia.org

Virginie Pichard
Atlantic Gene Therapies
INSERM UMR 1089
University Hospital Center
Nantes
France

virginie.pichard@univ-nantes.fr

Thomas C. Roberts
Department of Molecular and Experimental Medicine
The Scripps Research Institute
La Jolla, CA
USA

robertst@scripps.edu

Anca Savulescu
Gene Expression and Biophysics Group
CSIR Emerging Health Technologies Platform
Pretoria
South Africa

anca@mhlangalab.org

Elena Senís
Department of Infectious Diseases, Virology
Cluster of Excellence CellNetworks
Heidelberg University Hospital
Heidelberg
Germany

elena.senis@bioquant.uni-heidelberg.de

Adrian G. Torres
Institute for Research in Biomedicine (IRB)
Barcelona
Spain

adriangabriel.torres@irbbarcelona.org

Marc S. Weinberg
Department of Molecular and Experimental Medicine
The Scripps Research Institute
La Jolla, CA
USA; and
Antiviral Gene Therapy Research Unit
School of Pathology
University of the Witwatersrand
Johannesburg
South Africa

marc.weinberg@wits.ac.za

Matthew J.A. Wood
Department of Physiology Anatomy and Genetics
University of Oxford
Oxford
UK

matthew.wood@dpag.ox.ac.uk

Preface

Since discovery of the RNA interference (RNAi) pathway in 1998, research on the topic has advanced at an impressive pace. Small RNAs, and in particular microRNAs (miRNAs), play a fundamental role in gene regulation through activation of RNAi. Detailed insights into the mechanisms of RNAi have led to an improved understanding of gene regulation in normal and disease states, and also enabled exploitation of RNAi for a variety of applications. This book reviews some particularly interesting recent advances that have taken place in harnessing RNAi for medical research. The scope of the book is not intended to be comprehensive; rather it is a compilation to highlight exciting aspects of current RNAi research. The subjects cover a variety of themes. In the area of RNAi-based therapy, harnessing RNAi for treating viral infections, optimizing design of exogenous RNAi activators, blocking miRNA function with synthetic agents, somatic cell reprogramming and use of miRNA target sites for control of transgene expression are reviewed. The role of miRNAs in trinucleotide repeat disorders, HIV pathogenesis, epigenetic gene silencing, virus-related cancers as well as non-canonical miRNA biogenesis deal with more fundamental biological aspects of applied RNAi investigation. In addition to highlighting exciting developments in this ever-expanding field the articles also expose the challenges that are currently faced. We are grateful to the authors and are sure that their perceptive contributions will provide further stimulus to research on the exciting topic of RNAi application. We are also thankful to Caister Academic Press for providing us with the opportunity to edit a book on this important topic.

Patrick Arbuthnot and
Marc S. Weinburg

Current books of interest

Microarrays: Current Technology, Innovations and Applications	2014
Metagenomics of the Microbial Nitrogen Cycle: Theory, Methods and Applications	2014
Pathogenic *Neisseria*: Genomics, Molecular Biology and Disease Intervention	2014
Proteomics: Targeted Technology, Innovations and Applications	2014
Biofuels: From Microbes to Molecules	2014
Human Pathogenic Fungi: Molecular Biology and Pathogenic Mechanisms	2014
Phage Therapy: Current Research and Applications	2014
Halophiles: Genetics and Genomes	2014
Molecular Diagnostics: Current Research and Applications	2014
Bioinformatics and Data Analysis in Microbiology	2014
The Cell Biology of Cyanobacteria	2014
Pathogenic *Escherichia coli*: Molecular and Cellular Microbiology	2014
Campylobacter Ecology and Evolution	2014
Burkholderia: From Genomes to Function	2014
Myxobacteria: Genomics, Cellular and Molecular Biology	2014
Next-generation Sequencing: Current Technologies and Applications	2014
Omics in Soil Science	2014
Applications of Molecular Microbiological Methods	2014
Mollicutes: Molecular Biology and Pathogenesis	2014
Genome Analysis: Current Procedures and Applications	2014
Bacterial Membranes: Structural and Molecular Biology	2014
Bacterial Toxins: Genetics, Cellular Biology and Practical Applications	2013
Cold-Adapted Microorganisms	2013
Fusarium: Genomics, Molecular and Cellular Biology	2013
Prions: Current Progress in Advanced Research	2013
RNA Editing: Current Research and Future Trends	2013
Real-Time PCR: Advanced Technologies and Applications	2013
Microbial Efflux Pumps: Current Research	2013
Cytomegaloviruses: From Molecular Pathogenesis to Intervention	2013
Oral Microbial Ecology: Current Research and New Perspectives	2013
Bionanotechnology: Biological Self-assembly and its Applications	2013
Real-Time PCR in Food Science: Current Technology and Applications	2013
Bacterial Gene Regulation and Transcriptional Networks	2013
Bioremediation of Mercury: Current Research and Industrial Applications	2013

Full details at www.caister.com

Overview of Biogenesis and Applications of MicroRNA

1

Patrick Arbuthnot and Marc S. Weinberg

Abstract

Intensive research on the RNA interference (RNAi) pathway has provided detailed understanding of the intricacies of gene silencing by small RNAs. Insights into the regulation of microRNA (miRNA) biogenesis have revealed a complex interplay of controlling mechanisms. Efficacies of miRNAs are potentially influenced by modulation at all steps of their biogenesis: transcription of primary miRNAs, nuclear processing and export, cytoplasmic Dicer cleavage of precursor miRNAs (pre-miRNAs) and loading of mature miRNAs onto the RNA-induced silencing complex (RISC). In addition competing endogenous sequences, especially circular RNA sponges, may function to sequester miRNAs and attenuate their effects on mRNA targets. Disruption of miRNA function is implicated in disease processes such as cancer and modulating natural miRNA function has potential for treating diseases. Remarkably, miRNAs may be secreted from cells and detection of particular miRNA profiles in serum has diagnostic utility. Synthetic and expressed RNAi activators are being applied to therapeutic inhibition of pathology-causing genes. Also incorporating miRNA targets into untranslated sequences has provided a useful mechanism of regulating therapeutic transgene expression. RNAi research has matured and remains an exciting field of investigation. Improved understanding has now provided the means for powerful application.

Introduction

Since publication of the landmark article by Fire, Melo and colleagues (Fire *et al.*, 1998), which described RNA interference (RNAi) in *Chaenorhabditis elegans*, knowledge on the topic has expanded impressively. RNAi is mainly a gene silencing mechanism which includes effects of microRNAs (miRNA) and other small RNAs. The different types of small non protein-coding RNAs exert complex regulatory effects by engaging with Argonaute (Ago) proteins and directing sequence-dependent post transcriptional inhibition of gene expression. Currently, almost all known cellular processes, which include cell division, differentiation and development of tissue-specific cellular phenotypes, are subject to control by endogenous short RNAs that activate RNAi (reviewed in Bushati and Cohen, 2007). Also, miRNA dysfunction is implicated in the pathogenesis of many diseases such as cancer (reviewed in Garzon *et al.*, 2009).

The fundamental mechanisms that are responsible for formation of miRNAs are pertinent to a full appreciation of the intricacies of both natural and applied gene silencing. In this chapter, we review the essential features of RNAi and provide a general overview on how the pathway may be harnessed for various uses. This is intended to serve as a background to the interesting recent developments that are discussed in subsequent sections of this book.

Mammalian microRNA biogenesis

The initiating event of canonical miRNA biogenesis is Pol II-mediated transcription of 5′ capped and polyadenylated RNA that contains characteristic primary miRNA (pri-miR) hairpin motifs (Fig. 1.1) (Cai *et al.*, 2004; Lee *et al.*, 2004). Multiple individual hairpin structures, each representing precursors of mature miRNAs, may be present on one polycistronic transcript. These pri-miRNA sequences may be located within various positions of protein-coding mRNAs, non-protein coding intergenic transcripts, mirtrons or primary Pol III tRNA transcripts (reviewed in Miyoshi *et al.*, 2010; Rother and Meister, 2011). Apart from miRNAs, other cellular RNAs, such as piwi interacting RNAs (piRNAs) and endogenous short interfering RNAs (endo-siRNAs), also play a role in regulating cellular gene function.

Pri-miRNAs are subjected to processing by the nuclear microprocessor. The core components of this complex comprise the DiGeorge critical region 8 dsRNA-binding protein (DGCR8) and Drosha RNase III (Denli *et al.*, 2004; Gregory *et al.*, 2004). Accessory proteins, such as the DEAD box helicases DDX5 and DDX17 play an important role in the processing of subsets of miRNAs (Fukuda *et al.*, 2007). DGCR8 binds to the base of the pri-miRNA hairpin structure and directs Drosha to cleave the pri-miRNA sequence within its duplex at a position that is 11 bp from the stem base (Han *et al.*, 2006). This processing results in formation of precursor miRNAs (pre-miRNAs) that comprise 70–100 nts and have characteristic hairpin structures with 2 nt overhangs at their 3′ ends. Pre-miRNAs are exported from the nucleus via the exportin-5 karyopherin (Lund *et al.*, 2004; Yi *et al.*, 2003). This transmembrane protein achieves specificity through recognition of the 2 nt 3′ overhangs and non-sequence dependent detection of the double-stranded stem of pre-miRNAs (Okada *et al.*, 2009).

Further miRNA maturation occurs in the cytoplasm where Dicer, another RNase III, cleaves pre-miRNA (Chendrimada *et al.*, 2005; Zhang *et al.*, 2004). This function is carried out in conjunction with a dsRNA binding protein, which is HIV TAR RNA binding protein (TRBP) in humans. Dicer contains a PAZ (Piwi–Argonaute–Zwille) domain that is responsible for anchoring the 3′ end of pre-miR. Cleavage by the two RNase domains then typically occurs 22 bp from the 3′ end of the pre-miRNA stem base. Evidence also indicates that Dicer is capable of binding to the 5′ end of pre-miR, which may be required to enable discrimination between pre-miRNAs and other RNAs (Park *et al.*, 2011). The product of Dicer cleavage of pre-miRNA is a 22 nt duplex structure, which has 2 nt overhangs at both 3′ ends. In the final step of miRNA biogenesis, one of the strands is selected as a guide for gene silencing through association with a member of the Argonaute (Ago) family of proteins [reviewed in Joshua-Tor and Hannon (2011); also see below]. The other miRNA star (miRNA*) strand is usually degraded but may exert a silencing effect. The helicase domain of Dicer is thought to be important for selection of the guide strand. The strand within the duplex that has lower thermodynamic stability at its 5′ end is preferred for incorporation into the RNA-induced silencing complex (RISC), and properties of the helicase domain of Dicer are responsible for this bias (Noland *et al.*, 2011). Loading of RISC with the selected strand is accomplished through the concerted action of TRBP, Ago and Dicer.

Binding of the mature miRNA guide to its target is usually mediated by complementary base pairing at the 3′ untranslated region (3′ UTR) of the mRNA cognate (Fig. 1.1) (Friedman *et al.*, 2009). In mammals, it is very unusual for miRNAs to have a perfect match to the target. Rather it is partial base pairing that effects silencing by destabilization and translational suppression of the target mRNA. The seed region, comprising a minimum of six nucleotides spanning positions 2 to 7 from the 5′ end of the guide, is particularly important for silencing (Grimson *et al.*, 2007). Longer seed match sites of 7 and 8 complementary bases at the 5′ end of mature miRNA have also been described. A detailed study that entailed use of high-throughput sequencing to profile ribosome-bound RNA, together with mRNA sequencing and proteomic analysis revealed that target mRNA degradation is primarily responsible for inhibition of target gene expression (Guo *et al.*, 2010). Approximately 85% of the silencing effect was mediated by mRNA degradation and only

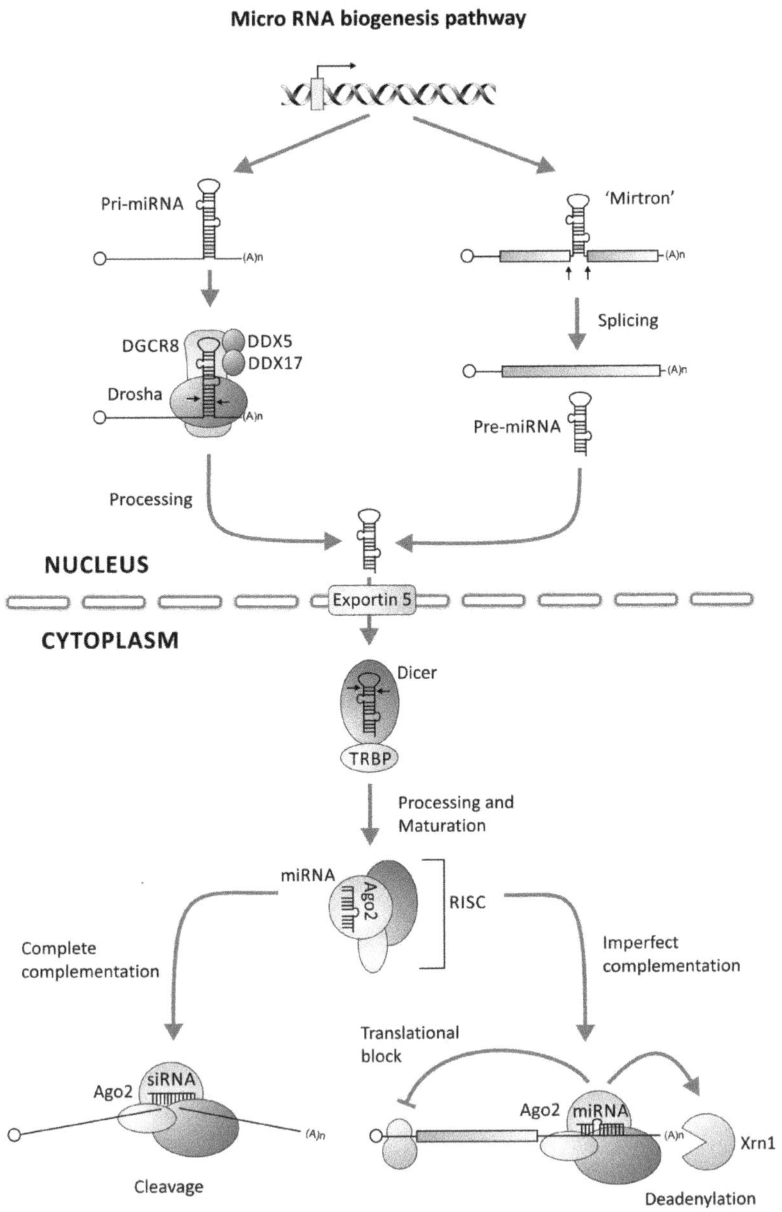

Figure 1.1 Mammalian miRNA biogenesis pathway. MicroRNAs are encoded in pri-miRNAs, which are ~100 nt inverted repeat structures usually found embedded once or multiple times within coding or non-coding RNA Pol II-derived transcripts. Pri-miRNAs are first processed in the nucleus where their hairpin-like structures are recognized and cleaved by the RNase III enzyme Drosha together with DiGeorge critical region 8 protein (DGCR8) and aided by DEAD box helicases DDX5 and DDX17 to produce shorter hairpin duplexes of 70–80 nt and known as pre-miRNAs. For a small minority of miRNAs, short intronic sequences, referred to 'mirtrons', can be directly processed by the spliceosome into pre-miRNA-like hairpins without requiring Drosha cleavage. Spliced lariats are de-branched and likely produce functional pre-miRNAs for export. Pre-miRNAs are exported from the nucleus to the cytoplasm by the exportin-5. This is followed by recognition and cleavage by a second RNase III enzyme, Dicer and its partner, TAR RNA-binding protein (TRBP), to produce a ~22 bp miRNA duplex with 2 nt 3′ overhangs. Dicer/TRBP, loads one of the strands, the 'guide strand', into a RISC consisting in its simplest form of Ago2. Ago2-directed target cleavage results from perfect sequence complementarity between miRNA and its cognate target. Several mechanisms result in translational suppression, including RNA degradation (by the Xrn1 exonuclease) or reduced translational initiation, all caused by miRNAs with an imperfect match to the target.

15% resulted from translational suppression. The mechanism of mRNA degradation is thought to involve shortening of the polyA tail, decapping and degradation by 5′ to 3′ exonucleases such as Xrn1. Translational suppression is likely to be a result of reduced translation initiation by ribosomes and increased drop off of ribosomes from mRNA.

Ago proteins

A common feature of the pathways that utilize small RNAs to effect gene silencing is their dependence on Ago proteins to exert target-specific effects (Joshua-Tor and Hannon, 2011). These proteins comprise a large and highly conserved family that was first implicated in RNAi when the protein affected in the RDE-1 (RNAi-defective-1) mutant of *C. elegans* was found to belong to the Ago family (Tabara *et al.*, 1999). The name Argonaute is derived from the appearance of an Ago1 mutant of *Arabidopsis*, which has foliage resembling the tentacles of the pelagic octopus, *Argonaute Argo* (Bohmert *et al.*, 1998).

Three clades of Argonaute proteins have been described: Ago-like, Piwi-like and Wagos (Joshua-Tor and Hannon, 2011; Tolia and Joshua-Tor, 2007). The Ago-like family is widely expressed in tissues and is found in animals, plants and fission yeast. Ago-like proteins function by causing transcriptional and post transcriptional gene silencing and typically mediate effects of miRNAs. Piwi proteins are exclusive to humans and are abundant in gonadal tissue. They cause silencing that is mediated by piRNAs and are important for countering effects of transposons. Wago proteins are specific to worms and are required for responses to exogenous RNAs, endogenous RNAs and changes to chromosome structure. In humans there are four Ago proteins: Ago 1, Ago 2, Ago 3 and Ago 4. As with Dicer, Ago proteins have a PAZ domain that is responsible for anchoring the 3′ end of the guide. The MID domain binds the 5′ end and has a high affinity for A or U residues, which accounts for the predominance of these bases located at the 5′ end of miRNA sequences (Frank *et al.*, 2010). Of the four Ago proteins, only Ago2 has a nuclease capability (Liu *et al.*, 2004; Meister *et al.*, 2004). This function is required for target cleavage when guide and target are completely complementary to each other.

Regulation during normal and potentially pathogenic microRNA biogenesis

Given that miRNAs regulate essential cellular processes, it is not surprising that control of miRNA biogenesis is subject to complex and subtle cellular influences. Studies have reported on a variety of mechanisms that influence miRNA biogenesis (recently reviewed in Treiber *et al.*, 2012). Several factors influence miRNA biogenesis: (1) transcription of pri-miR-containing sequences, (2) pri-miRNA processing by the Drosha–DGCR8 complex, (3) nuclear export of pre-miRNA sequences, (4) Dicer processing, (5) effects of rate limiting components of the RNAi pathway and (6) concentrations of cellular sequences that function as 'sponges' to diminish the concentrations of mature miRNAs.

Transcriptional regulation of miRNA biogenesis

The rate of transcription of pri-miR-containing sequences has a direct influence on the concentration of downstream mature miRNAs. Transcripts that encode proteins and also non-coding sequences are subject to transcription regulation, which in turn may influence intracellular miRNA concentrations. An example of transcriptional regulation of miRNA concentrations is that of induction of the miR-34 family in response to the effects of the tumour-suppressor protein, p53 (Bommer *et al.*, 2007; Chang *et al.*, 2007; Raver-Shapira *et al.*, 2007). Following genotoxic stress to cells, p53 activates expression of several protein-coding RNA sequences, as well as miR-34a, miR-34b and miR-34c, which are collectively involved with control of apoptosis, cell cycle arrest and senescence. The mature miR-34 sequences interact with specific target mRNAs to function as tumour suppressors themselves. miR-34 targets include cyclin E2, cyclin dependent kinase 4, the hepatocyte growth factor receptor and the anti-apoptotic protein BCL2 (reviewed in Yamakuchi and Lowenstein, 2009). An interesting feedback mechanism involving p53, miR-34a and SIRT1 (sirtuin 1 or silent mating type information regulation 2 homologue 1) has been proposed to play a role in regulating the cellular response to DNA damage (Yamakuchi and Lowenstein, 2009). DNA damage leads to p53 overexpression, increased transcription of miR-34a-containing sequences and resultant silencing of SIRT1 expression. SIRT1 deacetylates Lys382 of p53 to diminish transcriptional effects of the tumour suppressor. Through this mechanism, SIRT1 counters the effects of p53 on apoptosis to enhance cell survival during periods of severe oxidative stress.

The antineoplastic efficacy of augmenting miR-34 function has been explored in a recent interesting therapeutic approach (Bouchie, 2013). MRX34, which is a mimic of miR-34, has been used in to restore tumour-suppressor functions in cells. The duplex miR-34 mimic, formulated in lipid nanoparticles for targeted delivery to cancer cells, has now reached phase 1 clinical trial for the treatment of patients with liver cancer.

Nuclear processing of pri-miRNA in the control of miRNA biogenesis

Nuclear pre-miRNA formation is also subjected to control by miRNA regulatory proteins. For example, the Smad transcription regulatory proteins, which exert their effects in response to stimulation by bone morphogenic protein, influence maturation of a select group of miRNAs (Davis *et al.*, 2008). The mechanism involves recruitment of pri-miRNA sequences to the microprocessor complex and their facilitated processing. Interestingly, the Smad binding motifs within the pri-miRNAs comprise duplexes that also confer specificity on the processing of particular pri-miRNAs (Davis *et al.*, 2010). Smads therefore influence gene expression by two sequence-specific mechanisms that are the binding to DNA transcription regulatory elements and enhancement of pri-miRNA processing.

P53 is also involved in the regulation of microprocessor function, and has a mechanism that entails contact with the p68 DEAD box helicase (Suzuki *et al.*, 2009). This interaction is responsible for the facilitated processing of particular miRNAs, including miR-145, that are responsible for the suppression of potentially oncogenic genes such as c-*myc*. The mechanism for the specificity of action has not yet been elucidated, but does not entail a sequence-specific interaction with pri-miRNA duplex regions, as with Smad proteins.

Oestrogen receptor alpha (ER-α) inhibits nuclear processing of a group of 39 pri-miRNAs by binding to p68 and p72 DEAD box helicases (Yamagata *et al.*, 2009). Interaction with the miRNAs is thought to involve specific contact with tertiary structural motifs. Interestingly, the mRNA cognates of these down-regulated miRNAs are found in genes that are transcriptionally activated by ER-α. ER-α therefore acts at transcriptional as well as post transcriptional levels to augment target gene expression.

Influence of nuclear export on miRNA biogenesis

Although not yet elucidated, the export of pre-miRNAs from the nucleus by exportin-5 may have a role in regulating miRNA biogenesis. A study aimed at assessing efficacy of expressed pre-miRNA short hairpin RNA (shRNA) mimics targeted to hepatitis B virus (HBV), showed that serious toxicity was consequent to saturation of the exportin-5-mediated transport of pre-miRNAs from hepatocyte nuclei (Grimm *et al.*, 2006). Subsequently, it was shown that Ago proteins may play a more significant rate limiting role in miRNA maturation (Grimm *et al.*, 2010). Some malignancies that are characterized by microsatellite instability have been found to have C-terminal deletions of exportin-5 (Melo *et al.*, 2010). Resultant nuclear trapping of pre-miRNAs and global decrease in miRNA maturation are likely to contribute to malignant transformation.

The role of Dicer in modulating miRNA maturation

Currently the best characterized mechanism of regulating pre-miRNA processing by Dicer is that mediated by Lin-28. Binding of Lin-28 to lethal-7 (let-7) pre-miRNA has been characterized in detail (Bail *et al.*, 2010), and the interaction inhibits maturation of let-7 (reviewed in Rau *et al.*, 2011; Viswanathan and Daley, 2010). Let-7 miRNA is a tumour suppressor that targets cell proliferation genes, such as c-*myc* and K-*ras*, and is inactivated in several malignancies. Lin-28 and the related lin-28b, are overexpressed in ovarian cancer to inhibit let-7 maturation. As a result, let-7-mediated de-repression of growth regulatory genes contributes to the malignant phenotype (Chakraborty *et al.*, 2012).

In addition to causing malignant transformation, derangements of Lin-28 expression and disruption of miRNA maturation have been implicated in dystrophic cardiac disease (Rau *et al.*, 2011). The disease pathogenesis involves sequestration of the RNA binding protein, muscleblind-like 1 protein (MBNL1) by expanded CUG or CCUG repeats. Normally, MBNL1 competes with Lin-28 for binding to pre-miR-1, and counters the inhibitory effect of lin-28 on pre-miR-1 maturation. However, when MBNL1 is sequestered, lin-28 causes pre-miR-1 destabilization. Diminished miR-1 leads to changes in expression of genes that are responsible for ion channel maintenance, which in turn lead to abnormal cardiomyocyte contraction with dystrophy.

An interesting mechanism of regulating miRNA processing during Drosha/DGCR8 and Dicer cleavage involves conversion of adenosine to inosine residues by adenosine deaminase acting on RNA (ADAR). Action of ADARs on pri-miR-142 results in impaired microprocessor processing and diminished stability of the miRNA intermediate (Yang *et al.*, 2006). Tudor-SN, the nuclease recognizing inosine-containing duplex RNA, causes rapid degradation of pri-miR-142. Incorporation of inosine bases into pre-miR-151 inhibits cleavage of the pre-miRNA by Dicer and lower biogenesis of the mature miRNA results (Kawahara *et al.*, 2007). Although interesting, it is not yet clear how widespread this regulatory mechanism is.

Sequence determinants of miRNA biogenesis

Evidence indicates that sequence-specific mechanisms may affect miRNAs' biogenesis and degradation. Analysis carried out on miR-382 demonstrated that a sequence of seven nucleotides located at the 3′ end of the mature miRNA controls its stability (Treiber *et al.*, 2012). This region is positioned outside of the seed and mutation led to an increased half-life of miR-382 through a mechanism that primarily involves the exosome 3′ to 5′ exoribonuclease complex primarily and Xrn-1 to a lesser extent. HIV-1 latency is thought to be regulated by miR-382 and rapid decay of this regulatory miRNA may increase replication of the virus.

Other evidence implicating the importance of miRNA sequences in their maturation comes from studies on the tertiary structure of the miR-17-92 cluster (Kawahara *et al.*, 2007; Nam *et al.*, 2011). The pri-miR-17-92 sequence, encoding six miRNAs, folds into a tightly folded tertiary architecture. The miRNAs that are located at the centre of the structure are processed less efficiently than those positioned at the surface. Disruption of the tertiary structure of the miR-17-92 cluster resulted in increased miR-92 processing with functional repression of the *integrin* $\alpha 5$ mRNA target.

Control of miRNA function by sequestration

The recent competing endogenous RNA (ceRNA) hypothesis describes mechanisms that control RNA function and provides a persuasive explanation for a major role of non-protein-coding RNA in the regulation of gene expression (Salmena *et al.*, 2011). Central to the mechanism is the notion that coding and non-coding RNAs compete for sequestration of a limited pool of miRNAs and a regulatory network allows communication between different transcripts. The essential components of the ceRNA network are (1) miRNAs, (2) protein-encoding RNA, (3) RNA transcribed from pseudogenes, and (4) long non-coding RNAs. MiRNA response elements (MREs) are located on protein coding sequences, long non-coding (lnc) RNAs and pseudogenes. Non protein coding RNAs with miRNA targets serve as decoys to diminish the available pool of miRNAs that interact with MREs. In addition, mRNAs may influence each other's function by serving as decoys or sponges for miRNAs interacting with other mRNAs. Ultimately, functional consequences of effects of ceRNAs are dependent on the availability of inhibitory miRNAs to bind to protein-coding mRNAs.

Two recent reports on circular RNAs that function as miRNA sponges have been a particularly interesting development supporting the ceRNA hypothesis (Hansen *et al.*, 2013; Memczak *et al.*, 2013). Both studies examined a circular RNA derived from the cerebellar degeneration-related protein 1 (CDR1) locus. The circular structure was formed as a result of non-linear splicing of a transcript. An unusual feature of this circular RNA was that the sequence contains seventy three miRNA-7 seed targets and a complete complement to miR-671, and these targets are conserved in eutherial mammals (Hansen *et al.*, 2013). The miRNA-7 target sequences were typically mismatched in the central regions and the circular sequence, termed ciRS-7 by Hansen *et al.* (2013) and CDR1 by Memczak *et al.* (2013), was resistant to endonucleolytic cleavage. Enhanced stability conferred by the circular structure enhances the sponge effects of the non-linear splice RNA. In addition, high-throughput sequencing of RNA isolated by cross-linking immunoprecipitation (HITS-CLIP) confirmed that AGO2 associates with miR-7 at the non-linear splice junction (Hansen *et al.*, 2013). Interestingly, although ciRS-7 RNA was directed to P bodies after association with Ago2 and miR-7, the circular structure provided resistance to degradation. Colocalization of ciRS-7 with miR-7 *in vivo* in the mouse brain and attenuation of expression of endogenous miR-7 targets provided further support for a functional relationship between the two RNA species. ciRS-7-related altered gene expression may contribute to pathogenesis of Parkinson's disease and brain tumour development. Moreover, the role of circular RNAs in regulating miRNA function is likely to be a more general phenomenon than was originally thought. The sex determining region Y (Sry) is another example of circular RNA influencing miRNA function (Hansen *et al.*, 2013). As with ciRS-7, non-linear splicing of Sry leads to formation of a circular RNA that has multiple miRNA-138-sequestering sequences.

Harnessing the RNAi pathway for gene silencing

The demonstration by Elbashir *et al.* (2001) that synthetic RNAi intermediates can be used to effect gene knockdown was a major development in the field or gene silencing. This work provided evidence that RNAi may be harnessed with exogenous RNAi mimics to achieve specific silencing of almost any intended RNA target. This paved the way for the enormous strides that have been made in using RNAi for functional studies and therapeutic application.

Synthetic RNAi activators

Synthetic activators of RNAi are typically short interfering RNAs (siRNAs) and comprise double-stranded RNAs of about 21 bp. They usually have two nucleotide overhangs at their 3′ ends and resemble mature miRNA duplexes. siRNAs have been widely used for various gene silencing applications and they activate RNAi at the stage of guide strand incorporation into RISC (Elbashir *et al.*, 2001). Since siRNAs are small nucleic acids and have a cytoplasmic, not nuclear, site of action they may be delivered more easily than larger DNA expression cassettes. However siRNAs do have limitations of susceptibility to degradation, unfavourable biodistribution, non-specific immunostimulation and unintended silencing of host genes. To attenuate these drawbacks, siRNAs have been altered by chemical modification. Alteration of biological properties of siRNAs through the use of chemical modification has been a very active field of research and several comprehensive recent reviews have been published on the topic (Behlke, 2008; Engels, 2012; Rettig and Behlke, 2012). Changes have included incorporation of 2′-*O*-methyl (2′-*O*-Me), 2′-fluoro (2′-F), or 2-methoxyethyl (2′-O-MOE) moieties. Addition of moieties to the 2′ and 4′ positions of the ribose (Engels *et al.*, 2010), changes to the natural phosphodiester linkages, and substitution of the ribose for sugars such as hexitol were subsequently used (Bramsen *et al.*, 2009; Bramsen *et al.*, 2010; Fisher *et al.*, 2009; Fisher *et al.*, 2007). Interestingly, positioning of modifications at specific nucleotides within the sense or antisense strands of siRNAs influences their efficacy. For example, boranophosphonate-modified siRNAs show improved resistance to nucleases, but incorporation of this modification at the centre of the guide strand results in loss of silencing efficacy (Hall *et al.*, 2004).

Expressed RNAi activators

Fundamental to gene silencing with exogenous RNAi activators has been the use of short hairpin RNAs (shRNAs), which function as artificial mimics of pre-miRNA (Brummelkamp *et al.*, 2002; McManus *et al.*, 2002; Paddison *et al.*, 2002; Paul *et al.*, 2002; Sui *et al.*, 2002; Yu *et al.*, 2002). shRNAs comprise a ~22 bp duplexed dsRNA stem, which is flanked by defined 5′ and 3′ ends and a ssRNA loop. Several different shRNA structures exist and, combined with different modes of expression, there are now several options available for the design of these RNAi triggers. Useful design 'rules' have now been derived from an improved understanding of the function and potency of shRNAs. Some of these include adaptations to the structure and sequence of the ssRNA loop, the length of the dsRNA duplex stem, and the sequence composition and the thermodynamic properties of the duplex stem. Factors associated with the expression of the hairpin also play an important role (Fig. 1.2 and discussed later).

Typically, shRNAs are transcribed using RNA Pol III promoters, which are known to be capable of producing precisely defined short RNA sequences in abundance. Widely used Pol III promoters include the U6 small nuclear RNA (snRNA) (Bertrand *et al.*, 1997), 7SK RNA (Kruger and Benecke, 1987) and the RNase P H1 RNA (Baer *et al.*, 1990) transcriptional regulatory sequences. These promoters require a single purine (G for U6 and 7SK, A/G for H1) as a 5′ terminal nucleotide. They regulate production of RNAs with two to three uridine residues at their 3′ ends that are formed from a transcription termination signal comprising four or five thymidines on the sense strand of the DNA expression cassette. Distinct shRNAs may be generated with staggered 3′ end protrusions that resemble pre-miRNAs in structure. Several tRNA Pol III promoters have also been used to express shRNAs (Boden *et al.*, 2003; Kawasaki and Taira, 2003; Scherer *et al.*, 2007). tRNA-shRNA chimeras are initially processed by the nuclear tRNA 3′ processing enzyme tRNAse Z to release the clover-leaf tRNA motif from a functional shRNA (Scherer *et al.*, 2007). RNA Pol II promoters have also been used for expression of shRNA, but have been met with

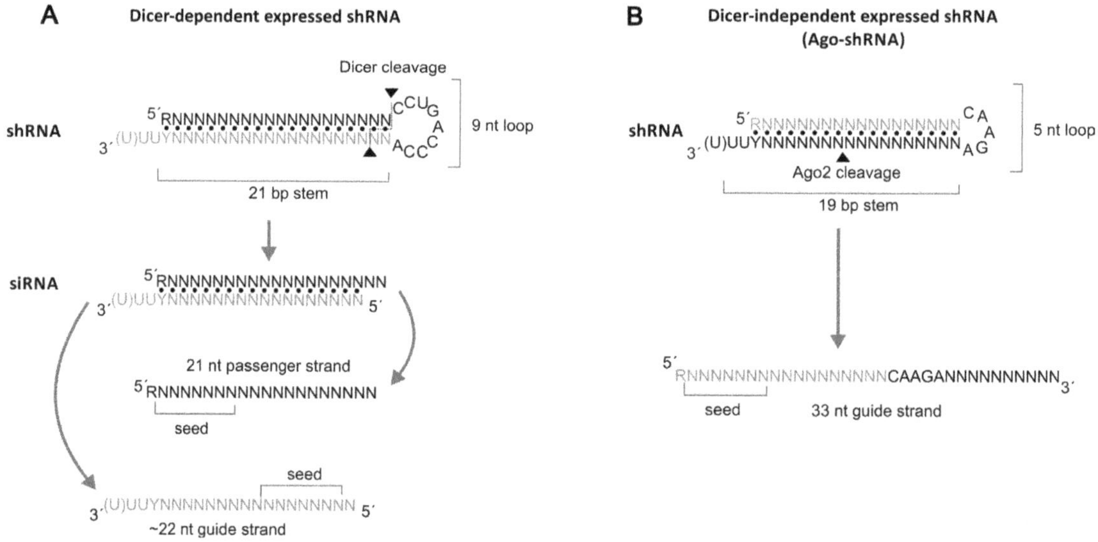

Figure 1.2 Design criteria for shRNA activators expressed off RNA Pol III promoters. (A) The optimal design for a Dicer-dependent shRNA includes a 21 bp stem duplex followed by a flexible loop >7 nt (the 9 nt miR-30 loop is shown). A +1 nt purine (R) is added to ensure transcriptional initiation (G for U6 shRNA promoters; G/A for H1 promoters). Precise and homogeneous Dicer cleavage occurs 2 nt from the loop on the guide strand producing a ~22 nt guide strand (grey) loaded into the RNA-induced Silencing Complex (RISC) (Gu *et al.*, 2012). In this configuration, the passenger strand (black) is also capable of loading into RISC and mediates off-target effects. (B) A Dicer-independent shRNA has the optimal configuration of a 19 bp stem duplex and a short <5 nt loop. The 5 nt loop CAAGA is derived from the Berkhout lab's optimal design (Liu *et al.*, 2013). The guide strand sequence (grey) is placed in the 5′ arm. Ago2 cleaves the shRNA to generate a 33 bp single-stranded RNA that is incorporated effectively into RISC. No passenger strand is produced.

limited success. This is because minimizing 5′ and 3′ terminal sequences on RNAi activators derived from Pol II promoters is difficult to achieve (Giering *et al.*, 2008; Xia *et al.*, 2002). The U1 snRNA Poll II promoter, which produces a minimally extended 5′ termini with defined 3′ termination sequences, has however proven to be useful for expression of functional shRNAs (Denti *et al.*, 2004).

Recently, to enhance production of effective guide strands, attention has been given to achieving a better understanding of the factors that influence shRNA processing. While design rules and algorithms for siRNAs are available and work reasonably well, generating potent shRNA expression cassettes is more difficult. Typically, design of short hairpins is based on an optimal 19 bp siRNA core that is extended (usually towards the loop), to result in shRNA stem lengths of 19 to 21 bp (Brummelkamp *et al.*, 2002; McIntyre *et al.*, 2011). Duplexes may, however, vary from 19 to 29 bp. Conflicting evidence about the factors that affect the potency of shRNAs, especially those with stems that are longer than 21 bp, has been presented. In a comprehensive study of shRNA function, McIntyre and colleagues examined the silencing effects of 91 expressed shRNAs, which varied in stem length and sequence composition. They found that duplex stem length correlated poorly with shRNA suppressive activity (McIntyre *et al.*, 2011). Related to this, Gu and colleagues used high-throughput sequencing data to show that processing of the long 24 bp shRNAs resulted in siRNAs comprising an unpredictable heterogeneous population (Gu *et al.*, 2012). Dicer-induced canonical and non-canonical cleavage events generated a range of active and inactive shRNA-derived siRNAs. Inactive siRNAs are likely to have reduced on-target efficiencies and an unintended bias for passenger strand selection. This property of shRNAs with longer duplexes is an important consideration for the development of shRNAs intended for therapeutic application, where off-target effects may cause toxicity. In fact, this may help explain why shRNAs with 25 bp duplexes were more toxic than 19 bp shRNAs when overexpressed in mice livers (Grimm *et al.*, 2006). Interestingly, efficacy of longer anti hepatitis C virus 29 bp shRNAs could be restored by placing an internal loop/bulge 2 nt from the site of Dicer cleavage (Gu *et al.*, 2012). This adjustment resulted in a precise and predictable cleavage pattern. The ssRNA bulge/loops are likely to secure Dicer to a fixed position for cleavage, and may be the reason for presence of internal loop/bulge sequences adjacent to the Dicer cleavage site in many natural pre-miRNAs. These data indicate that careful positioning of bulge/loop sequences in artificial shRNAs may be used to ensure optimal Dicer processing with accurate, predictable and homogeneous formation of gene silencers (Fig.1.2A).

shRNAs with stems that are 19 bp or shorter may also be used to achieve gene silencing (Ge *et al.*, 2010; Siolas *et al.*, 2005). Interestingly, these short shRNA (sshRNAs) are not substrates for Dicer cleavage since processing to form siRNAs is achieved in Dicer knockout cells. Instead, 19 bp shRNAs follow a similar biogenesis pathway to that of miR-451 (Liu *et al.*, 2013). This miRNA derives from a pre-miRNA that consists of a highly paired, unusually short 19–20 bp stem–loop region (Fig. 1.2B) (Cheloufi *et al.*, 2010; Cifuentes *et al.*, 2010; Yang *et al.*, 2010). Deep sequencing mapped the pre-miR-451 loop region to within the most commonly detected mature miRNA (Yang *et al.*, 2010). Also, mature miR-451 was lost in mice with catalytically inactive Ago2 (Cheloufi *et al.*, 2010). Biochemical and genetic studies indicate that pre-miR-451 is processed by Ago2 to form a 30 nt miRNA intermediate, which subsequently undergoes terminal uridylation and 3′–5′ exonuclease-mediated resection to yield a mature 23 or 24 nt miRNA (Cifuentes *et al.*, 2010; Yang *et al.*, 2010). Importantly, Yang *et al.* (2010) ingeniously reprogrammed a mir-451 backbone to generate miRNA species of different sizes. This process has been exploited to generate effective Dicer-independent sshRNAs (Ge *et al.*, 2010; Liu *et al.*, 2013). Engineered sshRNAs have distinct advantages of being active in the nucleus without a requirement for cytoplasmic processing by Dicer. In addition, sshRNA processing does not result in formation of a passenger strand, which improves the target specificity of these RNAi activators.

Another very effective approach to generating active siRNAs is through construction of artificial pri-miRNAs. Here, the guide strands are embedded into the backbone of known pri-miRNA sequences, but the structural motifs that define an active pri-miRNA are preserved. pri-miRNA mimics are efficiently processed by Drosha/DGCR8 to generate pre-miRNA precursors, with improved entry into the RNAi pathway (Boudreau *et al.*, 2009; Ely *et al.*, 2008, 2009; Maczuga *et al.*, 2012). Useful features of artificial pri-miRNAs make them a popular choice for many applications: pri-miRNAs may be expressed using Pol II promoters, allowing for inducible or tissue-specific expression; and individual pri-miRNAs may be multiplexed for combinatorial silencing. U6-promoter-driven shRNAs are known to produce toxicities *in vivo*, which are likely to be a result of off-target effects and saturation of the endogenous miRNA biogenesis pathway (An *et al.*, 2006; Ehlert *et al.*, 2010; Grimm *et al.*, 2006; Martin *et al.*, 2011; McBride *et al.*, 2008). These effects are largely avoided with artificial pri-miRNA cassettes, which appear to be much safer than expressed shRNAs (Boudreau *et al.*, 2009; Maczuga *et al.*, 2012; McBride *et al.*, 2008). Pri-miRNA mimics have been used in the development of safer knockdown technologies for therapy (Aagaard *et al.*, 2008; Ely *et al.*, 2008, 2009) and in the study of gene function (Premsrirut *et al.*, 2011).

Augmenting and antagonizing microRNA function

The study of miRNA function has been greatly facilitated by increasing or decreasing the cellular concentrations of specific miRNAs to augment or inhibit their regulatory effects. Such gain-of function and loss-of-function studies have enabled the identification of novel mRNA targets and aided in the detailed analysis of miRNAs' effects on target mRNA and protein levels. Moreover, the approach has led to development of novel therapeutic approaches that rely on targeted replacement or inhibition of miRNA activity to reverse a disease phenotype.

As with the use of artificial miRNAs to achieve targeted silencing of pathology-causing genes, supplementing intracellular concentrations of sequences that simulate function of natural miRNAs has therapeutic potential (Chang *et al.*, 2006; Zeng *et al.*, 2002, 2005). Similarly, chemically synthesized or expressed miRNAs have been applied to achieve this effect (Dickins *et al.*, 2005; Takeshita *et al.*, 2010). The targeted replacement of miRNA function has potential application in cancer where miRNA down-regulation is found in many tumour types (reviewed in Bader *et al.*, 2010). Examples include the evolutionary conserved regulatory miRNA families, such as let-7, miR-34 and miR-200, which control key pathways that may be disrupted during malignant transformation. Replacing these miRNAs has potential for cancer therapy (Bader, 2012; Kasinski and Slack, 2012; Kota *et al.*, 2009; Liu *et al.*, 2011). As indicated above, synthetic miR-34 delivered in a lipoplex formulation for liver cancer treatment has now reached an early stage of clinical trial (Bouchie, 2013).

MicroRNAs may also be inhibited with antisense sequences that are complementary to mature miRNAs. Typically, synthetic miRNA inhibitors have chemical modifications that improve their stability and binding affinity (Hutvagner *et al.*, 2004) (also see above). Examples include antagomirs and locked nucleic acids (LNA). Antagomirs are ~22 nt RNA molecules that have backbones containing phosphorothioates and 2′-*O*-methoxyethyl (2′-MOE) residues (Krützfeldt *et al.*, 2005). Antisense oligonucleotides with one or more LNA nucleotide monomers confer stronger binding affinity to target sequences (Elmen *et al.*, 2008; Obad *et al.*, 2011). Other approaches to inhibiting miRNA activity include the use of oligonucleotides to mask mRNA targets (Choi *et al.*, 2007). However, to be effective, chemically synthesized sequences require efficient delivery and repeat administration to induce long-lasting miRNA inhibition. MicroRNA sponge-like inhibitors expressed from DNA have the advantage of being compatible with viral vectors for highly efficient delivery. Ebert *et al.* (2007) showed that expressed transcripts encoding a series of imperfect complementary sequences effectively inhibited individual and whole families of miRNAs. MiRNA sponges often occur in nature as multiple-repeat motifs of imperfect miRNA target sequences (Hansen *et al.*, 2011, 2013; Memczak *et al.*, 2013; Poliseno *et al.*, 2010), and this topic has been discussed in the section of this chapter that deals with control of miRNA function by sequestration (see above).

MicroRNA targets as gene regulators and sensors

Inserting miRNA cognates into target genes has been shown to be a useful way of using endogenous miRNAs to control gene expression (Brown *et al.*, 2006). This is especially valuable for applications where gene expression needs to occur in a time-specific manner, be regulated in response to developmental cues, or where tissue specificity is required (Brown *et al.*, 2006; Mansfield *et al.*, 2004; Wu *et al.*, 2009). Several elegant studies have demonstrated that transgenes containing engineered miRNA targets are silenced by miRNAs to provide a means for improving specificity of gene expression (Brown *et al.*, 2006, 2007; Kelly *et al.*, 2008). This approach can also be exploited for cell- or tissue-specific attenuation of viral replication to achieve intended cytotoxicity or to improve safety of live vaccines (Barnes *et al.*, 2008; Cawood *et al.*, 2009). By ensuring targets are fully complementary to the miRNAs, Ago2 cleavage results in turnover and complete destruction of the RNA template and avoids potentially toxic saturation of endogenous miRNA pathways.

Cell-specificity of miRNA-regulated transgene expression may also be combined with selectivity of cell-specific promoters. Wu *et al* used miRNAs up-regulated in astrocytes and a tumour-specific glial fibrillary acidic protein (GFAP) gene promoter to control suicide gene expression in glial cells in the brains of mice (Wu *et al.*, 2009). This approach resulted in significantly improved tissue selectivity *in vivo* and will be useful to achieve the precise control required for suicide gene therapy of cancer.

What are the factors that enable optimal control of transgene expression by miRNAs? It is known that individual miRNAs interact with mRNAs to achieve subtle alterations of protein expression. Single miRNAs rarely exert dramatic effects on their own (reviewed in Bartel, 2009), and adding multiple fully complementary miRNA target sites adds versatility to control of transgene activity (Brown *et al.*, 2007). However, relative concentrations of miRNAs and their transgene mRNA targets requires careful control to ensure an optimal threshold of regulation (Mukherji *et al.*, 2011). Recently, a high-throughput, cell-based, assay called Sensor-seq was developed to monitor the relationship between cellular miRNA activity and target concentration (Mullokandov *et al.*, 2012). The study demonstrated that up to 60% of miRNAs do not exert any detectable activity against their targets. High miRNA concentrations were needed to affect the translation of targeted RNA decisively, and some miRNAs may not exert any effects at all, even when abundantly expressed (Mullokandov *et al.*, 2012).

One of the principal difficulties associated with miRNA sensors is that they function through an inhibitory, not stimulatory, output. To address this, Amendola *et al.* (2013) made use of an ingenious sensor vector comprising two OFF switches in series: a reporter gene under control of a tetracycline repressor (tTR) that was fused to the Kruppel-associated box (KRAB) domain, and a miRNA target regulating expression of the tTR-KRAB repressor. Using this system, the authors were able to monitor expression of miRNAs accurately in different stages of haematopoietic progenitor cell differentiation *in vivo*. A similar approach, coupled to use of liver-specific gene expression, was employed by Pichard *et al.* (2012) to modulate hepatic gene expression.

Lastly, one of the more exciting new applications of miRNAs as sensors is in the development of novel 'gene circuits' where miRNAs behave as switches or logic gates. Such systems can function both to detect and execute specific biological programs. These properties were employed to identify a cancerous cell state based on expression of multiple endogenous miRNA markers (Xie *et al.*, 2011).

MicroRNAs as prognostic and diagnostic markers

An exciting development in the field of RNAi has been the measurement of miRNAs for diagnostic purposes. MiRNAs may be detected in saliva, serum, plasma, fixed tissue and milk and their concentrations correlate with disease processes. It remains unclear whether miRNAs are secreted or whether their presence is a consequence of release during cell turnover or death. Several studies suggest miRNAs are actively secreted within extracellular vesicles (e.g. exosomes) (Gallo *et al.*, 2012; Valadi *et al.*, 2007) or as part of Ago2 complexes (Arroyo *et al.*, 2011). Interestingly, circulating miRNAs are highly stable (Mitchell *et al.*, 2008) and given the fact that blood can conveniently be obtained from patients, sensitive detection of miRNA biomarkers is potentially a powerful new diagnostic tool. This is especially useful for application in oncology diagnosis where comparison of miRNA and mRNA profiles provides information on the tissue of origin of poorly differentiated tumours (Lu *et al.*, 2005). Recently, discriminatory miRNA 'classifiers' have also been developed to assist with identification of tumours of unknown origin (Ferracin *et al.*, 2011). Moreover miRNA profiles may be used to correlate with prognosis or therapeutic outcome (reviewed in Pritchard *et al.*, 2012).

Conclusion

During the past 15 years, rapid and significant advances in understanding the RNAi pathway have taken place. Concerted efforts of investigators from a variety of backgrounds have provided detailed mechanistic insights into the important RNAi-induced gene regulation processes. As a result, several new and interesting applications for RNAi have been developed. An overview of the RNAi pathway has been provided in this chapter as a background to understanding how the naturally occurring process may be harnessed for a wide range of uses. Also, a summary of how RNAi may be harnessed has been provided as an introduction to the detailed explanations of specific uses of RNAi that follow in this book.

Applied RNAi research has matured as investigators have gained improved understanding of the advantages and potential pitfalls within the field. Initial enthusiasm for the power of RNAi has been tempered by a realization that unanticipated variables may have unexpected consequences. A particularly relevant example of this is in the use of RNAi for therapeutic application. Minimizing off target effects and overcoming difficulties with delivery of therapeutic RNAi activators to diseased tissues remain challenging. Although application of RNAi has encountered setbacks, the subject remains exciting and the field has considerable impetus. Coming years are likely to witness significant advances, and the following chapters of this book discuss a selection of the exciting topics of the field.

Acknowledgements

Work in the authors' laboratories is supported by financial assistance from the South African National Research Foundation (NRF, GUNs 81768, 81692, 68339, 85981 and 77954), Medical Research Council and Poliomyelitis Research Foundation.

References

Aagaard, L.A., Zhang, J., von Eije, K.J., Li, H., Saetrom, P., Amarzguioui, M., and Rossi, J.J. (2008). Engineering and optimization of the miR-106b cluster for ectopic expression of multiplexed anti-HIV RNAs. Gene Ther. *15*, 1536–1549.

Amendola, M., Giustacchini, A., Gentner, B., and Naldini, L. (2013). A double-switch vector system positively regulates transgene expression by endogenous microRNA expression (miR-ON Vector). Mol. Ther. *21*, 934–946.

An, D.S., Qin, F.X., Auyeung, V.C., Mao, S.H., Kung, S.K., Baltimore, D., and Chen, I.S. (2006). Optimization and functional effects of stable short hairpin RNA expression in primary human lymphocytes via lentiviral vectors. Mol. Ther. *14*, 494–504.

Arroyo, J.D., Chevillet, J.R., Kroh, E.M., Ruf, I.K., Pritchard, C.C., Gibson, D.F., Mitchell, P.S., Bennett, C.F., Pogosova-Agadjanyan, E.L., Stirewalt, D.L., *et al.* (2011). Argonaute2 complexes carry a population of circulating microRNAs independent of vesicles in human plasma. Proc. Natl. Acad. Sci. U.S.A. *108*, 5003–5008.

Bader, A.G. (2012). miR-34 – a microRNA replacement therapy is headed to the clinic. Front. Genet. *3*, 120.

Bader, A.G., Brown, D., and Winkler, M. (2010). The promise of microRNA replacement therapy. Cancer Res. *70*, 7027–7030.

Baer, M., Nilsen, T.W., Costigan, C., and Altman, S. (1990). Structure and transcription of a human gene for H1 RNA, the RNA component of human RNase P. Nucleic Acids Res. *18*, 97–103.

Bail, S., Swerdel, M., Liu, H., Jiao, X., Goff, L.A., Hart, R.P., and Kiledjian, M. (2010). Differential regulation of microRNA stability. RNA *16*, 1032–1039.

Barnes, D., Kunitomi, M., Vignuzzi, M., Saksela, K., and Andino, R. (2008). Harnessing endogenous miRNAs to control virus tissue tropism as a strategy for developing attenuated virus vaccines. Cell Host Microbe *4*, 239–248.

Bartel, D.P. (2009). MicroRNAs: target recognition and regulatory functions. Cell *136*, 215–233.

Behlke, M.A. (2008). Chemical modification of siRNAs for *in vivo* use. Oligonucleotides *18*, 305–319.

Bertrand, E., Castanotto, D., Zhou, C., Carbonnelle, C., Lee, N.S., Good, P., Chatterjee, S., Grange, T., Pictet, R., Kohn, D., *et al.* (1997). The expression cassette determines the functional activity of ribozymes in mammalian cells by controlling their intracellular localization. RNA *3*, 75–88.

Boden, D., Pusch, O., Lee, F., Tucker, L., Shank, P.R., and Ramratnam, B. (2003). Promoter choice affects the potency of HIV-1 specific RNA interference. Nucleic Acids Res. *31*, 5033–5038.

Bohmert, K., Camus, I., Bellini, C., Bouchez, D., Caboche, M., and Benning, C. (1998). AGO1 defines a novel locus of Arabidopsis controlling leaf development. EMBO J. *17*, 170–180.

Bommer, G.T., Gerin, I., Feng, Y., Kaczorowski, A.J., Kuick, R., Love, R.E., Zhai, Y., Giordano, T.J., Qin, Z.S., Moore, B.B., *et al.* (2007). p53-mediated activation of miRNA34 candidate tumor-suppressor genes. Curr. Biol. *17*, 1298–1307.

Bouchie, A. (2013). First microRNA mimic enters clinic. Nat. Biotechnol. *31*, 577.

Boudreau, R.L., Martins, I., and Davidson, B.L. (2009). Artificial microRNAs as siRNA shuttles: improved safety as compared to shRNAs *in vitro* and *in vivo*. Mol. Ther. *17*, 169–175.

Bramsen, J.B., Laursen, M.B., Nielsen, A.F., Hansen, T.B., Bus, C., Langkjaer, N., Babu, B.R., Hojland, T., Abramov, M., Van Aerschot, A., *et al.* (2009). A large-scale chemical modification screen identifies design rules to generate siRNAs with high activity, high stability and low toxicity. Nucleic Acids Res. *37*, 2867–2881.

Bramsen, J.B., Pakula, M.M., Hansen, T.B., Bus, C., Langkjaer, N., Odadzic, D., Smicius, R., Wengel, S.L., Chattopadhyaya, J., Engels, J.W., *et al.* (2010). A screen of chemical modifications identifies position-specific modification by UNA to most potently reduce siRNA off-target effects. Nucleic Acids Res. *38*, 5761–5773.

Brown, B.D., Venneri, M.A., Zingale, A., Sergi Sergi, L., and Naldini, L. (2006). Endogenous microRNA regulation suppresses transgene expression in hematopoietic lineages and enables stable gene transfer. Nat. Med. *12*, 585–591.

Brown, B.D., Gentner, B., Cantore, A., Colleoni, S., Amendola, M., Zingale, A., Baccarini, A., Lazzari, G., Galli, C., and Naldini, L. (2007). Endogenous microRNA can be broadly exploited to regulate transgene expression according to tissue, lineage and differentiation state. Nat. Biotechnol. *25*, 1457–1467.

Brummelkamp, T.R., Bernards, R., and Agami, R. (2002). A system for stable expression of short interfering RNAs in mammalian cells. Science *296*, 550–553.

Bushati, N., and Cohen, S.M. (2007). MicroRNA functions. Annu. Rev. Cell Dev. Biol. *23*, 175–205.

Cai, X., Hagedorn, C.H., and Cullen, B.R. (2004). Human microRNAs are processed from capped, polyadenylated transcripts that can also function as mRNAs. RNA *10*, 1957–1966.

Cawood, R., Chen, H.H., Carroll, F., Bazan-Peregrino, M., van Rooijen, N., and Seymour, L.W. (2009). Use of tissue-specific microRNA to control pathology of wild-type adenovirus without attenuation of its ability to kill cancer cells. PLoS Pathog. *5*, e1000440.

Chakraborty, S., Mehtab, S., Patwardhan, A., and Krishnan, Y. (2012). Pri-miR-17–92a transcript folds into a tertiary structure and autoregulates its processing. RNA *18*, 1014–1028.

Chang, K., Elledge, S.J., and Hannon, G.J. (2006). Lessons from Nature: microRNA-based shRNA libraries. Nat. Methods *3*, 707–714.

Chang, T.C., Wentzel, E.A., Kent, O.A., Ramachandran, K., Mullendore, M., Lee, K.H., Feldmann, G., Yamakuchi, M., Ferlito, M., Lowenstein, C.J., *et al.* (2007). Transactivation of miR-34a by p53 broadly influences gene expression and promotes apoptosis. Mol. Cell. *26*, 745–752.

Cheloufi, S., Dos Santos, C.O., Chong, M.M., and Hannon, G.J. (2010). A dicer-independent miRNA biogenesis pathway that requires Ago catalysis. Nature *465*, 584–589.

Chendrimada, T.P., Gregory, R.I., Kumaraswamy, E., Norman, J., Cooch, N., Nishikura, K., and Shiekhattar, R. (2005). TRBP recruits the Dicer complex to Ago2 for microRNA processing and gene silencing. Nature *436*, 740–744.

Choi, W.Y., Giraldez, A.J., and Schier, A.F. (2007). Target protectors reveal dampening and balancing of Nodal agonist and antagonist by miR-430. Science *318*, 271–274.

Cifuentes, D., Xue, H., Taylor, D.W., Patnode, H., Mishima, Y., Cheloufi, S., Ma, E., Mane, S., Hannon, G.J., Lawson, N.D., *et al.* (2010). A novel miRNA processing pathway independent of Dicer requires Argonaute2 catalytic activity. Science *328*, 1694–1698.

Davis, B.N., Hilyard, A.C., Lagna, G., and Hata, A. (2008). SMAD proteins control DROSHA-mediated microRNA maturation. Nature *454*, 56–61.

Davis, B.N., Hilyard, A.C., Nguyen, P.H., Lagna, G., and Hata, A. (2010). Smad proteins bind a conserved RNA sequence to promote microRNA maturation by Drosha. Mol. Cell. *39*, 373–384.

Denli, A.M., Tops, B.B., Plasterk, R.H., Ketting, R.F., and Hannon, G.J. (2004). Processing of primary microRNAs by the Microprocessor complex. Nature *432*, 231–235.

Denti, M.A., Rosa, A., Sthandier, O., De Angelis, F.G., and Bozzoni, I. (2004). A new vector, based on the PolII promoter of the U1 snRNA gene, for the expression of siRNAs in mammalian cells. Mol. Ther. *10*, 191–199.

Dickins, R.A., Hemann, M.T., Zilfou, J.T., Simpson, D.R., Ibarra, I., Hannon, G.J., and Lowe, S.W. (2005). Probing tumor phenotypes using stable and regulated synthetic microRNA precursors. Nat. Genet. *37*, 1289–1295.

Ebert, M.S., Neilson, J.R., and Sharp, P.A. (2007). MicroRNA sponges: competitive inhibitors of small RNAs in mammalian cells. Nat. Methods *4*, 721–726.

Ehlert, E.M., Eggers, R., Niclou, S.P., and Verhaagen, J. (2010). Cellular toxicity following application of adeno-associated viral vector-mediated

RNA interference in the nervous system. BMC Neuroscience *11*, 20.

Elbashir, S.M., Harborth, J., Lendeckel, W., Yalcin, A., Weber, K., and Tuschl, T. (2001). Duplexes of 21-nucleotide RNAs mediate RNA interference in cultured mammalian cells. Nature *411*, 494–498.

Elmen, J., Lindow, M., Schutz, S., Lawrence, M., Petri, A., Obad, S., Lindholm, M., Hedtjarn, M., Hansen, H.F., Berger, U., *et al.* (2008). LNA-mediated microRNA silencing in non-human primates. Nature *452*, 896–899.

Ely, A., Naidoo, T., Mufamadi, S., Crowther, C., and Arbuthnot, P. (2008). Expressed anti-HBV primary microRNA shuttles inhibit viral replication efficiently *in vitro* and *in vivo*. Mol. Ther. *16*, 1105–1112.

Ely, A., Naidoo, T., and Arbuthnot, P. (2009). Efficient silencing of gene expression with modular trimeric Pol II expression cassettes comprising microRNA shuttles. Nucleic Acids Res. *37*, e91.

Engels, J.W. (2012). Gene silencing by chemically modified siRNAs. N. Biotechnol. *30*, 302–307.

Engels, J.W., Odadzic, D., Smicius, R., and Haas, J. (2010). Chemical synthesis of 2′-O-alkylated siRNAs. Methods Mol. Biol. *623*, 155–170.

Ferracin, M., Pedriali, M., Veronese, A., Zagatti, B., Gafa, R., Magri, E., Lunardi, M., Munerato, G., Querzoli, G., Maestri, I., *et al.* (2011). MicroRNA profiling for the identification of cancers with unknown primary tissue-of-origin. J. Pathol. *225*, 43–53.

Fire, A., Xu, S., Montgomery, M.K., Kostas, S.A., Driver, S.E., and Mello, C.C. (1998). Potent and specific genetic interference by double-stranded RNA in *Caenorhabditis elegans*. Nature *391*, 806–811.

Fisher, M., Abramov, M., Van Aerschot, A., Xu, D., Juliano, R.L., and Herdewijn, P. (2007). Inhibition of MDR1 expression with altritol-modified siRNAs. Nucleic Acids Res. *35*, 1064–1074.

Fisher, M., Abramov, M., Van Aerschot, A., Rozenski, J., Dixit, V., Juliano, R.L., and Herdewijn, P. (2009). Biological effects of hexitol and altritol-modified siRNAs targeting B-Raf. Eur. J. Pharmacol. *606*, 38–44.

Frank, F., Sonenberg, N., and Nagar, B. (2010). Structural basis for 5′-nucleotide base-specific recognition of guide RNA by human AGO2. Nature *465*, 818–822.

Friedman, R.C., Farh, K.K., Burge, C.B., and Bartel, D.P. (2009). Most mammalian mRNAs are conserved targets of microRNAs. Genome Res. *19*, 92–105.

Fukuda, T., Yamagata, K., Fujiyama, S., Matsumoto, T., Koshida, I., Yoshimura, K., Mihara, M., Naitou, M., Endoh, H., Nakamura, T., *et al.* (2007). DEAD-box RNA helicase subunits of the Drosha complex are required for processing of rRNA and a subset of microRNAs. Nat. Cell Biol. *9*, 604–611.

Gallo, A., Tandon, M., Alevizos, I., and Illei, G.G. (2012). The majority of microRNAs detectable in serum and saliva is concentrated in exosomes. PLoS ONE *7*, e30679.

Garzon, R., Calin, G.A., and Croce, C.M. (2009). MicroRNAs in Cancer. Annu. Rev. Med. *60*, 167–179.

Ge, Q., Ilves, H., Dallas, A., Kumar, P., Shorenstein, J., Kazakov, S.A., and Johnston, B.H. (2010). Minimal-length short hairpin RNAs: the relationship of structure and RNAi activity. RNA *16*, 106–117.

Giering, J.C., Grimm, D., Storm, T.A., and Kay, M.A. (2008). Expression of shRNA from a tissue-specific pol II promoter is an effective and safe RNAi therapeutic. Mol. Ther. *16*, 1630–1636.

Gregory, R.I., Yan, K.P., Amuthan, G., Chendrimada, T., Doratotaj, B., Cooch, N., and Shiekhattar, R. (2004). The Microprocessor complex mediates the genesis of microRNAs. Nature *432*, 235–240.

Grimm, D., Streetz, K.L., Jopling, C.L., Storm, T.A., Pandey, K., Davis, C.R., Marion, P., Salazar, F., and Kay, M.A. (2006). Fatality in mice due to oversaturation of cellular microRNA/short hairpin RNA pathways. Nature *441*, 537–541.

Grimm, D., Wang, L., Lee, J.S., Schurmann, N., Gu, S., Borner, K., Storm, T.A., and Kay, M.A. (2010). Argonaute proteins are key determinants of RNAi efficacy, toxicity, and persistence in the adult mouse liver. J. Clin. Invest. *120*, 3106–3119.

Grimson, A., Farh, K.K., Johnston, W.K., Garrett-Engele, P., Lim, L.P., and Bartel, D.P. (2007). MicroRNA targeting specificity in mammals: determinants beyond seed pairing. Mol. Cell. *27*, 91–105.

Gu, S., Jin, L., Zhang, Y., Huang, Y., Zhang, F., Valdmanis, P.N., and Kay, M.A. (2012). The loop position of shRNAs and pre-miRNAs is critical for the accuracy of dicer processing *in vivo*. Cell *151*, 900–911.

Guo, H., Ingolia, N.T., Weissman, J.S., and Bartel, D.P. (2010). Mammalian microRNAs predominantly act to decrease target mRNA levels. Nature *466*, 835–840.

Hall, A.H., Wan, J., Shaughnessy, E.E., Ramsay Shaw, B., and Alexander, K.A. (2004). RNA interference using boranophosphate siRNAs: structure–activity relationships. Nucleic Acids Res. *32*, 5991–6000.

Han, J., Lee, Y., Yeom, K.H., Nam, J.W., Heo, I., Rhee, J.K., Sohn, S.Y., Cho, Y., Zhang, B.T., and Kim, V.N. (2006). Molecular basis for the recognition of primary microRNAs by the Drosha–DGCR8 complex. Cell *125*, 887–901.

Hansen, T.B., Wiklund, E.D., Bramsen, J.B., Villadsen, S.B., Statham, A.L., Clark, S.J., and Kjems, J. (2011). MiRNA-dependent gene silencing involving Ago2-mediated cleavage of a circular antisense RNA. EMBO J. 1*30*, 4414–4422.

Hansen, T.B., Jensen, T.I., Clausen, B.H., Bramsen, J.B., Finsen, B., Damgaard, C.K., and Kjems, J. (2013). Natural RNA circles function as efficient microRNA sponges. Nature *495*, 384–388.

Hutvagner, G., Simard, M.J., Mello, C.C., and Zamore, P.D. (2004). Sequence-specific inhibition of small RNA function. PLoS Biol. *2*, E98.

Joshua-Tor, L., and Hannon, G.J. (2011). Ancestral roles of small RNAs: an Ago-centric perspective. Cold Spring Harb. Perspect. Biol. *3*, a003772.

Kasinski, A.L., and Slack, F.J. (2012). MiRNA-34 prevents cancer initiation and progression in a therapeutically resistant K-ras and p53-induced mouse model of lung adenocarcinoma. Cancer Res. *72*, 5576–5587.

Kawahara, Y., Zinshteyn, B., Chendrimada, T.P., Shiekhattar, R., and Nishikura, K. (2007). RNA editing of the microRNA-151 precursor blocks cleavage by the Dicer–TRBP complex. EMBO Rep. *8*, 763–769.

Kawasaki, H., and Taira, K. (2003). Short hairpin type of dsRNAs that are controlled by tRNA(Val) promoter significantly induce RNAi-mediated gene silencing in the cytoplasm of human cells. Nucleic Acids Res. *31*, 700–707.

Kelly, E.J., Hadac, E.M., Greiner, S., and Russell, S.J. (2008). Engineering microRNA responsiveness to decrease virus pathogenicity. Nat. Med. *14*, 1278–1283.

Kota, J., Chivukula, R.R., O'Donnell, K.A., Wentzel, E.A., Montgomery, C.L., Hwang, H.W., Chang, T.C., Vivekanandan, P., Torbenson, M., Clark, K.R., *et al.* (2009). Therapeutic microRNA delivery suppresses tumorigenesis in a murine liver cancer model. Cell *137*, 1005–1017.

Kruger, W., and Benecke, B.J. (1987). Structural and functional analysis of a human 7 S K RNA gene. J. Mol. Biol. *195*, 31–41.

Krützfeldt, J., Rajewsky, N., Braich, R., Rajeev, K.G., Tuschl, T., Manoharan, M., and Stoffel, M. (2005). Silencing of microRNAs *in vivo* with 'antagomirs'. Nature *438*, 685–689.

Lee, Y., Kim, M., Han, J., Yeom, K.H., Lee, S., Baek, S.H., and Kim, V.N. (2004). MicroRNA genes are transcribed by RNA polymerase II. EMBO J. *23*, 4051–4060.

Liu, C., Kelnar, K., Liu, B., Chen, X., Calhoun-Davis, T., Li, H., Patrawala, L., Yan, H., Jeter, C., Honorio, S., *et al.* (2011). The microRNA miR-34a inhibits prostate cancer stem cells and metastasis by directly repressing CD44. Nat. Med. *17*, 211–215.

Liu, J., Carmell, M.A., Rivas, F.V., Marsden, C.G., Thomson, J.M., Song, J.J., Hammond, S.M., Joshua-Tor, L., and Hannon, G.J. (2004). Argonaute2 is the catalytic engine of mammalian RNAi. Science *305*, 1437–1441.

Liu, Y.P., Schopman, N.C., and Berkhout, B. (2013). Dicer-independent processing of short hairpin RNAs. Nucleic Acids Res. *41*, 3723–3733.

Lu, J., Getz, G., Miska, E.A., Alvarez-Saavedra, E., Lamb, J., Peck, D., Sweet-Cordero, A., Ebert, B.L., Mak, R.H., Ferrando, A.A., *et al.* (2005). MicroRNA expression profiles classify human cancers. Nature *435*, 834–838.

Lund, E., Guttinger, S., Calado, A., Dahlberg, J.E., and Kutay, U. (2004). Nuclear export of microRNA precursors. Science *303*, 95–98.

McBride, J.L., Boudreau, R.L., Harper, S.Q., Staber, P.D., Monteys, A.M., Martins, I., Gilmore, B.L., Burstein, H., Peluso, R.W., Polisky, B., *et al.* (2008). Artificial miRNAs mitigate shRNA-mediated toxicity in the brain: implications for the therapeutic development of RNAi. Proc. Natl. Acad. Sci. U.S.A. *105*, 5868–5873.

McIntyre, G.J., Yu, Y.H., Lomas, M., and Fanning, G.C. (2011). The effects of stem length and core placement on shRNA activity. BMC Mol. Biol. *12*, 34.

McManus, M.T., Petersen, C.P., Haines, B.B., Chen, J., and Sharp, P.A. (2002). Gene silencing using micro-RNA designed hairpins. RNA *8*, 842–850.

Maczuga, P., Koornneef, A., Borel, F., Petry, H., van Deventer, S., Ritsema, T., and Konstantinova, P. (2012). Optimization and comparison of knockdown efficacy between polymerase II expressed shRNA and artificial miRNA targeting luciferase and Apolipoprotein B100. BMC Biotechnol. *12*, 42.

Mansfield, J.H., Harfe, B.D., Nissen, R., Obenauer, J., Srineel, J., Chaudhuri, A., Farzan-Kashani, R., Zuker, M., Pasquinelli, A.E., Ruvkun, G., *et al.* (2004). MicroRNA-responsive 'sensor' transgenes uncover Hox-like and other developmentally regulated patterns of vertebrate microRNA expression. Nat. Genet. *36*, 1079–1083.

Martin, J.N., Wolken, N., Brown, T., Dauer, W.T., Ehrlich, M.E., and Gonzalez-Alegre, P. (2011). Lethal toxicity caused by expression of shRNA in the mouse striatum: implications for therapeutic design. Gene. Ther. *18*, 666–673.

Meister, G., Landthaler, M., Patkaniowska, A., Dorsett, Y., Teng, G., and Tuschl, T. (2004). Human Argonaute2 mediates RNA cleavage targeted by miRNAs and siRNAs. Mol. Cell. *15*, 185–197.

Melo, S.A., Moutinho, C., Ropero, S., Calin, G.A., Rossi, S., Spizzo, R., Fernandez, A.F., Davalos, V., Villanueva, A., Montoya, G., *et al.* (2010). A genetic defect in exportin-5 traps precursor microRNAs in the nucleus of cancer cells. Cancer Cell. *18*, 303–315.

Memczak, S., Jens, M., Elefsinioti, A., Torti, F., Krueger, J., Rybak, A., Maier, L., Mackowiak, S.D., Gregersen, L.H., Munschauer, M., *et al.* (2013). Circular RNAs are a large class of animal RNAs with regulatory potency. Nature *495*, 333–338.

Mitchell, P.S., Parkin, R.K., Kroh, E.M., Fritz, B.R., Wyman, S.K., Pogosova-Agadjanyan, E.L., Peterson, A., Noteboom, J., O'Briant, K.C., Allen, A., *et al.* (2008). Circulating microRNAs as stable blood-based markers for cancer detection. Proc. Natl. Acad. Sci. U.S.A. *105*, 10513–10518.

Miyoshi, K., Miyoshi, T., and Siomi, H. (2010). Many ways to generate microRNA-like small RNAs: non-canonical pathways for microRNA production. Mol. Genet. Genomics *284*, 95–103.

Mukherji, S., Ebert, M.S., Zheng, G.X., Tsang, J.S., Sharp, P.A., and van Oudenaarden, A. (2011). MicroRNAs can generate thresholds in target gene expression. Nat. Genet. *43*, 854–859.

Mullokandov, G., Baccarini, A., Ruzo, A., Jayaprakash, A.D., Tung, N., Israelow, B., Evans, M.J., Sachidanandam, R., and Brown, B.D. (2012). High-throughput assessment of microRNA activity and function using microRNA sensor and decoy libraries. Nat. Methods *9*, 840–846.

Nam, Y., Chen, C., Gregory, R.I., Chou, J.J., and Sliz, P. (2011). Molecular basis for interaction of let-7 microRNAs with Lin28. Cell *147*, 1080–1091.

Noland, C.L., Ma, E., and Doudna, J.A. (2011). siRNA repositioning for guide strand selection by human Dicer complexes. Mol. Cell. *43*, 110–121.

Obad, S., dos Santos, C.O., Petri, A., Heidenblad, M., Broom, O., Ruse, C., Fu, C., Lindow, M., Stenvang, J., Straarup, E.M., *et al.* (2011). Silencing of microRNA families by seed-targeting tiny LNAs. Nat. Genet. *43*, 371–378.

Okada, C., Yamashita, E., Lee, S.J., Shibata, S., Katahira, J., Nakagawa, A., Yoneda, Y., and Tsukihara, T. (2009). A high-resolution structure of the pre-microRNA nuclear export machinery. Science *326*, 1275–1279.

Paddison, P.J., Caudy, A.A., Bernstein, E., Hannon, G.J., and Conklin, D.S. (2002). Short hairpin RNAs (shRNAs) induce sequence-specific silencing in mammalian cells. Genes Dev.*16*, 948–958.

Park, J.E., Heo, I., Tian, Y., Simanshu, D.K., Chang, H., Jee, D., Patel, D.J., and Kim, V.N. (2011). Dicer recognizes the 5′ end of RNA for efficient and accurate processing. Nature 475, 201–205.

Paul, C.P., Good, P.D., Winer, I., and Engelke, D.R. (2002). Effective expression of small interfering RNA in human cells. Nat. Biotechnol. *20*, 505–508.

Pichard, V., Aubert, D., Boni, S., Battaglia, S., Ivacik, D., Nguyen, T.H., Arbuthnot, P., and Ferry, N. (2012). Specific microRNA-regulated TetR-KRAB transcriptional control of transgene expression in viral vector-transduced cells. PLoS One 7, e51952.

Poliseno, L., Salmena, L., Zhang, J., Carver, B., Haveman, W.J., and Pandolfi, P.P. (2010). A coding-independent function of gene and pseudogene mRNAs regulates tumour biology. Nature *465*, 1033–1038.

Premsrirut, P.K., Dow, L.E., Kim, S.Y., Camiolo, M., Malone, C.D., Miething, C., Scuoppo, C., Zuber, J., Dickins, R.A., Kogan, S.C., *et al.* (2011). A rapid and scalable system for studying gene function in mice using conditional RNA interference. Cell *145*, 145–158.

Pritchard, C.C., Cheng, H.H., and Tewari, M. (2012). MicroRNA profiling: approaches and considerations. Nat. Rev. Genet. *13*, 358–369.

Rau, F., Freyermuth, F., Fugier, C., Villemin, J.P., Fischer, M.C., Jost, B., Dembele, D., Gourdon, G., Nicole, A., Duboc, D., *et al.* (2011). Misregulation of miR-1 processing is associated with heart defects in myotonic dystrophy. Nat. Struct. Mol. Biol. *18*, 840–845.

Raver-Shapira, N., Marciano, E., Meiri, E., Spector, Y., Rosenfeld, N., Moskovits, N., Bentwich, Z., and Oren, M. (2007). Transcriptional activation of miR-34a contributes to p53-mediated apoptosis. Mol. Cell. *26*, 731–743.

Rettig, G.R., and Behlke, M.A. (2012). Progress toward *in vivo* use of siRNAs-II. Mol. Ther. *20*, 483–512.

Rother, S., and Meister, G. (2011). Small RNAs derived from longer non-coding RNAs. Biochimie *93*, 1905–1915.

Salmena, L., Poliseno, L., Tay, Y., Kats, L., and Pandolfi, P.P. (2011). A ceRNA hypothesis: the Rosetta Stone of a hidden RNA language? Cell *146*, 353–358.

Scherer, L.J., Frank, R., and Rossi, J.J. (2007). Optimization and characterization of tRNA-shRNA expression constructs. Nucleic Acids Res. *35*, 2620–2628.

Siolas, D., Lerner, C., Burchard, J., Ge, W., Linsley, P.S., Paddison, P.J., Hannon, G.J., and Cleary, M.A. (2005). Synthetic shRNAs as potent RNAi triggers. Nat. Biotechnol. *23*, 227–231.

Sui, G., Soohoo, C., Affar el, B., Gay, F., Shi, Y., and Forrester, W.C. (2002). A DNA vector-based RNAi technology to suppress gene expression in mammalian cells. Proc. Natl. Acad. Sci. U.S.A. *99*, 5515–5520.

Suzuki, H.I., Yamagata, K., Sugimoto, K., Iwamoto, T., Kato, S., and Miyazono, K. (2009). Modulation of microRNA processing by p53. Nature *460*, 529–533.

Tabara, H., Sarkissian, M., Kelly, W.G., Fleenor, J., Grishok, A., Timmons, L., Fire, A., and Mello, C.C. (1999). The rde-1 gene, RNA interference, and transposon silencing in C. elegans. Cell 99, 123–132.

Takeshita, F., Patrawala, L., Osaki, M., Takahashi, R.U., Yamamoto, Y., Kosaka, N., Kawamata, M., Kelnar, K., Bader, A.G., Brown, D., *et al.* (2010). Systemic delivery of synthetic microRNA-16 inhibits the growth of metastatic prostate tumors via down-regulation of multiple cell-cycle genes. Mol. Ther. *18*, 181–187.

Tolia, N.H., and Joshua-Tor, L. (2007). Slicer and the argonautes. Nat. Chem. Biol. 3, 36–43.

Treiber, T., Treiber, N., and Meister, G. (2012). Regulation of microRNA biogenesis and function. Thromb. Haemost. *107*, 605–610.

Valadi, H., Ekstrom, K., Bossios, A., Sjostrand, M., Lee, J.J., and Lotvall, J.O. (2007). Exosome-mediated transfer of mRNAs and microRNAs is a novel mechanism of genetic exchange between cells. Nat. Cell Biol. *9*, 654–659.

Viswanathan, S.R., and Daley, G.Q. (2010). Lin28, A MicroRNA Regulator with a Macro Role. Cell *140*, 445–449.

Wu, C., Lin, J., Hong, M., Choudhury, Y., Balani, P., Leung, D., Dang, L.H., Zhao, Y., Zeng, J., and Wang, S. (2009). Combinatorial control of suicide gene expression by tissue-specific promoter and microRNA regulation for cancer therapy. Mol. Ther. *17*, 2058–2066.

Xia, H., Mao, Q., Paulson, H.L., and Davidson, B.L. (2002). siRNA-mediated gene silencing *in vitro* and *in vivo*. Nat. Biotechnol. *20*, 1006–1010.

Xie, Z., Wroblewska, L., Prochazka, L., Weiss, R., and Benenson, Y. (2011). Multi-input RNAi-based logic circuit for identification of specific cancer cells. Science *333*, 1307–1311.

Yamagata, K., Fujiyama, S., Ito, S., Ueda, T., Murata, T., Naitou, M., Takeyama, K., Minami, Y., O'Malley, B.W., and Kato, S. (2009). Maturation of microRNA is hormonally regulated by a nuclear receptor. Mol. Cell. *36*, 340–347.

Yamakuchi, M., and Lowenstein, C.J. (2009). MiR-34, SIRT1 and p53, the feedback loop. Cell Cycle *8*, 712–715.

Yang, J.S., Maurin, T., Robine, N., Rasmussen, K.D., Jeffrey, K.L., Chandwani, R., Papapetrou, E.P., Sadelain, M., O'Carroll, D., and Lai, E.C. (2010). Conserved vertebrate mir-451 provides a platform for Dicer-independent, Ago2-mediated microRNA biogenesis. Proc. Natl. Acad. Sci. U.S.A. *107*, 15163–15168.

Yang, W., Chendrimada, T.P., Wang, Q., Higuchi, M., Seeburg, P.H., Shiekhattar, R., and Nishikura, K. (2006). Modulation of microRNA processing and expression through RNA editing by ADAR deaminases. Nat. Struct. Mol. Biol. *13*, 13–21.

Yi, R., Qin, Y., Macara, I.G., and Cullen, B.R. (2003). Exportin-5 mediates the nuclear export of pre-microRNAs and short hairpin RNAs. Genes Dev. *17*, 3011–3016.

Yu, J.Y., DeRuiter, S.L., and Turner, D.L. (2002). RNA interference by expression of short-interfering RNAs and hairpin RNAs in mammalian cells. Proc. Natl. Acad. Sci. U.S.A. *99*, 6047–6052.

Zeng, Y., Wagner, E.J., and Cullen, B.R. (2002). Both natural and designed microRNAs can inhibit the expression of cognate mRNAs when expressed in human cells. Mol. Cell. *9*, 1327–1333.

Zeng, Y., Cai, X., and Cullen, B.R. (2005). Use of RNA polymerase II to transcribe artificial microRNAs. Methods Enzymol. *392*, 371–380.

Zhang, H., Kolb, F.A., Jaskiewicz, L., Westhof, E., and Filipowicz, W. (2004). Single processing center models for human Dicer and bacterial RNase III. Cell *118*, 57–68.

Non-canonical MicroRNA Biogenesis and Function

2

Thomas C. Roberts and Matthew J.A. Wood

Abstract

MicroRNAs (miRNAs) are small, single-stranded non-coding RNAs that act primarily as post-transcriptional negative regulators of gene expression and are implicated in a wide range of cellular processes and pathological conditions. As such, miRNAs are being investigated both as disease biomarkers and as novel therapeutic targets for the treatment of a plethora of disease conditions including cancers, viral infections and genetic disorders. Here we consider non-canonical processing and functions of miRNAs. Alternative sources of miRNAs are described including Dicer-negative miRNAs, mirtrons, endo-shRNAs as well as miRNAs derived from snoRNAs and tRNAs. Non-miRNA small RNA biogenesis pathways are also briefly discussed including piRNAs, endo-siRNAs and tiRNAs. In addition, we discuss the causes and consequences of heterogeneity in processing and post-transcriptional editing of miRNAs. Lastly, we consider non-canonical miRNA functions including miRNAs that activate translation or regulate gene expression at the epigenetic level, interactions between miRNAs and long non-coding RNAs (lncRNAs), regulation of miRNA activity by endogenous miRNA sponges, and the role of miRNAs in the extracellular environment.

Canonical microRNA biogenesis and function

MicroRNAs (miRNAs) are genome-encoded, single-stranded, small (~21–23 nt) RNA molecules which regulate gene expression by acting as guides for ribonucleoprotein complexes. MiRNAs are thus considered endogenous effectors of the RNA interference (RNAi) pathway (Filipowicz *et al.*, 2005). Mature miRNA sequences are embedded within long primary-miRNA (pri-miRNA) transcripts. These transcripts are typically transcribed by RNA Polymerase II (RNAPII) (Lee *et al.*, 2004) and are capped, spliced and polyadenylated as with conventional messenger RNA processing. pri-miRNAs can be either derived from non-coding intergenic transcripts or embedded within the introns of coding genes (i.e. intragenic) (Kim and Kim, 2007). pri-miRNA transcripts contain one or more precursor-miRNA (pre-miRNA) hairpins which are sequentially processed in order to generate the mature single-stranded miRNA. Firstly, the pre-miRNA hairpin is excised from the primary transcript in the nucleus by the Microprocessor complex which consists of DGCR8 (DiGeorge Syndrome Critical Region 8, also known as Pasha) and the ribonuclease (RNase) III family enzyme Drosha (Gregory *et al.*, 2004; Denli *et al.*, 2004). DGCR8 is required for recognition of the miRNA stem–loop (Han *et al.*, 2006; Faller *et al.*, 2010) whereas cleavage by Drosha generates the pre-miRNA hairpin (Lee *et al.*, 2003). Nuclear export of the cleaved precursors is facilitated by the karyopherin Exportin-5 in a RAN-GTP-dependent process (Lund *et al.*, 2004; Yi *et al.*, 2003; Bohnsack *et al.*, 2004). The pre-miRNA is further processed in the cytoplasm by a second RNase III family enzyme, Dicer, which cleaves the hairpin to remove the loop sequence. The remaining RNA duplex consists of the mature miRNA and the complementary passenger strand and has 2 nt 3′ overhangs characteristic of RNase III cleavage (Bernstein *et al.*, 2001). In mammals, the duplex is generally imperfect with one or more mismatches between the strands.

Dicer forms part of the enzyme complex miRISC (miRNA-induced silencing complex) which also contains an RNA binding protein, Argonaute, as a core component. Dicer cleavage and Argonaute loading are coupled, such that the pre-miRNA species is immediately passed to Argonaute after processing by Dicer in a process facilitated by TRBP [trans-activation response (TAR) RNA-binding protein] (Wang *et al.*, 2009; Chendrimada *et al.*, 2005). The duplex RNA species generated by sequential Drosha and Dicer processing is analogous to a small interfering RNA (siRNA) such as that introduced exogenously to induce RNAi (Filipowicz *et al.*, 2005).

Once the dicer cleavage product is loaded into Argonaute, one strand of the duplex is preferentially retained and the other discarded. miRISC loading is asymmetric such that the strand with the most thermodynamically unstable 5′ end is preferentially incorporated. (Khorova *et al.*, 2003; Scharz *et al.*, 2003). The strand that becomes incorporated into miRISC constitutes the mature miRNA species whereas the complementary strand is degraded. The majority of miRNAs exhibit a strong strand bias although the minor strand (called the 'star strand' and denoted miRNA*) is sometimes incorporated and, in some cases, has been shown to be biologically important (Martin *et al.*, 2012).

Human genomes encode four Argonaute proteins (AGO1–4). AGO2 is the dominant Argonaute and has been described as the 'catalytic engine of RNAi' (Liu *et al.*, 2004). Only AGO2 retains the DDH motif at its active site which is a metal ion coordination site required for the catalytic 'slicer' endonuclease activity (Rivas *et al.*, 2005; Tolia and Joshua-Tor, 2007) (i.e. cleavage of the scissile phosphate group between positions 10 and 11 of the small RNA guide). Depletion of AGO2 can be compensated for by Argonautes-1, 3 and 4 illustrating that slicer functionality is dispensable for miRNA activity, and that AGO2 can mediate target repression by multiple mechanisms (Liu *et al.*, 2004).

Once loaded with a mature miRNA, the active miRISC complex is able to scan cytoplasmic target mRNAs for miRNA target sites (which are predominantly found in the 3′ untranslated region (UTR) of target mRNAs) (Tokumaru *et al.*, 2008) and induce gene silencing. The fate of an mRNA targeted by miRISC is determined by the degree of complementarity between the guide and target RNAs. Perfect complementarity leads to slicing of the target and release of miRISC (allowing it to search for its next target) (Hutvágner and Zamore, 2002; Yekta *et al.*, 2004). Conversely, imperfect complementarity leads to translational repression and mRNA decay which require that miRISC remains bound to the target (Zeng *et al.*, 2003). Whereas, plant miRNAs typically bind with complete complementarity to their target transcripts to induce slicing, the majority of mammalian miRNAs form imperfect miRNA-mRNA duplexes. Almost all mammalian miRNAs lack target complementarity at the 10/11 position, thus causing a bulge in the miRNA:mRNA duplex and inhibiting the slicer activity of AGO2 (Meister *et al.*, 2004).

The mature miRNA species is a single stranded ~23 nt RNA molecule with a 3′ hydroxyl group and a 5′ terminal phosphate group. The 5′ phosphate is required for interaction with Argonaute proteins and fits in a conserved basic binding pocket within the MID domain (Ma *et al.*, 2005). The first nucleotide of the RNA guide is unpaired and the following 7 nts (i.e. positions 2 to 8) are known as the miRNA 'seed' region, which is the principal determinant of target recognition. Classically, complete complementarity of the seed region with a cognate mRNA target has been considered the minimum requirement for target recognition although binding of the 3′ end of a miRNA, or central pairing, may compensate for imperfect binding in the seed region (Doench and Sharp, 2004; Kiriakidou *et al.*, 2004; Grimson *et al.*, 2007; Shin *et al.*, 2010). The canonical miRNA processing pathway is shown in Fig. 2.1.

Classically, miRNA-mediated gene silencing has been considered to be at the level of translation as early reports showed a decrease in protein levels in the absence of mRNA-level silencing (Leung and Sharp, 2010). However, recently it has been shown that the majority of miRNA targets are changed at the RNA level (Guo *et al.*,

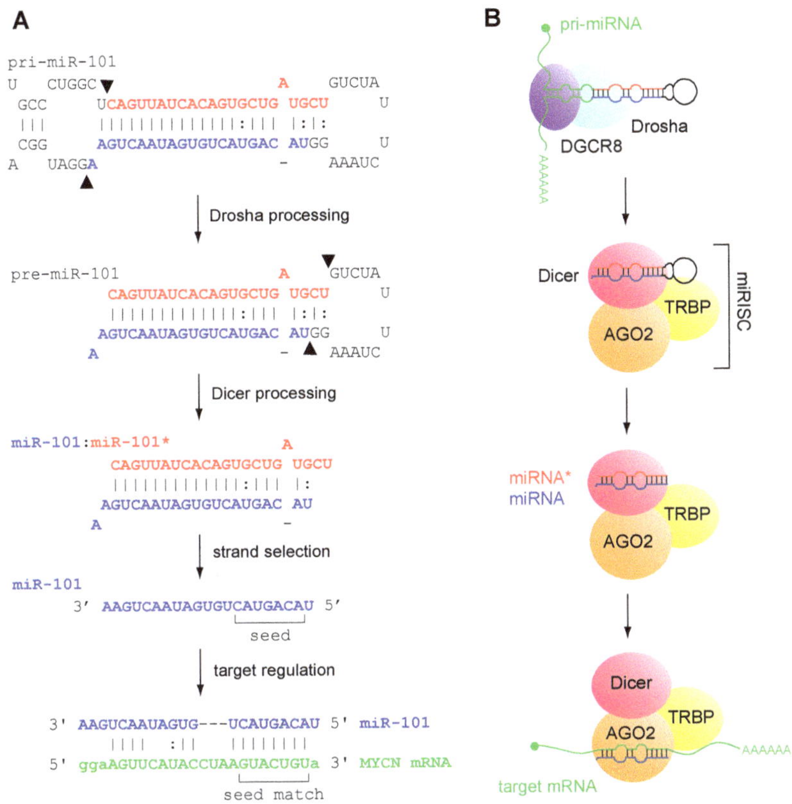

Figure 2.1 The canonical microRNA biogenesis pathway. Canonical miRNA processing is shown for the specific case of hsa-miR-101 (A), and in schematic for the general case (B). MiRNA hairpins are embedded within long pri-miRNA transcripts. The pri-miRNA is processed by Drosha and DGCR8 to produce the pre-miRNA which is exported from the nucleus in a process that is facilitated by Exportin-5. The loop sequence is removed by Dicer, part of the miRISC. TRBP mediates loading of the miRNA duplex into AGO2 where one strand is preferentially retained, and the other (miRNA*) is discarded. The loaded miRNA guides miRISC to partially complementary mRNA targets (e.g. MYCN is a predicted target of miR-101). The seed region is indicted in the case of the mature miR-101. Black triangles indicate RNase III cleavage sites.

2010). A discussion of the various mechanisms by which miRNAs repress the expression of their target transcripts is beyond the scope of this review (see Pillai *et al.*, 2007 for a discussion). To summarize, miRNAs primarily induce gene silencing by translational repression and slicer-independent mRNA decay (Doench and Sharp, 2004; Guo *et al.*, 2010; Olsen and Ambros, 1999; Ohler *et al.*, 2004).

A single miRNA may target hundreds or even thousands of mRNA transcripts. As a result, miRNAs can regulate families of transcripts with related functions and, consequently, a change in expression of a single miRNA may have a dramatic impact upon a cellular pathway or process (Grimson *et al.*, 2007). Conversely, a single mRNA may contain binding sites for multiple miRNAs such that the multiple miRNA regulatory signals are integrated, thereby allowing for additive or combinatorial control. MiRNAs form part of a gene regulatory network and, as such, have been implicated in control of a wide range of cellular functions and disease processes. It is now clear that there are multiple exceptions to the classical miRNA processing pathway and miRNAs have a variety of interesting functions. Here we present a summary of non-canonical miRNA biogenesis and function.

Other small RNA biogenesis pathways

A number of small RNA processing pathways exist in nature which are distinct from miRNA processing (both canonical and non-canonical) in that they produce small RNAs of non-standard size, or are not derived from hairpin precursors. Nevertheless, these small RNAs may utilize components of the miRNA processing pathway for their biogenesis, or may be involved in gene silencing processes. Although not the focus of this review, we consider each briefly below.

piRNAs

The activity of transposable elements in the germline can result in the accumulation of deleterious mutations and/or reduce fertility (Aravin *et al.*, 2007; Carmell *et al.*, 2007) thereby necessitating a means of host cell defence. PIWI-interacting RNAs (piRNAs) are small single-stranded RNA species (26–31 nt) that are expressed in germ cells and function primarily to silence intergenic repetitive regions and transposable elements by post-transcriptional and transcriptional gene silencing. piRNAs differ from miRNAs in a number of important respects. The 5′ terminal nucleotide of a primary piRNA is generally a uridine and the 3′ terminal nucleotide is methylated at the 2′ ribose position by the RNA methyltransferase, HEN1 (Ji and Chen, 2012). The biogenesis of piRNAs, which does not go through a hairpin intermediate, is different from that of miRNAs. piRNAs are generated by a mechanism known as a 'ping-pong' cycle first identified in *Drosophila* (Gunawardane *et al.*, 2007) but also operative in mammals (although with important differences) (Aravin *et al.*, 2008) (Fig. 2.2A). In mice, intergenic repeats, transposons or genomic piRNA clusters are transcribed to produce strand specific piRNA precursors. These are subsequently processed by an unknown cytoplasmic ribonuclease into short single-stranded sense piRNAs with a preference for uridine at the 5′ terminus. These piRNAs are loaded into a PIWI protein (PIWI2 or MILI, which are members of the Argonaute family) and guide it to cleave antisense transcripts thereby generating antisense secondary piRNAs with a preference for adenosine at the 10th nucleotide from the 5′ terminus (Fig. 2.2B). The antisense piRNAs then direct further cleavage of transposon transcripts to generate more sense primary piRNAs. piRNAs bound to PIWI2 facilitate transcriptional silencing of transposon loci in the nucleus. Both PIWI proteins bind to piRNAs, although MILI has a preference for primary piRNAs whereas PIWI2 has a preference for secondary piRNAs. The ping-pong cycle amplifies the production of piRNAs and serves to induce continuous transposon silencing. Tens of thousands of unique piRNA species are generated as a consequence of the ping-pong mechanism and, as a result, piRNAs constitute the largest class of small RNAs in higher organisms. Furthermore, in contrast with miRNAs, individual piRNA sequences are not evolutionarily conserved between mammals although they are derived from syntenic regions.

endo-siRNA

Endogenous-small interfering RNAs (endo-siRNAs) are ~21 nt RNAi effectors derived from long double-stranded RNAs that are generated from long hairpins or when overlapping transcripts hybridize (Fig. 2.2C). These RNA duplexes are subsequently processed by Dicer (Yang and Kazazian, 2006; Kawaji *et al.*, 2008). The resultant siRNAs are capable of silencing complementary transcripts and, like piRNAs, are believed to play a role in the cellular defence against transposable elements. In fact, in mouse embryonic stem cells the majority of endo-siRNAs were shown to originate from a single locus consisting of inverted short interspersed nuclear elements (SINEs) (Babiarz *et al.*, 2008).

tiRNA

Transcription initiation RNAs (tiRNAs or tiny RNAs) are 18–22 nt small single-stranded RNA molecules that are found in the nucleus and map to the region surrounding RNAPII transcription start sites (positions −60 to +120) (Taft *et al.*, 2009a; 2010). Additionally, tiRNAs are the same sense with respect to the associated RNAPII gene. tiRNAs were detected in small RNA libraries prepared from *Dicer-2* negative *Drosophila* cells, suggesting that they are likely not generated by RNAse III cleavage. Instead, it is possible that tiRNAs are generated by polymerase backtracking

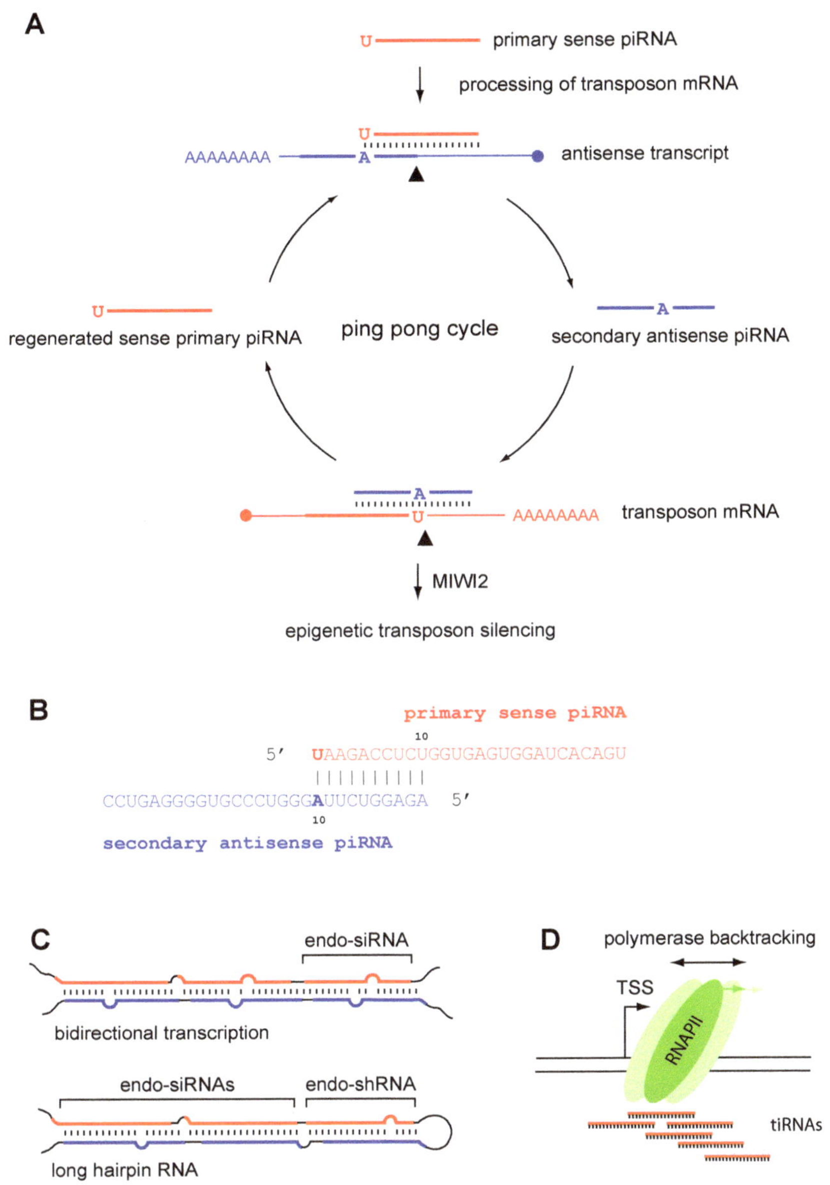

Figure 2.2 Small RNA biogenesis pathways. (A) The mammalian ping-pong cycle of piRNA biogenesis and amplification. Primary sense piRNAs are processed from intergenic repeats, transposons or genomic piRNA precursor clusters by an unknown nuclease. These piRNAs guide PIWI2 or MILI to cleave complementary antisense transcripts leading to secondary antisense piRNAs. Secondary piRNAs cleave further precursor transcripts to generate further primary piRNAs. Black triangles indicate the locations of slicer cleavage. piRNAs silence transposons and repetitive elements by post-transcriptional and transcriptional gene silencing pathways. (B) Example of characteristic piRNAs generated by ping-pong cycle amplification. The primary sense piRNA and secondary antisense piRNAs are offset by 10 nts as a result of slicer-mediated cleavage. Primary sense piRNAs have a preference for U at position one thus accounting for the complementary A at position 10 of the secondary antisense piRNA. (Adapted from Aravin *et al.*, 2008). (C) Generation of endo-siRNAs by Dicer processing of overlapping bidirectional transcripts or long hairpin RNAs. The terminal hairpin of long hairpin RNAs are similarly processed by Dicer to form endo-shRNAs but can also be considered *bona fide* miRNAs on account of their ~21 nt length and the fact that they are processed from hairpin precursors. (D) Generation of tiRNAs (in the sense orientation only) as a result of RNA polymerase II (RNAPII) backtracking in the region surrounding transcription start sites (TSS).

(Fig. 2.2D) (Taft *et al.*, 2009a). Small RNAs with similar properties were also identified which mapped to splice junctions and CTCF (CCCTC-motif binding factor) binding sites. CTCF regulates chromatin macro-structure by facilitating gene loop formation (Ling *et al.*, 2006). A recent study by Taft *et al.* (2011) showed that there is an inverse relationship between tiRNA activity and CTCF binding, as determined by chromatin immunoprecipitation, suggesting that tiRNAs act as epigenetic regulators.

Non-canonical microRNA biogenesis

It is now clear that there are multiple alternative routes to generate a mature miRNA species. These typically bypass one or more processing steps of the canonical miRNA pathway or are derived from alternative RNA sources.

Dicer-independent microRNAs

Two recent reports showed that maturation of the miR-451 (involved in erythroid differentiation) was independent of Dicer processing (Cheloufi *et al.*, 2010; Yang *et al.*, 2010). Instead, pre-miR-451 is cleaved by the endonucleolytic slicer activity of Argonaute-2 which generates a long single-stranded RNA which is subsequently trimmed by exonucleases in order to generate the mature miR-451 sequence (Fig. 2.3). As a result, miR-451 has a well-defined 5′ end (from Drosha cleavage) and a more variable 3′ end that forms part of the precursor loop sequence. The longest templated sequencing reads showed that the 3′ end of miR-451 extended as far as the Argonaute-2 cleavage site. Transgenic mice in which the endonuclease activity of Argonaute-2 was ablated were found to die prematurely and were visibly anaemic as a result of impaired miR-451 processing (Cheloufi *et al.*, 2010). Maturation of miR-451 required Drosha processing but was independent of Dicer cleavage as the expression of miR-451 was unaffected in Dicer knockout cells. Interestingly, the unique processing pathway of this essential miRNA may explain why the catalytic slicer activity of human Argonaute-2 has been retained by evolution despite the vast majority of miRNAs functioning by slicer-independent mechanisms. Interestingly, Argonaute-2 has recently been shown to associate with pre-miRNAs and long unstructured RNAs independent of Dicer suggesting a potential new role for Argonaute proteins in miRNA processing (Tan *et al.*, 2009a).

Mirtrons

Mirtrons are miRNAs derived from introns where the 5′ and 3′ ends of the miRNA are defined by cleavage by the spliceosome (Sibley *et al.*, 2012). In other words, mirtrons are introns which consist entirely of a pre-miRNA hairpin. Processing of mirtrons is dependent on Dicer, splicing and intron lariat debranching, but independent of Drosha cleavage. Mirtrons differ from other intron-encoded miRNAs that can be Drosha processed before the splicing reaction takes place (Kim and Kim, 2007). Mirtrons were first identified in *Drosophila* where deep sequencing approaches detected small RNAs with the potential to form hairpin structures that originated from short introns (the size of a pre-miRNA) (Ruby *et al.*, 2007; Okamura *et al.*, 2007). After generation

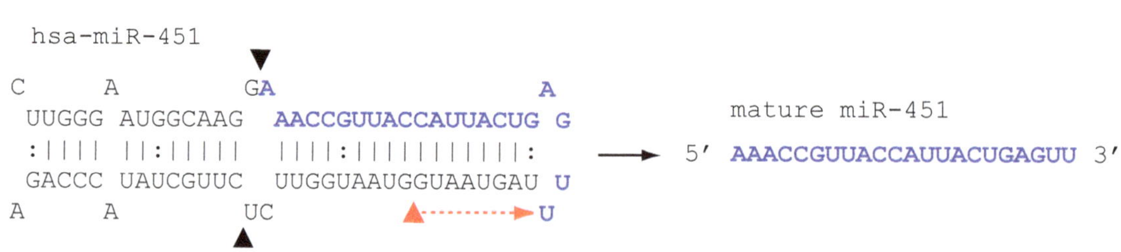

Figure 2.3 Dicer-independent processing of miR-451. The miR-451 primary transcript is cleaved by Drosha at the sites indicated by back triangles. The hairpin is loaded into AGO2 which cleaves the duplex at the site indicated by the red triangle. The resulting RNA is then trimmed by exonuclease activity (red dotted line) in order to generate the mature miR-451 sequence.

of the pre-miRNA by splicing and debranching, mirtron-derived miRNAs can be further processed by the canonical miRNA biogenesis pathway (Fig. 2.4). Mirtrons have also been identified in mammalian cells. For example, Barbiaz *et al.* (2008) identified mirtrons by small RNA sequencing of Dicer and Drosha knockout mouse embryonic stem cells. However, in higher organisms introns are typically large (in the kilobase range) and so mirtrons are relatively rare and represent only a small fraction of vertebrate miRNAs. Another class of mirtron-related miRNAs is the tailed mirtron (also known as semi-mirtron). These miRNAs are also produced by the splicing reaction but, unlike conventional mirtrons, the 5′ or 3′ terminus consists of a tail sequence between the miRNA hairpin and the corresponding splice site. This tail sequence is trimmed by the RNA exosome or other unknown nucleases in order to produce the pre-miRNA after splicing has taken place (Flynt *et al.*, 2010).

Recently, Havens *et al.* (2012) identified a new class of mirtrons which they term 'simtrons' (splicing-independent mirtron-like miRNAs). The predicted mirtrons miR-1225 and miR-1228 were found to have the potential to elicit gene silencing in HEK293T cells and immunoprecipitated with Argonaute proteins suggesting they are bone fide RNAi effectors. However, processing of these miRNAs was independent of splicing and did not require the RNAi components DGCR8, Dicer, Exportin-5 or AGO2. Interestingly, Drosha was required for simtron biogenesis suggesting a novel function for this protein (Havens *et*

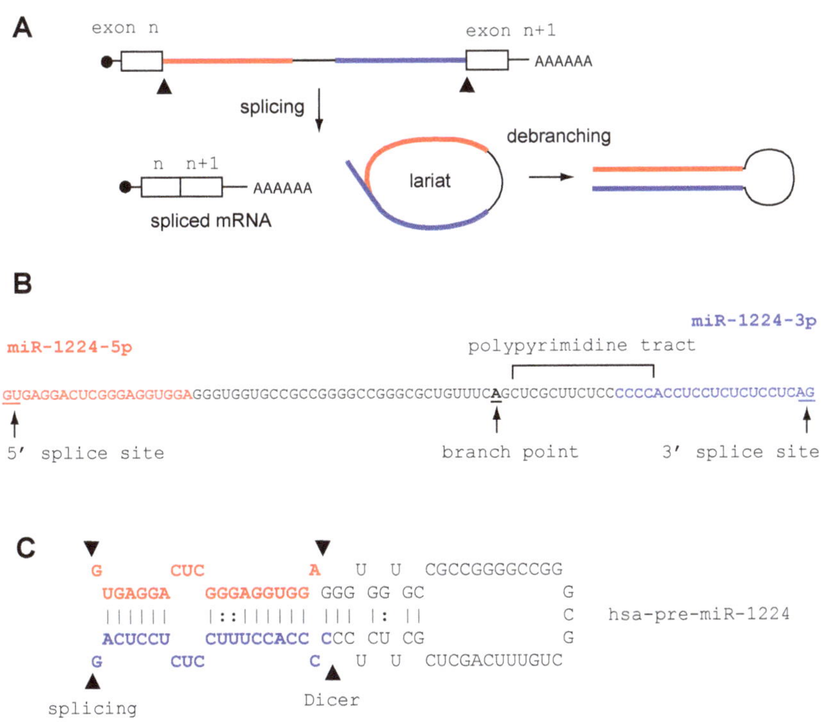

Figure 2.4 Mirton biogenesis. (A) Schematic of mirtron processing. The sequence of the mirtron precursor hairpin is excised by the spliceosome (cleavage points indicated by black triangles). Splicing generates the mature mirtron-containing mRNA and a mirtron lariat. The lariat is debranched by the action of RNA lariat debranching enzyme (DBR1) and refolded to form the pre-miRNA hairpin which is subsequently processed by Dicer as per the canonical miRNA processing pathway. (B) Linear sequence of the mirtron hsa-miR-1224. The splice acceptor sites, branch point and polypyrimidine tract are indicated by arrows. hsa-miR-1224 5′ and 3′ arms are indicated in red and blue respectively. (C) hsa-miR-1224 precursor hairpin structure. Mature miRNA species are indicated in red and blue. Black triangles indicate the sites of spliceosome processing (at the base of the stem) and Dicer processing (at the loop).

al., 2012) although a detailed understanding of simtron biogenesis remains to be elucidated.

endo-shRNA

Endogenous-short hairpin RNAs (endo-shRNAs) are derived from short transcripts (typically from Pol III promoters and terminating with a terminal stretch of uridines) that are processed by Dicer to generate mature miRNA species but, like mirtrons, are independent of processing by Drosha (e.g. miR-320 and miR-484). A unique Drosha-independent miRNA, miR-1980, is derived from the 3′ end of a double hairpin structure. This structure is first cleaved by an unknown nuclease in order to remove the 5′ hairpin which does not generate a mature miRNA, whereas the 3′ hairpin is processed by Drosha (Babiarz *et al.*, 2008). In *Drosophila*, miRNAs have been identified which are derived from the terminal hairpin of an endo-siRNA precursor. In this instance, endo-siRNAs are generated by Dicer cleavage of a long hairpin RNA. The terminus of this hairpin is processed to yield a ~21 nt endo-shRNA (Miyoshi *et al.*, 2010).

snoRNA-derived microRNAs

Small nucleolar RNAs (snoRNAs) are RNA species which are found predominantly in the nucleoli and cajal bodies of eukaryotic cells and primarily act as part of ribonucleoprotein complexes to guide the editing of ribosomal RNAs. snoRNAs fall into two classes, H/ACA box snoRNAs form a dual hairpin structure and guide the position-specific isomerization of a uridine nucleotide to pseudouridine, whereas C/D box snoRNAs act to guide the methylation of 2′ O-ribose (Bratkovič and Rogelj, 2011).

snoRNAs with miRNA-like functions were initially identified in human cells and in the protozoan *Giardia lamblia* (Saraiya and Wang, 2008; Ender *et al.*, 2008). In the case of humans, snoRNA-derived small RNAs were identified by high-throughput sequencing of AGO1 and AGO2 immunoprecipitates. A miRNA derived from the ACA45 snoRNA required processing by Dicer but not Drosha/DGCR8 and was found to target CDC2L6 mRNA (Ender *et al.*, 2008) (Fig. 2.5A). Similarly, in the *Giardia* study, snoRNA-derived miRNAs were identified by small RNA cloning. When *Giardia* cells were transformed with a synthetic snoRNA-derived miRNA the translation of a predicted target reporter gene was repressed. Furthermore, FISH analysis was used to identify the cellular locations of the snoRNA and respective derived miRNA. Whereas the snoRNA was found to be located in the nucleus in only one of the two *Giardia* nuclei the snoRNA-derived miRNA was found to be located in the cytoplasm (Saraiya and Wang, 2008).

miRNAs have been demonstrated to be derived from H/ACA box snoRNAs (Scott *et al.*, 2009; Alawi and Lin, 2010) and C/D box snoRNAs (Brameier *et al.*, 2011; Ono *et al.*, 2011). Analysis of sequencing data of cells from human, mouse, chicken, fission yeast, *Arabidopsis* and *Drosophila* revealed a multitude of snoRNA-derived RNAs (sdRNAs) that showed conservation of both position and size. sdRNAs show distinct size distributions and positions depending on the class of host snoRNA. H/ACA box snoRNAs produced small RNA species of 20–24 nt in length originating at the 3′ end of the host transcript, whereas two kinds of RNA species were produced from C/D box snoRNA (17–19 nt and > 27 nt). Both of these species were primarily derived from the 5′ end of the host snoRNA (Fig. 2.5B). Interestingly, this study also showed that many highly abundant sdRNAs were derived from lowly expressed host snoRNAs. Taken together these observations suggest that sdRNAs are the result of a specific regulated processing pathway rather than degradation of snoRNA transcripts (Taft *et al.*, 2009b). Analysis of deep sequencing data from Dicer knockout cells suggests that, in contrast with H/ACA box snoRNAs, C/D box snoRNAs are not processed by Dicer (Langenberger *et al.*, 2013). A number of snoRNAs have no known target ribosomal RNAs (so-called 'orphan snoRNAs'). An intriguing possibility is that these orphan snoRNAs function primarily as carriers for their miRNAs.

tRNA-derived microRNAs

Deep sequencing studies have identified a plethora of abundant small RNA species which map to transfer RNA (tRNA) genes (transcribed by RNA Pol III) (Schopman *et al.*, 2010). tRNA-derived RNA fragments (tRFs) were found to originate from both the 5′ and 3′ tRNA termini,

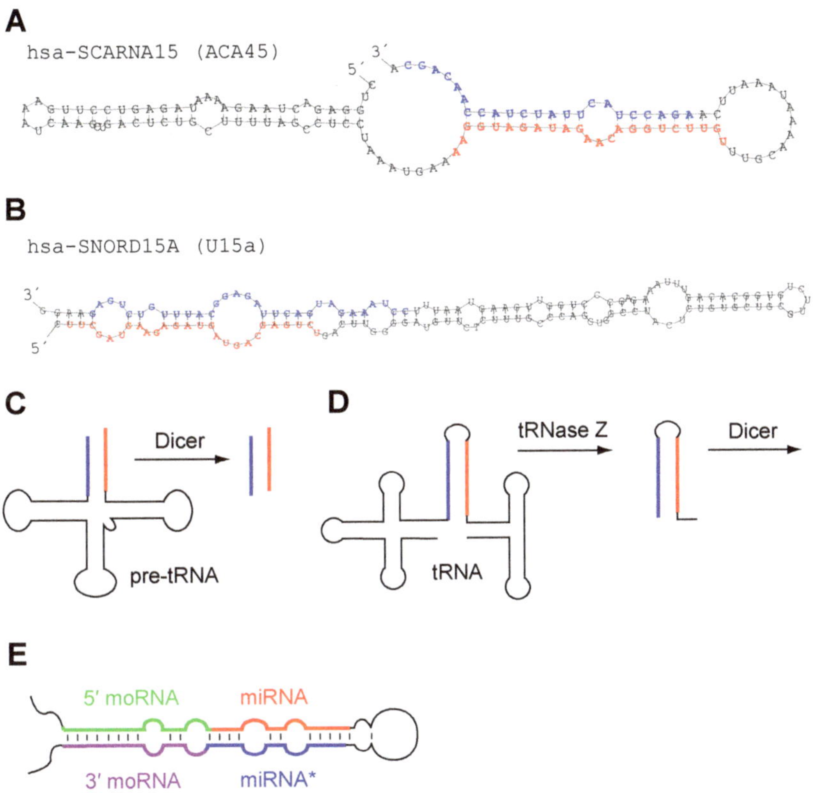

Figure 2.5 Non-canonical microRNA processing. (A) miRNA sequences derived from the 3′ end of a H/ACA box snoRNA (SCARNA15). (B) miRNA sequences derived from C/D box snoRNA (SNORD15A). (C) Cleavage of pre-tRNA structures by Dicer produces miRNA-like sequences. (D) Processing of tRNAs by tRNase Z liberates the pre-miRNA hairpin which is subsequently processed by Dicer to produce mature miRNA sequences. (E) MicroRNA-offset RNAs (moRNAs) are processed from the sequences immediately adjacent to some pre-miRNA hairpins.

and from the 3′ pre-tRNA trailer sequences (Lee *et al.*, 2009). Whether these small RNAs qualify as miRNAs is a matter of debate. Processing of 5′ tRFs required Dicer (but not Drosha) (Fig. 2.5C) (Cole *et al.*, 2009) whereas 3′ tRFs were cleaved by tRNase Z. tRFs immunoprecipitate with Argonaute proteins and were capable of exerting gene silencing effects in luciferase reporter assays. Taken together, these observations suggest that at least some of these fragments have miRNA-like functionality. Similarly, processing of tRNA tailer sequences by tRNase Z and tRNase P results in 5′ and 3′ fragments of ~18–25 nt that have the potential to hybridize and act as RNAi effectors (Lee *et al.*, 2009). The murine γ-herpesvirus 68 (MHV68) encodes miRNAs that are immediately down-stream of tRNA sequences. The pre-miRNA hairpins are liberated by tRNase Z (instead of Drosha) and the resulting hairpin is then cleaved by Dicer (Reese *et al.*, 2010) (Fig. 2.5D). MiRNA-like small RNAs have also been identified which are derived from other cellular RNAs including vault RNA (vRNA) (Persson *et al.*, 2009), Y RNA (Meiri *et al.*, 2010), small nuclear RNA (snRNA) (Langenberger *et al.*, 2010) and the snaR family of small non-coding RNAs (Langenberger *et al.*, 2013). It is likely that this list of non-canonical miRNA sources is incomplete and further investigation may identify others.

MicroRNA-offset RNAs

A further class of miRNAs was identified in the sea squirt *Ciona intestinalis*. Small RNA sequencing identified multiple miRNA species generated at ~50% of miRNA loci. These variants appeared to be the products of RNase III cleavage and were generated from the sequence immediately adjacent to the pre-miRNA sequence on one or both strands such that up to 4 variants were produced from around the same stem loop (Shi *et al.*, 2009) (Fig. 2.5E). These variants were named miRNA-offset RNAs (moRNAs) and were found to be stable and expressed at levels comparable with their immediately adjacent 'expected' miRNAs. moRNAs have also been detected in mouse hippocampus, ES cells and iPS cells although their biogenesis and biological significance are currently not well understood (Babiarz *et al.*, 2008; Zhou *et al.*, 2012).

Virus-encoded microRNAs

As alluded to above, the genomes of some viruses encode miRNAs that regulate the expression of both viral and host cell genes (Takane and Kanai, 2011; Pfeffer *et al.*, 2005). Similarly, some viruses are dependent on host cell miRNAs (e.g. replication of hepatitis C virus (HCV) requires miR-122 expression (Jopling *et al.* 2005), discussed below). Interestingly, in infected cells, adenovirus expressed high levels of a 160 nt non-coding RNA, VA1, which acts to globally inhibit miRNA activity. VA1 competes with pre-miRNAs for Exportin-5 binding thereby interfering with miRNA processing and nuclear export whereas gene silencing by siRNAs was unaffected. Global inhibition of miRNA regulation resulted in enhanced viral replication and infectivity (Lu and Cullen, 2004).

IsomiRs and microRNA editing

Currently, miRNAs are assigned a single annotated sequence in miRBase (Griffiths-Jones *et al.*, 2008). However, next generation sequencing of small RNAs (small RNA-seq) has revealed that many miRNAs are heterogeneous in length or sequence, and that the annotated sequence represents the dominant isoform. MiRNA isomers are collectively referred to as isomiRs which can be classified into three types: (1) 5′ isomiRs, (2) 3′ isomiRs, and (3) polymorphic isomiRs (Fig. 2.6). The first two types have termini which differ with respect to the annotated sequence. These can arise for a number of reasons. Firstly, heterogeneity in Drosha/Dicer processing can result in mature miRNA sequences of variable length or terminal identity (Wu *et al.*, 2009). Similarly, TRBP can also alter the processing fidelity of pre-miRNAs (Lee and Doudna, 2012). Secondly, the mature miRNA can be subject to exonucleolytic degradation post-Drosha/Dicer processing. The 3′ ends of mature miRNAs, protrude from the PAZ domain of Argonaute and are therefore susceptible to nucleolytic trimming or 'nibbling,' whereas the 5′ miRNA termini are bound in an internal pocket within the MID domain and are therefore protected from exonucleolytic attack. As a result, 3′ isomiRs are the most common type of isomiRs.

Whereas heterogeneity in Dicer/Drosha processing or differential exonucleolytic processing yield isomiRs which align with the genomic pre-miRNA locus (so called templated isomiRs), other types of 5′ or 3′ isomiRs contain nucleotides which do not (i.e. non-templated isomiRs). Typically, this is due to the activity of nucleotidyl transferases. Seven nucleotidyl transferases have been identified in humans which catalyse the template independent addition of adenosine or uridine nucleotides to the 3′ ends of mature miRNA (Burroughs *et al.*, 2010; Wyman *et al.*, 2011). In plants miRNA uridylation and adenylation have the opposing effects of promoting degradation and stabilization respectively, although in animals the consequences of these 3′ additions are less clear. Given the non-specific nature of nucleotide addition to miRNAs, adenylation or uridylation may be mechanisms of influencing global miRNA activity.

Polymorphic isomiRs are miRNA isomers that contain internal mismatches to the canonical miRNA sequence (and therefore contain non-templated nucleotides). Typically this is due to post-transcriptional RNA editing. RNA editing is a widespread phenomenon which adds to the complexity of the mammalian transcriptome and proteome. Although much investigation has focused on modified nucleotides in coding regions which alter translation it is now also clear that extensive RNA editing also occurs in non-coding transcripts, including miRNAs. Editing of miRNA results in a variety of functional outcomes, although the degree to which miRNA

A

```
       hsa-miR-99b                             hsa-miR-99b*                          Count    RPM
....CCCACCCGUAGAACCGACCUUGC..........................................  17       71.2
....CCCACCCGUAGAACCGACCUUGCG.........................................  16       69.9
....CCCACCCGUAGAACCGACCUUG...........................................  3        17.2
....CCCACCCGUAGAACCGACCUU............................................  1        7.02
.....CCACCCGUAGAACCGACCUUGCG.........................................  13       1.61
......CACCCGUAGAACCGACCUUGCG.........................................  18148    624
......CACCCGUAGAACCGACCUUGC..........................................  3121     109
......CACCCGUAGAACCGACC..............................................  368      5.74
......CACCCGUAGAACCGACCUUG...........................................  175      34.9
......CACCCGUAGAACCGACCUUGCGG........................................  95       1.87
......CACCCGUAGAACCGACCUU............................................  31       0.674
......CACCCGUAGAACCGACCU.............................................  20       0.359
......CACCCGUAGAACCGACCUUGCGGG.......................................  6        0.1
.......ACCCGUAGAACCGACCUUGCG.........................................  122      2.79
.......ACCCGUAGAACCGACCUUGC..........................................  19       0.3
........CCCGUAGAACCGACCUUGCG.........................................  2        7.75
.........CCGUAGAACCGACCUUGCG.........................................  1        0.227
..........CGUAGAACCGACCUUGCG.........................................  8        0.56
...........................................CAAGCUCGUGUCUGUGGGUCCGU...  408      8.98
...........................................CAAGCUCGUGUCUGUGGGUCCG....  380      12.6
...........................................CAAGCUCGUGUCUGUGGGUC......  52       1.21
...........................................CAAGCUCGUGUCUGUGGGUCC.....  29       1.91
...........................................CAAGCUCGUGUCUGUGGGU.......  8        0.106
...........................................CAAGCUCGUGUCUGUGGG........  2        0.0282
GGCACCCACCCGUAGAACCGACCUUGCGGGCCUUCGCCGCACACAAGCUCGUGUCUGUGGGUCCGUGUC
(((((..(((((((((..(((.((((.(((((....))).).).)))).)))..)))))))))..)))))
```

B **schematic of hypothetical isomiRs**

```
......CACCCGUIGAACCGACCUUGCG.......  polymorphic isomiR
.......ACCCGUAGAACCGACCUUGCG.......  templated 5′ isomiR
.....ACACCCGUAGAACCGACCUUGCG.......  non-templated 5′ isomiR
....CCCACCCGUAGAACCGACCUUGCG.......  templated 5′ isomiR
......CACCCGUAGAACCGACCUUGC........  templated 3′ isomiR
......CACCCGUAGAACCGACCUUGCGUU.....  non-templated 3′ isomiR
......CACCCGUAGAACCGACCUUGCGGG.....  templated 3′ isomiR
......CACCCGUAGAACCGACCUUGCG.......  annotated mature miRNA
GGCACCCACCCGUAGAACCGACCUUGCGGGGCCUU  pri-miRNA sequence
(((((..(((((((((..(((.((((.(((((...  hairpin structure
```

Figure 2.6 MicroRNA isomiRs. (A) Heterogeneity in miRNA processing in the case of hsa-miR-99b. Data are taken from miRBase (Griffiths-Jones *et al.* 2008). Linear sequence of pri-miR-99b is show at the bottom of the panel and miRNA variants derived the hairpin are shown in the top of the panel. Read counts and reads per million (RPM) summed from small RNA sequencing experiments (36267 total reads across 39 experiments) for each variant are shown on the right of the panel and reflect their relative abundances. Canonical miR-99b and miR-99b* sequences are indicated in red and the highlighted sequencing statistics indicate that these are the dominant variants derived from each arm of the hairpin. The 5′ arm dominates over the 3′ 'star-form' arm. Multiple 5′ and 3′ isomiRs are generated from the miR-99b hairpin. Note that some isomiRs derived from the 5′ arm are more abundant than the dominant species generated from the 3′ arm. (B) Schematic of hypothetical isomiRs. Hypothetical examples of templated, non-templated, 5′, 3′ and polymorphic isomiRs are shown. Important differences are highlighted in red.

editing is a stochastic or regulated process is currently unknown.

The ADAR (adenosine deaminase acting on RNA) family of enzymes catalyse the hydrolytic deamination of adenosine to form inosine (Fig. 2.7A). Inosine is read as guanine by the translation machinery and so ADAR editing adds further complexity to the proteome. For example, A-to-I editing leading to a glutamine (CAG) to arginine (CIG) conversion in the mammalian AMPA receptor subunit (GluR-B) renders the receptor impermeable to Ca^{2+} ions (Higuchi *et al.*, 1993; Seeberg and Hartner, 2003). ADARs recognize double-stranded RNA substrates that adopt fold-back structures and, as a result, frequently target miRNA hairpin structures (Nishikura 2010). ADAR-mediated miRNA editing was first demonstrated for miR-22 in human brain and testis (Luciano *et al.*, 2004). Studies that have tried to determine the frequency of A-to-I editing in

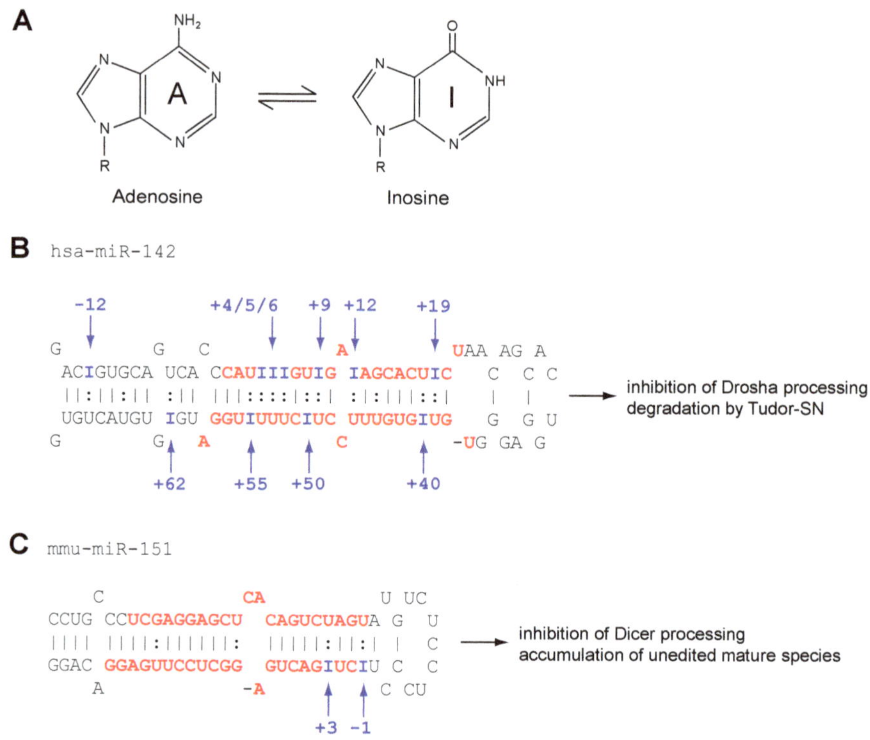

Figure 2.7 Adenosine-to-inosine editing of microRNAs. (A) Conversion of adenosine to inosine by deamination. R indicates position of ribose ring. (B) Multiple A-to-I editing events occur in the hsa-miR-142 precursor hairpin. Editing at the base of the stem loop inhibits processing by Drosha and leads to degradation by Tudor-SN. (C) A-to-I editing adjacent to the loop region of mmu-miR-151 inhibits Dicer processing leading to accumulation of the unedited mature species.

human tissues have estimated between 6% (Blow *et al.*, 2006) and 16% (Kawahara *et al.*, 2008) by direct sequencing of RT-PCR products compared with genomic DNA from the same individual. Both pri-miRNAs and mature miRNAs are subject to A-to-I editing with UAG trinucleotides being preferentially targeted in ~50% of cases (Blow *et al.*, 2006; Yang *et al.*, 2006). The editing of pri-miRNAs suggests that the editing machinery is active early in miRNA biogenesis.

A-to-I editing can have a multitude of effects on both the biogenesis and function of miRNAs and has been implicated in human disease. [Some cancers exhibit changes in ADAR activity and RNA editing (Paz *et al.*, 2007) and editing has also been reported for some viral miRNAs (Pfeffer *et al.*, 2005).] ADAR-mediated miRNA editing has been shown to modulate Drosha processing. In the case of pri-miR-142, U·I and I·U wobble base pairs introduced in the stem loop structure inhibit cleavage by Drosha (Yang *et al.*, 2006) (Fig. 2.7B). Inosine containing stem loop structures are subsequently degraded by the RISC component Tudor-SN (Tudor staphylococcal nuclease) (Scadden, 2005). Similarly, A-to-I editing has also been shown to affect processing by Dicer. The introduction of 2 U·I wobbles in the miR-151 primary hairpin did not affect Drosha processing but instead inhibited processing by Dicer (Fig. 2.7C). As a result accumulation of edited mature species was inhibited (Blow *et al.*, 2006; Kawahara *et al.*, 2007a). In summary, introduction of inosine nucleotides into the primary miRNA stem loop can result in reduced generation of the mature miRNA species, changes in the ratio of edited:unedited miRNA molecules, or alternatively target the hairpin for degradation. The activity of ADARs and Tudor-SN can thus be considered antagonistic with respective to the RNAi pathway.

Editing of the mature miRNA sequence also affects the targeting properties of the respective miRNA. Editing of sequences important for target recognition can lead to either impaired miRNA-mediated regulation or re-targeting of a miRNA such that it regulates a different set of target transcripts (Kawahara *et al.*, 2007b). For example, miR-376 is subject to A-to-I editing in its seed region which redirects it to target new mRNAs (e.g. PRPS1). Similarly, if a miRNA is edited such that it starts with a non-canonical 5′ nucleotide the seed sequence will be shifted leading to dramatic changes in miRNA targeting or strand selection preference (Humphreys *et al.*, 2012). It should be noted that 5′ isomiRs constitute the minority of edited miRNAs. The reason for this is probably due to the deleterious effects of inducing a shift in the seed sequence given the importance of this region for target recognition. Conversely, canonical editing sites are enriched at miRNA target sites in 3′ UTRs suggesting that mRNA target modification has a regulatory function by altering the preference for specific cognate miRNA binding partners (Gu *et al.*, 2012). Investigation of A-to-I editing events in target transcripts using a computational approach identified edited nucleotides that disrupted miRNA target sites. However, these sites were found to occur at frequencies below what is expected by chance, thus suggesting an evolutionary pressure to preserve the fidelity of miRNA recognition. Furthermore, ~200 editing events generated novel miRNA seed recognition motifs (Liang and Landweber, 2007).

Detection of isomiR variants occurs at frequencies above the error rates of sequencing technologies (Lee *et al.*, 2010) and association of isomiRs with RISC components and polysomes (Cloonan *et al.*, 2011) suggests that they are functional. However, the extent to which they are biologically important remains to be determined. A 5′ variant of miR-101 is highly abundant and was shown to participate in RISC mediated silencing suggesting it is a functional miRNA (Llorens *et al.*, 2013). A major unanswered question is the degree to which the generation of isomiR biogenesis is a regulated process. A recent study by Fernandez-Valverde *et al.* (2010) showed that isomiRs are dynamically regulated during *Drosophila* development. Similarly, miRNA editing has been shown to be tissue specific, although this may simply reflect the differential ADAR expression (Kawahara *et al.*, 2007b). An interesting possibility is that specific isomiRs exhibit differential loading into Argonaute proteins (Burroughs *et al.*, 2011; Azuma-Mukai *et al.*, 2008). In *Drosophila* and plants the various Argonaute proteins have non-redundant functions (Hutvagner and Simard, 2008; Mallory and Vaucheret, 2010) although in mammals this distinction is less clear. The functional outcome of differential loading of isomiRs into Argonaute proteins is currently unknown.

Other RNA modifications

Several other covalent modifications of RNA sequences have recently been identified with the potential to modulate miRNA biogenesis or function. A study by Xhemalce *et al.* (2012) showed that an RNA methyltransferase BCDIN3D can regulate miRNA processing in human cells. BCDIN3D dimethylates the 5′ terminal phosphate group of miRNA precursors leading to an inhibition of Dicer processing and reduced accumulation of the mature miRNA species. Dimethylation of the terminal phosphate reduces its negative charge, thereby impairing its recognition by Dicer.

In a separate study, Meyer *et al.* (2012) utilized methyl RNA immunoprecipitation and RNA sequencing (MeRIP-seq) in order to profile the abundance of N^6-methyladenosine (m6A) modifications in the human and mouse transcriptomes. N^6-methylation of adenosine residues was found to be widespread and reversible (via the action of the RNA demethylase, FTO). Methylated adenosines were enriched at termination codons and in 3′ UTRs of mRNA transcripts. Interestingly, an inverse correlation was observed between miRNA binding sites and m6A modification suggesting a possible relationship between RNA editing and miRNA function. Additionally, the predicted target transcripts of highly abundant miRNAs in the brain were enriched for m6A modifications which raises the intriguing possibility that miRNAs may direct adenosine methylation (Meyer *et al.*, 2012).

Non-canonical microRNA function

Classically, mammalian miRNAs have been considered to repress gene expression at the post-transcriptional level through imperfect binding in the 3′ UTRs of their cognate mRNA targets. MiRNA target sites are principally found in 3′UTR regions for two reasons. Firstly, there is evidence that active translation inhibits miRNA activity. Extending the ORF into the 3′UTR by deletion of a stop codon abrogated miRNA-mediated repression of reporter transcripts, whereas, inclusion of rarer stop codons was able to rescue silencing (Gu *et al.,* 2009). Secondly, untranslated regions are under less selective pressure than coding regions. Consequently, mutations can accumulate in the 3′ UTR leading to subtle inherited changes in gene expression without affecting the translation of the primary amino acid coding sequence.

Although artificial siRNAs can direct RISC-mediated slicer cleavage of an mRNA through perfect complementarity with targets throughout the transcript, early work suggested that miRNA sites within ORFs and 5′ UTR regions were poor targets, possibly as repressive miRISC protein-complexes are displaced by ribosome procession (Grimson *et al.,* 2007). However, Saxena *et al.* (2003) showed that siRNAs with mismatches to their cognate coding region targets can behave as miRNAs and induce translational arrest and mRNA destabilization, thus suggesting that there could be functional miRNA sites within ORFs. This was first confirmed in mammals for miR-148 which targets the mRNA coding region of the DNA methyltransferase, DNMT3B (Duursma *et al.,* 2008). Similarly, computational prediction of conserved miRNA seed matches in coding regions identified Dicer as a target of the let-7 miRNA. As Dicer plays a central role in the maturation of miRNAs this interaction constitutes a negative feedback loop (Forman and Coller, 2008). Furthermore, analysis of proteomics data and predicted let-7 targets identified multiple functional coding region targets (Forman and Coller, 2010). high-throughput sequencing of RNA-cross-linking immunoprecipitation (HITS-CLIP) analysis has revealed that Argonaute–miRNA complexes exhibit substantial binding in coding regions, introns and to non-coding RNA suggesting further complexity in miRNA-mediated gene regulation (Chi et a., 2009) and a plethora of additional coding region targets have since been identified (Shen *et al.,* 2008; Lal *et al.,* 2008; Tay *et al.,* 2008).

5′ UTR targeting

The application of miRNA target-finding algorithms miRanda and TargetScan to both coding and untranslated regions of human, mouse and *Drosophila* genes found that conserved miRNA target sites are abundant in 5′ UTR regions. Furthermore, the density of target sites was higher than in 3′ UTR or coding regions (Zhou *et al.,* 2009). Similarly, in *Drosophila* embryo extracts, binding of miR-2 in the 5′UTR and coding regions was shown to induce translational repression of reporter constructs analogous with interactions in the 3′UTR (Moretti *et al.,* 2010). Using luciferase reporter constructs Lytle *et al.* (2007) showed that miRNAs can repress transcripts with 5′UTR or 3′ UTR targets equally in human cells. Morreti *et al.* (2010) used an immunoprecipitation protocol to pulldown mRNAs that differentially associate with AGO2 after transfection with mimics of either miR-1 or miR-124. A substantial proportion of mRNAs recovered contained seed matches for miR-1/miR-124 in the 5′ UTR or coding region thus providing empirical evidence that non-3′ UTR directed activity of miRNAs is a general phenomenon (Hendrickson *et al.,* 2008). Similarly, Grey at al. showed that a human cytomegalovirus miRNA, miR-US25-1, induces silencing of multiple host genes involved in cell cycle progression through interactions with their 5′UTRs as determined by RISC-immunoprecipitation (Grey *et al.,* 2010).

Interestingly, Lee and co-workers showed that some miRNAs can simultaneously target both untranslated regions. In these cases the classical 3′ UTR site targeted by the 5′ (seed) end of the miRNA sequence was complemented by an interaction in the 5′ UTR targeted by the 3′ end of the miRNA. The presence of target sequences in both UTRs had the effect of enhancing target repression (Lee *et al.,* 2009). The targeting of mRNA transcripts in the coding region and 5′ UTR has been largely overlooked for the reasons discussed above. Given the growing number of studies reporting miRNA regulation of non-canonical regions this area warrants further study. A number of bioinformatics tools have been developed to assist researchers in this area (Kumar *et al.*, 2012; Sacco and Masotti, 2012; Dweep *et al.*, 2011).

MicroRNA-mediated translation activation

miRNAs can also function to activate translation by a variety of mechanisms. Perhaps the most well understood example of miRNA-mediated translational activation (on account of its pharmacological significance) is in the case of the hepatitis C virus (HCV) RNA. Unusually, the liver-specific miRNA, miR-122, is required for HCV replication as cells which do not express miR-122 (i.e. HepG2) are refractory to infection (Jopling *et al.*, 2005). Inhibition of miR-122 using antisense oligonucleotides has been shown to control infection (Lanford *et al.*, 2010). The HCV viral RNA contains three miR-122 target sites; two in the 5′ UTR upstream of the IRES site and one in the 3′ UTR (Jopling *et al.*, 2008). Binding of miR-122 at the 5′ UTR sites does not act to reduce mRNA levels, but rather the AGO–miR-122 complexes protect the viral RNA from exonucleolytic degradation (Shimakami *et al.*, 2012). This stabilization of the mRNA leads to an apparent increase in translation of viral proteins (Henke *et al.*, 2008).

Vasudevan and co-workers showed that differential miRNA-mediated silencing and activation is context dependent. In the case of tumour necrosis factor–alpha (TNF-α), an AU-rich element (ARE) found in the 3′ UTR becomes a translational activation motif following cell cycle arrest. Association of activating factors with the TNFα mRNA was directed by miR-369-3. Similarly, let-7 was shown to mediate gene silencing in proliferating cell cultures but induced up-regulation of translation after cell cycle arrest (Vasudevan *et al.*, 2007). An exogenous small RNA was also able to switch between silencing and activation depending on whether the cells were grown in the presence or absence of serum respectively, thereby demonstrating the generality of this phenomenon. A complex of AGO2 and FXR1 (fragile X mental retardation syndrome-related protein 1) was found to mediate this activation phenomenon. The remarkable findings of this study suggest that the widely reported gene repressive functionality of miRNAs may be a property of proliferating cells.

Several studies have demonstrated that miRNAs can enhance the translation of their cognate mRNA targets through interaction with their 5′ UTRs. For example, miR-10a up-regulates the translation of a number of ribosomal proteins by binding to conserved miRNA 5′ UTR target sites. The authors of this study propose a mechanism by which binding of miR-10a and associated protein factors precludes binding of a hypothetical negative regulator with consequent activation of translation (Ørom *et al.*, 2008). A separate study identified a miR-346 binding site in the 5′ UTR of RIP140 (receptor-interacting protein 140) and showed that miR-346 activates translation of RIP140 but did not affect mRNA levels. Interestingly, RIP140 activation was shown to occur via an AGO2-indpendent mechanism (Tsai *et al.*, 2009). MiRNAs have also been shown to induce activation of gene expression at the epigenetic level (discussed below).

MicroRNA-mediated epigenetic regulation

In addition to their role as post-transcriptional regulators, miRNAs can also influence chromatin structure and epigenetic control of gene expression in the nucleus (Fig. 2.8). Small RNA-mediated transcriptional control has been well described in fission yeast, *Drosophila* and *Arabidopsis* (Creamer and Partridge, 2011; Verdel *et al.*, 2009; Grewal and Elgin, 2007; Matzke *et al.*, 2001) but has only been described in mammals relatively recently. In 2004 Morris and co-workers showed that siRNAs complementary to a promoter sequence can induce transcriptional gene silencing (TGS) of the targeted gene in human cells (Morris *et al.*, 2004). This TGS effect was shown to be epigenetic, involving alterations in chromatin structure (Weinberg *et al.*, 2006; Hawkins *et al.*, 2009) and promoter DNA methylation (Kin *et al.*, 2006; Castanotto *et al.*, 2005). It was later shown that siRNAs can also induce transcriptional gene activation (TGA) (Li *et al.*, 2006, Janowski *et al.*, 2006) through a related, although less well understood, epigenetic mechanism (Morris *et al.*, 2008). Reports of miRNAs

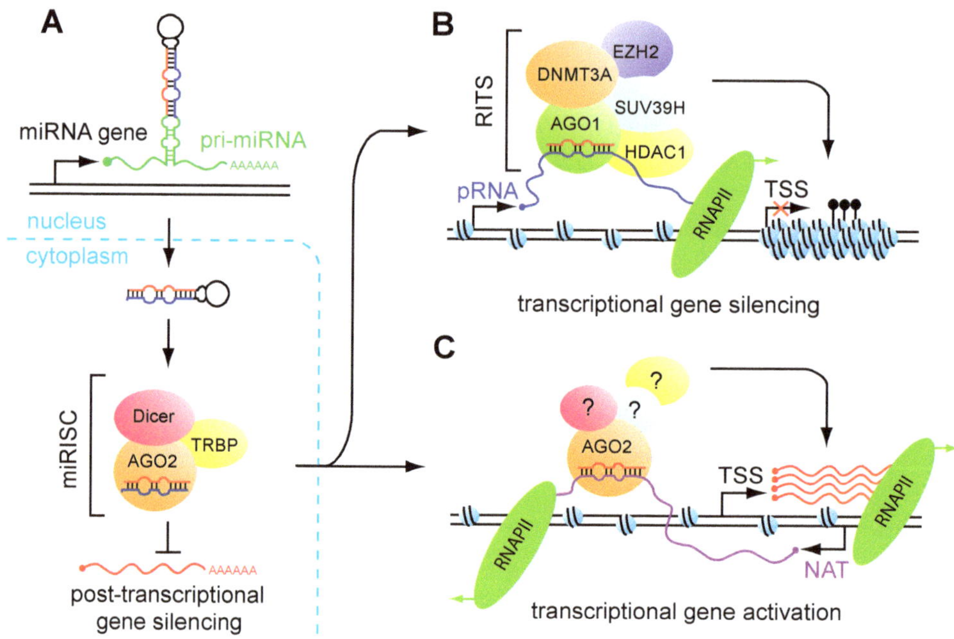

Figure 2.8 MicroRNA-mediated transcriptional gene regulation. (A) Canonical miRNA processing and post-transcriptional gene silencing. The mature miRNA sequence guides miRISC (miRNA Induced Silencing Complex), which consists of AGO2 (Argonaute-2), Dicer and TRBP (Trans-activation response (TAR) RNA Binding Protein), to complementary target transcripts in the cytoplasm and facilitates gene silencing. (B) Mature miRNAs are imported back into the nucleus where they mediate transcriptional gene silencing. The mature miRNA guides the RITS (RNA-Induced Transcriptional Silencing) complex to complementary promoter-associated RNAs (pRNAs) which are transcribed from upstream of the transcription start site (TSS) and remain tethered to their respective genomic loci through association with RNA polymerase II (RNAPII). The RITS complex consists of AGO1 (Argonaute-1), The histone methyltransferases EZH2 and SUV39H, the histone deacetylase HDAC1 and the DNA methyltransferase DNMT3A. RITS recruitment therefore induces silent-state chromatin formation and, in some cases, DNA methylation. (C) Nuclear miRNAs can also induce transcriptional gene activation by targeting natural antisense transcripts (NATs). Recruitment of AGO2 and other unknown factors results in loss of endogenous epigenetic control and concomitant activation of sense transcription.

that induce TGS and TGA followed soon after suggesting the existence of endogenous pathways of small RNA-mediated epigenetic regulation. One such example is miR-320 which was shown to silence *POLR3D* by TGS by recruiting AGO1 and EZH2 to the target promoter and inducing chromatin remodelling (Kim *et al.*, 2008). Similarly, miR-10a induces epigenetic remodelling at the *HOXD* promoter (Tan *et al.*, 2009b).

Several recent studies have shown a role for miRNA-mediated TGS in controlling cellular processes. Benhamad *et al.* (2012) demonstrated a functional link between miRNA-mediated TGS and cellular senescence. ChIP-on-chip analysis was performed using a pan-AGO antibody and RIP-seq with pan-AGO and H3Kme2 antibodies in order to determine potential Argonaute–promoter interactions in senescent and presenescent cells. The miRNA let-7f was shown to physically interact with the promoters of RB1/E2F-target genes via binding with AGO2. let-7 induced TGS of target promoters, and inhibition of let-7 with an antagomiR perturbed senescence (Benhamed *et al.*, 2012).

Similarly, Adilakshmi *et al.* (2012) showed that, following experimentally induced sciatic nerve injury in a mouse model, a group of 22 miRNAs were differentially expressed which mediate the repair response by modulating the expression of genes involved in neuronal differentiation. Notably, miR-709 triggered TGS by recruiting Argonaute-1 to the *Egr2* promoter and inducing silent state chromatin formation. Notably, this study demonstrated that miRNA-mediated TGS and PTGS cooperate in order to coordinate cellular processes (Adilakshmi *et al.*, 2012). Similarly, miR-233 regulates granulopoiesis by transcriptionally silencing the NF1-A promoter (Zardo *et al.*, 2012).

Conversely, miR-373 has been shown to activate transcription of *CDH1* (E-cadherin) and *CSDC2* (cold-shock domain-containing protein C2) in PC3 cells (Place *et al.*, 2008) although the mechanistic details are less understood than for miRNA-mediated TGS. A separate study found that miR-205 was able to induce transcriptional activation of the interleukins *IL24* and *IL32* with only partial complementarity with the target promoter (Majid *et al.*, 2010) suggesting that miRNA-mediated transcriptional modulation may be a widespread phenomenon. Specific mature and precursor miRNAs (e.g. miR-206) have also been found to be enriched in the nucleolus of rat myoblasts suggesting other possible nuclear functions of miRNAs (Politz *et al.*, 2009).

Extracellular microRNA

A surprising finding of recent years is that miRNAs are readily detectable in the extracellular environment (namely blood serum/plasma, urine, cerebrospinal fluid, saliva and seminal fluid) (Mitchell *et al.*, 2008). These bodily fluids are rich in RNase activity, and so the apparent stability of extracellular miRNAs was a surprise [the half-life of an unmodified oligonucleotide in serum is on the order of seconds (Tsui *et al.*, 2002)]. The stability of miRNAs in the extracellular space can be explained by their encapsulation in lipid bound vesicles [such as exosomes, microvesicles (Kosaka *et al.*, 2010; Valadi *et al.*, 2007) and apoptotic bodies (Zernecke *et al.*, 2009)], or by complexation with proteins [e.g. Argonautes (Turchinovich *et al.*, 2011, Turchinovich and Burwinkel, 2012) or nucleophosmin-1 (Wang *et al.*, 2010)] and lipoproteins (e.g. high-density lipoprotein, HDL) (Vickers *et al*, 2011). Given their abundance and stability, serum miRNAs are promising biomarkers for a number of disease conditions, including cancer. For example, tumour-derived microvesicle-bound miRNAs can be used to detect gliomas by a relatively simple, non-invasive blood test (Skog *et al.*, 2008). Serum biomarkers are especially promising in the case of muscle wasting disorders such as Duchenne muscular dystrophy (DMD). Muscle comprises ~50% of the total body mass meaning that, in contrast with cancers, the amount of tissue releasing miRNAs into the circulation is large and, as a result, 50- to 100-fold changes in muscle-specific miRNAs are typically seen in DMD patients and dystrophic animal models relative to healthy controls (Mizuno *et al.*, 2011; Cacchiarelli *et al.*, 2011). Furthermore, miRNA profiling has revealed that the pattern of serum miRNA expression does not parallel that seen in muscle, which suggests that miRNA release in specific, rather than passive, and points to a potential role of serum miRNAs in the control of muscle homeostasis (Roberts *et al.*, 2012). Similarly, miRNA profiling of serum, plasma and placental tissue revealed that certain placenta-specific (i.e. miR-498 family) miRNAs were preferentially abundant in serum, again suggesting the possibility of specific miRNA release (Williams *et al.*, 2013).

In a landmark study, Zhang *et al.* (2012) showed that the plant-specific miR-168a is highly enriched in the sera of Chinese subjects. As this miRNA has no mammalian counterpart the authors reasoned that dietary intake was responsible for serum miR-168a in humans. Remarkably, miR-168a was shown to regulate expression of low-density lipoprotein receptor adaptor protein (LDLRAP1) in mouse liver and human cells. Inhibition of LDLRAP1 resulted in decreased LDL uptake. This study is therefore the first demonstration of cross-kingdom miRNA regulation (Zhang *et al.*, 2012).

Endogenous microRNA sponges

A recent development is the regulation of miRNA functionality is the discovery that long non-coding RNA transcripts can act as miRNA sponges (also called competing endogenous RNAs, ceRNAs). MiRNA sponges are RNA molecules which typically contain multiple binding sites for one or more miRNAs. Exogenous miRNA sponges were first utilized as tools to study miRNA gene function and were subsequently found to exist in nature (Franco-Zorrilla *et al.*, 2007). Sponges compete with miRNA target transcripts for miRNA binding and so effectively inhibit miRNA regulation. For example, a pseudogene of PTEN (PTENP1) is expressed and acts as a sponge for miRNAs which regulate PTEN expression. Expression of PTENP1 therefore acts to de-repress expression of PTEN (Poliseno *et al.*, 2010). Similarly, a miRNA sponge has also been implicated in the regulation of muscle differentiation (Cesana *et al.*, 2011).

A study by Hansen *et al.* (2011) identified a novel nuclear function of miR-671. This miRNA guides nuclear AGO2 to cleave a circular non-coding RNA (circRNA) transcript antisense to the *CDR1* gene. Slicing of this circular RNA resulted in destabilization of both the antisense and CDR1 coding mRNA and thus constitutes a novel post-transcriptional gene silencing pathway active in the nucleus (Hansen *et al.*, 2011). Following up on this study, two groups showed that the CDR1-associated circRNA also acts as a miRNA sponge for miR-7. The circRNA contains >70 binding sites for miR-7, was resistant to RNAi-mediated degradation and greatly suppressed miR-7 activity. Similarly, a separate testis-specific circRNA was also found to act as a sponge for miR-138 (Hansen *et al.*, 2013; Memczak *et al.*, 2013). The modulation of miRNA activity by endogenous miRNA sponges and circRNAs adds another level of complexity to miRNA-mediated gene regulation.

Concluding remarks

Over the last 10 years there has been an explosion of interest in miRNA-mediated gene regulation. Understanding the link between the expression of miRNAs and the regulation of their cognate targets is still a major challenge and an area of active research. Importantly, the vast majority of work in this area has been undertaken within a canonical paradigm of miRNA biogenesis and function and, as such, makes a number of assumptions: (1) that miRNAs are represented by a single sequence, (2) miRNAs bind in 3′ UTRs and (3) miRNAs negatively regulate their mRNA targets at the post-transcriptional level. In this review we have summarized the many caveats to these assumptions. Specifically, (1) heterogeneity in processing and post-transcriptional editing contribute to the flexibility of miRNA sequence, (2) miRNAs can target mRNAs in coding regions and 5′ UTRs, or alternatively target non-coding regions such as promoters, and (3) miRNAs can act as post-transcriptional and transcriptional regulators that can both silence or activate gene expression. These exceptions-to-the-rules greatly expand the functional repertoire of miRNAs. The many ways of producing mature miRNA species reflects their exaptation from diverse RNA substrates (e.g. snoRNAs, tRNAs, long hairpin RNAs etc.). Conversely, the unusual processing requirements of certain miRNAs appear to have constrained the evolution of components of the miRNA processing pathway (as in the case of miR-451 and AGO2-mediated slicer activity). The plasticity of miRNA sequences adds a further layer of regulation to the control of gene expression, which in turn has permitted increased organismal complexity.

References

Adilakshmi, T., Sudol, I., and Tapinos, N. (2012). Combinatorial action of miRNAs regulates transcriptional and post-transcriptional gene silencing following *in vivo* PNS injury. PLoS ONE *7*, e39674.

Alawi, F., and Lin, P. (2010). Loss of dyskerin reduces the accumulation of a subset of H/ACA snoRNA-derived miRNA. Cell Cycle *9*, 2467–2469.

Aravin, A.A., Sachidanandam, R., Girard, A., Fejes-Toth, K., and Hannon, G.J. (2007). Developmentally regulated piRNA clusters implicate MILI in transposon control. Science *316*, 744–747.

Aravin, A.A., Sachidanandam, R., Bourc'his, D., Schaefer, C., Pezic, D., Toth, K.F., Bestor, T., and Hannon, G.J. (2008). A piRNA pathway primed by individual transposons is linked to de novo DNA methylation in mice. Mol. Cell. *31*, 785–799.

Azuma-Mukai, A., Oguri, H., Mituyama, T., Qian, Z.R., Asai, K., Siomi, H., and Siomi, M.K. (2008). Characterization of endogenous human Argonautes and their miRNA partners in RNA silencing. Proc. Natl. Acad. Sci. U.S.A. *105*, 7964–7969.

Babiarz, J.E., Ruby, J.G., Wang, Y., Bartel, D.P., and Blelloch, R. (2008). Mouse ES cells express endogenous shRNAs, siRNAs, and other Microprocessor-independent, Dicer-dependent small RNAs. Genes Dev. *22*, 2773–2785.

Benhamed, M., Herbig, U., Ye, T., Dejean, A., and Bischof, O. (2012). Senescence is an endogenous trigger for microRNA-directed transcriptional gene silencing in human cells. Nat. Cell Biol. *14*, 266–275.

Bennasser, Y., Chable-Bessia, C., Triboulet, R., Gibbings, D., Gwizdek, C., Dargemont, C., Kremer, E.J., Voinnet, O., and Benkirane, M. (2011). Competition for XPO5 binding between Dicer mRNA, pre-miRNA and viral

RNA regulates human Dicer levels. Nat. Struct. Mol. Biol. *18*, 323–327.

Bernstein, E., Caudy, A.A., Hammond, S.M., and Hannon, G.J. (2001). Role for a bidentate ribonuclease in the initiation step of RNA interference. Nature *409*, 363–366.

Blow, M.J., Grocock, R.J., van Dongen, S., Enright, A.J., Dicks, E., Futreal, P.A., Wooster, R., and Stratton, M.J. (2006). RNA editing of human microRNAs. Genome Biol. *7*, R27.

Bohnsack, M.T., Czaplinski, K., and Gorlich, D. (2004). Exportin-5 is a RanGTP-dependent dsRNA-binding protein that mediates nuclear export of pre-miRNAs. RNA. *10*, 185–191.

Brameier, M., Herwig, A., Reinhardt, R., Walter, L., and Gruber, J. (2011). Human box C/D snoRNAs with miRNA like functions: expanding the range of regulatory RNAs. Nucleic Acids Res. *39*, 675–686.

Bratkovič, T., and Rogelj, B. (2011). Biology and applications of small nucleolar RNAs. Cell. Mol. Life Sci. *68*, 3843–3851.

Burroughs, A.M., Ando, Y., de Hoon, M.J.L., Tomaru, Y., Nishibu, T., Ukekawa, R., Funakoshi, T., Kurokawa, T., Hayashizaki, Y., and Daub, C.O. (2010). A comprehensive survey of 3′ animal miRNA modification events and a possible role for 3′ adenylation in modulating miRNA targeting effectiveness. Genome Res. *20*, 1398–1410.

Burroughs, A.M., Ando, Y., de Hoon, M.J.L., Tomaru, Y., Suzuki, H., Hayashizaki, Y., and Daub, C.O. (2011). Deep-sequencing of human Argonaute-associated small RNAs provides insight into miRNA sorting and reveals Argonaute association with RNA fragments of diverse origin. RNA Biol. *8*, 158–177.

Cacchiarelli, D., Legnini, I., Martone, J., Cazzella, V., D'Amico, A., Bertini, E., and Bozzoni, I. (2011). MiRNAs as serum biomarkers for Duchenne muscular dystrophy. EMBO Mol. Med. *3*, 258–265.

Carmell, M.A., Girard, A., van de Kant, H.J.G., Bourc'his, D., Bestor, T.H., de Rooij, D.G., and Hannon, G.J. (2007). MIWI2 is essential for spermatogenesis and repression of transposons in the mouse male germline. Dev. Cell. *12*, 503–514.

Castanotto, D., Tommasi, S., Li, M., Li, H., Yanow, S., Pfeifer, G.P., and Rossi, J.J. (2005). Short hairpin RNA-directed cytosine (CpG) methylation of the RASSF1A gene promoter in HeLa cells. Mol. Ther. *12*, 179–83.

Cesana, M., Cacchiarelli, D., Legnini, I., Santini, T., Sthandier, O., Chinappi, M., Tramontano, A., and Bozzonni, I. (2011). A long noncoding RNA controls muscle differentiation by functioning as a competing endogenous RNA. Cell; *147*, 358–369.

Cheloufi, S., Dos Santos, C.O., Chong, M.M.W., and Hannon, G.J. (2010). A dicer-independent miRNA biogenesis pathway that requires Ago catalysis. Nature *465*, 584–589.

Chendrimada, T.P., Gregory, R.I., Kumaraswamy, E., Norman, J., Cooch, N., Nishikura, K., and Shiekhattar, R. (2005). TRBP recruits the Dicer complex to Ago2 for microRNA processing and gene silencing. Nature *436*, 740–744.

Chi, S.W., Zang, J.B., Mele, A. and Darnell, R.B. (2009). Argonaute HITS-CLIP decodes microRNA–mRNA interaction maps. Nature *460*, 479–486.

Cloonan, N., Wani, S., Xu, Q., Gu, J., Lea, K., Heater, S., Barbacioru, C., Steptoe, A.L., Martin, H.C., Nourbakhsh, E., *et al.* (2011). MicroRNAs and their isomiRs function cooperatively to target common biological pathways. Genome Biol. *12*, R126.

Cole, C., Sobala, A., Lu, C., Thatcher, S.R., Bowman, A., Brown, J.W.S., Green, P.J., Barton, G.J., and Hutvagner, G. (2009). Filtering of deep sequencing data reveals the existence of abundant Dicer-dependent small RNAs derived from tRNAs. RNA; *15*, 2147–2160.

Creamer, K.M., and Partridge, J.F. (2011). RITS-connecting transcription, RNA interference, and heterochromatin assembly in fission yeast. Wiley Interdiscip. Rev. RNA. *2*, 632–646.

Denli, A.M., Tops, B.B.J., Plasterk, R.H.A., Ketting, R.F., and Hannon, G.J. (2004). Processing of primary microRNAs by the Microprocessor complex. Nature *432*, 231–235.

Doench, J.G., and Sharp, P.A. (2004). Specificity of microRNA target selection in translational repression. Genes Dev. *18*, 504–511.

Duursma, A.M., Kedde, M., Schrier, M., le Sage, C., and Agami, R. (2008). miR-148 targets human DNMT3b protein coding region. RNA *14*, 872–877.

Dweep, H., Sticht, C., Pandey, P., and Gretz, N. (2011). miRWalk--database: prediction of possible miRNA binding sites by 'walking' the genes of three genomes. J. Biomed. Inform. *44*, 839–847.

Ender, C., Krek, A., Friedländer, M.R., Beitzinger, M., Weinmann, L., Chen, W., Pfeffer, S., Rajewsky, N., and Mesiter, G. (2008). A human snoRNA with microRNA-like functions. Mol. Cell. *32*, 519–528.

Faller, M., Toso, D., Matsunaga, M., Atanasov, I., Senturia, R., Chen, Y., Zhou, Z.H., and Guo, F. (2010). DGCR8 recognizes primary transcripts of microRNAs through highly cooperative binding and formation of higher-order structures. RNA *16*, 1570–1583.

Fernandez-Valverde, S.L., Taft, R.J., and Mattick, J.S. (2010). Dynamic isomiR regulation in Drosophila development. RNA *16*, 1881–1888.

Filipowicz, W., Jaskiewicz, L., Kolb, F.A., and Pillai, R.S. (2005). Post-transcriptional gene silencing by siRNAs and miRNAs. Curr. Opin. Struct. Biol. *15*, 331–341.

Flynt, A.S., Greimann, J.C., Chung, W.-J., Lima, C.D., and Lai, E.C. (2010). MicroRNA biogenesis via splicing and exosome-mediated trimming in Drosophila. Mol. Cell *38*, 900–907.

Forman, J.J., and Coller, H.A. (2010). The code within the code: microRNAs target coding regions. Cell Cycle *9*, 1533–1541.

Forman, J.J., Legesse-Miller, A., and Coller, H.A. (2008). A search for conserved sequences in coding regions reveals that the let-7 microRNA targets Dicer within its coding sequence. Proc. Natl. Acad. Sci. U.S.A. *105*, 14879–14884.

Franco-Zorrilla, J.M., Valli, A., Todesco, M., Mateos, I., Puga, M.I., Rubio-Somoza, I., Leyva, A., Weigel, D., García, J.A., and Paz-Ares, J. (2007). Target mimicry provides a new mechanism for regulation of microRNA activity. Nat. Genet. *39*, 1033–1037.

Gilad, S., Meiri, E., Yogev, Y., Benjamin, S., Lebanony, D., Yerushalmi, N., Benjamin, H., Kushnir, M., Cholakh, H., Melamed, N., *et al.* (2008). Serum microRNAs are promising novel biomarkers. PLoS ONE 3, e3148.

Gregory, R.I., Yan, K.-P., Amuthan, G., Chendrimada, T., Doratotaj, B., Cooch, N., and Shiekhattar, R. (2004). The Microprocessor complex mediates the genesis of microRNAs. Nature *432*, 235–240.

Grewal, S.I.S. and Elgin, S.C.R. (2007). Transcription and RNA interference in the formation of heterochromatin. Nature *447*, 399–406.

Grey, F., Tirabassi, R., Meyers, H., Wu, G., McWeeney, S., Hook, L., and Nelson, J.A. (2010). A viral microRNA down-regulates multiple cell cycle genes through mRNA 5′UTRs. PLoS Pathog. *6*, e1000967.

Griffiths-Jones, S., Saini, H.K., van Dongen, S., and Enright, A.J. (2008). miRBase: tools for microRNA genomics. Nucleic Acids Res. *36*, D154–158.

Grimson, A., Farh, K.K.-H., Johnston, W.K., Garrett-Engele, P., Lim, L.P., and Bartel, D.P. (2007). MicroRNA targeting specificity in mammals: determinants beyond seed pairing. Mol. Cell. *27*, 91–105.

Gu, S., Jin, L., Zhang, F., Sarnow, P., and Kay, M.A. (2009). Biological basis for restriction of microRNA targets to the 3′ untranslated region in mammalian mRNAs. Nat. Struct. Mol. Biol. *16*, 144–150.

Gu, T., Buaas, F.W., Simons, A.K., Ackert-Bicknell, C.L., Braun, R.E., and Hibbs, M.A. (2012). Canonical A-to-I and C-to-U RNA editing is enriched at 3′UTRs and microRNA target sites in multiple mouse tissues. PLoS ONE 7, e33720.

Gunawardane, L.S., Saito, K., Nishida, K.M., Miyoshi, K., Kawamura, Y., Nagami, T., Siomi, H., and Siomi, M.C. (2007). A slicer-mediated mechanism for repeat-associated siRNA 5′ end formation in Drosophila. Science *315*, 1587–1590.

Guo, H., Ingolia, N.T., Weissman, J.S., and Bartel, D.P. (2010). Mammalian microRNAs predominantly act to decrease target mRNA levels. Nature *466*, 835–840.

Han, J., Lee, Y., Yeom, K.-H., Nam, J.-W., Heo, I., Rhee, J.-K., Sohn, S.J., Cho, Y., Zhang, B-T., and Kim, V.N. (2006). Molecular basis for the recognition of primary microRNAs by the Drosha–DGCR8 complex. Cell *125*, 887–901.

Hansen, T.B., Wiklund, E.D., Bramsen, J.B., Villadsen, S.B., Statham, A.L., Clark, S.J., and Kjems, J. (2011). MiRNA-dependent gene silencing involving Ago2-mediated cleavage of a circular antisense RNA. EMBO J. *30*, 4414–4422.

Hansen, T.B., Jensen, T.I., Clausen, B.H., Bramsen, J.B., Finsen, B., Damgaard, C.K., and Kjems, J. (2013). Natural RNA circles function as efficient microRNA sponges. Nature *495*, 384–388.

Havens, M.A., Reich, A.A., Duelli, D.M., and Hastings, M.L. (2012). Biogenesis of mammalian microRNAs by a non-canonical processing pathway. Nucleic Acids Res. *40*, 4626–4640.

Hawkins, P.G., Santoso, S., Adams, C., Anest, V., and Morris, K.V. (2009). Promoter targeted small RNAs induce long-term transcriptional gene silencing in human cells. Nucleic Acids Res. *37*, 2984–2995.

Hendrickson, D.G., Hogan, D.J., Herschlag, D., Ferrell, J.E., and Brown, P.O. (2008). Systematic identification of mRNAs recruited to argonaute 2 by specific microRNAs and corresponding changes in transcript abundance. PLoS ONE *3*, e2126.

Henke, J.I., Goergen, D., Zheng, J., Song, Y., Schüttler, C.G., Fehr, C., Jünermann, C., and Niepmann, M. (2008). MicroRNA-122 stimulates translation of hepatitis C virus RNA. EMBO J. *27*, 3300–3310.

Higuchi, M., Single, F.N., Köhler, M., Sommer, B., Sprengel, R., and Seeburg, P.H. (1993). RNA editing of AMPA receptor subunit GluR-B: a base-paired intron–exon structure determines position and efficiency. Cell *75*, 1361–1370.

Humphreys, D.T., Hynes, C.J., Patel, H.R., Wei, G.H., Cannon, L., Fatkin, D., Suter, C.M., Clancy, J.L., and Preiss, T. (2012). Complexity of murine cardiomyocyte miRNA biogenesis, sequence variant expression and function. PLoS ONE 7, e30933.

Hutvagner, G., and Simard, M.J. (2008). Argonaute proteins: key players in RNA silencing. Nat. Rev. Mol. Cell. Biol. *9*, 22–32.

Hutvágner, G., and Zamore, P.D. (2002). A microRNA in a multiple-turnover RNAi enzyme complex. Science *297*, 2056–2060.

Janowski, B.A., Younger, S.T., Hardy, D.B., Ram, R., Huffman, K.E., and Corey, D.R. (2007). Activating gene expression in mammalian cells with promoter-targeted duplex RNAs. Nat. Chem. Biol. *3*, 166–73.

Ji, L., and Chen, X. (2012). Regulation of small RNA stability: methylation and beyond. Cell Res. *22*, 624–636.

Jopling, C.L., Yi, M., Lancaster, A.M., Lemon, S.M., and Sarnow, P. (2005). Modulation of hepatitis C virus RNA abundance by a liver-specific MicroRNA. Science *309*, 1577–1581.

Jopling, C.L., Schütz, S., and Sarnow, P. (2008). Position-dependent function for a tandem microRNA miR-122-binding site located in the hepatitis C virus RNA genome. Cell Host Microbe *4*, 77–85.

Kawahara, Y., Zinshteyn, B., Sethupathy, P., Iizasa, H., Hatzigeorgiou, A.G., and Nishikura, K. (2007a). Redirection of silencing targets by adenosine-to-inosine editing of miRNAs. Science *315*, 1137–1140.

Kawahara, Y., Zinshteyn, B., Chendrimada, T.P., Shiekhattar, R., and Nishikura, K. (2007b). RNA editing of the microRNA-151 precursor blocks cleavage by the Dicer–TRBP complex. EMBO Rep.. *8*, 763–769.

Kawahara, Y., Megraw, M., Kreider, E., Iizasa, H., Valente, L., Hatzigeorgiou, A.G., and Nishikura, K. (2008). Frequency and fate of microRNA editing in human brain. Nucleic Acids Res. *36*, 5270–5280.

Kawaji, H., Nakamura, M., Takahashi, Y., Sandelin, A., Katayama, S., Fukuda, S., Daub, C.O., Kai, C., Kawai, J., Yasuda, J., *et al.* (2008). Hidden layers of human small RNAs. BMC Genomics 9, 157.

Khvorova, A., Reynolds, A., and Jayasena, S.D. (2003). Functional siRNAs and miRNAs exhibit strand bias. Cell *115*, 209–216.

Kim, D.H., Villeneuve, L.M., Morris, K.V., and Rossi, J.J. (2006). Argonaute-1 directs siRNA-mediated transcriptional gene silencing in human cells. Nat. Struct. Mol. Biol. *13*, 793–797.

Kim, D.H., Saetrom, P., Snøve, O., and Rossi, J.J. (2008). MicroRNA-directed transcriptional gene silencing in mammalian cells. Proc. Natl. Acad. Sci. U.S.A. *105*, 16230–16235.

Kim, Y.-K., and Kim, V.N. (2007). Processing of intronic microRNAs. EMBO J. *26*, 775–783.

Kiriakidou, M., Nelson, P.T., Kouranov, A., Fitziev, P., Bouyioukos, C., Mourelatos, Z., and Hatzigeorgiou, A. (2004). A combined computational-experimental approach predicts human microRNA targets. Genes Dev. *18*, 1165–1178.

Kosaka, N., Iguchi, H., Yoshioka, Y., Takeshita, F., Matsuki, Y., and Ochiya, T. (2010). Secretory mechanisms and intercellular transfer of microRNAs in living cells. J. Biol. Chem. *285*, 17442–17452.

Kumar, A., Wong, A.K.-L., Tizard, M.L., Moore, R.J., and Lefèvre, C. (2012). MiRNA_Targets: a database for miRNA target predictions in coding and non-coding regions of mRNAs. Genomics *100*, 352–356.

Lal, A., Kim, H.H., Abdelmohsen, K., Kuwano, Y., Pullmann, R., Jr, Srikantan, S., Subrahmanyam, R., Martindale, J.L., Yang, X., Ahmed, F., *et al.* (2008). p16(INK4a) translation suppressed by miR-24. PLoS ONE. *3*, e1864.

Lanford, R.E., Hildebrandt-Eriksen, E.S., Petri, A., Persson, R., Lindow, M., Munk, M.E., Kauppinen, S., and Ørum, H. (2010). Therapeutic silencing of microRNA-122 in primates with chronic hepatitis C virus infection. Science *327*, 198–201.

Langenberger, D., Bermudez-Santana, C.I., Stadler, P.F., and Hoffmann, S. (2010). Identification and classification of small rnas in transcriptome sequence data. Pac. Symp. Biocomput. 80–87.

Langenberger, D., Çakir, M.V., Hoffmann, S., and Stadler, P.F. (2013). Dicer-processed small RNAs: rules and exceptions. J. Exp. Zool. B. Mol. Dev. Evol. *320*, 35–46.

Lee, H.Y., and Doudna, J.A. (2012). TRBP alters human precursor microRNA processing *in vitro*. RNA *18*, 2012–2019.

Lee, I., Ajay, S.S., Yook, J.I., Kim, H.S., Hong, S.H., Kim, N.H., *et al.* (2009). New class of microRNA targets containing simultaneous 5′-UTR and 3′-UTR interaction sites. Genome Res. *19*, 1175–1183.

Lee, L.W., Zhang, S., Etheridge, A., Ma, L., Martin, D., Galas, D., and Wang, K. (2010). Complexity of the microRNA repertoire revealed by next-generation sequencing. RNA *16*, 2170–2180.

Lee, Y., Ahn, C., Han, J., Choi, H., Kim, J., Yim, J., Lee, J., Provost, P., Rådmark, O., Kim, S., *et al.* (2003). The nuclear RNase III Drosha initiates microRNA processing. Nature. *425*, 415–419.

Lee, Y., Kim, M., Han, J., Yeom, K.-H., Lee, S., Baek, S.H., and Kim, V.N. (2004). MicroRNA genes are transcribed by RNA polymerase II. EMBO J. *23*, 4051–4060.

Lee, Y.S., Shibata, Y., Malhotra, A., and Dutta, A. (2009). A novel class of small RNAs: tRNA-derived RNA fragments (tRFs). Genes Dev. *23*, 2639–2649.

Leung, A.K.L., and Sharp, P.A. (2010). MicroRNA Functions in Stress Responses. Mol. Cell. *40*, 205–215.

Li, L.-C., Okino, S.T., Zhao, H., Pookot, D., Place, R.F., Urakami, S., Enokida, H., and Dahiya, R. (2006). Small dsRNAs induce transcriptional activation in human cells. Proc. Natl. Acad. Sci. U.S.A. *103*, 17337–42.

Liang, H., and Landweber, L.F. (2007). Hypothesis: RNA editing of microRNA target sites in humans? RNA. *13*, 463–467.

Ling, J.Q., Li, T., Hu, J.F., Vu, T.H., Chen, H.L., Qiu, X.W., Cherry, A.M., and Hoffman, A.R. (2006). CTCF mediates interchromosomal colocalization between Igf2/H19 and Wsb1/Nf1. Science. *312*, 269–272.

Liu, J., Carmell, M.A., Rivas, F.V., Marsden, C.G., Thomson, J.M., Song, J.-J., Hammond, S.M., Joshua-Tor, L., and Hannon, G.J. (2004). Argonaute2 is the catalytic engine of mammalian RNAi. Science. *305*, 1437–1441.

Llorens, F., Bañez-Coronel, M., Pantano, L., Del Río, J.A., Ferrer, I., Estivill, X., and Martí, E. (2013). A highly expressed miR-101 isomiR is a functional silencing small RNA. BMC Genomics. *14*, 104.

Lu, S., and Cullen, B.R. (2004). Adenovirus VA1 noncoding RNA can inhibit small interfering RNA and MicroRNA biogenesis. J. Virol. *78*, 12868–12876.

Luciano, D.J., Mirsky, H., Vendetti, N.J., and Maas, S. (2004). RNA editing of a miRNA precursor. RNA *10*, 1174–1177.

Lund, E., Güttinger, S., Calado, A., Dahlberg, J.E., and Kutay, U. (2004). Nuclear export of microRNA precursors. Science *303*, 95–98.

Lytle, J.R., Yario, T.A., and Steitz, J.A. (2007). Target mRNAs are repressed as efficiently by microRNA-binding sites in the 5′ UTR as in the 3′ UTR. Proc. Natl. Acad. Sci. U.S.A. *104*, 9667–9672.

Ma, J.-B., Yuan, Y.-R., Meister, G., Pei, Y., Tuschl, T., and Patel, D.J. (2005). Structural basis for 5′-end-specific recognition of guide RNA by the A. fulgidus Piwi protein. Nature *434*, 666–670.

Majid, S., Dar, A.A., Saini, S., Yamamura, S., Hirata, H., Tanaka, Y., Deng, G., and Dahiya, R. (2010). MicroRNA-205-directed transcriptional activation of tumor suppressor genes in prostate cancer. Cancer *116*, 5637–5649.

Mallory, A., and Vaucheret, H. (2010). Form, function, and regulation of ARGONAUTE proteins. Plant Cell. *22*, 3879–3889.

Martin, E.C., Elliott, S., Rhodes, L.V., Antoon, J.W., Fewell, C., Zhu, Y., Driver, J.L., Jodari-Karimi, M., Taylor, C.W., Flemington, E.K., *et al.* (2012). Preferential star strand biogenesis of pre-miR-24–2 targets PKC-alpha and suppresses cell survival in MCF-7 breast cancer cells. Mol. Carcinog. *53*, 38–48.

Matzke, M.A., Matzke, A.J., Pruss, G.J., and Vance, V.B. (2001). RNA-based silencing strategies in plants. Curr. Opin. Genet. Dev. *11*, 221–227.

Meiri, E., Levy, A., Benjamin, H., Ben-David, M., Cohen, L., Dov, A., Dromi, N., Elyakim, E., Yerushalmi, N., Zion, O., *et al.* (2010). Discovery of microRNAs and other small RNAs in solid tumors. Nucleic Acids Res. *38*, 6234–6246.

Meister, G., Landthaler, M., Patkaniowska, A., Dorsett, Y., Teng, G., and Tuschl, T. (2004). Human Argonaute2 mediates RNA cleavage targeted by miRNAs and siRNAs. Mol. Cell. *15*, 185–197.

Memczak, S., Jens, M., Elefsinioti, A., Torti, F., Krueger, J., Rybak, A., Maier, L., Mackowiak, S.D., Gregersen, L.H., Munschauer, M., *et al.* (2013). Circular RNAs are a large class of animal RNAs with regulatory potency. Nature *495*, 333–338.

Meyer, K.D., Saletore, Y., Zumbo, P., Elemento, O., Mason, C.E., and Jaffrey, S.R. (2012). Comprehensive analysis of mRNA methylation reveals enrichment in 3′ UTRs and near stop codons. Cell. *149*, 1635–1646.

Mitchell, P.S., Parkin, R.K., Kroh, E.M., Fritz, B.R., Wyman, S.K., Pogosova-Agadjanyan, E.L., Petersen, A., Noteboom, J., O'Briant, K.C., Allen, A., *et al.* (2008). Circulating microRNAs as stable blood-based markers for cancer detection. Proc. Natl. Acad. Sci. U.S.A. *105*, 10513–10518.

Miyoshi, K., Miyoshi, T., Hartig, J.V., Siomi, H., and Siomi, M.C. (2010). Molecular mechanisms that funnel RNA precursors into endogenous small-interfering RNA and microRNA biogenesis pathways in Drosophila. RNA. *16*, 506–515.

Mizuno, H., Nakamura, A., Aoki, Y., Ito, N., Kishi, S., Yamamoto, K., Sekiguchi, M., Takeda, S., and Hashido, K. (2011). Identification of muscle-specific microRNAs in serum of muscular dystrophy animal models: promising novel blood-based markers for muscular dystrophy. PLoS ONE. *6*, e18388.

Moretti, F., Thermann, R., and Hentze, M.W. (2010). Mechanism of translational regulation by miR-2 from sites in the 5′ untranslated region or the open reading frame. RNA *16*, 2493–2502.

Morris, K.V., Chan, S.W.-L., Jacobsen, S.E., and Looney, D.J. (2004). Small interfering RNA-induced transcriptional gene silencing in human cells. Science *305*, 1289–92.

Morris, K.V., Santoso, S., Turner, A.-M., Pastori, C., and Hawkins, P.G. (2008). Bidirectional transcription directs both transcriptional gene activation and suppression in human cells. PLoS Genet. *4*, e1000258.

Nishikura, K. (2010). Functions and Regulation of RNA Editing by ADAR Deaminases. Annu. Rev. Biochem. *79*, 321–349.

Ohler, U., Yekta, S., Lim, L.P., Bartel, D.P., and Burge, C.B. (2004). Patterns of flanking sequence conservation and a characteristic upstream motif for microRNA gene identification. RNA *10*, 1309–1322.

Okamura, K., Hagen, J.W., Duan, H., Tyler, D.M., and Lai, E.C. (2007). The mirtron pathway generates microRNA-class regulatory RNAs in Drosophila. Cell *130*, 89–100.

Olsen, P.H., and Ambros, V. (1999). The lin-4 regulatory RNA controls developmental timing in *Caenorhabditis elegans* by blocking LIN-14 protein synthesis after the initiation of translation. Dev. Biol. *216*, 671–680.

Ono, M., Scott, M.S., Yamada, K., Avolio, F., Barton, G.J., and Lamond, A.I. (2011). Identification of human miRNA precursors that resemble box C/D snoRNAs. Nucleic Acids Res. *39*, 3879–3891.

Ørom, U.A., Nielsen, F.C., and Lund, A.H. (2008). MicroRNA-10a binds the 5′UTR of ribosomal protein mRNAs and enhances their translation. Mol. Cell. *30*, 460–471.

Paz, N., Levanon, E.Y., Amariglio, N., Heimberger, A.B., Ram, Z., Constantini, S., Barbash, Z.S., Adamsky, K., Safran, M., Hirschberg, A., *et al.* (2007). Altered adenosine-to-inosine RNA editing in human cancer. Genome Res. *17*, 1586–1595.

Persson, H., Kvist, A., Vallon-Christersson, J., Medstrand, P., Borg, A., and Rovira, C. (2009). The non-coding RNA of the multidrug resistance-linked vault particle encodes multiple regulatory small RNAs. Nat. Cell Biol. *11*, 1268–1271.

Pfeffer, S., Sewer, A., Lagos-Quintana, M., Sheridan, R., Sander, C., Grässer, F.A., van Dyk, L.F., Ho, C.K., Schuman, S., Chien, M., *et al.* (2005). Identification of microRNAs of the herpesvirus family. Nat. Methods *2*, 269–276.

Pillai, R.S., Bhattacharyya, S.N., and Filipowicz, W. (2007). Repression of protein synthesis by miRNAs: how many mechanisms? Trends Cell Biol. *17*, 118–126.

Place, R.F., Li, L.-C., Pookot, D., Noonan, E.J., and Dahiya, R. (2008). MicroRNA-373 induces expression of genes with complementary promoter sequences. Proc. Natl. Acad. Sci. U.S.A. *105*, 1608–1613.

Poliseno, L., Salmena, L., Zhang, J., Carver, B., Haveman, W.J., and Pandolfi, P.P. (2010). A coding-independent function of gene and pseudogene mRNAs regulates tumour biology. Nature *465*, 1033–1038.

Politz, J.C.R., Hogan, E.M., and Pederson, T. (2009). MicroRNAs with a nucleolar location. RNA *15*, 1705–1715.

Reese, T.A., Xia, J., Johnson, L.S., Zhou, X., Zhang, W., and Virgin, H.W. (2010). Identification of novel microRNA-like molecules generated from herpesvirus and host tRNA transcripts. J. Virol. *84*, 10344–10353.

Rivas, F.V., Tolia, N.H., Song, J.-J., Aragon, J.P., Liu, J., Hannon, G.J., *et al.* (2005). Purified Argonaute2 and an siRNA form recombinant human RISC. Nat. Struct. Mol. Biol. *12*, 340–349.

Roberts, T.C., Blomberg, K.E.M., McClorey, G., Andaloussi, S.E., Godfrey, C., Betts, C., Coursindel, T., Gait, M.J., Smith, C.I.E., and Wood, M.J. (2012). Expression analysis in multiple muscle groups and serum reveals complexity in the microRNA transcriptome of the mdx mouse with implications for therapy. Mol. Ther. Nucleic Acids *1*, e39.

Ruby, J.G., Jan, C.H., and Bartel, D.P. (2007). Intronic microRNA precursors that bypass Drosha processing. Nature; *448*, 83–86.

Russo, F., Di Bella, S., Nigita, G., Macca, V., Laganà, A., Giugno, R., *et al.* (2012). miRandola: extracellular circulating microRNAs database. PLoS ONE 7, e47786.

Sacco, L.D., and Masotti, A. (2012). Recent Insights and Novel Bioinformatics Tools to Understand the Role of MicroRNAs Binding to 5′ Untranslated Region. Int. J. Mol. Sci. *14*, 480–495.

Saraiya, A.A., and Wang, C.C. (2008). snoRNA, a novel precursor of microRNA in Giardia lamblia. PLoS Pathog. *4*, e1000224.

Saxena, S., Jónsson, Z.O., and Dutta, A. (2003). Small RNAs with imperfect match to endogenous mRNA repress translation. Implications for off-target activity of small inhibitory RNA in mammalian cells. J. Biol. Chem. *278*, 44312–44319.

Scadden, A.D.J. (2005). The RISC subunit Tudor-SN binds to hyper-edited double-stranded RNA and promotes its cleavage. Nat. Struct. Mol. Biol. *12*, 489–496.

Schopman, N.C.T., Heynen, S., Haasnoot, J., and Berkhout, B. (2010). A miRNA-tRNA mix-up: tRNA origin of proposed miRNA. RNA Biol. *7*, 573–576.

Schwarz, D.S., Hutvágner, G., Du, T., Xu, Z., Aronin, N., and Zamore, P.D. (2003). Asymmetry in the assembly of the RNAi enzyme complex. Cell *115*, 199–208.

Scott, M.S., Avolio, F., Ono, M., Lamond, A.I., and Barton, G.J. (2009). Human miRNA precursors with box H/ACA snoRNA features. PLoS Comput. Biol. *5*, e1000507.

Seeburg, P.H., and Hartner, J. (2003). Regulation of ion channel/neurotransmitter receptor function by RNA editing. Curr. Opin. Neuro. Biol. *13*, 279–283.

Shen, W.-F., Hu, Y.-L., Uttarwar, L., Passegue, E., and Largman, C. (2008). MicroRNA-126 regulates HOXA9 by binding to the homeobox. Mol. Cell. Biol. *28*, 4609–4619.

Shi, W., Hendrix, D., Levine, M., and Haley, B. (2009). A distinct class of small RNAs arises from pre-miRNA-proximal regions in a simple chordate. Nat. Struct. Mol. Biol. *16*, 183–189.

Shimakami, T., Yamane, D., Jangra, R.K., Kempf, B.J., Spaniel, C., Barton, D.J., and Lemon, S.M. (2012). Stabilization of hepatitis C virus RNA by an Ago2–miR-122 complex. Proc. Natl. Acad. Sci. U.S.A. *109*, 941–946.

Shin, C., Nam, J.-W., Farh, K.K.-H., Chiang, H.R., Shkumatava, A., and Bartel, D.P. (2010). Expanding the microRNA targeting code: functional sites with centered pairing. Mol. Cell. *38*, 789–802.

Sibley, C.R., Seow, Y., Saayman, S., Dijkstra, K.K., El Andaloussi, S., Weinberg, M.S., and Wood, M.J. (2012). The biogenesis and characterization of mammalian microRNAs of mirtron origin. Nucleic Acids Res. *40*, 438–448.

Skog, J., Würdinger, T., van Rijn, S., Meijer, D.H., Gainche, L., Sena-Esteves, M., Curry, W.T., Carter, B.S., Krichevsky, A.M., and Breakefield, X.O. (2008). Glioblastoma microvesicles transport RNA and proteins that promote tumour growth and provide diagnostic biomarkers. Nat. Cell Biol. *10*, 1470–1476.

Taft, R.J., Glazov, E.A., Cloonan, N., Simons, C., Stephen, S., Faulkner, G.J., Lassmann, T., Forrest, A.R.R., Grimmond, S.M., Schroder, K., *et al.* (2009a). Tiny RNAs associated with transcription start sites in animals. Nat. Genet. *41*, 572–578.

Taft, R.J., Glazov, E.A., Lassmann, T., Hayashizaki, Y., Carninci, P., and Mattick, J.S. (2009b). Small RNAs derived from snoRNAs. RNA *15*, 1233–1240.

Taft, R.J., Simons, C., Nahkuri, S., Oey, H., Korbie, D.J., Mercer, T.R., Holst, J., Ritchie, W., Wong, J.J-L., Rasko, J.E.J., *et al.* (2010). Nuclear-localized tiny RNAs are associated with transcription initiation and splice sites in metazoans. Nat. Struct. Mol. Biol. *17*, 1030–1034.

Taft, R.J., Hawkins, P.G., Mattick, J.S., and Morris, K.V. (2011). The relationship between transcription initiation RNAs and CCCTC-binding factor (CTCF) localization. Epigenetics Chromatin. *4*, 13.

Takane, K., and Kanai, A. (2011). Vertebrate virus-encoded microRNAs and their sequence conservation. Jpn. J. Infect. Dis. *64*, 357–366.

Tan, G.S., Garchow, B.G., Liu, X., Yeung, J., Morris, J.P., Cuellar, T.L., McManus, M.T., and Kiriakidou, M. (2009a). Expanded RNA-binding activities of mammalian Argonaute 2. Nucleic Acids Res. *37*, 7533–7545.

Tan, Y., Zhang, B., Wu, T., Skogerbø, G., Zhu, X., Guo, X., He, S., and Chen, R. (2009b). Transcriptional inhibiton of Hoxd4 expression by miRNA-10a in human breast cancer cells. BMC Mol. Biol. *10*, 12.

Tay, Y., Zhang, J., Thomson, A.M., Lim, B., and Rigoutsos, I. (2008). MicroRNAs to Nanog, Oct4 and Sox2 coding regions modulate embryonic stem cell differentiation. Nature *455*, 1124–1128.

Tokumaru, S., Suzuki, M., Yamada, H., Nagino, M., and Takahashi, T. (2008). Let-7 regulates Dicer expression and constitutes a negative feedback loop. Carcinogenesis *29*, 2073–2077.

Tolia, N.H., and Joshua-Tor, L. (2007). Slicer and the argonautes. Nat. Chem. Biol. *3*, 36–43.

Tsai, N.-P., Lin, Y.-L., and Wei, L.-N. (2009). MicroRNA mir-346 targets the 5′-untranslated region of receptor-interacting protein 140 (RIP140) mRNA and up-regulates its protein expression. Biochem. J. *424*, 411–418.

Tsui, N.B.Y., Ng, E.K.O., and Lo, Y.M.D. (2002). Stability of endogenous and added RNA in blood specimens, serum, and plasma. Clin. Chem. *48*, 1647–1653.

Turchinovich, A., and Burwinkel, B. (2012). Distinct AGO1 and AGO2 associated miRNA profiles in human cells and blood plasma. RNA Biol. *9*, 1066–1075.

Turchinovich, A., Weiz, L., Langheinz, A., and Burwinkel, B. (2011). Characterization of extracellular circulating microRNA. Nucleic Acids Res. *39*, 7223–7233.

Valadi, H., Ekström, K., Bossios, A., Sjöstrand, M., Lee, J.J., and Lötvall, J.O. (2007). Exosome-mediated transfer of mRNAs and microRNAs is a novel mechanism of genetic exchange between cells. Nat. Cell Biol. *9*, 654–659.

Vasudevan, S., Tong, Y., and Steitz, J.A. (2007). Switching from repression to activation: microRNAs can up-regulate translation. Science *318*, 1931–1934.

Verdel, A., Vavasseur, A., Le Gorrec, M., and Touat-Todeschini, L. (2009). Common themes in siRNA-mediated epigenetic silencing pathways. Int. J. Dev. Biol. *53*, 245–257.

Vickers, K.C., Palmisano, B.T., Shoucri, B.M., Shamburek, R.D., and Remaley, A.T. (2011). MicroRNAs are transported in plasma and delivered to recipient cells by high-density lipoproteins. Nat. Cell Biol. *13*, 423–433.

Wang, H.-W., Noland, C., Siridechadilok, B., Taylor, D.W., Ma, E., Felderer, K., Doudna, J.A., and Nogales, E. (2009). Structural insights into RNA processing by the human RISC-loading complex. Nat. Struct. Mol. Biol. *16*, 1148–1153.

Wang, K., Zhang, S., Weber, J., Baxter, D., and Galas, D.J. (2010). Export of microRNAs and microRNA-protective protein by mammalian cells. Nucleic Acids Res. *38*, 7248–7259.

Weinberg, M.S., Villeneuve, L.M., Ehsani, A., Amarzguioui, M., Aagaard, L., Chen, Z.-X., Riggs, A.D., Rossi, J.J., and Morris, K.V. (2006). The antisense strand of small interfering RNAs directs histone methylation and transcriptional gene silencing in human cells. RNA *12*, 256–262.

Williams, Z., Ben-Dov, I.Z., Elias, R., Mihailovic, A., Brown, M., Rosenwaks, Z., and Tuschl, T. (2013). Comprehensive profiling of circulating microRNA via small RNA sequencing of cDNA libraries reveals biomarker potential and limitations. Proc. Natl. Acad. Sci. U.S.A. *110*, 4255–4260.

Wu, H., Ye, C., Ramirez, D., and Manjunath, N. (2009). Alternative processing of primary microRNA transcripts by Drosha generates 5′ end variation of mature microRNA. PLoS ONE. *4*, e7566.

Wyman, S.K., Knouf, E.C., Parkin, R.K., Fritz, B.R., Lin, D.W., Dennis, L.M., Krouse, M.A., Webster, P.J., and Tewari, M. (2011). Post-transcriptional generation of miRNA variants by multiple nucleotidyl transferases contributes to miRNA transcriptome complexity. Genome Res. *21*, 1450–1461.

Xhemalce, B., Robson, S.C., and Kouzarides, T. (2012). Human RNA methyltransferase BCDIN3D regulates microRNA processing. Cell *151*, 278–288.

Yang, J.-S., Maurin, T., Robine, N., Rasmussen, K.D., Jeffrey, K.L., Chandwani, R., Papapetrou, E., Sadelain, M., O'Carroll, D., and Lai, E.C. (2010). Conserved vertebrate mir-451 provides a platform for Dicer-independent, Ago2-mediated microRNA biogenesis. Proc. Natl. Acad. Sci. U.S.A. *107*, 15163–15168.

Yang, N., and Kazazian, H.H., Jr. (2006). L1 retrotransposition is suppressed by endogenously encoded small interfering RNAs in human cultured cells. Nat. Struct. Mol. Biol. *13*, 763–771.

Yang, W., Chendrimada, T.P., Wang, Q., Higuchi, M., Seeburg, P.H., Shiekhattar, R., and Nishikura, K. (2006). Modulation of microRNA processing and expression through RNA editing by ADAR deaminases. Nat. Struct. Mol. Biol. *13*, 13–21.

Yekta, S., Shih, I.-H., and Bartel, D.P. (2004). MicroRNA-directed cleavage of HOXB8 mRNA. Science; *304*, 594–596.

Yi, R., Qin, Y., Macara, I.G., and Cullen, B.R. (2003). Exportin-5 mediates the nuclear export of pre-microRNAs and short hairpin RNAs. Genes Dev. *17*, 3011–3016.

Zardo, G., Ciolfi, A., Vian, L., Starnes, L.M., Billi, M., Racanicchi, S., Maresa, C., Fazi, F., Travaglini, L., Noguera, N., *et al.* (2012). Polycombs and microRNA-223 regulate human granulopoiesis by transcriptional control of target gene expression. Blood; *119*, 4034–4046.

Zeng, Y., Yi, R., and Cullen, B.R. (2003). MicroRNAs and small interfering RNAs can inhibit mRNA expression by similar mechanisms. Proc. Natl. Acad. Sci. U.S.A. *100*, 9779–9784.

Zernecke, A., Bidzhekov, K., Noels, H., Shagdarsuren, E., Gan, L., Denecke, B., Hristov, M., Köppel, T., Jahantigh, M.N., Lutgens, E., *et al.* (2009). Delivery of microRNA-126 by apoptotic bodies induces CXCL12-dependent vascular protection. Sci. Signal. *2*, Ra81.

Zhang, L., Hou, D., Chen, X., Li, D., Zhu, L., Zhang, Y., Li, J., Bian, Z., Liang, X., and Caj, X. (2012). Exogenous plant MIR168a specifically targets mammalian LDLRAP*1*, evidence of cross-kingdom regulation by microRNA. Cell Res. *22*, 107–126.

Zhou, H., Arcila, M.L., Li, Z., Lee, E.J., Henzler, C., Liu, J., Rana, T.M., and Kosik, K.S. (2012). Deep annotation of mouse iso-miR and iso-moR variation. *Nucleic Acids Res. 40*, 5864–5875.

Zhou, X., Duan, X., Qian, J., and Li, F. (2009). Abundant conserved microRNA target sites in the 5′-untranslated region and coding sequence. Genetica; *137*, 159–164.

Non-coding RNAs and the Epigenetic Control of Gene Expression

3

Kevin V. Morris

Abstract

A paradigm shift is occurring in molecular biology with the realization that much of the genome is transcribed as non-coding RNA and that these non-coding transcripts are biologically relevant. Many examples have now been provided whereby long non-coding RNAs that are antisense to their protein-coding counterpart are actually involved in epigenetically regulating the protein coding genes expression. Importantly, small non-coding RNAs can be utilized to either take advantage of these endogenous non-coding pathways of gene regulation and silence gene expression or to target regulatory non-coding RNAs and de-repress particular protein-coding genes, essentially turning on a genes expression. Knowledge of this emerging RNA based epigenetic regulatory network and our ability to control gene expression has deep implications in the development of entirely new areas of therapeutics.

Long non-coding RNAs epigenetic regulators in human cells

Long non-coding RNAs (ncRNA) have now been estimated to make upwards to ~70% of the transcripts in the cell (Banfai *et al.*, 2012; Derrien *et al.*, 2012; Rosenbloom *et al.*, 2012). While originally considered 'junk DNA' with no known function, these transcripts are beginning to cause consternation within modern molecular biology and our collective understanding of the cell. Emerging evidence suggests that one mechanism of action for long ncRNAs is to drive epigenetic forms of regulation and fundamentally transcriptional processes in human cells. Epigenetic changes can range from histone modifications, such as acetylation and methylation of core histones to the methylation of genomic DNA. The result of these forms of epigenetic states can be chromatin compaction, which leads ultimately to reduced transcriptional output due to the compaction of the targeted genomic loci, ultimately reducing the ability of RNA Polymerase II (RNAPII) to interact with the particular locus. Recent observations suggest that long antisense ncRNAs are active modulators of epigenetic states in human cells (Hawkins and Morris 2010; Poliseno 2010; Johnsson *et al.*, 2013a). Several different models have been put forth as to how these long ncRNAs modulate epigenetic states in human cells, many of which could be endogenously active (reviewed in Morris 2011). Structural indications suggest that the larger a long ncRNA is the greater the complexity instilled in forming various secondary structures and interacting with different protein binding partners (Novikova *et al.*, 2013). Collectively, an important insight gained from many of those studies on ncRNA directed mechanisms suggests that ncRNAs provide an avenue to directing protein payloads around the cell and affecting various aspects of the cell, in the case of epigenetic states transcription is affected (Fig. 3.1A–C). While clearly long ncRNAs play a plethora of roles in the cell the fact that epigenetic marks are instilled in target loci by the action of ncRNAs suggests a complex mechanism of gene regulation that has the potential to be long lasting if not permanent due fundamentally to the targeted recruitment of epigenetic marks such as DNA methylation.

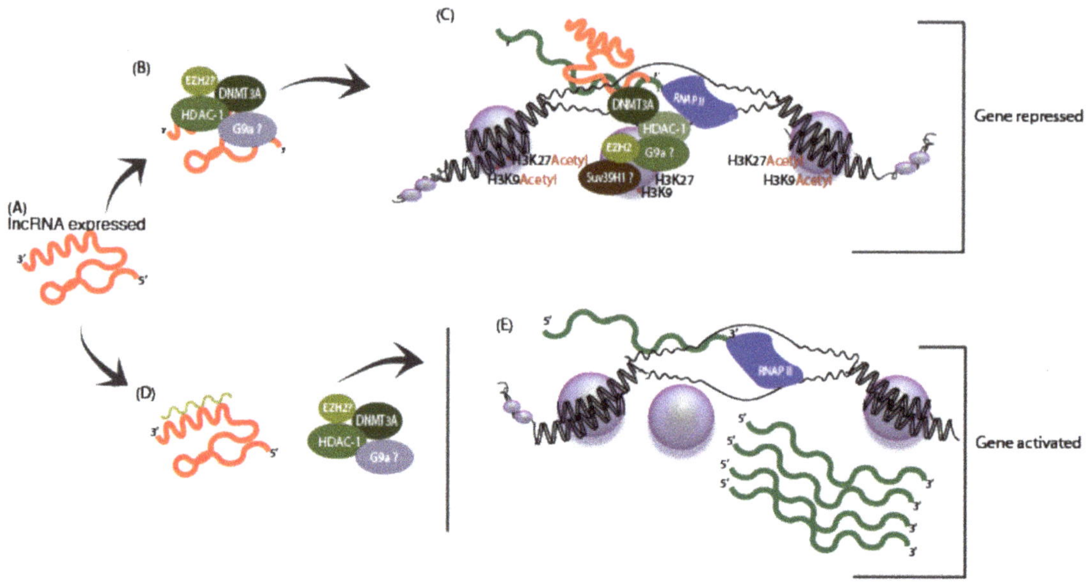

Figure 3.1 Long antisense ncRNA directed gene regulation in human cells. (A) Long antisense non-coding RNAs (lncRNAs) are expressed in trans. (B) These lncRNAs interact with epigenetic remodelling complexes consisting of DNA methyltransferase 3A (DNTM3a), Enhancer of Zeste (EZH2), and Histone deacetylase 1 (HDAC-1)(Jeffery and Nakielny, 2004; Weinberg *et al.*, 2006; Suzuki *et al.*, 2008; Turner *et al.*, 2009; Hawkins and Morris, 2010; Johnsson, 2013). (C) This lncRNA-Epigenetic remodelling complex can then target homology-containing loci, resulting in transcriptional gene silencing of the targeted locus. When this pathway is targeted with (D) antisense oligonucleotides, or RNAi, there is a loss of the lncRNA directed targeting resulting in (E) transcriptional activation of the lncRNA-targeted locus. This interpretation of the lncRNA regulatory pathway in human cells is based on several bodies of work (reviewed in Knowling and Morris, 2011) as well as experimentation ongoing in the Morris laboratory.

Unlike short ncRNAs, such as short interfering RNA (siRNA), which predominantly silence genes by cutting their messenger RNA transcripts, long ncRNAs appear capable of (1) binding directly to targeted transcribed regions in the nucleus and (2) recruiting epigenetic remodelling complexes to these target loci (Fig. 3.1B and C) (reviewed in Morris, 2009a). Some of these long ncRNAs have been found to bind in antisense, or on the opposite strand to the one that codes for a protein-coding gene, thereby regulating the sense counterpart gene's transcription (Morris *et al.*, 2008; Yu *et al.*, 2008; Hawkins and Morris 2010; Johnsson *et al.*, 2013a)(Fig. 3.1A–C). Mechanistically, these antisense long ncRNAs affect their targeted loci by binding to a promoter-associated transcript (Han *et al.*, 2007; Schmitz *et al.*, 2010) at the ncRNA target locus and simultaneously recruiting chromatin remodelling proteins which then modify the local chromatin and DNA (Fig. 3.1C). In the case of DNA modification, the DNA methyltransferase 3a (DNMT3a) may direct modifications of the target sequence by adding a methyl group (Weinberg *et al.*, 2006; Johnsson *et al.*, 2013a). Whereas silencing via histone modification is thought to occur mainly on the tail of histone 3 (at position H3K27me3 for instance). The methyl group, be it on H3K27 or DNA, can silence the long ncRNA targeted locus. Notably, methylated DNA has been found to be passed down from one cell to the next and to be long lasting with regards to suppression of transcription. One observation found that the duration of effect of ncRNA directed epigenetic regulation can be drastic with gene expression reduced long-term, and in this one example silencing was conferred out to a month following ncRNA targeted epigenetic silencing (Hawkins *et al.*, 2009). These observations along with reports from others strongly implicate ncRNAs as drivers of diverse gene expression states and that these states can be retained in a long-term and stable manner.

Disruption of long antisense ncRNA networks

The observations to date suggest that long ncRNAs act to instil an epigenetic brake on the particular ncRNA targeted gene. Several functional long antisense ncRNAs have been observed to have a regulatory role in different genes, ranging from those genes associated with cancer (Morris *et al.*, 2008; Yu *et al.*, 2008; Tahira *et al.*, 2011; Johnsson *et al.*, 2013b), to stem cell genesis (Hawkins and Morris 2010), to autism (Kerin *et al.*, 2012), and neurodegenerative disease (Lipovich *et al.*, 2012; Modarresi *et al.*, 2012). Notably, when these ncRNAs are targeted with small antisense RNAs, siRNAs, or even antisense oligonucleotides (Fig. 3.1D and E), the result is de-repression of the long antisense ncRNA target locus. This de-repression leads to activation of the long antisense ncRNA targeted locus, and if a protein-coding gene increased activation of the particular protein-coding genes expression. This activation of gene expression has been observed to be physiologically relevant and in the case of PTEN affecting cell cycling (Johnsson *et al.*, 2013a). *In vivo* studies demonstrated functional targeting of a Brain-derived neurotrophic factor (BDNF)-associated antisense long ncRNA that resulted in distinct physiologically relevant changes (Modarresi *et al.*, 2012). In this study Modarresi and colleagues found that targeting of the BDNF associated antisense lncRNA using either chemically modified oligonucleotides or siRNAs resulted in upwards to seven-fold increased expression of BDNF in the mouse brain. Notably, the observed increased BDNF expression had a direct effect on downstream BDNF targets (Modarresi *et al.*, 2012). These observations suggest that the knowledge of the endogenous long ncRNA pathways and their targets can lead to a methodology whereby gene activation can be instilled in particular genes, namely those that are actively modulated by the action of a long antisense ncRNA. This knowledge can lead directly to siRNA targeting or antisense oligonucleotide based applications of targeted gene activation/de-repression.

The therapeutic potential of targeting antisense ncRNAs

The long ncRNA pathway described here (Fig. 3.1) can be taken advantage of in two distinct and therapeutically relevant ways to modulate gene expression epigenetically; either turns a gene on or off. To activate a particular gene of interest knowledge of the endogenous long antisense ncRNA that targets or regulates the particular gene of interest is required. Once this is known one can then target this particular ncRNA using siRNAs or antisense oligonucleotides (as depicted in Fig. 3.1D and E). Such a method allows in essence for an ability to activate gene expression and is of profound interest with regards to new therapeutics, as currently very few compounds are specific activators of gene expression.

The same endogenous lncRNA pathway can however also be usurped with small antisense ncRNAs (Fig. 3.2A–C). Early studies showed that siRNAs designed to target gene promoters were capable of directing transcriptional gene silencing (TGS) in human cells (reviewed in Turner and Morris, 2010). In this study, siRNAs were designed to target a genes promoter instead of the mRNA/coding region, which is typically targeted with siRNAs (Morris *et al.*, 2004). The result of this siRNA targeting of a gene promoter was the suppression of gene expression at the transcriptional level, well upstream of the well documented PTGS pathway (Morris *et al.*, 2004). This mode of RNA based TGS involved epigenetic components, as Trichostatin A and 5′ Aza-cytadine suppressed the observed gene silencing. This observed suppression was similar to the first observation of RNA directed TGS carried out in plants and yeast (Matzke *et al.*, 2004).

Later studies by Weinberg and colleagues discovered, quite serendipitously, that only the antisense strand of the siRNA was required for TGS in human cells and that DNMT3a is required for the antisense RNA directed targeting (Weinberg *et al.*, 2006). Other studies with small RNA directed TGS have further validated the requirement for DNMT3a, Histone deacetylase 1 (HDAC1) and Enhancer of Zeste (EZH2) (Han *et al.*, 2007; Lim *et al.*, 2008; Suzuki *et al.*, 2008; Hawkins *et al.*, 2009; Turner *et al.*, 2009, 2012), suggesting that the endogenous mechanism in

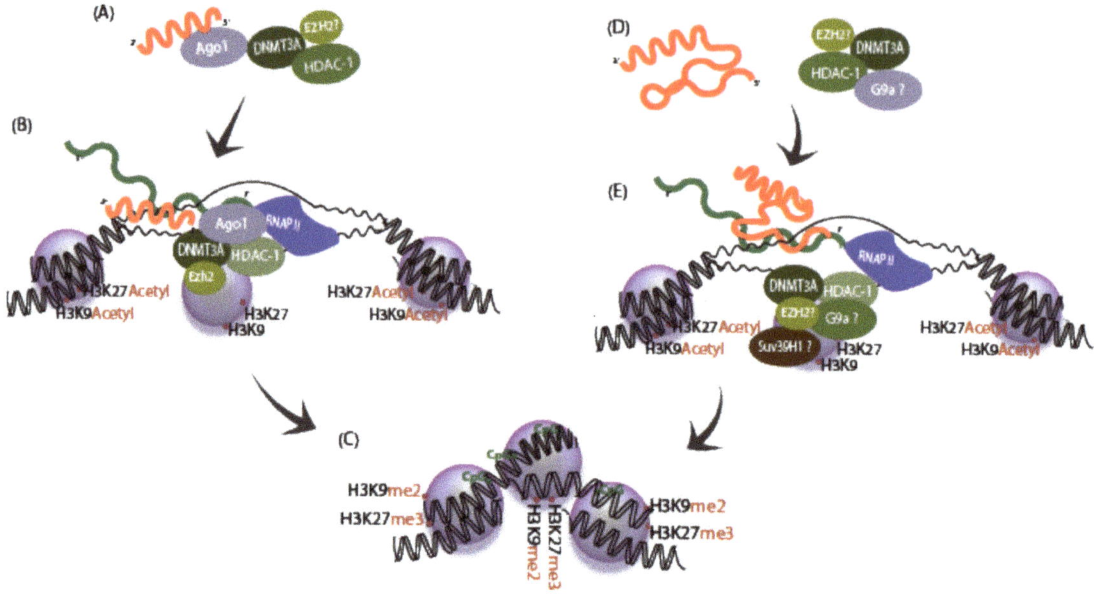

Figure 3.2 The parallels of small and long antisense ncRNA directed transcriptional gene silencing. (A) Small antisense RNAs generated synthetically and designed to target a promoter of choice can (B) interact with promoter-associated RNAs at the transcribed promoter (Han *et al.*, 2007; Hawkins *et al.*, 2009; Napoli *et al.*, 2009) and facilitate the recruitment of epigenetic remodelling complexes (Suzuki, 2005; Kim *et al.*, 2006; Weinberg *et al.*, 2006; Han *et al.*, 2007; Suzuki *et al.*, 2008; Hawkins *et al.*, 2009; Turner *et al.*, 2009) to the targeted loci. (C) The result of small antisense ncRNA targeting is the epigenetic remodelling and stable silencing of transcription from the targeted promoter. (D) Long antisense ncRNAs can also interact with DNMT3a, and possibly also HDAC1, G9a and EZH2 (direct evidence supporting this has not yet been published) and (E) remodel the targeted loci leading to (C) the heterochromatinization and transcriptional gene silencing.

human cells whereby non-coding RNAs regulate gene transcription may involve antisense ncRNAs directing epigenetic complexes (Fig. 3.2D–F). While at the time of these discoveries it was unknown as to what ncRNAs, if any, were active endogenous epigenetic modulators of gene expression in human cells the data collectively suggested that antisense ncRNAs could function in this manner. We now suspect that those small antisense ncRNAs used in earlier studies were in essence mimicking the action of the endogenous long antisense ncRNA pathway (Fig. 3.2). Collectively, the emerging notion from studies on both small and long antisense ncRNAs involved in epigenetic regulation of gene expression is that small ncRNAs can be used in two ways to control a genes transcriptional expression. For activating gene expression these small antisense ncRNAs can be targeted to endogenous long antisense ncRNAs and either degrade the targeted transcripts (RNAi) or bind to the long antisense ncRNA and obstruct secondary structures (antisense oligonucleotides) leading ultimately to a loss of this non-coding transcripts ability to epigenetically regulate its target gene(s), e.g. de-repression of the long antisense ncRNA targeted gene (Fig. 3.1D and E). Alternatively, small antisense ncRNAs can be generated to mimic long antisense ncRNAs and target epigenetic silencing to particular loci (Fig. 3.2). While clearly a path exists whereby small ncRNAs can be used to differentially modulate gene expression such modulation is not without its pitfalls and problems.

Issues with using small antisense ncRNAs therapeutically

The use of small antisense ncRNAs or siRNAs as therapeutics has several issues that require understanding and remain to be resolved. One major issue that might be encountered from the use of small antisense ncRNAs or siRNAs to target gene silencing is that this form of targeting, within the nucleus, is fundamentally epigenetic (Hawkins *et al.*, 2009). For instance siRNA targeting of gene promoters results in epigenetic remodelling of the targeted promoter. This form of targeting however is not specific to gene promoters and any siRNA will have this effect when one assesses the epigenetic footprint at the homology containing loci of the siRNA within the context of the nucleus (demonstrated in Hawkins *et al.*, 2009). These observations suggest that siRNAs, regardless of their intended target, be it a particular coding transcript, also have effects within the nucleus of the cell, e.g. they direct epigenetic remodelling to homology containing transcribed loci in the nucleus of the treated cell. Such observations also suggest that the mechanism of ncRNA directed epigenetic remodelling is ubiquitous and ncRNAs load into this pathway and guide epigenetic complexes to homology containing target loci, regardless of where this loci is in the chromatin (provided it is actually transcribed, e.g. transcription is needed to get transcriptional gene silencing). At least this is the notion based on published observations (Weinberg *et al.*, 2006). Exactly how this targeting localizes to the transcribed locus remains largely unknown but does appear to involve active transcription at the homology containing target locus (Weinberg *et al.*, 2006; Hawkins *et al.*, 2009; Napoli *et al.*, 2009).

Another confounding issue with using small antisense ncRNAs to target epigenetic based regulation is that the target loci can become stably silenced and that this form of silencing can be passed on to daughter cells (reviewed in Morris, 2009b). While this is an added benefit in treating diseases such as HIV, where long-term stable silencing could be beneficial, the ramifications of using small ncRNAs in other disease settings whereby long-term stable changes are not desired, could be disastrous. Indeed, many tumour-suppressor genes become epigenetically silenced in a long-term and stable manner, presumably via the action of long antisense ncRNAs regulating the corresponding tumour-suppressor gene (reviewed in Morris and Vogt, 2010). Such long-term silencing may not be advantageous in certain disease settings and thus make the choice of targeted regulation of epigenetic states by small antisense ncRNAs less than desirable. One method around this might be to utilize antisense oligonucleotides with various chemical modifications. Some studies with antisense oligonucleotides targeted to gene promoters found that the antisense oligonucleotides could block transcription in the absence of recruiting silent state epigenetic marks to the targeted locus (Knowling *et al.*, 2012). In this study only fluorinated antisense transcripts were capable of recruiting silent state epigenetic marks to the promoter of HIV, supporting the notion that in human cells antisense ncRNAs are drivers of epigenetic-based forms of gene regulation (Fig. 3.1A–C).

Lastly, the activation of gene expression, by targeting the regulatory long antisense ncRNAs might also prove disadvantageous in some regards. For instance the removal of the epigenetic brake, by the targeting of the modulatory long antisense ncRNA, could also be relatively long-term and stable. This could occur if the small antisense ncRNAs targeted to the regulatory long antisense ncRNAs results in directing transcriptional gene silencing (Fig. 3.2A–C) to the particular long antisense ncRNA locus. Early experimental indications suggest that many long antisense ncRNAs are RNAPII expressed transcripts (Naganuma and Hirose, 2013), supporting the notion that these transcripts are also susceptible to the same mechanism of ncRNA based regulation which is instilled by these long antisense ncRNAs at other targeted loci (Figs. 3.1A–C and 3.2D and E). As such one can surmise that targeting regulatory long antisense ncRNAs can lead to their stable suppression, which ultimately results in the stable activation of their particular targeted locus. Collectively, the notion from these studies on both small and long forms of antisense ncRNAs is that ncRNA may in fact be driving the epigenetic footprint of the cell (Mattick *et al.*, 2009; Morris, 2009b). Indeed, ncRNAs may be drivers of natural selection (Morris, 2009b) and as such if this is the case then utilizing small ncRNAs as therapeutics, while beneficial to controlling disease and pathological states, may also place a selective force on the particular exposed organism. While clearly an exciting time exists with regards to non-coding RNA Biol. the relatively unknown aspects of utilizing molecules to regulate these new species of RNAs with regards to downstream eventualities remains unknown. One has to wonder if one day targeted natural selection will be possible whereby antisense ncRNAs are used to epigenetically modify organisms; ranging from better producers of biofuels to resistance to Dengue virus, HIV, or cancer.

Acknowledgements

The project was supported by the National Institute of Allergy and Infectious Disease (NIAID) R56 AI096861-01, P01 AI099783-01 and the National Cancer Institute (NCI) R01 CA151574, National Institutes of Health (NIH) R01 CA153124 to KVM.

References

Banfai, B., Jia, H., Khatun, J., Wood, E., Risk, B., Gundling, W.E. Jr., Kundaje, A., Gunawardena, H.P., Yu, Y., Xie, L., *et al.* (2012). Long noncoding RNAs are rarely translated in two human cell lines. Genome Res. *22*, 1646–1657.

Derrien, T., Johnson, R., Bussotti, G., Tanzer, A., Djebali, S., Tilgner, H., Guernec, G., Martin, D., Merkel, A., Knowles, D.G., Lagarde, J., Veeravalli, L., *et al.* (2012). The GENCODE v7 catalog of human long noncoding RNAs: Analysis of their gene structure, evolution, and expression. Genome Res. *22*, 1775–1789.

Han, J., Kim, D., and Morris, K.V. (2007). Promoter-associated RNA is required for RNA-directed transcriptional gene silencing in human cells. Proc. Natl. Acad. Sci. U.S.A. *104*, 12422–12427.

Hawkins, P.G., and Morris K.V. (2010). Transcriptional regulation of Oct4 by a long non-coding RNA antisense to Oct4-pseudogene 5. Transcription *1*, 165–175.

Hawkins, P.G., Santoso, S., Adams, C., Anest, V., and Morris, K.V. (2009). Promoter targeted small RNAs induce long-term transcriptional gene silencing in human cells. Nucleic Acids Res. *37*, 2984–2995.

Jeffery, L., and Nakielny S. (2004). Components of the DNA methylation system of chromatin control are RNA-binding proteins. J. Biol. Chem. *279*, 49479–49487.

Johnsson, P., Ackley, A., Vidarsdottir, L., Lui, W., Corcoran, M., Grandér, D., and Morris K.V. (2013). A pseudogene long non-coding RNA network regulates PTEN transcription and translation in human cells. Nat. Struct. Mol. Biol. *20*, 440–446.

Kerin, T., Ramanathan, A., Rivas, K., Grepo, N., Coetzee, G.A., and Campbell, D.B. (2012). A noncoding RNA antisense to moesin at 5p14.1 in autism. Sci. Transl. Med. *4*, 128ra140.

Kim, D.H., Villeneuve, L.M., Morris, K.V., and Rossi, J.J. (2006). Argonaute-1 directs siRNA-mediated transcriptional gene silencing in human cells. Nat. Struct. Mol. Biol. *13*, 793–797.

Knowling, S., and Morris, K.V. (2011). Epigenetic regulation of gene expression in human cells by noncoding RNAs. Prog. Mol. Biol. Transl. Sci *102*, 1–10.

Knowling, S., Stapleton, K., Turner, A.M., Uhlmann, E., Lehmann, T., Vollmer, J., and Morris, K.V. (2012). Chemically modified oligonucleotides modulate an epigenetically varied and transient form of transcription silencing of HIV-1 in human cells. Mol. Ther. Nucleic Acids *1*, e16.

Lim, H.G., Suzuki, K., Cooper, D.A., and Kelleher, A.D. (2008). Promoter-targeted siRNAs induce gene silencing of simian immunodeficiency virus (SIV) infection *in vitro*. Mol. Ther. *16*, 565–570.

Lipovich, L., Dachet, F., Cai, J., Bagla, S., Balan, K., Jia, H., and Loeb, J.A. (2012). Activity-dependent Human Brain Coding/Non-coding Gene Regulatory Networks. Genetics *192*, 1133–1148.

Mattick, J.S., Amaral, P.P., Dinger, M.E., Mercer, T.R., and Mehler, M.F. (2009). RNA regulation of epigenetic processes. Bioessays *31*, 51–59.

Matzke, M., Aufsatz, W., Kanno, T., Daxinger, L., Papp, I., Mette, M.F., and Matzke, A.J. (2004). Genetic analysis of RNA-mediated transcriptional gene silencing. Biochim. Biophys. Acta *1677*, 129–141.

Modarresi, F., Faghihi, M.A., Lopez-Toledano, M.A, Fatemi, R.P., Magistri, M., Brothers, S.P., van der Brug, M.P., and Wahlestedt, C. (2012). Inhibition of natural antisense transcripts *in vivo* results in gene-specific transcriptional up-regulation. Nat. Biotechnol. *30*, 453–459.

Morris, K.V. (2009a). Long antisense non-coding RNAs function to direct epigenetic complexes that regulate transcription in human cells. Epigenetics *4*, 296–301.

Morris, K.V. (2009b). Non-coding RNAs, epigenetic memory, and the passage of information to progeny. RNA Biol. *6*, 242–247.

Morris, K.V. (2011). The emerging role of RNA in the regulation of gene transcription in human cells. Semin. Cell Dev. Biol. *22*, 351–358.

Morris, K.V., and Vogt, P.K. (2010). Long antisense non-coding RNAs and their role in transcription and oncogenesis. Cell Cycle *9*, 2544–2547.

Morris, K.V., Chan, S.W., Jacobsen, S.E., and Looney, D.J. (2004). Small interfering RNA-induced transcriptional gene silencing in human cells. Science *305*, 1289–1292.

Morris, K.V., Santoso, S., Turner, A.-M., Pastori, C., and Hawkins, P.G. (2008). Bidirectional transcription directs both transcriptional gene activation and suppression in human cells. PLoS Genet. *4*, e1000258.

Naganuma, T., and Hirose, T. (2013). Paraspeckle formation during the biogenesis of long non-coding RNAs. RNA Biol. *10*, 456–461.

Napoli, S., Pastori, C., Magistri, M., Carbone, G.M., and Catapano, C.V. (2009). Promoter-specific transcriptional interference and c-myc gene silencing by siRNAs in human cells. EMBO J. *28*, 1708–1719.

Novikova, I.V., Hennelly, S.P., Tung, C.S., and Sanbonmatsu, K.Y. (2013). Rise of the RNA machines: exploring the structure of long non-coding RNAs. J. Mol. Biol. *425*, 3731–3746.

Poliseno, L., Salmena, L., Zhang, J., Carver, B., Haveman, W.J., and Pandolfi, P.P. (2010). A coding-independent function of gene and pseudogene mRNAs regulates tumour biology. Nature *465*, 1033–1038.

Rosenbloom, K.R., Dreszer, T.R., Long, J.C., Malladi, V.S., Sloan, C.A., Raney, B.J., Cline, M.S., Karolchik, D., Barber, G.P., Clawson, H., *et al.* (2012). ENCODE whole-genome data in the UCSC Genome Browser: update 2012. Nucleic Acids Res. *40*(Database issue), D912–917.

Schmitz, K.M., Mayer, C., Postepska, A., and Grummt, I. (2010). Interaction of noncoding RNA with the rDNA promoter mediates recruitment of DNMT3b and silencing of rRNA genes. Genes Dev. *24*, 2264–2269.

Suzuki, K., Shijuuku, T., Fukamachi, T., Zaunders, J., Guillemin, G., Cooper, D., and Kelleher., A. (2005). Prolonged transcriptional silencing and CpG methylation induced by siRNAs targeted to the HIV-1 promoter region. J. RNAi Gene Silencing *1*, 66–78.

Suzuki, K., Juelich, T., Lim, H., Ishida, T., Watanebe, T., Cooper, D.A., Rao, S., and Kelleher, A.D. (2008). Closed chromatin architecture is induced by an RNA duplex targeting the HIV-1 promoter region. J. Biol. Chem. *283*, 23353–23363.

Tahira, A.C., Kubrusly, M.S., Faria, M.F., Dazzani, B., Fonseca, R.S., Maracaja-Coutinho, V., Verjovski-Almeida, S., Machado, M.C.C., and Reis, E.M. (2011). Long noncoding intronic RNAs are differentially expressed in primary and metastatic pancreatic cancer. Mol. Cancer *10*, 141.

Turner, A.M., and Morris, K.V. (2010). Controlling transcription with noncoding RNAs in mammalian cells. Biotechniques *48*, ix–xvi.

Turner, A.M., De La Cruz, J., and Morris, K.V. (2009). Mobilization-competent lentiviral vector-mediated sustained transcriptional modulation of HIV-1 expression. Mol. Ther. *17*, 360–368.

Turner, A.M., Ackley, A.M., Matrone, M.A., and Morris, K.V. (2012). Characterization of an HIV-targeted transcriptional gene-silencing RNA in primary cells. Hum. Gene. Ther. *23*, 473–483.

Weinberg, M.S., Villeneuve, L.M., Ehsani, A., Amarzguioui, M., Aagaard, L., Chen, Z.X., Riggs, A.D., Rossi, J.J., and Morris, K.V. (2006). The antisense strand of small interfering RNAs directs histone methylation and transcriptional gene silencing in human cells. RNA *12*, 256–262.

Yu, W., Gius, D., Onyango, P., Muldoon-Jacobs, K., Karp, J., Feinberg, A.P., and Cui, H. (2008). Epigenetic silencing of tumour suppressor gene p15 by its antisense RNA. Nature *451*, 202–206.

From Mice to Men: Towards the Clinical Translation of MicroRNA Technologies for Somatic Cell Reprogramming

4

Elena Senís and Dirk Grimm

Abstract

The 2006/2007 discovery that somatic cells can be reprogrammed to induced pluripotent stem cells through the introduction of a defined class of transcription factors has revolutionized the field of regenerative medicine. More recently, the additional finding that the RNAi machinery and in particular miRNAs are fundamentally involved in pluripotency has sparked a flurry of efforts to develop new technologies for miRNA-regulated reprogramming of human cells. Here, we summarize and critically review the latest literature describing the three most prominent advances in this field, with a specific focus on the underlying methodologies and their clinical relevance. First, we compare the different experimental approaches – forward versus reverse screening – to identify endogenous miRNAs that are inherently involved in the induction and/or maintenance of pluripotency. Second, we comprehensively discuss the principal strategies for deliberate dysregulation of these miRNAs in order to reprogramme somatic cells that have been tested thus far, from viral vector-mediated miRNA expression along with classical pluripotency factors, to exclusive reprogramming through delivery of 'naked' miRNAs. Third and last, we highlight novel innovative concepts that exploit the inherent alterations in miRNA expression profiles between pluripotent and differentiated cells, with the aim to mark and track these distinct subpopulations, or to purify one cell type from contaminations with the other. Finally, we conclude with our outlook into the possible future of human cell reprogramming and with suggestions for improvements to existing miRNA technologies that may further foster their clinical translation.

Introduction

Embryonic stem cells (ESCs) are derived from the inner cell mass of the mammalian blastocyst (Evans and Kaufman, 1981) and are characterized by two main features: they can be cultured indefinitely and they are pluripotent, which means that they have the capacity to generate all types of somatic cells that make up an organism. Indeed, when transferred back into a blastocyst, they can give rise to differentiated cells that will form tissues belonging to the three embryonic germ layers: endoderm (such as the digestive track and some of the adjacent organs, like the liver and the pancreas; the urinary bladder and the lungs), mesoderm (comprising the heart and the circulatory system, the skeletal muscle, the cartilage, the dermis and the kidney) and ectoderm (such as neural tissue including the brain, the epidermis, hair, nails, etc.) (Fig. 4.1). To some extent, ESC can also be differentiated *ex vivo* into specific progeny cell types, by culturing them with certain hormones and/or growth factors. This property makes pluripotent stem cells highly attractive for the field of regenerative medicine, especially when combined with the latest advances in cell transplantation technologies. Their appeal is further increased by the option to genetically modify ESC prior to differentiation using genome engineering tools such as zinc finger nucleases, TALEns or the CRISPR system (Gaj *et al.*, 2013), in order to correct genetic defects or to introduce exogenous

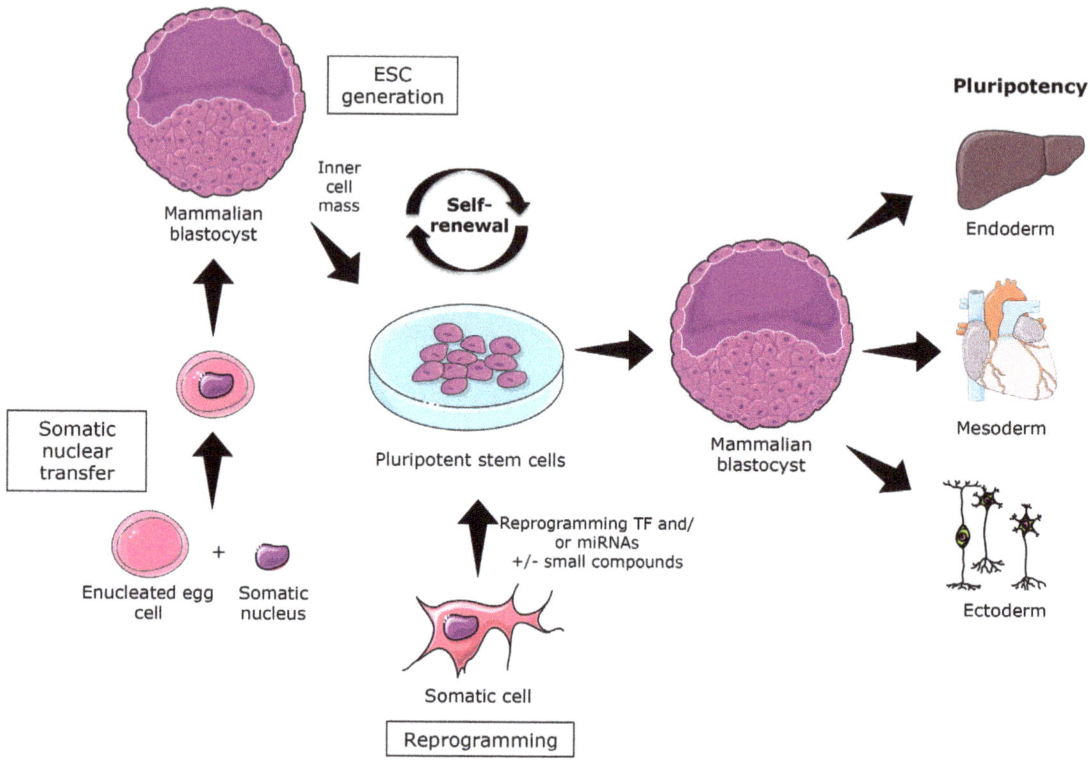

Figure 4.1 Methods to generate pluripotent stem cells. In the classical method of ESC generation, cells from the inner cells mass of the mammalian blastocyst are extracted and maintained in appropriate culture conditions. The cell lines generated are able to self-renew and are pluripotent, meaning that they can differentiate into cells belonging to the three embryonic germ layers: endoderm, mesoderm and ectoderm. In the somatic cell nuclear transfer method, a somatic nucleus is transferred to a enucleated egg cell. The resulting cell divides and is allowed to develop until the blastocyst stage from which the same procedure as for the classical method to generate ESC is followed. The reprogramming method circumvents the blastocyst stage by introducing defined transcription factors (TF) and/or miRNAs ± small molecules into somatic cells. The induced pluripotent stem cells generated with this method maintain the ESC characteristics, i.e. self-renewal and pluripotency. This figure was produced using Servier Medical Art (www.servier.com).

sequences for therapeutic benefit. Nonetheless, the clinical translation of ESC technologies is hampered by a number of practical and theoretical concerns, including the risk of immune rejection of the transplanted/engrafted cells due to their derivation from a genetically different donor. Moreover, for obvious reasons, the isolation of human embryonic stem cells from early embryos continues to be a source of intense ethical and political debate.

Consequently, these problems have tempted many labs to undertake extensive efforts to develop alternative technologies that can circumvent these issues. One surrogate strategy that emerged is somatic cell nuclear transfer, whereby a nucleus is transferred from a differentiated cell into an enucleated unfertilized egg. If successful, the latter will subsequently develop further and give rise to proper offspring (Wilmut *et al.*, 1997). Remarkably, despite the enormous challenges that are inherent to this technique, one group has recently accomplished its application to derive human embryonic stem cells (Tachibana *et al.*, 2013) (Fig. 4.1). Perhaps even more notable are, however, the seminal 2006/2007 reports by Takahashi and Yamanaka which demonstrated the possibility to reprogramme mouse somatic cells back to an immature, pluripotent state. The only manipulation required to obtain these so-called induced pluripotent stem cells (iPSC) was to ectopically express four transcription factors, Oct3/4 (O), Klf4 (K), Sox2 (S) and c-Myc (M), in the somatic cells (Takahashi and Yamanaka, 2006) (Fig. 4.1). Shortly after these initial publications, the same group then also managed to reprogramme human cells with the same four transcription factors (Takahashi *et al.*, 2007). Moreover, the Thompson lab was likewise successful using Oct3/4 and Sox2 in combination with Lin28 (an inhibitor of the let7 miRNA cluster) and Nanog, instead of Klf4 and c-Myc (Yu *et al.*, 2007). As a whole, these reports represented a major revolution in the stem cell field as they illustrated, for the first time, the option to derive pluripotent cells from somatic cells rather than embryos, which fully overcomes the ethical problems indicated above. Concurrently, this modification relieves concerns about immunological rejection since donor and host can in principle be identical. This ground breaking work thus entirely changed our perception of cell development and specialization, and at the same time significantly expanded our options to study diseases and to devise new therapeutic and diagnostic modalities. In view of this enormous relevance of his discoveries for biological and medical research, Shinya Yamanaka was awarded the Nobel Prize for Physiology and Medicine in 2012 (together with John Gurdon, for somatic cell nuclear transfer).

Today, the field of somatic cell reprogramming is flourishing, with an ever extending wealth of groups employing the original Yamanaka factors (OKSM) to generate iPSC from a large variety of mouse or human somatic cells. Still, a number of lingering issues continue to prevent the further exploitation of this technology for routine application in human patients. One is that the gene delivery vectors used to transfer the OKSM factors in the original work were derived from retro- or lentiviruses, which are notoriously known for their propensity to randomly integrate into the genome and potentially cause cellular transformation. A related second concern is that one of the four factors, c-Myc, is oncogenic *per se*. Furthermore, the necessity to co-deliver four individual factors to the same cell, and at a very specific stoichiometry that mediates best efficiencies, puts substantial demands on the design of appropriate vector constructs.

The sum of these concerns has therefore sparked tremendous additional efforts to further improve reprogramming technology, towards an ideal system that (i) is non-integrating and thus inherently safer, and that (ii) requires only a minimal amount of exogenous factors. To achieve the first aim, a large diversity of alternative gene delivery vehicles has already been tested, including non-integrating viral vectors (replication-incompetent adenoviral vectors, Adeno-associated viral (AAV) vectors or Sendai viral vectors), DNA-based methods (plasmids, minicircle vectors, piggyBac transposons or episomal plasmids) as well as DNA-free strategies (protein or mRNA delivery). The common goal of these alternatives was to minimize or completely eliminate any traces of the vectors carrying the reprogramming factors, without compromising their overall efficiency. Towards the second aim, many other groups have attempted to reprogramme cells with different factors, delivered either alone or in combination with all or some of the OKSM factors. The list of surrogates tested thus far is long and diverse, and includes short hairpin RNAs (against specific cellular genes whose products hamper reprogramming), as well as small compounds, such as valproic acid (Huangfu *et al.*, 2008) or ascorbic acid (Stadtfeld *et al.*, 2012). In several reported cases, the use of these alternative factors indeed helped to increase the efficiency of the reprogramming process and/or the stability of the progeny cells.

In this chapter, we will focus on yet another class of small molecules that have exceptional potential to spark a new revolution in the stem cell field, namely, miRNAs. As detailed in other chapters of this issue, miRNAs are small, roughly 21 to 23 nt long regulatory RNAs that are endogenously expressed in a highly cell type- and developmental stage-specific manner. Following their Drosha- and Dicer-mediated processing, they are loaded into the RNA-induced silencing complex RISC and then typically repress gene expression. Therefore, the RISC/miRNA complex binds to partially complementary sequences in the 3'UTR of an mRNA and subsequently triggers destabilization of this transcript via decapping and deadenylation, followed by mRNA degradation (note that other mechanisms are also discussed in the literature).

Because miRNA–mRNA association relies on imperfect complementarity, a single miRNA can in principle regulate hundreds of mRNA targets; in contrast, a single mRNA is typically targeted by several different miRNAs. Altogether, this creates a highly complex regulatory network through which the currently known over 2500 human miRNAs probably control at least 60% of all human genes. Considering this intricacy, it is not surprising that miRNAs are involved in a vast array of fundamental physiological and pathological cellular processes, including cell division, tumorigenesis, viral infection and apoptosis (Berezikov, 2011; Fabian and Sonenberg, 2012).

Important in the present context, a large body of evidence that has accumulated over the past decade clearly proves that miRNAs are also essential players in cellular pluripotency and (de) differentiation. This raises considerable hope that a better understanding of the natural role of miRNAs in these processes will eventually result in novel technologies and applications that can improve or even replace our current reprogramming and differentiation strategies, and that will thereby accelerate the clinical translation of this exciting field. In the following, we will critically review the latest advances in iPSC technologies that could pave the way towards this important goal, from (i) methods for identification of relevant miRNAs and (ii) strategies to exploit these miRNAs for reprogramming, to (iii) recent innovations that use stem cell-specific miRNAs for cell tracking or purging. Our particular focus is on the underlying methodologies and on their relevance for the application in human iPSC, which we will also summarize in our outlook at the end of this chapter. We will not specifically review the biology of miRNAs in pluripotency, such as their mRNA targets and associated pathways, as this topic has already been covered extensively in a flurry of excellent and comprehensive reviews to which we refer the interested reader (Table 4.1).

Table 4.1 List of selected reviews on miRNA biology in reprogramming

Year	Reference
2009	Dang (2009)
	Feng *et al.* (2009)
2010	Sun *et al.* (2010)
	Mallanna and Rizzino (2010)
	Tiscornia and Izpisua Belmonte (2010)
2011	Onder and Daley (2011)
	Lai *et al.* (2011)
	Lin (2011)
	Chang and Gregory (2011)
	Sridharan and Plath (2011)
2012	Kuo *et al.* (2012)
	Lin *et al.* (2012)
	Heinrich and Dimmeler (2012)
	Anokye-Danso *et al.* (2012)
	Lipchina *et al.* (2012)
	Leonardo *et al.* (2012)
	Kuo and Ying (2012)
	Underbayev *et al.* (2012)
	Li and He (2012)
2013	Bao *et al.* (2013)
	Luningschror *et al.* (2013)
	Lamouille *et al.* (2013)
	Yang and Rana (2013)
	Rosa and Brivanlou (2013)
	Huo and Zambidis (2013)

Listed are selected reviews that focus on the biology of miRNAs in pluripotency rather than on technological aspects, since the latter are the topic of the current article. We apologize to all authors whose work was not included here due to space constraints.

General evidence for a role of miRNAs in pluripotency and differentiation

The search for miRNAs that regulate pluripotency and/or differentiation started about a decade ago, soon after the initial report of RNAi in nematodes (Fire *et al.*, 1998) and its subsequent confirmation in other species including mammals (He and Hannon, 2004). One of the first general clues that miRNAs are critically involved in mammalian stem cell functionality was provided in 2003 by Bernstein and colleagues, when they reported their inability to generate homozygous ESC from a Dicer knock-out mouse model (Bernstein *et al.*, 2003). Albeit this study could not make a direct link to miRNAs due to technical limitations at the time, the knowledge that Dicer is a key regulator of miRNA processing strongly hinted at their possible involvement in the maintenance of a stem cell population in the early mouse embryo.

The role of Dicer and hence the RNAi machinery was subsequently solidified in a number of studies, including a 2004 report by Kanellopoulou *et al.* (2005) in which the authors used a more sophisticated strategy (conditional gene targeting and stringent selection) to create homozygous Dicer knock-out ESC. Curiously, these cells were viable despite their inability to produce mature miRNAs. In addition, they displayed the typical ESC morphology when grown on MEF feeders, and also expressed Oct4 and other ESC markers. However, the cells grew more slowly than wildtype ESC, and, even more notably, they did not differentiate properly *in vitro* and *in vivo*. Evidence was that the Dicer knock-out ESC formed embryoid bodies (three-dimensional pluripotent cell aggregates from which cells can differentiate into cells belonging to the three germ layers – endoderm, mesoderm and ectoderm) that were poorly differentiated and eventually ceased growing. Moreover, the cells failed to yield detectable teratomas upon subcutaneous injection into nude mice (another typical assay to measure pluripotency and the ability to differentiate). Interestingly, these effects were partially attributed to Dicer knock-out-dependent epigenetic changes in centromeric heterochromatin structure, satellite repeat methylation and centromeric silencing.

Similar findings on the role of Dicer for ESC proliferation were made in a 2005 study by the Hannon lab (Murchison *et al.*, 2005), which had already reported the initial observations in the Dicer knock-out mice (Bernstein *et al.*, 2003). When this group now also used a conditional strategy to ablate Dicer, they likewise managed to obtain homozygous and vital Dicer knock-out ESC. Akin to the Kanellopoulou study (Kanellopoulou *et al.*, 2005), these cells lacked mature miRNAs and exhibited a significant proliferation lag; however, this newer report could not reproduce the same DNA methylation defects. Despite this unresolved controversy, the two studies together suggest a very complex role of Dicer and the RNAi machinery in pluripotency and differentiation, involving miRNA-related mechanisms, but probably also others.

Further proof for a seminal role of miRNAs came in 2007 when Wang and colleagues created mice with a knock-out of DGCR8, a double-stranded RNA-binding protein that assists Drosha in nuclear processing of primary to precursor miRNAs (one of the first steps in miRNA biogenesis) (Wang *et al.*, 2007). Similar to the Dicer-depleted cells, miRNA-deficient ESC from these mice grew more slowly and accumulated more in G1 as compared to their wildtype counterparts, further supporting the role of miRNAs in ESC proliferation and cell cycle progression. The knock-out cells also failed to differentiate properly *in vitro* and *in vivo* since they were unable to fully silence pluripotency markers including Oct4, even under stringent differentiation conditions. Altogether, this implied that differentiating ESC normally express miRNAs that inhibit cellular genes/proteins which maintain ESC pluripotency, extending the findings on the role of miRNAs from the previous Dicer studies.

Further support for this function of miRNAs comes from a 2011 study by Shekar *et al.* (2011), who knocked out Argonaute 2 (Ago2), a key component of RISC. Notably, Ago2 knock-out only reduces but does not completely eliminate the biogenesis of mature miRNAs. It also does not entirely impair miRNA activity due to the expression of three other Ago proteins in mammals (Ago1, 3, and 4) that exert redundant functions in the miRNA pathway (Schurmann *et al.*, 2013; Su *et al.*, 2009). Nonetheless, Ago2-deficient ESC displayed retarded growth and G1 accumulation, reminiscent of the data in DGCR8 knock-out ESC from Wang and colleagues (Wang *et al.*, 2007). Moreover similarly, ESC lacking Ago2 had a defect in differentiation, as evidenced by only partial suppression of the Oct4 and Nanog pluripotency markers, as well as by delayed and weaker expression of differentiation markers in Ago2-deficient embryoid bodies. Yet, as compared to the corresponding previous findings in the Dicer and DGCR8 knock-out ESC, these effects were rather subtle which may reflect the aforementioned redundancy among the four mammalian Ago proteins.

Finally, Li *et al.* (2011) further validated the role of miRNA-mediated post-transcriptional gene regulation for reprogramming by knocking down Ago2, Drosha or Dicer in Oct4-Gfp MEFs using specific short hairpin RNAs, and by showing that this resulted in a dramatic drop in the number of iPSC colonies following transduction with OKSM-expressing retroviral vectors.

Strategies for the identification of pluripotency-associated miRNAs

Collectively, the discoveries highlighted above (and others not mentioned here) show the relevance of the RNAi machinery for the regulation of the three basic pillars of stem cells: self-renewal, pluripotency maintenance and differentiation. They hence provide a solid and strong rationale to search for the actual miRNAs that are the potential effectors in all these different processes. As noted before, their characterization is not only interesting from a fundamental biological standpoint, but their identification and subsequent use as tools also promises to improve reprogramming technologies. Ideally, they will even permit to replace some of the conventional factors, especially the oncogenic c-Myc. In the recent literature, one can distinguish two principal strategies towards these goals, which are illustrated in Fig. 4.2 and will be described in more detail in the following.

The first is based on the reasonable assumption that miRNAs which favour the pluripotent state and/or suppress differentiation will be inherently up-regulated in stem cells as compared to their parental cells or differentiated progeny, respectively. Accordingly, a large number of groups have pursued a forward genetics approach in which they have initially applied various methods to determine the 'microRNA-ome' of pluripotent versus differentiated cells, and then subsequently tested single lead candidates for their ability to enhance reprogramming in combination with the classical factors (Fig. 4.2, top). Because the profiling platforms and cell types used thus far are highly complex and diverse, they will not be reviewed in detail; instead, representative examples are summarized in Table 4.2. Despite this diversity, it is striking to note that nearly all studies conducted so far congruently identified a subset of miRNAs that are significantly up-regulated in both m/hESC as well as hiPSC, implying their critical involvement in the induction and/or maintenance of pluripotency. The three most prominent and most consistently identified candidates are miRNAs from the miR-302-367, miR-371-373 as well as miR-17-92 clusters (highlighted in red, green and blue in Table 4.2). Indeed, as will be described in more detail in the next chapter, in particular the miR-302-367 cluster has recently been validated as a very potent reprogramming factor. In turn, this verifies the power of the overall approach and should fuel the further exploitation of other leading miRNAs that have emerged from these profiling studies.

In contrast to this 'top-down' or 'educated guesses' tactic, the second strategy to identify pluripotency-associated miRNAs relies on the largely unbiased deliberate up-regulation of numerous miRNAs in parallel, followed by phenotypic inspection of the treated cells. Typically, this 'bottom-up' or reverse genetics approach is pursued in the presence of at least a subset of the conventional reprogramming factors, in order to increase chances to identify miRNAs that foster pluripotency by cooperating with these factors (Fig. 4.2, bottom). A notable example for the unique power of such a high-throughput miRNA screening was provided by Pfaff and colleagues in a 2011 study (Pfaff *et al.*, 2011). The group first transduced MEFs with a lentivirus expressing human OKS which they had previously shown to induce iPSC within 10 days, split the cells and 12 hours later transfected them with a commercial library of 379 murine precursor miRNAs (one per well). Evaluation of colonies expressing Gfp (encoded in the MEFs under the Oct4 promoter) then allowed for the identification of miRNAs that enhanced or blocked iPSC generation. Strikingly, this approach not only re-identified several previously reported candidates (such as the miR-290 and miR-302 clusters, see Table 4.2) but also an additional 14 miRNAs that were unknown to contribute to reprogramming at the time of this study. Validation experiments confirmed their essential role, by showing that transfection of these newly found miRNAs along with the OKS vector increases iPSC colony numbers, sizes and Oct4-Gfp fluorescence intensities. The colonies expressed typical stem cell markers and had silenced the exogenous reprogramming factors,

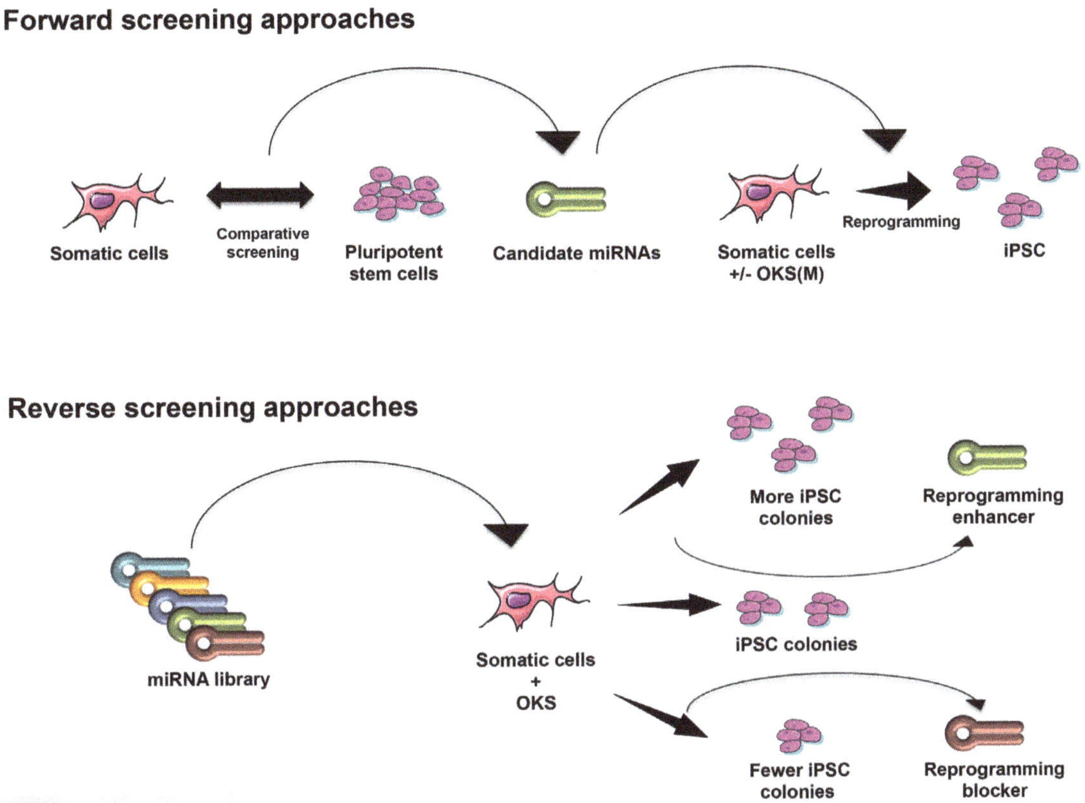

Figure 4.2 Screening approaches to identify reprogramming-associated miRNAs. In forward miRNA screening approaches (top), somatic versus pluripotent cells are profiled and the miRNAs that are expressed preferentially in pluripotent stem cells are chosen. These candidate miRNAs are then used in reprogramming experiments alone or in combination with OKS(M) to test whether they enhance the reprogramming process or whether they are even able to reprogramme somatic cells by themselves. In reverse screening approaches (bottom), a miRNA library is tested directly in reprogramming experiments in combination with OKS, with the aim to identify candidates which enhance or block the reprogramming process. In hybrid forward-reverse approaches, a pre-selected library of miRNAs which play a role in pluripotency (identified through forward screening) is tested in functional reprogramming assays together with OKS (i.e. reverse screening). The overall aim is again to recognize and validate miRNAs that enhance reprogramming efficiency. This figure was produced using Servier Medical Art (www.servier.com).

indicative of their pluripotent nature. Lastly, the group also reported evidence that one of the main targets of their miRNAs might be the homeobox transcription factor Meox2, a regulator of TGF-β signalling and p21, which might provide a mechanistic link to reprogramming.

As a whole, this study impressively demonstrates the potential of library-based reverse screening to identify new miRNAs with a role in reprogramming. Most notable in this respect is that RT-PCR analyses of endogenous miRNA expression during the first six days after MEF infection with the OKS vector showed no up-regulation of one of the lead candidates in this study, miR-721. This clearly exemplifies the unique strength of reverse approaches to also identify relevant miRNAs that are inconspicuous in profiling studies and hence may have been missed in forward screenings.

Finally, a notable recent study by Ruiz *et al.* (2012) reported a hybrid forward-reverse genetics approach in which the group performed a miRNA screen akin to Pfaff and colleagues (Pfaff *et al.*, 2011), but using a smaller pre-selected library of miRNAs already known to be crucial for reprogramming (Ruiz *et al.*, 2012). Moreover

Table 4.2 Identification of stem cell-specific miRNAs

Method[a]	Cell types[b]	Up-regulated stem cell-specific miRNAs[c]	Reference
cDNA cloning 53 miRNAs	mESC	First report of miR-290 cluster, **miR-302** (15 new mESC-specific miRNAs)	Houbaviy *et al.* (2003)
cDNA cloning 36 miRNAs	hESC	First report of **miR-302**/367 cluster, ***miR-371*** cluster, miR-200c (14 new hESC-specific miRNAs)	Suh *et al.* (2004)
RT-PCR 220 miRNAs	mESC	miRNA profiling on single cell level, data confirm prior literature, including miR-290 cluster	Tang *et al.* (2006)
Microarray miRBase n.d. 312 (?) miRNAs	hESC	**miR-302a/d**, ***miR-371-373*** cluster, miR-200c; also confirmed let7a down-regulation	Lakshmipathy *et al.* (2007)
Deep sequencing 323 miRNAs	mESC	miR-290 cluster, miR-17-92 cluster; also found 46 novel miRNAs	Calabrese *et al.* (2007)
Microarray miRBase n.d. miRNAs n.d.	hESC	**miR-302** cluster, miR-17-92 cluster	Josephson *et al.* (2007)
Microarray miRBase n.d. 436 miRNAs	hESC	**miR-302** cluster, ***miR-372*** cluster, miR-17-92 cluster, miR-520 cluster	Tzur *et al.* (2008)
Deep sequencing 260 miRNAs	hESC	**miR-302** cluster, miR-17-92 cluster; also found 13 novel miRNAs	Bar *et al.* (2008)
Deep sequencing 194 miRNAs	hESC	**miR-302** cluster, found miR-372 down-regulated; also found 23 novel miRNAs	Morin *et al.* (2008)
Microarray miRBase 9.0 397 miRNAs	hESC	Confirmed 37 previously reported miRNAs, including **miR-302** cluster, ***miR-371-373*** cluster, miR-17-92 cluster	Laurent *et al.* (2008)
Microarray miRBase 10.0 697 miRNAs	hiPSC, hESC	**miR-302** cluster, miR-17-92 cluster Small differences hiPSC<>hESC	Wilson *et al.* (2009)
Microarray miRBase n.d. 326 miRNAs	hiPSC, hESC	**miR-302** cluster, miR-17-92 cluster Small differences hiPSC<>hESC	Chin *et al.* (2009)
Microarray miRBase 9.1 827 miRNAs	hESC	**miR-302** cluster, miR-520 cluster, miR-200c, ***miR-372***	Ren *et al.* (2009)
Deep sequencing 733 miRNAs	hESC	**miR-302** cluster, predicted nearly 300 new miRNAs and other small RNAs	Goff *et al.* (2009)
RT-PCR 466 miRNAs	hiPSC, hESC	**miR-302** cluster, ***miR-371*** cluster; specifically confirmed miR-302 as a crucial marker	Neveu *et al.* (2010)
Microarray miRBase n.d. 470 miRNAs	hiPSC, hESC	**miR-302** cluster, miR-17-92 cluster, ***miR-371-373*** cluster	Stadler *et al.* (2010)

[a]Listed (where named in the respective paper) are the miRBase version (www.mirbase.org) for the microarray-based profiling studies and the number of miRNAs that were assessed in these papers. n.d., not determined.
[b]Listed are the stem cell lines that were profiled. Not shown are the differentiated controls that were used for comparison (typically either fibroblasts from the same species or embryoid bodies derived from the stem cells).
[c]Listed are only selected findings from these papers with respect to major miRNAs or clusters that were either originally discovered or validated. The three major candidates are highlighted in the table using bold, bold italic and underlined text. 'Novel' or 'predicted' miRNAs were new at the time of the publication; by now, they have been incorporated into miRBase or discarded.

remarkably, this latest screen was more clinically relevant as it was conducted in human cells. To this end, the authors established an intricate reprogramming system in which they first introduced an Oct4 promoter-driven Gfp reporter into human fibroblasts. These cells were then infected with doxycycline-inducible lentiviral vectors individually expressing the classical Yamanaka factors. Importantly, these vectors were selected so that the resulting cell lines would miss one of the four factors, in order to allow screening for miRNAs that could functionally substitute for this particular factor. Specifically, the cell lines expressed combinations of SKM, OKM or OSK, respectively. These cells could be differentiated into fibroblasts, and, upon addition of doxycycline and infection with a retrovirus supplying the missing fourth factor, reprogrammed back into pluripotency, altogether proving their usefulness.

To harness this system to identify miRNAs with a role in human cell reprogramming, the authors next profiled the expression of 667 miRNAs amongst 18 human somatic or iPS cell types/lines and isolated 26 miRNAs that were strongly expressed in the iPSC or down-regulated in the somatic cells, respectively. When 19 of these 26 miRNAs were individually expressed from a lentivirus in doxycycline-treated OSK fibroblasts, five of them greatly improved reprogramming efficiency (as measured via appearance of Gfp-positive colonies), including the previously unreported miR-519a. This miRNA, which is only found in primates, shares an almost identical seed sequence with several ES cell-specific cell cycle-regulating (ESCC) miRNAs, targets TGFβRII and enhances the proliferation levels of cells subjected to reprogramming, providing first clues to its possible functions in iPSC generation.

Collectively, these and other related studies clearly prove the power of both strategies – forward or reverse screening – to recognize and corroborate miRNAs with a key role in reprogramming. With respect to the looming clinical translation of iPSC technologies, it is particularly remarkable that both are applicable to, and relevant for, human cells. As noted, this has in fact already been proved experimentally by a series of studies that have successfully used these miRNAs, especially miR-302, to improve the generation of iPSC from human somatic ancestors. Nonetheless, it could still be rewarding to more thoroughly mine the existing data sets and to continue screening for additional miRNAs, for a number of reasons. First is that forward approaches are inherently restricted to identifying miRNAs that are up-regulated in stem cells, which includes a risk to miss other interesting candidates which are not increased *per se*, but still function as a reprogramming enhancer upon deliberate delivery/expression. In other words, it is possible that many more important miRNAs have escaped their discovery in forward genetics approaches and hence still await their identification through reverse screening or other experimental means. A good example for this class of naturally inconspicuous miRNAs is the aforementioned miR-721 (see above) which only showed up in a reverse screen, but was never a prominent hit in any of the published forward assays (to our knowledge). Second and third, it is likewise conceivable that some useful miRNAs were missed due to the enormous complexity of typical data sets resulting from miRNA profiling studies, as well as due to underlying technical demands and the ensuing error-prone nature of these methodologies. Examples for critical parameters whose slight variation can already largely influence the screening outcome are the selection of the reference cells, contaminations of stem cell cultures with spontaneously differentiated cells, or biases between individual miRNAs during library preparation for deep sequencing. Fourth, pluripotent cells moreover likely exhibit large changes in the expression of many miRNAs which are not mechanistically linked to reprogramming, and which thus create a high background noise that hampers the identification of truly interesting candidates. Lastly, it is a well known phenomenon and problem in the miRNA profiling field (also beyond iPSC) that results can differ substantially between the various technology platforms (Git *et al.*, 2010). As a whole, these considerations clearly highlight the necessity for very careful experimental planning and subsequent independent validation of forward approaches for miRNA identification, in order to minimize the number of false-positives.

The same of course also applies to the reverse strategies which are likewise inherently prone to yield flawed data if not meticulously planned and executed. In fact, these assays might be even more critical as they require extensive manipulation of the cells under investigation, including their bulk transfection with miRNA candidates and subsequent prolonged cultivation until faithful reprogramming becomes apparent. Hence, there is a wealth of parameters than can affect and falsify the outcome of such reverse screening strategies, and that have indeed been confirmed as critical in the respective studies. Examples are the concentration of doxycycline used to induce stably integrated reprogramming factors, or the selection of supporting feeder MEFs versus matrigel or other substrates. Equally important and error-prone is the choice of the specific endpoint that determines successful reprogramming and thus relevance of the tested miRNA, and that must be selected carefully to maximize the stringency of selection. For instance, the time point after miRNA delivery when positive colonies are scored can be crucial since the individual miRNAs that are compared in parallel might not only differentially affect reprogramming efficiencies, but also kinetics. Further potential limitations of these assays are their strict dependence on available (commercial or self-made) tools for miRNA delivery or expression, as well as on the individual lab's capacity and equipment to perform and analyse massive parallel screenings of large miRNA collections. A clear advantage over the physical screening methods outlined above is, however, that reverse strategies represent a functional screen designed to straightforwardly identify miRNAs that actually play a role in reprogramming, rather than being dysregulated during pluripotency or differentiation.

Delivery or expression of miRNAs as novel tools to improve reprogramming

Over the past 4 years, the miRNAs identified using the means outlined in the previous chapter have been tested extensively for their capability to induce or enhance the reprogramming of various mouse and human cells. From all these studies, several parameters have emerged whose profound analysis and comparison is particularly relevant with respect to the potential clinical use of reprogramming technologies: (i) miRNA expression from vectors versus delivery of 'naked' miRNAs, (ii) the miRNA expression/delivery vehicle, (iii) the number of miRNAs that need to be altered concurrently, and (iv) the requirement for co-expression/delivery with conventional reprogramming factors versus the option to reprogramme solely with miRNAs. Because the original literature describing the different underlying approaches is already vast and diverse, we have summarized a selection of the most important and influential reports on the use of miRNAs to reprogramme mouse or human somatic cells in Tables 4.3 and 4.4, respectively. These two tables list the different cell sources, delivery methods, conventional co-factors (if any), as well as corresponding efficiencies and kinetics (i.e. days of appearance of the first colonies). In the following, we will additionally review some of the hallmark papers in more detail with a focus on the different parameters noted above.

A first seminal study in this field came from Judson *et al.* (2009) who reported the feasibility to reprogramme MEFs with a combination of OKS and miRNAs from the miRNA miR-290-295 cluster. Notably, over 70% of all miRNAs expressed in mouse ESC belong to this cluster whose members are master regulators of pluripotency. They up-regulate Lin28, Sall4 and c-Myc (Melton *et al.*, 2010), promote cell cycle progression (Wang *et al.*, 2008) and down-regulate differentiation genes (Lichner *et al.*, 2011) as well as pro-apoptotic genes, such as Caspase-2 and Ei24, under stress conditions (Zheng *et al.*, 2011). These functions have been particularly assigned to miR-291-3p, miR-294 and miR-295, the so-called ESCC miRNAs. Probably due to this unique role, this was the first miRNA cluster used to reprogramme somatic cells into iPSCs. Specifically, MEFs were infected with retroviruses encoding OKS and then transfected twice with miR-291-3p, miR-294 or miR-295 mimics, one and six days later. As compared to OKS alone, adding either of these three miRNAs increased the efficiency of reprogramming up to ten-fold. The same was observed for miR-302d, a

Table 4.3 miRNAs used to reprogramme mouse somatic cells

miRNA	Reference	Delivery method	Cell source	Co-factors used	First colony	Reprogramming efficiency
miR-290/295	Judson *et al.* (2009)	miRNA mimics (transfections at days 0 and 6)	MEFs	OKS	Day 10	0.1–0.3% (miR-294)
	Lu *et al.* (2012)	PiggyBac vector	MEFs	OKSM	Day 16	2 × more than OKSM control
miR-302d	Judson *et al.* (2009)	miRNA mimics (transfections at days 0 and 6)	MEFs	OKS	Day 10	0.1–0.3%
miR-200c	Samavarchi-Tehrani *et al.* (2010)	miRNA mimics (transfection)	MEFs	OKSM ± miR-200b	Not shown	Increased compared to mock transfected control but no clear numbers shown
	Miyoshi *et al.* (2011)	Transfection of mature miRNAs (four transfections every 2 days)	mASC and MEFs	miR-302c and miR-369	Day 15	0.01% (mASC) and 0.002% (MEFs)
miR-200b	Samavarchi-Tehrani *et al.* (2010)	miRNA mimics (transfection)	MEFs	OKSM ± miR-200c	Not shown	Increased compared to mock transfected control but no clear numbers shown
miR-302/367	Anokye-Danso *et al.* (2011)	Lentiviral vector	MEFs	VPA	Day 6–8	Approx. 50%
	Liao *et al.* (2011)	Retroviral vector	MEFs	OKS ± vitamin C or OKSM ± vitamin C	Day 15 (OKS) and day 11 (OKSM)	0.25% (OKS–), 3.5% (OKS+), 0.075% (OKSM–), 0.75% (OKSM+)
	Miyoshi *et al.* (2011)	Transfection of mature miRNAs (4 transfections every 2 days)	mASC and MEFs	miR-200c and miR-369	Day 15	0.01% (mASC) and 0.002% (MEFs)
	Lu *et al.* (2012)	PiggyBac vector	MEFs	OKSM	Day 16	4 × more than OKSM control
	Sohn *et al.* (2013)	Polyketal nanoparticles (every 2 days for 1 week)	Bone marrow mononuclear cells	VPA	Day 8	Not shown and no iPSC lines established
miR-106a/363	Liao *et al.* (2011)	Retroviral vector	MEFs	OKS ± vitamin C or OKSM ± vitamin C	Day 15 (OKS) and day 11 (OKSM)	0.05% (OKS–), 1.5% (OKS+), 0.025% (OKSM–), 0.5% (OKSM+)
miR-369	Miyoshi *et al.* (2011)	Transfection of mature miRNAs (4 transfections every 2 days)	mASC and MEFs	miR-200c and miR-302	Day 15	0.01% (mASC) and 0.002% (MEFs)

Table 4.3 continued

miRNA	Reference	Delivery method	Cell source	Co-factors used	First colony	Reprogramming efficiency
miR-93	Li *et al.* (2011)	miRNA mimics (transfections at days 0 and 5)	MEFs	OKS or OKSM	Days 15–20 (OKS) and day 11 (OKSM)	0.5% (OKS) and 0.3% (OKSM)
miR-103b	Li *et al.* (2011)	miRNA mimics (transfections at days 0 and 5)	MEFs	OKS or OKSM	Days 15–20 (OKS) and day 11 (OKSM)	0.5% (OKS and OKSM)
miR-17	Li *et al.* (2011)	miRNA mimics (transfections at days 0 and 5)	MEFs	OKSM?	Day 11	0.35%
miR-106a	Li *et al.* (2011)	miRNA mimics (transfections at days 0 and 5)	MEFs	OKSM?	Day 11	0.40%
miR-130/301/721	Pfaff *et al.* (2011)	Transfection of precursor miRNAs (12 h post OKS transduction)	MEFs	OKS and VPA	Days 7–8	Approx. 0.6%
miR-25	Lu *et al.* (2012)	PiggyBac vector and miRNA mimic	MEFs	OKSM	Day 16	3 × more than OKSM control
miR-298	Lu *et al.* (2012)	PiggyBac vector	MEFs	OKSM	Day 16	2.5 × more than OKSM control

mASC, mouse adipose stem cells; VPA, valproic acid. In some cases, the reprogramming efficiency was not specifically mentioned. If possible, it was thus calculated by us from the data provided in the text and figures (number of iPSC colonies divided by number of cells initially seeded).

Table 4.4 miRNAs used to reprogramme human somatic cells

miRNA	Reference	Delivery method	Cell source	Co-factors used	First colony	Reprogramming efficiency
miR-302/367	Lin *et al.* (2008)	Transfection of Pol II- based intronic miRNA expression system	Skin cancer cells		Not shown	2–5%
	Anokye-Danso *et al.* (2011)	Lentiviral vector	Foreskin and dermal fibroblasts		Day 18	Approx. 10%
	Liao *et al.* (2011)	Retroviral vector	Fibroblasts	OKS	Not shown	0.01%
	Miyoshi *et al.* (2011)	Direct transfection of mature miRNAs (four transfections every 2 days)	hASC and dermal fibroblasts	miR-200c and miR-369	Day 20	0.002%
	Lin *et al.* (2011)	Electroporation of a doxycycline-inducible vector	Hair follicle cells		Day 5–6	Data not shown

Table 4.4 continued

miRNA	Reference	Delivery method	Cell source	Co-factors used	First colony	Reprogramming efficiency
	Lee *et al.* (2012)	Lentiviral vector	Partially reprogrammed cells from CD34+ cells from cord blood	OKSM to reprogramme human CB-CD34+	Day 15 (conversion, no reprogramming)	16.4% conversion efficiency (not reprogramming efficiency)
miR-302 (alone)	Subramanyam *et al.* (2011)	Mimic of mature miR-302b (transfections at days 3 and 10)	Foreskin and lung fibroblasts	OKS ± miR-372 or OKSM ± miR-372	Day 31 (OKS) and day 21 (OKSM)	Increased compared to mock transfected control but no clear numbers shown
	Lee *et al.* (2012)	Lentiviral vector	Partially reprogrammed cells from CD34+ cells from cord blood	OKSM to reprogramme human CB-CD34+	Day 15 (conversion, no reprogramming)	16.4% conversion efficiency (not reprogramming efficiency)
	Hu *et al.* (2013)	Lentiviral vector	hASC and dermal fibroblasts	OKSM	Day 12	0.143% (vs. 0.035% only OKSM)
miR-200c	Miyoshi *et al.* (2011)	Direct transfection of mature miRNAs (four transfections every 2 days)	hASC and dermal fibroblasts	miR-302 and miR-369	Day 20	0.002%
miR-369	Miyoshi *et al.* (2011)	Direct transfection of mature miRNAs (four transfections every 2 days)	hASC and dermal fibroblasts	miR-200c and miR-302	Day 20	0.002%
miR-372	Subramanyam *et al.* (2011)	Mimic of mature miR-372 (transfections at days 3 and 10)	Foreskin and lung fibroblasts	OKS ± miR-302b or OKSM ± miR-302b	Day 31 (OKS) and day 21 (OKSM)	Increased compared to mock transfected control but no clear numbers shown

hASC, human adipose stem cells. In some cases, the reprogramming efficiency was not specifically mentioned. If possible, it was calculated by us from the data provided in the text and figures (number of iPSC colonies divided by number of cells initially seeded).

member of another miRNA family with an important role in pluripotency, the miR-302-367 cluster. As will be described below, the miR-302-367 cluster was later used alone to reprogramme mouse and human somatic cells (Anokye-Danso *et al.*, 2011). Interestingly, mixing the three different miRNAs did not further enhance reprogramming efficiencies, and adding c-Myc to the cocktail also provided no benefit, suggesting that the ESCC miRNAs act redundantly between each other and substitute for, but do not enhance, c-Myc. Finally important to note is that iPSC colonies derived from OKS plus miR-294 were more uniform with respect to morphology and expression of an Oct4 promoter-driven Gfp reporter than those generated with OKSM, and that they were also less invasive upon teratoma formation in mice, indicating that the effects of miR-294 differ from those of c-Myc.

In conclusion, this early study already provided key evidence for the feasibility to employ naked miRNAs for reprogramming, using a commercial transfection reagent for delivery and two rounds of transfection (needed because miRNA mimics function only transiently, resulting in a 10,000-fold drop in activity over 6 days). It also showed that a single miRNA can already suffice to substantially enhance reprogramming and to eliminate the problematic c-Myc from the cocktail, albeit the best potencies of OKS/miR-294 reprogramming reached only ~75% of those with OKSM.

Two years later, Anokye-Danso and colleagues published their findings that the miR-302-367 cluster, comprising the miR-302d miRNA that had been evaluated in the Judson study (Judson *et al.*, 2009), is able to reprogramme mouse and human somatic cells in the absence of the OKSM factors (Anokye-Danso *et al.*, 2011). To do so, they created a lentiviral vector expressing a 690 base pair region including the mouse miR-302-367 sequences and then used it to infect MEFs derived from Oct4-Gfp mice. When subsequently cultured in the presence of valproic acid (VPA), a histone deacetylase (Hdac) inhibitor, the cells formed fully pluripotent iPSC colonies even faster than controls treated with OKSM-expressing viruses. Moreover, the efficiency of miR-302-367 reprogramming surpassed that of OKSM by two orders of magnitude. Notably, the miR-302-367-expressing lentivirus was also capable of reprogramming human foreskin and dermal fibroblasts, again with greatly increased efficiency. VPA was not required in the human cells, most likely due to the fact that these cells express low levels of Hdac(2) to begin with. Lastly, they dissected the miR-302-367 cluster and showed that miR-367 as well as a coordinated action of miR-302a/b/c/d are critically required for reprogramming, as neither miR-302a/b/c/d nor miR-367 alone were able to yield iPSC.

Like the preceding work by Judson and colleagues (Judson *et al.*, 2009), this study is highly notable in the field as it introduced important new scientific principles and technologies. First, it demonstrated the feasibility to reprogramme with miRNAs that are expressed from a lentiviral vector. As compared to the mimic approach tested before, this represents a much more efficient and less transient delivery option; however, it concurrently also increases concerns about vector integration. Second, this was the first report to show that a single miRNA cluster not only suffices for reprogramming, but also increases efficiencies by up to 100-fold as opposed to the conventional OKSM strategy. Yet, it must be pointed out that the results from this study are debated very controversially to date, as several groups subsequently failed to reproduce somatic cell reprogramming with the miR-302-367 cluster in the absence of additional classical factors (see below).

Others did, however, succeed in reprogramming cells with cocktails of naked miRNAs. One example is a 2011 study by Miyoshi *et al.* (2011) who first screened for miRNAs that were two-fold up-regulated in miPSC and mESC in comparison to mouse adipose stromal cells (mASC). When the identified miRNAs were then transfected as mimics into mASC, the combination of miR-200c, miR-302 and miR-369 miRNA clusters proved to be very potent, as it reached reprogramming efficiencies akin to retroviral-mediated OSKM delivery. Interestingly, the group showed that all three miRNA families were necessary to activate the expression of *Nanog*, since mASC from *Nanog* promoter-Gfp reporter mice only gave rise to Gfp-positive colonies when the cells were transfected with the three miRNA families, and not with only one or combinations of two of them. Importantly, the same cocktail of three miRNA clusters was also able to reprogramme human ASC, albeit at a relatively low efficiency (two out of 100,000 cells, i.e. 0.002%; see Table 4.4). Also notable from a clinical perspective is again the higher safety of this approach as compared to retroviral miRNA expression, as the mimics are not supposed to integrate into the iPSC genome. Last but not least, it should be pointed out that this is one of the studies noted above which directly contradict the work from Anokye-Danso and colleagues (Anokye-Danso *et al.*, 2011), for reasons not resolved thus far.

In contrast to this ongoing controversy, there is no doubt in the field that the use of miRNA mimics in combination with OKS(M), as originally reported by the Blelloch lab (Judson *et al.*, 2009), provides a powerful and promising option with high clinical relevance. In fact, this lab itself as well as multiple other groups were able to recapitulate the principal finding, using further combinations of miRNAs and the Yamanaka factors. For instance, Li and co-workers found (or confirmed, respectively) that the miR-17-92, miR-106b-25, miR-106a-363 and miR-302-367 clusters are highly expressed in mESC (Li *et al.*, 2011). Next, they showed that transfection of mimics for miR-93 or miR-106b, two members of the miR-106-25 cluster, together with retroviral vector-mediated OKS expression, promoted a four- to six-fold increase in iPSC formation from MEFs. The role of these miRNAs as enhancers of reprogramming was further validated through the reverse experiment in which their levels were decreased using transfected synthetic miRNA inhibitors, resulting in drops in iPSC colony numbers. Similar findings were finally also made with miR-17 and miR-106a mimics, implying that numerous combinations of miRNAs from the three clusters studied in this work with OKS hold the potential to boost reprogramming efficiencies. Remarkably, these clusters are conserved between mouse and human, providing a rationale for future work to try and recapitulate the data from this report in human cells.

Another study that should be briefly highlighted in this context is again from the Blelloch lab who, in 2011, translated their original 2009 findings into human cells. Specifically, Subramanyam *et al.* (2011) reported that transduction of human foreskin or lung fibroblasts with OKSM-expressing retroviruses followed by transduction with miR-302b and/or miR-372 mimics resulted in an increase in the number of ensuing colonies with iPSC characteristics. Important to note is that human miR-372/373 are orthologous to mouse miR-291, miR-294 and miR-295 that have been used successfully by Judson *et al.* (2009) in the preceding mouse study. This confirms the relevance of these miRNA families and their putative targets (whose discussion is beyond the scope of this article, but see the reviews in Table 4.1 for details) across species.

In all studies described thus far, miRNAs were either delivered as naked molecules through RNA transfection, or encoded by – and expressed from – viral vectors. As indicated above and moreover summarized in Tables 4.3 and 4.4 as well as in Fig. 4.3, both methods have their pros and cons with respect to delivery, efficiency, kinetics and safety which will all have to be considered for their application in human cells. Noteworthy in this context is thus a recent study by Sohn *et al.* (2013) who encapsulated miRNA mimics in nanoparticles. More specifically, they packaged the miR-302-367 cluster into a polyketal copolymer termed PK3 [polyketals are synthetic polymers whose backbone contains ketal bonds (Heffernan and Murthy, 2005)]. These nanoparticles were then used to reprogramme mouse bone marrow-derived mononuclear cells (BM-MNC), although it must be noted that the only endpoints were activation of pluripotency markers, and that true iPSC lines were not established. Nonetheless, this latest work is interesting and promising for several reasons. First, (partial) reprogramming was again achieved only with multiple miRNAs in the absence of the conventional factors and without genomic integration, confirming previous studies (see above). Second, BM-MNC are a very attractive cell type for reprogramming because they are easily accessible and less likely to have been affected by environmental mutagens, as compared to other typically used cell types such as human dermal fibroblasts. Besides, the polymers used are generally intriguing as they are highly suitable as a drug delivery system, considering that their degradation products are biocompatible and that the nanoparticles and microparticles derived thereof can encapsidate and deliver molecules with different sizes.

Next to these and other studies that demonstrated the possibility to improve reprogramming through miRNA up-regulation, it is finally intriguing that a few reports also succeeded with the opposite strategy, i.e. deliberate inhibition of specific miRNAs. A first example is the work by Melton and colleagues who hypothesized that inhibition of the let-7 family would enhance somatic reprogramming (Melton *et al.*, 2010), based on knowledge that this miRNA cluster is very abundant in MEFs but down-regulated in stem cells (Marson *et al.*, 2008). Indeed, when Oct4-Gfp MEFs were transduced with

Delivery method	Viral vectors	Mimics/Antagomirs	Nanoparticles
	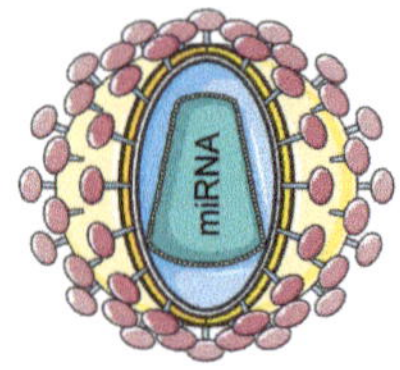		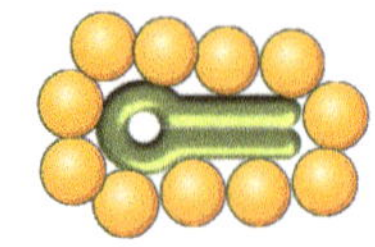
Expression	Long-term	Short-term	Short-term
Transfections/ Transductions	One	Several	Several
Integration into the cellular genome	Yes	No	No
Target cells	Wide range	Only cells susceptible to transfection	Hard-to-transfect cells (*e.g.*, macrophages)
Expertise	Medium	Company	High
Biosafety	Level 2	Level 1	Level 1
Others			Biocompatible degradation products

Figure 4.3 Comparison of miRNA delivery methods. Depicted are the pros (green) and cons (red) of the different principal miRNA delivery methods that have been used thus far to reprogramme somatic cells: retro- or lentiviral vectors, transfection of miRNA mimics or antagomirs, as well as nanoparticles for delivery of miRNA mimics. This figure was produced using Servier Medical Art (www.servier.com).

OKS-expressing retroviral vectors, subsequent transfection with a let-7 antisense inhibitor on days 0 and 6 improved reprogramming by 4.3-fold. Notably, when c-Myc was also co-delivered via retroviral vectors, the increase due to let-7 inhibition was reduced to only about 1.75-fold. The authors thus concluded that inhibition of the let-7 family promotes somatic cell reprogramming, and that c-Myc activity is one, but not the only important target of these miRNAs. More generally, they proposed that let-7 and ESCC miRNAs exert opposing functions in stabilizing the differentiated versus self-renewing phenotypes, perhaps akin to their function in mammalian cancers.

As a second example, Choi and colleagues focused on the p53-induced miR-34 miRNAs (miR-34a/b/c) (Choi *et al.*, 2011). This was based on prior findings that p53 suppression improves reprogramming, as shown (among others) by Utikal and colleagues (Utikal *et al.*, 2009). Moreover, miR-34 as a p53 target induces apoptosis (Chang *et al.*, 2007; Raver-Shapira *et al.*, 2007) and cell cycle arrest (He *et al.*, 2007), and its expression was up-regulated in MEFs when reprogramming was induced, altogether tempting the authors to reason that miR-34 might be involved in the regulatory reprogramming circuits downstream of p53. Indeed, they were then able to show that MEFs from miR-34a knock-out mice could be reprogrammed (using OKS- or OKSM-expressing retroviral vectors) with roughly four-fold higher efficiencies than wildtype MEFs. This effect could also be recapitulated in the wildtype cells by transient miR-34a suppression using locked nucleic acid-based inhibitors. Interestingly, MEFs from mice lacking all three miR-34 members could be reprogrammed even more efficiently, suggesting a cooperative effect among miR-34a/b/c. Also remarkable is that albeit miR-34a ablation was not as efficient as p53 deficiency in terms of improving reprogramming, the iPSC lines resulting from miR-34a knock-out were genetically more stable and readily gave rise to healthy adult chimeras (which was difficult with the p53-deficient lines). Similar to the let-7/c-Myc connection noted above (Melton *et al.*, 2010), these different phenotypes suggest that miR-34 miRNAs acts as a critical barrier for reprogramming, but is certainly not the only p53 target with a key role in pluripotency.

Another 2011 study by Yang *et al.* (2011) provides the third and final example for this category of strategies to improve somatic cell reprogramming through miRNA inhibition rather than up-regulation. Similar to the other two studies, the authors first identified two miRNAs that were highly abundant in MEFs as compared to mESC, namely, miR-21 and miR-29a. Further akin, transfection of MEFs with inhibitors against these two miRNAs and transduction with OKS- or OKSM-expressing retroviruses resulted in an up to three-fold boost in reprogramming efficiency. From a mechanistic standpoint, it was notable that the authors proposed a model whereby (i) miR-21 and miR-29a are repressed by c-Myc, and (ii) induce p53 protein levels, likely through a regulatory double negative loop involving inhibitors of p53 that are controlled by these miRNAs. This is interesting as it may not only explain how miR-21 and miR-29a act as reprogramming barriers, similar to let-7 and miR-34, but also as it ties in well with the known roles of c-Myc and p53 during the induction and maintenance of pluripotency (see above).

MiRNA-based systems to monitor or control pluripotency and differentiation

In the previous two sections, we have outlined recent technical advances that aid in the identification of iPSC-specific miRNAs and in their deliberate use to improve reprogramming methodologies. Even with these improvements, a bottleneck that remains and that continues to hamper the clinical translation of iPSC technologies is the need for powerful and high-throughput-compatible strategies for selection of faithfully reprogrammed clonal cell lines. This is critical as the heterogeneous cell populations that typically result from iPSC approaches contain an assortment of cell intermediates that have partially reprogrammed or differentiated, respectively. This not only lowers the quality of these preparations but concomitantly raises a major safety threat, since residual pluripotent cells among a culture of differentiated cells pose a risk of tumour formation upon transplantation. To date, the sole methodology to distinguish living reprogrammed from differentiated cells relies on morphological criteria whose accurate application requires experience and time, factors that prevent their routine use in the clinics. Fortunately though, pioneering work from Brian Brown and the Naldini lab has paved the way for a novel cell segregation technology that is based on differential miRNA expression in specific cell types, and that has the potential to revolutionize the iPSC field as well.

Briefly, this technology's hallmark is that it exploits the fact that many miRNAs have a distinct development- and tissue-specific expression pattern that frequently defines cell identity. In a string of seminal papers, Brown and colleagues provided proofs-of-principle that it is possible to incorporate binding sites for such a specific miRNA into a gene transfer vector, in order to suppress vector activity in cells that express this miRNA at sufficient amounts, without perturbing vector expression in other cell types. Importantly, one of their examples included human ESC and their progeny (Brown *et al.*, 2007). In an attempt to segregate gene expression between these two cell types, the group created lentiviral vectors encoding a *gfp* reporter gene whose 3′ untranslated region (3′UTR) was tagged with binding sites for miR-302a or miR-372, both of which are highly expressed in hESC lines. Notably, when hESC were grown on mouse embryonic fibroblasts (MEF) and then infected with these vectors, only the MEFs became fluorescent, while Gfp expression was reduced up to 150-fold in the hESC due to their inherent high miR-302a/372 levels. Moreover remarkably, differentiation of these transduced hESC into embryoid bodies reactivated Gfp expression, consistent with the down-regulation of the two hESC-specific miRNAs during the differentiation process. Finally, the opposite was observed for a neuron-specific miRNA (miR-124a), whose incorporation into the Gfp vector resulted in bright fluorescence in the pluripotent hESC cells (which lack this miRNA) and its subsequent shutdown upon differentiation of a subset of cells from the embryoid bodies to neurons.

Recently, a number of groups have now started to adapt this concept of miRNA-regulated reporter vectors to the field of iPSC as well. In principle, one can distinguish two approaches in the iPSC literature that recapitulate the two original findings in hESC from the Naldini lab: (i) an OFF system where a fluorescent reporter is tagged with binding sites for miRNAs that are abundant in iPSC, resulting in silencing of this reporter in these cells, versus (ii) an ON system, where the reporter is controlled by miRNAs that are down-regulated in iPSC, causing an increase in fluorescence in these cells (Fig. 4.4).

The OFF system has been exemplified in a 2010 paper by Kamata and colleagues (Kamata *et al.*, 2010). This group generated a bi-cistronic lentiviral vector expressing two reporter genes that were tagged in their 3′UTR with binding sites for miRNAs that are specific for either stem cells (fused to *egfp*) or differentiated cells (fused to *mCherry*). More specifically, the miRNAs chosen were miR-302a/d (characteristic of stem cells, see above) as well as miR-223, -155 and -142-3p, all of which are enriched in differentiated cells (in particular in haematopoietic cells, but miR-155 also in fibroblasts and cancers). Accordingly, it was expected that eGfp but not mCherry would be silenced (=eGfp OFF state) in pluripotent cells, whereas mCherry but not eGfp would be ablated in differentiated cells expressing one or more of the three selected miRNAs (Fig. 4.4, top). Vector sensitivity and functionality were principally verified in a human ESC line (eGfp OFF, mCherry ON) as well as in CD34+ haematopoietic stem cells (eGfp ON, mCherry OFF).

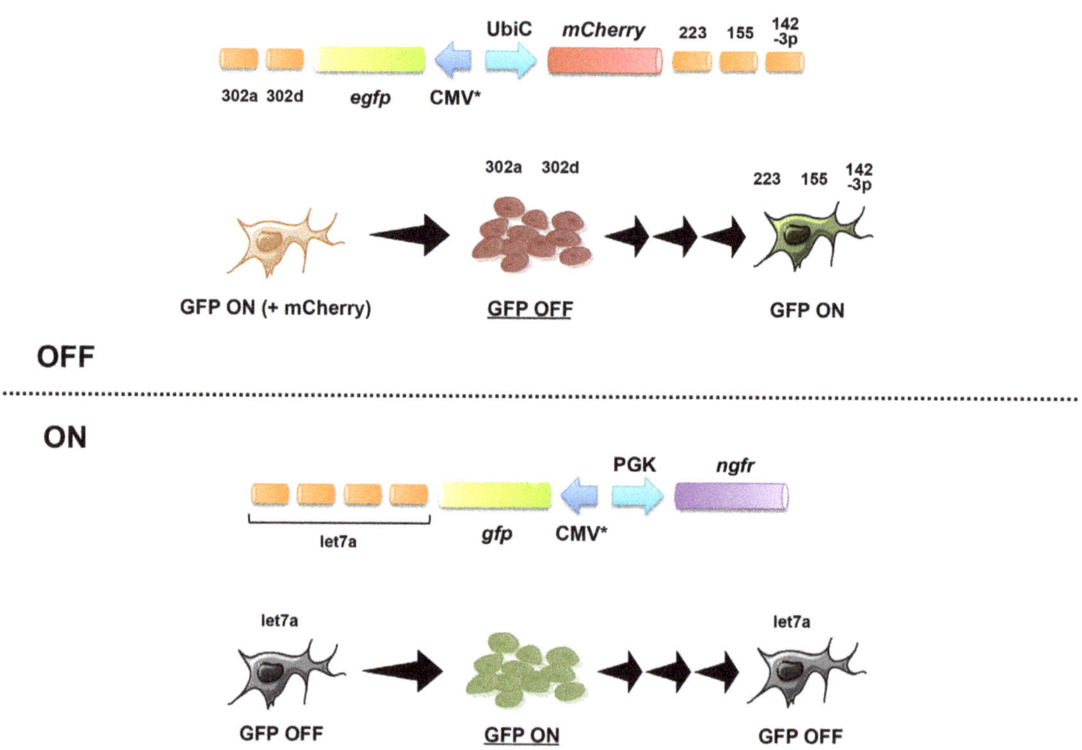

Figure 4.4 Comparison of miRNA ON/OFF sensor strategies for iPSC tracking. Note that 'ON' and 'OFF' refer to the status of reporter gene expression in the pluripotent stem cells. Shown are the two vector designs described in the main text and their activities in somatic (single cell) versus pluripotent (cell population) cells. Numbers denote binding sites for specific miRNAs (see text for details). CMV*, minimal cytomegalovirus promoter; ngfr, nerve growth factor receptor (reporter gene); PGK, phosphoglycerol kinase promoter; UbiC, ubiquitin C promoter. This figure was produced using Servier Medical Art (www.servier.com).

More interesting and relevant was then to test the capacity of this vector to monitor reprogramming of human fibroblasts to iPSC. Indeed, over a course of four weeks after transduction of the fibroblasts with the dual reporter vector as well as a cocktail of OKSM-expressing lentiviral vectors, the authors observed the expected shift in colours: While the initial fibroblast culture expressed eGfp (but also mCherry, since miR-155 expression is relatively low in these cells), this reporter decreased during the reprogramming phase presumably due to induction of miR-302a/d. Conversely, over the same period, mCherry increased probably due to the loss of residual miR-155 in the dedifferentiating cells (Fig. 4.4, top).

An intriguing side observation was that more than 50% of the cells expressed miR-302a/d at around two weeks (calculated based on loss of eGfp signals), yet only a small percentage – 0.03% – of the starting fibroblasts eventually resulted in iPSC colonies. Moreover, many colonies displayed a transformed phenotype and were eGfp-negative/mCherry-positive, suggesting that they were partially reprogrammed. These findings are notable as they support the recently emerging view that expression of miR-302 itself is not sufficient for reprogramming; in turn, they moreover imply that this miRNA should also not be used solely as a marker for truly reprogrammed cells.

Importantly, the group additionally verified that iPSC clones selected via their dual reporter indeed expressed the typical markers of pluripotent cells, indirectly confirming that their vector does not grossly affect iPSC induction or marker expression. Finally, they also demonstrated the ability of their vector to monitor the differentiation process in real-time, by showing a gain in eGfp expression during embryoid body formation (presumably reflecting the loss in miR-302a/d) and differentiation into neural lineages.

In another 2010 publication, Sachdeva and colleagues took this concept – to shut down a reporter in a fraction of a heterogeneous iPSC cell population – one step further, by demonstrating its usefulness to also actively select for a certain desired subpopulation (Sachdeva *et al.*, 2010). Akin to Kamata *et al.* (2010), this group created lentiviral vectors expressing a *gfp* gene tagged with binding sites for a stem cell-specific miRNA, in this case miR-292 (as they were predominantly working with mouse pluripotent cells in which this miRNA is highly abundant). Accordingly, Sachdeva *et al.* (2010) could show that Gfp expression was specifically silenced (= OFF state) in mESC, miPSC and also hES (due to their expression of miR-371 which is the human homologue of miR-292), whereas Gfp became detectable as the various stem cells differentiated towards the neural lineage. Importantly, they were subsequently able to exploit this segregated Gfp expression as a means to enrich and purify neuronal progenitors via FACS-based cell sorting. Following transplantation and engraftment of these purified differentiated cells in mouse brains, the authors noted a marked reduction in tumour formation and an increased survival of the neuronal progenitors, as compared to control groups that had received a non-FACS-selected heterogeneous cell population. This result clearly validates the principal clinical risks of using impure iPSC progeny that is contaminated with immature cells and that can result in overgrowth up to tumour formation, and/or in transplant rejection through an immune reaction. As demonstrated in this study, the use of an OFF reporter system in combination with FACS-based cell sorting represents one of several possible solutions to these problems that could be widely applied in many labs, provided they possess FACS sorting capabilities.

In 2011, Di Stefano *et al.* (2011) reported an example of the reverse strategy, i.e. the ON system which is characterized by an <u>in</u>crease of the fluorescent reporter as a result of a stem cell-specific <u>de</u>crease of a certain miRNA. The miRNA that was chosen for this purpose was let7a, based on its known ablation in pluripotent cells as a consequence of the expression of the negative let7 family regulator Lin28. Accordingly, the group generated a lentiviral vector encoding *gfp* tagged with four let7a binding sites in its 3′UTR, expecting Gfp to be exclusively expressed (= Gfp ON state) in cells that lack this miRNA, such as iPSC. Indeed, in a mixed culture of murine ESC and MEF, the let7a-regulated vector strongly suppressed Gfp expression in the MEFs but not the ESC. The Gfp signal was lost upon differentiation to cardiomyocytes or neurons, further proving the ability of the vector to direct reporter gene expression specifically to the pluripotent cell compartment (Fig. 4.4, bottom). Using a co-expressed and non-tagged second reporter for normalization, the degree of Gfp repression in cells encoding let7a was estimated to be about 30-fold, providing a high level of sensitivity to this ON vector. Notably, in a direct side-by-side comparison, the let7a reporter vector faithfully traced pluripotent cells derived from MEFs with a specificity comparable to that of a conventional marker, an Oct4 promoter-driven Gfp.

Importantly, the authors then also validated the usefulness of this ON vector to monitor reprogramming of human cells, akin to the work by Kamata and colleagues for their OFF system. Therefore, they transduced a variety of human fibroblasts, including primary cells from skin biopsies of Parkinson's disease and Rett syndrome patients, with the let7a-sensitive vector. They then used the co-encoded second reporter (NGFR) to enrich a homogeneous population of cells carrying about 10 vector genomes per cells and exhibiting good reporter expression. Subsequent transduction of these cells with OKSM-encoding retroviral vectors resulted in hiPSC lines which started to express Gfp from day 14 onward, concomitant with the TRA-1-60 pluripotency marker, but preceded by the SSEA3 and AP markers. Eventually, 88% of the Gfp-positive clones proliferated and generated stable iPSC lines. Interestingly and reminiscent of the Kamata findings (Kamata *et al.*, 2010), a large fraction of AP-positive clones were Gfp-negative and failed to produce stable iPSC lines, suggesting that they had also arrested at a partially reprogrammed state at which they still expressed let7a. Also in line with the former study, Di Stefano *et al.* (2011) obtained equal numbers of iPSC lines with or without the reporter and hence concluded that lentiviral marking had not compromised overall cell functionality. In fact, vector expression was maintained for up to 32 passages, without affecting the cells' capability to differentiate into embryoid bodies or to form teratomas. Remarkably, the latter were again Gfp-negative, congruent with let7a up-regulation during *in vivo* lineage cell differentiation.

Last but not least, this study demonstrated the feasibility to also exploit the ON vector design for selective enrichment of a certain subpopulation of cells in a heterogeneous iPSC culture. Due to the nature of the ON system, it will produce the opposite result of the OFF-based selection describe above, i.e. it will enrich pluripotent cells that express a specific gene, rather than the differentiated progeny. One option to use this feature is to FACS sort for pluripotent cells that express a fluorescent reporter, akin to the proof-of-concept demonstration for differentiated neuronal progenitors in the Sachdeva study (Sachdeva *et al.*, 2010). In an alternative scenario that was pursued by Di Stefano *et al.* (2011), the gene that is expressed in a stem cell-specific fashion protects the pluripotent cell population from drug-induced cytotoxicity, while concurrently resulting in eradication or purging of spontaneously differentiated cells from the iPSC culture. To this end, the group replaced their let7a-controlled *gfp* with a neomycin resistance gene, transduced mouse iPSC lines that had already started to differentiate with the ensuing vector and then grew them in the presence of G418. As hoped for, this resulted in efficient elimination of contaminating differentiated cells from the culture (due to their G418-mediated killing, facilitated by the let7a-induced suppression of the neomycin resistance cassette) and yielded a homogeneous population of pluripotent cells (that were protected from G418 toxicity).

As a whole, these studies including the work from Brown *et al.* (2007) clearly illustrate the high versatility and enormous potential of miRNA sensor technology to (i) track iPSC generation and differentiation, and (ii) to produce pure cultures of immature or mature cells, respectively. Particularly remarkable is that all reports consistently demonstrated the great flexibility of this strategy, as the reporter vector can be readily adapted depending on the miRNA profile in a specific differentiation lineage. *Vice versa*, the same vector design can also be harnessed as a biological tool, to measure the temporal and cell-specific expression of selected miRNAs during reprogramming or differentiation. All studies highlighted above are moreover congruent in that neither found any evidence for vector-associated inhibition of cell proliferation or differentiation capacity, suggesting that lentiviral marking and tracking are well tolerated.

Which of the two strategies – ON or OFF – is ultimately preferred may depend on the specific application and a number of parameters. For instance, a potential benefit of the OFF system as described is that it permits to mark and thus distinguish both cell populations simultaneously, since pluripotent cells will fluoresce red, while differentiated progeny will be green (Fig. 4.4). On the other hand, the ON system may exhibit faster kinetics (the long Gfp half-life may compromise the OFF strategy) and is more compatible with mono-cistronic vectors (as it is independent of a second reporter for counter-staining or normalization of transduction efficiencies).

One remaining and very critical disadvantage in both cases is that either strategy relies on lentiviral vectors which raises concerns about insertional mutagenesis, especially considering that the Di Stefano *et al.* (2011) study reported up to 10 integration events per cell. Adverse integration might be particularly problematic with iPSC selection/purging strategies based on either the OFF or the ON system, at least in the way they have been used thus far (Fig. 4.4). In the OFF system, as outlined, an iPSC-specific miRNA silences a reporter gene in pluripotent cells, allowing for isolation of differentiated cells that continue to express the reporter and that are hence selected for actively transcribed vector integrants. Likewise, the ON system as reported will only work efficiently when the enriched pluripotent cells permanently express the neomycin resistance gene which will protect them from G418 toxicity, which likely also favours vector genome integration. As another adverse event, it is conceivable that the persistent expression of abundant binding sites for a given endogenous miRNA might sequester this miRNA away from its natural targets and hence perturb its functionality. Finally, the need to select under neomycin resistance pressure for multiple passages may impair iPSC viability and/or differentiation capacity.

Outlook: the road towards clinical translation of human iPSC technologies

About 7 years ago, the discovery that ordinary somatic cells can be coaxed into becoming pluripotent stem cells, which have the capacity to differentiate into any cell or tissue of the body, has jump-started an exciting new era of stem cell research that has unprecedented potential to revolutionize regenerative medicine and transplantation therapy. Fuelled by these prospects, amazing scientific progress has been achieved in dissecting the intricate molecular mechanisms underlying the reprogramming and differentiation events, most recently culminating in the identification of endogenous miRNAs as essential regulators that either act as barriers or helpers in these two processes. Meanwhile, we have witnessed how a growing number of labs have realized the practical implications of these fascinating biological findings and have started to deliberately exploit the unique role of miRNAs to improve our repertoire of iPSC technologies, from new methods for iPSC generation to innovative concepts for cell tracking and purging, as outlined above. Already, it is evident that the promise of miRNAs to spark a second revolution in the stem cell field is enormous, yet it is also becoming increasingly clear that further improvements are needed to realize their full potential in future clinical applications.

A first area that we have reviewed in this chapter and that should benefit from additional refinements concerns the identification of miRNAs with a central role in pluripotency. In the future, it might be promising and worthwhile to focus even more on hybrid approaches that combine the advantages of forward and reverse screens akin to the study by Ruiz *et al.* (2012), i.e. an experimental pre-selection of a group of potentially interesting miRNAs followed by their immediate functional testing in reprogramming studies in a high-throughput/high-content format. The power of this avenue has also been confirmed independently in work from Miyoshi and colleagues who likewise first identified a subset of ESC/iPSC-specific miRNAs and then tested all individual candidates in parallel for their effect on reprogramming (Miyoshi *et al.*, 2011). Additionally, fuelling our confidence that this strategy will lead to successful identification of interesting and clinically relevant new miRNAs are the latest advances in the required technologies. This particularly comprises recent developments in automated, high-throughput and live-cell-compatible platforms and devices, as well as the continuous progress in next generation sequencing technologies, such as miRNA-seq (Garmire and Subramaniam, 2012), that are more comprehensive and have less of an initial selection bias than current genomics methods. As the use of microarrays and other present techniques gives way to these more global transcriptomic survey methods, and as these are rapidly becoming more practical and more affordable, this raises considerable hope that novel key miRNAs which have been overlooked before will be captured and help to complete our picture of the small RNA landscape that regulates pluripotency and differentiation.

We likewise predict substantial advances in the second field of miRNA applications in reprogramming, i.e. their use as enhancers or exclusive mediators of pluripotency. It may still be too early to reach a firm conclusion about the eventual clinical utility of miRNAs for generating translation-grade iPSC as this area remains in its infancy, and as the interpretation of individual findings is complicated by significant controversies. The latter probably reflects the inherent intricacy of the reprogramming process and its sensitivity to even marginal alterations in key parameters, which largely hampers the reconciliation of many recent studies, considering that the authors typically used different miRNAs, doses, time and end points, cell types, miRNA administration methodologies or combinations of miRNAs with other factors, such as OKSM or chemical compounds. Nevertheless, the potential of miRNAs to boost reprogramming, and to thus help overcome the low efficiency of this process which remains an impediment to adapting it to routine and high-throughput approaches, is certainly undisputed. To further foster the clinical translation and therapeutic application of miRNA-based reprogramming strategies, we first and foremost expect that pertinent issues related to the safety of miRNAs and their vectors will need to be carefully addressed. In this respect, the current generation of integrating lentiviral vectors does not provide an optimal choice for making translation-grade iPSC, as an ideal reprogramming strategy would leave no trace of the process itself and would function in the complete absence of genetic modifications. Accordingly, it will be necessary to try and alleviate any concerns about genotoxicity, by either replacing the integrating lentiviruses with episomal vector systems, such as adeno-associated viruses (AAVs) (Grimm and Kay, 2003), or by including safety measures allowing later excision of the stably integrated genomes. As a third alternative, one could exploit the recent advances in genome engineering, in particular the latest generations of TALENs and CRISPR systems (Gaj *et al.*, 2013), in order to direct the miRNA sensors to safe harbours in the human genome. This should not only increase the success rates of integration into chromosomal regions that tolerate insertions, but concurrently also mitigate concerns about neoplastic transformation.

Notably, these concerns do not apply to the alternative strategy whereby miRNAs are administered in the form of mimics, i.e. double-stranded RNA molecules that comprise only the actual 21- to 23-nt-long target binding site without any additional regulatory sequences, such as promoters or enhancers. At present, this is the only method that functions in the absence of vector-based gene transfer and that does not use oncogenic genes, transcripts or proteins, thus ameliorating the risks associated with conventional reprogramming technologies. On the other hand, the lack of regulatory sequences prevents any option for exogenous or endogenous control over miRNA expression, and moreover, current generations of miRNA mimics require repeated delivery owing to the intracellular instability of these synthetic molecules. It should thus be highly rewarding to attempt and design new variants with an extended half-life long enough to rival the persistent expression of miRNAs by viral vector integrants. In parallel, it will be important to improve the miRNA mimic delivery options in order to adapt this technology to other somatic cell types that represent interesting sources for reprogramming, such as keratinocytes or peripheral blood cells. One promising option in this respect are nanoparticles, such as those based on polyketals that we have outlined above. Next to these more technical advances, we also envisage a variety of conceptual improvements that could further enhance the efficiency and/or kinetics of the reprogramming process, such as combinatorial strategies whereby miRNA mimics are juxtaposed with small chemical compounds that modulate relevant cellular factors or pathways. Alternatively or in addition, it could be worthwhile to develop sequential paradigms in which specific miRNAs are delivered at a certain time point during the reprogramming phase, dictated by when they are most important. Indeed, a recent paper has demonstrated the power of such a sequential approach for the classical Yamanaka factors (Liu *et al.*, 2013), providing optimism that this concept will work for miRNAs as well. Finally, it could be beneficial to devise novel tools to activate or repress transcription of endogenous miRNAs that act as reprogramming enhancers or barriers, respectively, as another means to regulate their intra-cellular levels. A most promising option in this respect could again be provided by the CRISPR system, considering very recent reports of its successful development as a specific enhancer or repressor of gene expression (Gilbert *et al.*, 2013; Maeder *et al.*, 2013; Qi *et al.*, 2013).

Similar considerations about safety and necessary improvements towards clinical use are also important with regards to the third field of miRNA applications, i.e. their exploitation for cell tracking and purging. Here, it will also be essential to optimize any respective vector or delivery systems for minimal genotoxicity while providing the highest efficiency and widest applicability in many different cell types. In particular AAV vectors might represent an ideal vector of choice for these applications, as these vectors rarely or not at all integrate, are very potent in a large array of cultured cells, including various classes of stem cells, and can readily be engineered for targeting of almost any further cell type that is not amenable to infection with wildtype viruses (Grimm and Kay, 2003; Grimm *et al.*, 2008; Kienle *et al.*, 2012). Regardless, it will certainly remain essential to carefully monitor cells that have been marked with a vector-encoded fluorescence reporter for adverse effects on their general functionality, particularly possible perturbations of cellular differentiation.

Next to advancing the vector itself, key to success will also be to further improve the design of selection/purging strategies, considering the aforementioned potential problems with the current approaches (i.e. dependency on FACS sorting capabilities or adverse effects from neomycin pressure). Fortunately, there are already a number of promising alternatives that are worth exploring, such as the inducible Caspase-9 system which has already been used successfully as a tool for selective cell elimination *ex vivo* and even in human patients (Di Stasi *et al.*, 2011). As compared to the neomycin strategy, a benefit of such a suicide gene-based selection may be that transgene expression will ultimately be silenced in the surviving cells (which express a miRNA that shuts off suicide gene expression). The latter can be either pluripotent or differentiated, depending on the use of an OFF or ON strategy, respectively (Fig. 4.4). This lack of exogenous gene expression may help to stabilize the reprogrammed or differentiated state in the final cell population, and may generally be more acceptable from a clinical standpoint.

In conclusion, the sum of the recent miRNA-related technological advances that have been propelled by the limitations of the original reprogramming method raise considerable hope that the derivation of patient-specific iPSC with improved quality and safety for regenerative medicine and transplantation therapy is a realistic goal. Considering further possible miRNA applications such as their use for controlled transdifferentiation, as well as the emerging field of long non-coding RNAs as alternative regulators of many cellular processes, we believe that we have probably only witnessed the dawn of the field of small RNAs in pluripotency, and are largely optimistic that many more exciting novel avenues and breakthroughs lie ahead.

References

Anokye-Danso, F., Trivedi, C.M., Juhr, D., Gupta, M., Cui, Z., Tian, Y., Zhang, Y., Yang, W., Gruber, P.J., Epstein, J.A., *et al.* (2011). Highly efficient miRNA-mediated reprogramming of mouse and human somatic cells to pluripotency. Cell Stem Cell *8*, 376–388.

Anokye-Danso, F., Snitow, M., and Morrisey, E.E. (2012). How microRNAs facilitate reprogramming to pluripotency. J. Cell Sci. *125*, 4179–4187.

Bao, X., Zhu, X., Liao, B., Benda, C., Zhuang, Q., Pei, D., Qin, B., and Esteban, M.A. (2013). MicroRNAs in somatic cell reprogramming. Curr. Opin. Cell Biol. 25, 208–214.

Bar, M., Wyman, S.K., Fritz, B.R., Qi, J., Garg, K.S., Parkin, R.K., Kroh, E.M., Bendoraite, A., Mitchell, P.S., Nelson,

A.M., *et al.* (2008). MicroRNA discovery and profiling in human embryonic stem cells by deep sequencing of small RNA libraries. Stem Cells *26*, 2496–2505.

Berezikov, E. (2011). Evolution of microRNA diversity and regulation in animals. Nat. Rev. Genet. *12*, 846–860.

Bernstein, E., Kim, S.Y., Carmell, M.A., Murchison, E.P., Alcorn, H., Li, M.Z., Mills, A.A., Elledge, S.J., Anderson, K.V., and Hannon, G.J. (2003). Dicer is essential for mouse development. Nat. Genet. *35*, 215–217.

Brown, B.D., Gentner, B., Cantore, A., Colleoni, S., Amendola, M., Zingale, A., Baccarini, A., Lazzari, G., Galli, C., and Naldini, L. (2007). Endogenous microRNA can be broadly exploited to regulate transgene expression according to tissue, lineage and differentiation state. Nat. Biotechnol. *25*, 1457–1467.

Calabrese, J.M., Seila, A.C., Yeo, G.W., and Sharp, P.A. (2007). RNA sequence analysis defines Dicer's role in mouse embryonic stem cells. Proc. Natl. Acad. Sci. U.S.A. *104*, 18097–18102.

Chang, H.M., and Gregory, R.I. (2011). MicroRNAs and reprogramming. Nat. Biotechnol. *29*, 499–500.

Chang, T.C., Wentzel, E.A., Kent, O.A., Ramachandran, K., Mullendore, M., Lee, K.H., Feldmann, G., Yamakuchi, M., Ferlito, M., Lowenstein, C.J., *et al.* (2007). Transactivation of miR-34a by p53 broadly influences gene expression and promotes apoptosis. Mol. Cell. *26*, 745–752.

Chin, M.H., Mason, M.J., Xie, W., Volinia, S., Singer, M., Peterson, C., Ambartsumyan, G., Aimiuwu, O., Richter, L., Zhang, J., *et al.* (2009). Induced pluripotent stem cells and embryonic stem cells are distinguished by gene expression signatures. Cell Stem Cell *5*, 111–123.

Choi, Y.J., Lin, C.-P., Ho, J.J., He, X., Okada, N., Bu, P., Zhong, Y., Kim, S.Y., Bennett, M.J., Chen, C., *et al.* (2011). miR-34 miRNAs provide a barrier for somatic cell reprogramming. Nat. Cell Biol. *13*, 1353–1360.

Dang, C.V. (2009). Have you seen..?: Micro-managing and restraining pluripotent stem cells by MYC. EMBO J. *28*, 3065–3066.

Di Stasi, A., Tey, S.K., Dotti, G., Fujita, Y., Kennedy-Nasser, A., Martinez, C., Straathof, K., Liu, E., Durett, A.G., Grilley, B., *et al.* (2011). Inducible apoptosis as a safety switch for adoptive cell therapy. N. Engl. J. Med. *365*, 1673–1683.

Di Stefano, B., Maffioletti, S.M., Gentner, B., Ungaro, F., Schira, G., Naldini, L., and Broccoli, V. (2011). A microRNA-based system for selecting and maintaining the pluripotent state in human induced pluripotent stem cells. Stem Cells *29*, 1684–1695.

Evans, M.J., and Kaufman, M.H. (1981). Establishment in culture of pluripotential cells from mouse embryos. Nature *292*, 154–156.

Fabian, M.R., and Sonenberg, N. (2012). The mechanics of miRNA-mediated gene silencing: a look under the hood of miRISC. Nat. Struct. Mol. Biol. *19*, 586–593.

Feng, B., Ng, J.H., Heng, J.C., and Ng, H.H. (2009). Molecules that promote or enhance reprogramming of somatic cells to induced pluripotent stem cells. Cell Stem Cell *4*, 301–312.

Fire, A., Xu, S., Montgomery, M.K., Kostas, S.A., Driver, S.E., and Mello, C.C. (1998). Potent and specific genetic interference by double-stranded RNA in *Caenorhabditis elegans*. Nature *391*, 806–811.

Gaj, T., Gersbach, C.A., and Barbas, C.F., 3rd (2013). ZFN, TALEN, and CRISPR/Cas-based methods for genome engineering. Trends Biotechnol. *31*, 397–405.

Garmire, L.X., and Subramaniam, S. (2012). Evaluation of normalization methods in mammalian microRNA-Seq data. RNA *18*, 1279–1288.

Gilbert, L.A., Larson, M.H., Morsut, L., Liu, Z., Brar, G.A., Torres, S.E., Stern-Ginossar, N., Brandman, O., Whitehead, E.H., Doudna, J.A., *et al.* (2013). CRISPR-mediated modular RNA-guided regulation of transcription in eukaryotes. Cell *154*, 442–451.

Git, A., Dvinge, H., Salmon-Divon, M., Osborne, M., Kutter, C., Hadfield, J., Bertone, P., and Caldas, C. (2010). Systematic comparison of microarray profiling, real-time PCR, and next-generation sequencing technologies for measuring differential microRNA expression. RNA *16*, 991–1006.

Goff, L.A., Davila, J., Swerdel, M.R., Moore, J.C., Cohen, R.I., Wu, H., Sun, Y.E., and Hart, R.P. (2009). Ago2 immunoprecipitation identifies predicted microRNAs in human embryonic stem cells and neural precursors. PLoS One *4*, e7192.

Grimm, D., and Kay, M.A. (2003). From virus evolution to vector revolution: use of naturally occurring serotypes of adeno-associated virus (AAV) as novel vectors for human gene therapy. Curr. Gene Ther. *3*, 281–304.

Grimm, D., Lee, J.S., Wang, L., Desai, T., Akache, B., Storm, T.A., and Kay, M.A. (2008). *In vitro* and *in vivo* gene therapy vector evolution via multispecies interbreeding and retargeting of adeno-associated viruses. J. Virol. *82*, 5887–5911.

He, L., and Hannon, G.J. (2004). MicroRNAs: small RNAs with a big role in gene regulation. Nat. Rev. Genet. *5*, 522–531.

He, L., He, X., Lim, L.P., de Stanchina, E., Xuan, Z., Liang, Y., Xue, W., Zender, L., Magnus, J., Ridzon, D., *et al.* (2007). A microRNA component of the p53 tumour suppressor network. Nature *447*, 1130–1134.

Heffernan, M.J., and Murthy, N. (2005). Polyketal Nanoparticles: A new pH-sensitive biodegradable drug delivery vehicle. Bioconjugate Chem. *16*, 1340–1342.

Heinrich, E.M., and Dimmeler, S. (2012). MicroRNAs and stem cells: control of pluripotency, reprogramming, and lineage commitment. Circ. Res. *110*, 1014–1022.

Houbaviy, H.B., Murray, M.F., and Sharp, P.A. (2003). Embryonic stem cell-specific MicroRNAs. Dev. Cell. *5*, 351–358.

Hu, S., Wilson, K.D., Ghosh, Z., Han, L., Wang, Y., Lan, F., Ransohoff, K.J., Burridge, P., and Wu, J.C. (2013). MicroRNA-302 increases reprogramming efficiency via repression of NR2F2. Stem Cells *31*, 259–268.

Huangfu, D., Maehr, R., Guo, W., Eijkelenboom, A., Snitow, M., Chen, A.E., and Melton, D.A. (2008). Induction of pluripotent stem cells by defined factors is greatly improved by small-molecule compounds. Nat. Biotechnol. *26*, 795–797.

Huo, J.S., and Zambidis, E.T. (2013). Pivots of pluripotency: the roles of non-coding RNA in regulating embryonic and induced pluripotent stem cells. Biochim. Biophys. Acta *1830*, 2385–2394.

Josephson, R., Ording, C.J., Liu, Y., Shin, S., Lakshmipathy, U., Toumadje, A., Love, B., Chesnut, J.D., Andrews, P.W., Rao, M.S., *et al.* (2007). Qualification of embryonal carcinoma 2102Ep as a reference for human embryonic stem cell research. Stem Cells *25*, 437–446.

Judson, R.L., Babiarz, J.E., Venere, M., and Blelloch, R. (2009). Embryonic stem cell-specific microRNAs promote induced pluripotency. Nat. Biotech. *27*, 459–461.

Kamata, M., Liang, M., Liu, S., Nagaoka, Y., and Chen, I.S. (2010). Live cell monitoring of hiPSC generation and differentiation using differential expression of endogenous microRNAs. PLoS One *5*, e11834.

Kanellopoulou, C., Muljo, S.A., Kung, A.L., Ganesan, S., Drapkin, R., Jenuwein, T., Livingston, D.M., and Rajewsky, K. (2005). Dicer-deficient mouse embryonic stem cells are defective in differentiation and centromeric silencing. Genes Dev. *19*, 489–501.

Kienle, E., Senis, E., Borner, K., Niopek, D., Wiedtke, E., Grosse, S., and Grimm, D. (2012). Engineering and evolution of synthetic adeno-associated virus (AAV) gene therapy vectors via DNA family shuffling. J. Vis. Exp. *2*, 3819.

Kuo, C.H., and Ying, S.Y. (2012). Advances in microRNA-mediated reprogramming technology. Stem Cells Int. *2012*, 823709.

Kuo, C.H., Deng, J.H., Deng, Q., and Ying, S.Y. (2012). A novel role of miR-302/367 in reprogramming. Biochem. Biophys. Res. Commun. *417*, 11–16.

Lai, M.I., Wendy-Yeo, W.Y., Ramasamy, R., Nordin, N., Rosli, R., Veerakumarasivam, A., and Abdullah, S. (2011). Advancements in reprogramming strategies for the generation of induced pluripotent stem cells. J. Assist. Reprod. Genet. *28*, 291–301.

Lakshmipathy, U., Love, B., Goff, L.A., Jornsten, R., Graichen, R., Hart, R.P., and Chesnut, J.D. (2007). MicroRNA expression pattern of undifferentiated and differentiated human embryonic stem cells. Stem Cells Dev. *16*, 1003–1016.

Lamouille, S., Subramanyam, D., Blelloch, R., and Derynck, R. (2013). Regulation of epithelial-mesenchymal and mesenchymal-epithelial transitions by microRNAs. Curr. Opin. Cell Biol. *25*, 200–207.

Laurent, L.C., Chen, J., Ulitsky, I., Mueller, F.J., Lu, C., Shamir, R., Fan, J.B., and Loring, J.F. (2008). Comprehensive microRNA profiling reveals a unique human embryonic stem cell signature dominated by a single seed sequence. Stem Cells *26*, 1506–1516.

Lee, M.R., Prasain, N., Chae, H.D., Kim, Y.J., Mantel, C., Yoder, M.C., and Broxmeyer, H.E. (2012). Epigenetic regulation of NANOG by miR-302 cluster-MBD2 completes induced pluripotent stem cell reprogramming. Stem Cells *31*, 666–681.

Leonardo, T.R., Schultheisz, H.L., Loring, J.F., and Laurent, L.C. (2012). The functions of microRNAs in pluripotency and reprogramming. Nat. Cell Biol. *14*, 1114–1121.

Li, M.A., and He, L. (2012). MicroRNAs as novel regulators of stem cell pluripotency and somatic cell reprogramming. Bioessays *34*, 670–680.

Li, Z., Yang, C.-S., Nakashima, K., and Rana, T.M. (2011). Small RNA-mediated regulation of iPS cell generation. EMBO J. *30*, 823–834.

Liao, B., Bao, X., Liu, L., Feng, S., Zovoilis, A., Liu, W., Xue, Y., Cai, J., Guo, X., Qin, B., *et al.* (2011). MicroRNA cluster 302–367 enhances somatic cell reprogramming by accelerating a mesenchymal-to-epithelial transition. J. Biol. Chem. *286*, 17359–17364.

Lichner, Z., Páll, E., Kerekes, A., Pállinger, É., Maraghechi, P., Bősze, Z., and Gócza, E. (2011). The miR-290-295 cluster promotes pluripotency maintenance by regulating cell cycle phase distribution in mouse embryonic stem cells. Differentiation *81*, 11–24.

Lin, C.P., Choi, Y.J., Hicks, G.G., and He, L. (2012). The emerging functions of the p53-miRNA network in stem cell biology. Cell Cycle *11*, 2063–2072.

Lin, S.L. (2011). Concise review: Deciphering the mechanism behind induced pluripotent stem cell generation. Stem Cells *29*, 1645–1649.

Lin, S.L., Chang, D.C., Chang-Lin, S., Lin, C.H., Wu, D.T., Chen, D.T., and Ying, S.Y. (2008). Mir-302 reprograms human skin cancer cells into a pluripotent ES-cell-like state. RNA *14*, 2115–2124.

Lin, S.L., Chang, D.C., Lin, C.H., Ying, S.Y., Leu, D., and Wu, D.T. (2011). Regulation of somatic cell reprogramming through inducible mir-302 expression. Nucleic Acids Res. *39*, 1054–1065.

Lipchina, I., Studer, L., and Betel, D. (2012). The expanding role of miR-302-367 in pluripotency and reprogramming. Cell Cycle *11*, 1517–1523.

Liu, X., Sun, H., Qi, J., Wang, L., He, S., Liu, J., Feng, C., Chen, C., Li, W., Guo, Y., *et al.* (2013). Sequential introduction of reprogramming factors reveals a time-sensitive requirement for individual factors and a sequential EMT-MET mechanism for optimal reprogramming. Nat. Cell Biol. *15*, 829–838.

Lu, D., Davis, M.P., Abreu-Goodger, C., Wang, W., Campos, L.S., Siede, J., Vigorito, E., Skarnes, W.C., Dunham, I., Enright, A.J., *et al.* (2012). MiR-25 regulates Wwp2 and Fbxw7 and promotes reprogramming of mouse fibroblast cells to iPSCs. PLoS One *7*, e40938.

Luningschror, P., Hauser, S., Kaltschmidt, B., and Kaltschmidt, C. (2013). MicroRNAs in pluripotency, reprogramming and cell fate induction. Biochim. Biophys. Acta *1833*, 1894–1903.

Maeder, M.L., Linder, S.J., Cascio, V.M., Fu, Y., Ho, Q.H., and Joung, J.K. (2013). CRISPR RNA-guided activation of endogenous human genes. Nat. Methods *10*, 977–979.

Mallanna, S.K., and Rizzino, A. (2010). Emerging roles of microRNAs in the control of embryonic stem cells and the generation of induced pluripotent stem cells. Dev. Biol. *344*, 16–25.

Marson, A., Levine, S.S., Cole, M.F., Frampton, G.M., Brambrink, T., Johnstone, S., Guenther, M.G., Johnston, W.K., Wernig, M., Newman, J., *et al.* (2008). Connecting microRNA genes to the core transcriptional regulatory circuitry of embryonic stem cells. Cell *134*, 521–533.

Melton, C., Judson, R.L., and Blelloch, R. (2010). Opposing microRNA families regulate self-renewal in mouse embryonic stem cells. Nature *463*, 621–626.

Miyoshi, N., Ishii, H., Nagano, H., Haraguchi, N., Dewi, D.L., Kano, Y., Nishikawa, S., Tanemura, M., Mimori, K., Tanaka, F., *et al.* (2011). Reprogramming of mouse and human cells to pluripotency using mature microRNAs. Cell Stem Cell *8*, 633–638.

Morin, R.D., O'Connor, M.D., Griffith, M., Kuchenbauer, F., Delaney, A., Prabhu, A.L., Zhao, Y., McDonald, H., Zeng, T., Hirst, M., *et al.* (2008). Application of massively parallel sequencing to microRNA profiling and discovery in human embryonic stem cells. Genome Res. *18*, 610–621.

Murchison, E.P., Partridge, J.F., Tam, O.H., Cheloufi, S., and Hannon, G.J. (2005). Characterization of Dicer-deficient murine embryonic stem cells. Proc. Natl. Acad. Sci. U.S.A. *102*, 12135–12140.

Neveu, P., Kye, M.J., Qi, S., Buchholz, D.E., Clegg, D.O., Sahin, M., Park, I.H., Kim, K.S., Daley, G.Q., Kornblum, H.I., *et al.* (2010). MicroRNA profiling reveals two distinct p53-related human pluripotent stem cell states. Cell Stem Cell *7*, 671–681.

Onder, T.T., and Daley, G.Q. (2011). MicroRNAs become macro players in somatic cell reprogramming. Genome Med. *3*, 40.

Pfaff, N., Fiedler, J., Holzmann, A., Schambach, A., Moritz, T., Cantz, T., and Thum, T. (2011). MiRNA screening reveals a new miRNA family stimulating iPS cell generation via regulation of Meox2. EMBO Rep. *12*, 1153–1159.

Qi, L.S., Larson, M.H., Gilbert, L.A., Doudna, J.A., Weissman, J.S., Arkin, A.P., and Lim, W.A. (2013). Repurposing CRISPR as an RNA-guided platform for sequence-specific control of gene expression. Cell *152*, 1173–1183.

Raver-Shapira, N., Marciano, E., Meiri, E., Spector, Y., Rosenfeld, N., Moskovits, N., Bentwich, Z., and Oren, M. (2007). Transcriptional activation of miR-34a contributes to p53-mediated apoptosis. Mol. Cell. *26*, 731–743.

Ren, J., Jin, P., Wang, E., Marincola, F.M., and Stroncek, D.F. (2009). MicroRNA and gene expression patterns in the differentiation of human embryonic stem cells. J. Transl. Med. *7*, 20.

Rosa, A., and Brivanlou, A.H. (2013). Regulatory Non-Coding RNAs in Pluripotent Stem Cells. Int. J. Mol. Sci. *14*, 14346–14373.

Ruiz, S., Panopoulos, A.D., Montserrat, N., Multon, M.C., Daury, A., Rocher, C., Spanakis, E., Batchelder, E.M., Orsini, C., Deleuze, J.F., *et al.* (2012). Generation of a drug-inducible reporter system to study cell reprogramming in human cells. J. Biol. Chem. *287*, 40767–40778.

Sachdeva, R., Jonsson, M.E., Nelander, J., Kirkeby, A., Guibentif, C., Gentner, B., Naldini, L., Bjorklund, A., Parmar, M., and Jakobsson, J. (2010). Tracking differentiating neural progenitors in pluripotent cultures using microRNA-regulated lentiviral vectors. Proc. Natl. Acad. Sci. U.S.A. *107*, 11602–11607.

Samavarchi-Tehrani, P., Golipour, A., David, L., Sung, H.K., Beyer, T.A., Datti, A., Woltjen, K., Nagy, A., and Wrana, J.L. (2010). Functional genomics reveals a BMP-driven mesenchymal-to-epithelial transition in the initiation of somatic cell reprogramming. Cell Stem Cell *7*, 64–77.

Schurmann, N., Trabuco, L.G., Bender, C., Russell, R.B., and Grimm, D. (2013). Molecular dissection of human Argonaute proteins by DNA shuffling. Nat. Struct. Mol. Biol. *20*, 818–826.

Shekar, P.C., Naim, A., Sarathi, D.P., and Kumar, S. (2011). Argonaute-2-null embryonic stem cells are retarded in self-renewal and differentiation. J. Biosci. *36*, 649–657.

Sohn, Y.-D., Somasuntharam, I., Che, P.-L., Jayswal, R., Murthy, N., Davis, M.E., and Yoon, Y.-S. (2013). Induction of pluripotency in bone marrow mononuclear cells via polyketal nanoparticle-mediated delivery of mature microRNAs. Biomaterials *34*, 4235–4241.

Sridharan, R., and Plath, K. (2011). Small RNAs loom large during reprogramming. Cell Stem Cell *8*, 599–601.

Stadler, B., Ivanovska, I., Mehta, K., Song, S., Nelson, A., Tan, Y., Mathieu, J., Darby, C., Blau, C.A., Ware, C., *et al.* (2010). Characterization of microRNAs involved in embryonic stem cell states. Stem Cells Dev. *19*, 935–950.

Stadtfeld, M., Apostolou, E., Ferrari, F., Choi, J., Walsh, R.M., Chen, T., Ooi, S.S., Kim, S.Y., Bestor, T.H., Shioda, T., *et al.* (2012). Ascorbic acid prevents loss of Dlk1-Dio3 imprinting and facilitates generation of all-iPS cell mice from terminally differentiated B cells. Nat. Genet. *44*, 398–*405*, S391–392.

Su, H., Trombly, M.I., Chen, J., and Wang, X. (2009). Essential and overlapping functions for mammalian Argonautes in microRNA silencing. Genes Dev. *23*, 304–317.

Subramanyam, D., Lamouille, S., Judson, R.L., Liu, J.Y., Bucay, N., Derynck, R., and Blelloch, R. (2011). Multiple targets of miR-302 and miR-372 promote reprogramming of human fibroblasts to induced pluripotent stem cells. Nat. Biotechnol. *29*, 443–448.

Suh, M.R., Lee, Y., Kim, J.Y., Kim, S.K., Moon, S.H., Lee, J.Y., Cha, K.Y., Chung, H.M., Yoon, H.S., Moon, S.Y., *et al.* (2004). Human embryonic stem cells express a unique set of microRNAs. Dev. Biol. *270*, 488–498.

Sun, X., Fu, X., Han, W., Zhao, Y., and Liu, H. (2010). Can controlled cellular reprogramming be achieved using microRNAs? Ageing Res. Rev. *9*, 475–483.

Tachibana, M., Amato, P., Sparman, M., Gutierrez, N.M., Tippner-Hedges, R., Ma, H., Kang, E., Fulati, A., Lee, H.-S., Sritanaudomchai, H., *et al.* (2013). Human embryonic stem cells derived by somatic cell nuclear transfer. Cell *153*, 1–11.

Takahashi, K., and Yamanaka, S. (2006). Induction of pluripotent stem cells from mouse embryonic and adult fibroblast cultures by defined factors. Cell *126*, 663–676.

Takahashi, K., Tanabe, K., Ohnuki, M., Narita, M., Ichisaka, T., Tomoda, K., and Yamanaka, S. (2007). Induction of pluripotent stem cells from adult human fibroblasts by defined factors. Cell *131*, 861–872.

Tang, F., Hajkova, P., Barton, S.C., Lao, K., and Surani, M.A. (2006). MicroRNA expression profiling of single whole embryonic stem cells. Nucleic Acids Res. *34*, e9.

Tiscornia, G., and Izpisua Belmonte, J.C. (2010). MicroRNAs in embryonic stem cell function and fate. Genes Dev. *24*, 2732–2741.

Tzur, G., Levy, A., Meiri, E., Barad, O., Spector, Y., Bentwich, Z., Mizrahi, L., Katzenellenbogen, M., Ben-Shushan, E., Reubinoff, B.E., *et al.* (2008). MicroRNA expression patterns and function in endodermal differentiation of human embryonic stem cells. PLoS One *3*, e3726.

Underbayev, C., Kasar, S., Yuan, Y., and Raveche, E. (2012). MicroRNAs and induced pluripotent stem cells for human disease mouse modeling. J. Biomed. Biotechnol. *2012*, 758169.

Utikal, J., Polo, J.M., Stadtfeld, M., Maherali, N., Kulalert, W., Walsh, R.M., Khalil, A., Rheinwald, J.G., and Hochedlinger, K. (2009). Immortalization eliminates a roadblock during cellular reprogramming into iPS cells. Nature *460*, 1145–1148.

Wang, Y., Medvid, R., Melton, C., Jaenisch, R., and Blelloch, R. (2007). DGCR8 is essential for microRNA biogenesis and silencing of embryonic stem cell self-renewal. Nat. Genet. *39*, 380–385.

Wang, Y., Baskerville, S., Shenoy, A., Babiarz, J.E., Baehner, L., and Blelloch, R. (2008). Embryonic stem cell-specific microRNAs regulate the G1-S transition and promote rapid proliferation. Nat. Genet. *40*, 1478–1483.

Wilmut, I., Schnieke, A.E., McWhir, J., Kind, A.J., and Campbell, K.H. (1997). Viable offspring derived from fetal and adult mammalian cells. Nature *385*, 810–813.

Wilson, K.D., Venkatasubrahmanyam, S., Jia, F., Sun, N., Butte, A.J., and Wu, J.C. (2009). MicroRNA profiling of human-induced pluripotent stem cells. Stem Cells Dev. *18*, 749–758.

Yang, C.S., and Rana, T.M. (2013). Learning the molecular mechanisms of the reprogramming factors: let's start from microRNAs. Mol. Biosyst. *9*, 10–17.

Yang, C.S., Li, Z., and Rana, T.M. (2011). MicroRNAs modulate iPS cell generation. RNA *17*, 1451–1460.

Yu, J., Vodyanik, M.A., Smuga-Otto, K., Antosiewicz-Bourget, J., Frane, J.L., Tian, S., Nie, J., Jonsdottir, G.A., Ruotti, V., Stewart, R., *et al.* (2007). Induced pluripotent stem cell lines derived from human somatic cells. Science *318*, 1917–1920.

Zheng, G.X., Ravi, A., Calabrese, J.M., Medeiros, L.A., Kirak, O., Dennis, L.M., Jaenisch, R., Burge, C.B., and Sharp, P.A. (2011). A latent pro-survival function for the mir-290-295 cluster in mouse embryonic stem cells. PLoS Genet. *7*, e1002054.

Systems Biology Tools to Understand the Role of Host MicroRNAs in Infection: a Closer Look at HIV

5

Jerolen Naidoo, Robyn Brackin, Rethabile Khutlang, Anca Savulescu and Musa M. Mhlanga

Abstract

The discovery of mammalian microRNAs (miRNAs) has greatly enhanced our appreciation for the complexity associated with the regulation of the mammalian transcriptional landscape. Endogenous miRNA pathways mediate the targeted and subtle variations in gene expression required to drive complex biological processes that must be coordinated both spatially and temporally within cells and tissues. It is not surprising then that the dysregulation of miRNA function has been implicated in various models of disease and pathogenesis. Increasing interest in miRNA function has facilitated the transfer of many existing technologies to miRNA-based formats. Expression-based tools like RNAseq and qPCR microarray technologies, as well as the use of synthetic molecules to inhibit or enhance miRNA functions have been employed to identify and characterize distinct miRNA expression profiles in various models of infection. In this chapter we take a closer look at the application of some of the existing tools for miRNA-based analyses with a focus on host–pathogen interactions. Aspects pertinent to high-content miRNA-based screens are also discussed using an HIV screening workflow as a backdrop to address the important considerations associated with miRNA-based studies.

Background information

Intracellular pathogens and particularly viruses, rely heavily on the discrete control of host gene expression in order to 'hijack' host cellular machinery and complete their life cycles (Ahlquist *et al.*, 2003; Chisari, 2005; Konig *et al.*, 2008). Many viruses have evolved mechanisms to be able to avoid recognition by the host and to create a pro-viral environment within host cells (Santhakumar *et al.*, 2010; Hayes *et al.*, 2011). In the case of human immunodeficiency virus (HIV), the viral genome encodes only 15 proteins (Frankel *et al.*, 2008) and the virus must exploit multiple, specific host cell functions for successful infection (Goff, 2008). There is thus a growing interest in exploiting viral host-dependency factors and the inherent redundancy within host cellular processes in order to develop more effective antiviral therapies. Additionally, given their comparatively low mutation rate, host factors provide much more stable and reliable targets for drug development in comparison to viral factors. The prevalence of viraemic controllers, whose genetic backgrounds allow them to naturally restrict HIV infection, has also contributed to an increasing interest in tapping the anti-HIV potential within the human genome (Clavel and Hance, 2004; reviewed by Blankson, 2010; Witwer *et al.*, 2012). The identification of suitable host-dependency factors has, however, previously been limited by the lack of technologies that allow for the efficient and systematic scrutinization of host factors at the scale of the human genome. Recent advances in gene sequencing technologies, high-throughput screening techniques and RNA interference (RNAi) have led to the development of a number of research tools which allow for a systems biology approach to understanding the role of host factors in various models of infection, including HIV.

The discovery of mammalian miRNAs has provided an entirely new class of molecules that have yet to be classified in terms of their potential as viral host-dependency factors. As discussed in previous chapters, miRNAs represent one of the endogenous mechanisms of post-transcriptional gene regulation. The sequences of all known human miRNAs (miRNome) are clearly characterized and their biological effects are mapped not only at the level of individual genes, but also in regards to specific functional pathways (Gennarino *et al.*, 2012). MiRNAs have been shown to regulate many aspects of cellular physiology and have been closely linked to various models of infection (Santhakumar *et al.*, 2010; Gupta *et al.*, 2011; Witwer *et al.*, 2012). Indeed, numerous studies, most of which relied on expression profiling or miRNA target prediction, identified subsets of mammalian miRNAs that are involved in diverse viral infections, including those capable of actively modulating HIV replication (Huang *et al.*, 2007; Umbach and Cullen, 2009; Parameswaran *et al.*, 2010). Therefore the interrogation of the host miRNome holds great potential for the reliable identification of novel host-dependency factors and pathways. This chapter will discuss some of the currently available technologies for miRNome-wide studies with regard to HIV infection and will focus on the use of a high content screening (HCS) approach.

Systems biology tools to investigate differential miRNA expression in HIV infection

The role that miRNAs play in the defence against viral infection in mammalians as well as the influence of the virus on the expression profiles of miRNAs within the host cell, have been the object of many studies in the past few years. Various miRNA expression profile studies have demonstrated the differential expression of host miRNAs in HIV-1 infection while others have focused on the intriguing ability of HIV to encode its own miRNAs (Bennasser *et al.*, 2004; Houzet and Jeang, 2011; Narayanan *et al.*, 2011; Sun *et al.*, 2012; Witwer *et al.*, 2012; Chang *et al.*, 2013). The susceptibility of physiologically relevant cell lines like macrophages or monocytes to HIV-1 infection, along with their miRNA expression profiles have been evaluated by numerous studies (Swaminathan *et al.*, 2009; Wang *et al.*, 2009a). These studies have identified many host miRNAs as candidate biomarkers of HIV infection and some have even identified host miRNAs that may potentially regulate viral transcripts (Huang *et al.*, 2007). Recent studies have also sought to probe the expression levels of miRNAs in peripheral blood mononuclear cells derived HIV-1 seropositive individuals in order better understand the role of host miRNAs in context of infection (Houzet *et al.*, 2008).

MiRNA expression profiling approaches

Available technologies

The three principal methods used to measure miRNA expression levels in miRNA profiling are: real-time reverse-transcription PCR (RT-PCR) (Gupta *et al.*, 2011; Wang *et al.*, 2009b), microarray hybridization (Sisk *et al.*, 2012; Sun *et al.*, 2012) and next generation sequencing (NGS) (Bignami *et al.*, 2012; Chang *et al.*, 2011, 2013). All three methods have been employed to investigate differential miRNA expression in context of HIV-1 infection (Chang *et al.*, 2013; Gupta *et al.*, 2011; Sun *et al.*, 2012). However, each method faces challenges, among which, are the limited length of the mature miRNA, potentially low expression levels, miRNAs (~22 nucleotides), variable and tissue-specific expression, patterns of specific miRNAs as well as the effect of developmental stage and disease state effects on miRNA expression (Kong *et al.*, 2009). Other challenges arise from the existence of miRNA families where miRNAs that belong to one family exhibit differential expression patterns but may only differ by a single base (Telenti, 2009). The considerations to use RT-PCR, microarrays or sequencing to profile miRNA expression levels, based on the merits or drawbacks of each technology, is similar to their use for mRNA or genomic DNA profiling. RT-PCR suffers from the ability to analyse miRNAs in a high throughput manner, at the genomic scale, and is best suited for validation studies. The use of microarray technologies and NGS technologies are typically considered to be better suited as discovery tools for miRNA expression profiling.

Microarray miRNA profiling overcomes the limitations of RT-PCR miRNA expression profiling by assessing miRNAs levels at a global scale in a single experiment (Yin *et al.*, 2008). However, there are a few factors to be considered when applying this technique to profile miRNA expression, mainly the difference between mature miRNAs and their precursors, as well as single nucleotide differences (Git *et al.*, 2010). A typical miRNA array experiment consists of three stages: (i) design of the microarray which includes probe and control probe design, and spotting the probes onto the array surface (Barad *et al.*, 2004; Wang *et al.*, 2011); (ii) purification and enrichment of mature miRNAs from the sample of choice, sample labelling and hybridization to array (Shingara *et al.*, 2005; Yin *et al.*, 2008); and (iii) analysis of array results by quantification, normalization and validation of the readout data (Houzet *et al.*, 2008; Yeung *et al.*, 2005). Despite the many advantages of miRNA arrays, deep sequencing technologies have recently been favoured for miRNA expression profiling over microarray technologies. Firstly, sequencing approaches do not require an *a priori* knowledge of the genome (Wilhelm and Landry, 2009). Second, sequencing is not dependent on hybridization, thus reducing the possibility of cross-hybridization (Malone and Oliver, 2011). Third, NGS has greater sensitivity allowing detection of less abundant transcripts giving a more accurate value for abundance levels (Mardis, 2008).

Expression profiling of miRNAs by next-generation sequencing

NGS allows the detection of virtually all miRNAs expressed in the sample of interest allowing investigation into changes occurring in HIV-infected cells. The advantages of NGS for miRNA expression profiling are as follows: the ability to count the number of transcripts of miRNAs as a measure for expression, being able to multiplex, being species independent and having high sensitivity for low abundant miRNA transcripts (Buermans *et al.*, 2010; Creighton *et al.*, 2009; Wilhelm and Landry, 2009). This is highlighted in the ability to assess both the differential expression of host miRNAs as well as viral miRNAs. There are two varieties of NGS technologies: (i) sequencing by synthesis, which is employed by the Genome Analyser (Illumina, http://www.illumina.com) and 454 sequencing (454 Sequencing/Roche, http://454.com); and (ii) sequencing by ligation employed by the SOLiD system sequencing (Applied Biosystems, http://www.appliedbiosystems.com)(all reviewed in Mardis, 2008). The cost, read length and sequencing depth for miRNA profiling determine the sequencing platform of choice.

An overview of the strategy used for miRNA expression profiling using the Genome Analyser (Illumina) will be highlighted. The key aspects of the sequencing pipeline will be discussed in this section, however detailed descriptions of miRNA expression profiling using Illumia sequencing are available (Creighton *et al.*, 2009; Kadri *et al.*, 2011). Sequencing protocols are also available for 454 sequencing (Mizuguchi *et al.*, 2011) and SOLiD sequencing (Wu *et al.*, 2011).

The first step in all sequencing experiments entails the generation of a small RNA library. This begins with extracting the total RNA from the sample of interest, followed by enrichment of small RNAs, typically those < 200 bases in length. Adapters are then ligated to the 3′ and 5′ end of these small RNAs followed by a reverse transcription step and amplification by PCR, whereby one of the primers contains an index that tags each cDNA. The final step consists of sequencing the library using sequencing by synthesis technology employed by the Genome Analyser (Mardis, 2008; Wilhelm and Landry, 2009).

The output from the miRNA sequencing experiment contains millions of short reads and several basic data processing steps are required to extract the relevant information from the raw data. Processing begins with quality control analysis of the sequencing data using FastQC (FastQC, http://www.bioinformatics.bbsrc.ac.uk/projects/fastqc) (Creighton *et al.*, 2009; Wilhelm and Landry, 2009). The average length of the Illumina sequencing read is 36 nt in length, and as the length of the miRNAs of interest are between 17 and 25 nt, the reads are shortened to remove the 3′ sequencing adapters. This is followed by another quality control assessment to ensure the final set of sequence reads are between 17 nt and 25 nt in length. The final step in the data processing pipeline requires the reads to be aligned to a reference genome, each sequence should originate from the genome of the sequenced organism. The small RNAs that have passed the above-mentioned control steps are then aligned against reference miRNAs in the latest version of miRBase (Griffiths-Jones *et al.*, 2008; Motameny *et al.*, 2010).

The number of small RNA sequence reads from the filtered data can be used to estimate expression profiles of each identified miRNA in two or more samples. The expression level for each unique sequence can be represented as the number of times it occurs among all the reads for a given sample, and can be normalized by computing the Reads Per Million (RPM) value. The RPM is represented as the number of reads for a unique sequence in the sample divided by the total number of reads for the whole sample multiplied by 1 million. These normalized reads should reflect the abundance of each unique sequence in the sample that can then be related as its expression level. Performing this normalization allows for the comparison of expression levels between multiple samples as the absolute read count of the sample from the sequencing experiment is significantly altered by the RNA extraction and library preparation (Linsen *et al.*, 2009). The differential expression of miRNAs between samples of interest can be determined using various software packages available. A few of these include DeSeq package from Bioconductor (http://www.bioconductor.org) (Gentleman *et al.*, 2004), miR-Express (http://mirexpress.mbc.nctu.edu.tw) (Wang *et al.*, 2009a), and web-based platforms MirTools (http://http://centre.bioinformatics.zj.cn/mirtools) (Zhu *et al.*, 2010) and deepBase (http://deepbase.sysu.edu.cn) (Yang *et al.*, 2010). Differential expression profiles of miRNAs of interest can be verified using RT-PCR methods. Once a set of miRNAs have been identified as being differentially expressed between two test states (e.g. HIV-infected cells versus uninfected cells), the next step is to predict putative mRNAs being targeted by the miRNA. This can be done using either empirical or *in silico* methodology and further discussed in this chapter.

Expression-based approaches versus functional assays

Many studies have utilized comparative miRNome-wide (all known miRNAs) expression profiling and miRNA target prediction approaches in order to infer the significance of miRNAs within infection models (reviewed by Umbach and Cullen, 2009; Gupta *et al.*, 2011). Often microRNA expression profiling relies on the use of commercially available technologies to quantify the differential expression differences, especially observed with RT-PCR expression profiling and microarray expression profiling described above (Gupta *et al.*, 2011; Hayes *et al.*, 2011). As with miRNA sequencing approaches, qRT-PCR arrays have been useful in the identification of miRNAs as candidate biomarkers of infection and the observed global changes in miRNA expression profiles in response to infection have supported the critical role of miRNAs in infection biology. It should be noted that widely used expression profiling methods such as qRT-PCR arrays and RNA sequencing are susceptible to the conflation of variable phenotypes within cell populations. This can result in the propagation of distorted representations of miRNA function within a given model. The use of single cell profiling approaches such as the nCounter® miRNA assay (nanoString®) can overcome this issue and provide greater sensitivity and confidence in the evaluation of host miRNA responses to infection. Given the complex nature of miRNA functionality however, the differential expression of a specific miRNAs may not translate to a functional phenotype for a given fold change in its expression levels. Functional assays may therefore represent a much more definitive approach to fully elucidate the functional relevance of individual miRNAs in a given model of infection.

MiRNA libraries and functional screening approaches

There are currently a number of miRNA libraries available for functional miRNome-wide studies, as highlighted in Table 5.1. Commercially available miRNA libraries are regularly released in new iterations to ensure that they are up to date with the ever-growing complement of known miRNAs (miRBase, http://www.mirbase.org).

Table 5.1 Selected commercially available miRNA libraries

Company	Product	Coverage
Thermo Scientific™	miRIDIAN™ miRNA Libraries	Mimics and inhibitors
Exiqon	miRCURY LNA™ microRNA Inhibitors	Inhibitors only
Life Technologies™	Ambion mirVana™ Libraries	Mimics and inhibitors
Sigma-Aldrich®	MISSION® Human miRNA Mimics	Mimics (inhibitors available on request)

Lentiviral or plasmid based libraries allow for the overexpression of specific miRNAs via the canonical miRNA biogenesis pathways. These approaches are extremely useful in the generation of sustained overexpression phenotypes but can also result in broader effects due to integration at biologically relevant sites as well saturation of miRNA processing machinery. Another option is the use of miRNA libraries that are comprised of mature mimic and inhibitor molecules that feed into the miRNA pathway at the point of RISC-mediated functionality. Plasmid and lentiviral miRNA libraries are also available from various vendors for the stable overexpression of miRNAs, however this chapter will focus on the use of mimic and inhibitor molecule libraries. MiRNA mimic molecules are synthetic RNA duplexes that imitate mature endogenous miRNAs (Schoolmeester *et al.*, 2009; Santhakumar *et al.*, 2010). The introduction of miRNA mimics into a cell, results in the increased intracellular levels of a particular miRNA and thus augments the function of a specific miRNA. mimic molecules are readily incorporated into RISC complexes and are able to mediate the enhanced suppression of all mRNA targets specific to a particular endogenous miRNA (Fig. 5.1). MiRNA inhibitor molecules alternatively are single stranded RNA structures that are; (i) designed to have a high binding affinity to a specific target miRNA and (ii) chemically modified to be resistant to RISC-mediated degradation (Vermeulen *et al.*, 2007). Inhibitor molecules are thus able to function as miRNA 'sponges' that saturate target miRNA–RISC complexes and attenuate the repressive

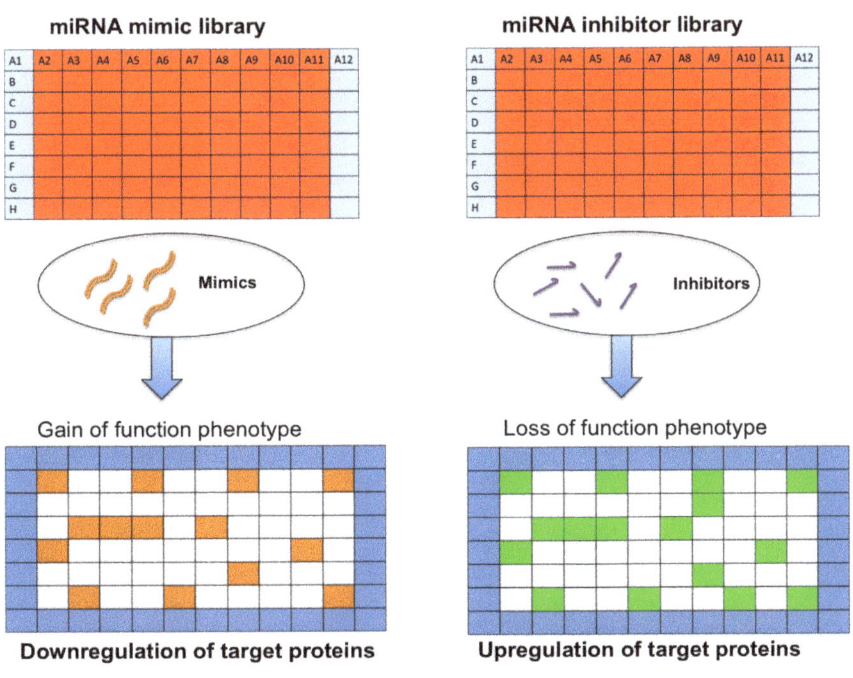

Figure 5.1 Agonist–antagonist approach using miRNA mimic and inhibitor molecules (adapted from http://www.thermoscientific.com/miRIDIAN).

effects of a specific miRNA (Fig. 5.1). This combined agonist-antagonist approach to evaluating miRNA function has an added advantage over conventional RNAi screens in that it allows for the parallel assessment of both loss-of-function (inhibitors) and gain-of-function (mimics) phenotypes for a given miRNA.

A high-content screening approach to evaluate the role of host miRNAs in HIV infection

The availability of miRNA libraries has enhanced the researchers ability to perform functional High-Throughput (HT) and high-content (HC) screening. A study by Santhakumar *et al.* (2010) was the first to combine a HT screen and the agonist-antagonist approach to evaluating global miRNA function in viral infection. The combination of HT experimental design and automated imaging systems has heralded the era of high content screening (HCS). These approaches represent a very powerful technique to obtain multi-parametric biological data at a genome-wide experimental scale, thus facilitating more robust and data-rich studies compared with classical HT approaches (Zanella *et al.*, 2010; Genovesio *et al.*, 2011; Xia and Wong, 2012). This chapter will focus on a recently conducted to-be-published visual-based HC screen for HIV infection as a backdrop to address some of the salient aspects related to functional miRNA screens. A summary of the experimental workflow that will be discussed is presented in Fig. 5.2.

Initial considerations

To conduct a successful and reproducible HCS there are certain components that need to be realized namely; an optimized cell-based assay, biologically relevant markers or readouts, image acquisition, image analysis and data management. Each component plays an important role and careful consideration needs to be made to ensure the outcome is in a format that is representative and relevant. A carefully designed assay will aid in the decision of which imaging method is best suited to extract the required information from the data. The following sections will highlight the importance of each component with regards to developing and performing a HCS to elucidate host miRNAs that play a role in HIV infection. The HCS can be applied to different formats from micro-well plates to micro-array slides and even microfluidic chips (Brass *et al.*,

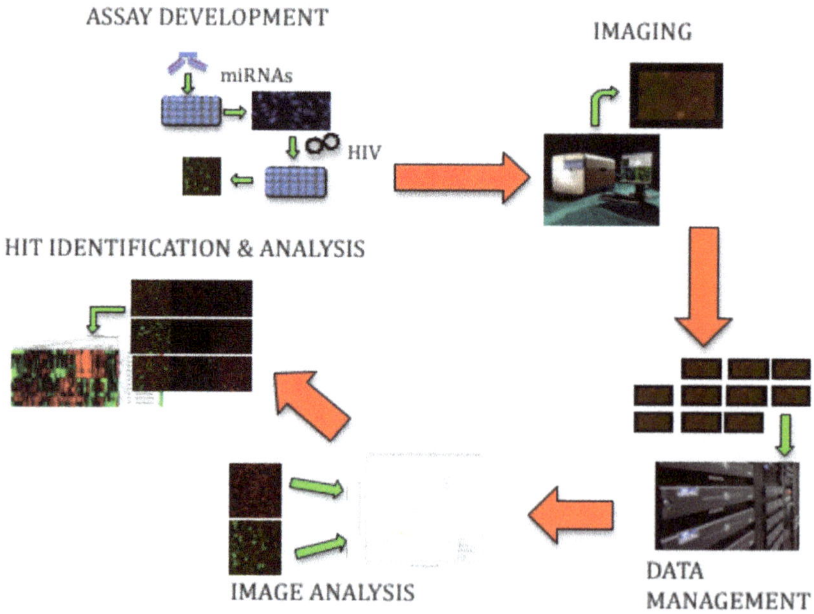

Figure 5.2 Summary of visual-based screen workflow.

2008; Santhakumar *et al.*, 2010; Genovesio *et al.*, 2011). In the case of micro-well plates, specialized microscopy-compatible, clear-bottom plates are required for HC fluorescent microscopy-based approaches. The choice of micro-well plate format also has downstream implications in terms of (i) the amount of library reagents that will be used per screen, (ii) the range of reagent volumes that need to be dispensed, (iii) whether automation is necessary or possible and (iv) the level of expertise required. Therefore the choice of the screen format is highly dependent on the availability of specialized equipment and expertise. The use of the 96-well micro-well plates represents a robust screening platform that can be adapted to both automated and mechanical approaches and strikes a balance between the degree of high-throughput application (experimental miniaturization) and the available data points per treatment (number of cells per well).

'Edge effects' refers to the phenomenon whereby evaporation of liquids can occur at an expedited rate in the outer wells of a micro-well plate in a humidified environment. This can result in altered concentrations of reagents and may compromise the reproducibility of experimental conditions across the wells of a micro-well plate. In order to diminish the impact of edge effects in experimental micro-well plates, the outer wells are generally filled with a sterile solution such as phosphate buffer saline solution (PBS). In the case of a 96-well micro-well plate for example, the outer 36 wells should be filled with 200–250 µl of PBS and only the 60 inner wells should be used for experimental reactions. The use of moated micro-well plates such as Nunc Edge 96-well plates (Thermo Scientific™, USA) can also be employed in order to make use of all of the micro-wells on each plate. It is however important to first evaluate the use of micro-well plates with regard to the microscopy-based readouts required for downstream analysis (e.g. sufficient fluorescent signal and identifiable subcellular features).

Assay development

To be able to relate the functional relevance of a specific miRNA to HIV infection, the miRNA mimic or inhibitor molecules must be shown to have a significant effect on HIV replication. There are many ways to assay HIV replication, in the context of host infection, which can be adapted for HCS. Immunofluorescence-based methods targeting specific HIV proteins may be adapted for HCS analysis (Brass *et al.*, 2008). The immunofluorescence approach is advantageous in that it does not necessitate the use of genetically altered viruses or host cells but requires extensive optimization, technical expertise and additional sample preparation steps (Pache *et al.*, 2011). Reporter viruses are genetically modified to encode fluorescent markers and can be used to directly assay HIV replication by the quantification of fluorescent signal (Santhakumar *et al.*, 2010). While this approach is readily applicable to HCS, the genetic modifications to these viruses may result in aberrant interactions with host factors and thus a distorted representation of the functional relevance of certain miRNAs.

The use of an HIV reporter cell line represents a technically undemanding approach to evaluate HIV replication. HIV reporter cell lines are routinely employed in order to quantitatively and qualitatively evaluate HIV infection (Brass *et al.*, 2008; Zhou *et al.*, 2008; Genovesio *et al.*, 2011). These cell lines are genetically modified to stably express a fluorescent or biochemical reporter signal that is inducible by a specific HIV factor or protein. A variety of both HIV reporter cell lines and reporter viruses can be obtained from the NIH AIDS Research and Reference Reagent Program (http://www.aidsreagent.org).

For the purposes of this chapter we will consider an HIV-1 screen utilizing the adherent GHOST(3) X4/R5 cell line. This cell line is available from the NIH AIDS Research and Reference Reagent Program (catalogue # 3942) and stably expresses an HIV-TAT sensitive Long Terminal Repeat (LTR) driving humanized GFP (hGFP) expression. The use of similar tat-responsive LTR-GFP reporter systems in the evaluation of HIV-1 infection has been previously described by Karroy *et al.* (2000) and Genovesio *et al.* (2011).

Understanding the limitations of the intended read-out assay is an important initial consideration. The use of a TAT-sensitive reporter system, as in the GHOST(3) cell line for example, can restrict the scope of a screen to miRNAs that regulate the early stages of infection (viral entry to

TAT protein production). The assay is thus naïve to miRNAs that may impact HIV replication processes that occur after TAT protein production (e.g. budding and viral packaging). A method to include the evaluation of late acting factors while using the same reporter system requires a secondary infection screen as described by Brass *et al.* (2008) and Zhou *et al.* (2008). This approach does not require the use of additional miRNA library reagents but does require use of a replication competent viral strain, additional miRNA control reagents and additional analyses. The scope of the study should thus be very carefully considered in conjunction with the experimental design.

Automated versus manual approaches

The use of commercially available systems like the JANUS liquid handling system (Perkin-Elmer, France) represents a fully automated, robotics approach to liquid handling solutions. Automated approaches are highly advantageous in that they are physically undemanding and offer superior efficiency along with quantifiably low error rates. These systems however require a substantial initial investment and extensive training and optimization before use. The use of automated systems can be applied to all aspects of screening protocols from the reconstitution and daughtering of library plates to cell seeding, transfections and sample preparation. While automated approaches are extremely useful with regard to reliability and reproducibility they are also limited in comparison to the robustness afforded by conducting experimental work by hand. A simple example of this is the routine practice of angling a micro-well plate in order to remove a solution from wells while minimizing the disruption to adherent cells. Similarly, the addition of liquids to micro-wells containing cells is routinely carried out by dispensing a solution directly against the walls of the micro-well in order to minimize wash effects. While these tasks can be readily achieved using a hand held multi-channel pipette, the adaptation of such methods to robotic systems will require a heavy investment of time, expertise and costly specialized components and consumables. Likewise, the incorporation of sterile tissue culture practices into robotics-based approaches may also require specialized sterile rooms or instrument casings that mimic the appropriate bio-safety level environments. Manually conducted screens while more robust are not without their drawbacks, these approaches demand tremendous levels of focus and dexterity, attention to detail and a high level of expertise in a number of advanced cell biology techniques. Manually conducting a screen at the miRNome-wide or genome-wide experimental scale is also subject to increased variability resulting from human error, pipetting errors and various other factors. Therefore the choice of whether automation is required or not, be it for all screen protocols or specific procedures, is dependent on the availability of experienced personnel, financial resources, the scale of experimentation and future prospects with regard to HT approaches.

Library management

Aspects of library management will be discussed with regard to the miRIDIAN miRNA library (Thermo Scientific™, USA) however these principles should be broadly applicable to miRNA libraries from different vendors. MiRNA libraries are generally shipped as lyophilized pellets in 96 well plates. These plates are stable at room temperature for 2–4 weeks but should be stored at −20°C or −80°C immediately upon receipt. Reconstituted miRNA libraries are reported to have a shelf life of ~1 year under these storage conditions however these libraries can be stored and used for longer periods. It is recommended that the efficacy and integrity of miRNA molecules be monitored regularly once stored for longer than a year. The number of plates in a complete library will vary from vendor to vendor and is dependent on the coverage of the library; the latest miRIDIAN miRNA library is based on miRBase v.19.0 (miRBase, http://www.mirbase.org) and accounts for 2044 mature miRNAs. Libraries are generally shipped as separate mimic and inhibitor components and are accompanied by electronic information describing the miRNA molecule annotations, well locations per library plate and further information on the miRNA molecules.

Library reconstitution

miRNA libraries (both mimic and inhibitors) can be reconstituted in either RNAse-free water or RNA stabilizing buffers used for siRNA reagents. Thermo Scientific recommends reconstitution of miRNA reagents with 1x siRNA buffer (cat #: B-002000-UB-100) at all concentration ranges, for long-term storage. RNAse-free water may be used to reconstitute miRNA pellets but only at high concentrations (20–100 μM) and only for short-term storage applications (less than 6 months). Detailed protocols on miRNA library/reagent reconstitution are available at http://www.thermo.com/eThermo/CMA/PDFs/Various/File_5516.pdf.

The concentration at which the original library plates (Master plates) are re-suspended must be carefully considered. Reconstitution at higher concentrations is generally favoured and if the downstream library is intended for use in microarray-based printing technologies then a minimum reconstitution concentration of 20 μM is advisable. There are two important factors that must be considered in terms of the downstream aliquoting and dilution (daughtering) of master plates for use in screen applications. The first of these is the final concentration and volume of library reagents that will be required for single screening applications and the second is the maximum number of freeze–thaw cycles that the reconstituted library reagents may undergo. The freeze–thaw cycle limitation is more reference point than law and refers to the number of times that the reconstituted library may be defrosted and frozen again before this process will negatively impact RNA integrity (Thermo Scientific recommends a maximum of five freeze–thaw cycles for miRIDIAN miRNA reagents).

The rationale behind the miRNA library daughtering process will be explained using an example of a library that has been reconstituted at a 20 μM final concentration and will be used in a 96-well micro-well plate based screen. The optimal final concentration of miRNA mimics and inhibitors that will be used in the experimental assay should first be determined as discussed in this chapter. Assuming that the final concentration of miRNA mimics and inhibitors to be used in each experimental well of the screen is 25 nM, the volume of mimic or inhibitor that will need to be pipetted to a 100 μl transfection reaction from different daughter library concentrations are summarized in Table 5.2 below.

Assuming that the transfection reactions will be prepared by hand using a multi-channel micropipette, the accuracy and reproducibility with which volumes of liquid can be handled is directly proportional to the volume. The production of miRNA daughter plates should facilitate the storage of miRNA library reagents at a concentration that facilitates the use of readily pipettable volumes of library and at a volume that takes into account the maximum number of freeze thaw cycles recommended by the manufacturer. An additional volume of miRNA library reagent should be included per well in order to compensate for reagent volume lost due to pipetting error and storage. The volume of reagent that is aliquoted per well of a daughter plate can be calculated as follows:

$$\text{Total volume} = (A \times B) + A$$

where A = volume of reagent required per transfection reaction and B = number of permitted freeze–thaw cycles.

Library storage

In order to minimize sample loss during storage and to prevent cross well contamination, all daughter and master plates should be sealed using sterile adhesive or thermal micro-well plate seals (e.g. Aluma Seal II™, Rotilabo®). These seals offer superior resistance against reagent loss during long term storage are suitable for storage at temperatures as low as –70°C if shipping of reconstituted library is required. Specialized storage plates are also required for the long-term storage

Table 5.2 miRNA reagent concentration and related volumes

Concentration of miRNA reagent (μM)	Required volume for transfection reaction (μl)
20	0.25
4	0.625
2	1.25
1	2.5

(−20°C to −70°C) of miRNA library reagents and the use of v-bottomed plates are recommended for the optimal recovery of low volumes of liquid (e.g. Corning®, Costar®).

Cell seeding

The primary advantage of HC visual-based screens is the rich amount data provided from single cell analysis using multiple cellular features or descriptors. In order to fully exploit this, downstream image analysis must be able to readily distinguish individual cells within each well. Ideally, multiple seeding densities should be evaluated within the experimental parameters intended for the screen (e.g. experimental time frame and all treatments). The optimal cell seeding density range should allow for the maximum number of cells per well that facilitates subsequent single cell segmentation by image analysis (Jones *et al.*, 2008). The optimization of cell seeding for subsequent image processing is thus an important factor that must be considered for experimental design (Jones *et al.*, 2008).

Transfection

In contrast to siRNA approaches where lower siRNA concentrations are favoured in order to reduce off-target effects (Brass *et al.*, 2008; Konig *et al.*, 2008), the promiscuous nature of miRNA targeting permits the use of maximal concentrations of miRNA molecules to fully understand the systemic range of miRNA function. Most commercially available lipid-mediated siRNA transfection reagents can also be used to introduce miRNA mimic and inhibitor molecules into mammalian cells (Lipofectamine® RNAiMax, Invitrogen, USA; DharmaFECT, Thermo Scientific™, USA). Different transfection conditions must be evaluated in order to determine the optimal transfection conditions for a particular experimental assay. Previous studies by Santhakumar *et al.* (2011) and Schoolmeester *et al.* (2009) have shown that a final concentration of 25 nM was the maximal miRNA molecule concentration that did not result in broad cytotoxic effects in mammalian cells. While the final concentration of miRNA mimic and inhibitor molecules should be optimized for each cell line and transfection reagent used, the reported 25 nM concentration can be exploited as a useful reference point in subsequent optimization steps.

Fluorescent dye-labelled, non-targeting miRNA and siRNA transfection controls used in combination with fluorescent nuclear and cytoplasmic markers represent a useful tool to evaluate transfection efficiency by fluorescent microscopy (Fig. 5.3). Using the fluorescent nuclear markers it is possible to calculate the average number of cells imaged per well. By quantitatively assessing the number of cells per well in which the signal from the fluorescent miRNA/siRNA controls overlap with either the nuclear or cytoplasmic markers, it is then possible to determine the average number of cells in each transfection condition that have been

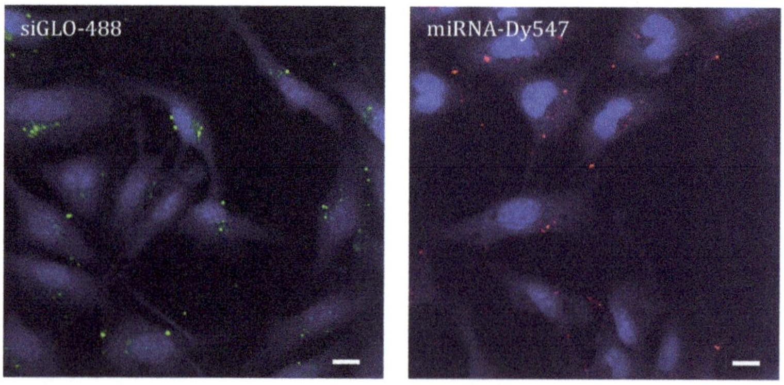

Figure 5.3 Fluorescent microscopy images of GHOST(3) cells transfected with dye-labelled non-targeting siRNAs (SiGLO-488) and non-targeting miRNA molecules (miRNA-Dy547) at a final concentration of 25 nM using Lipofectamine RNAiMax. Nuclei are stained with DAPI (blue). Overlap of dye channels with nuclear and cytoplasmic stain is indicative of positive transfection. Scale bar = 10 μM.

successfully transfected (Fig. 5.4). Nuclear markers can also be exploited to monitor cell viability across various experimental conditions. Comparisons of cell counts from mock-transfected control wells to test wells can be used to identify treatments that result in significant cytotoxic effects (cell counts <20% than control wells). Both transfection efficiency and cell viability can be determined by automated image analysis making use of cell counting/segmentation and spot detection algorithms. Alternatively, these values can be determined manually by visual inspection or via a semi-automated approach using open-source image analysis packages such as ICY (http://icy.bioimageanalysis.org/), FIJI (http://fiji.sc/Fiji) and CellProfiler (http://www.cellprofiler.org/).

The experimental optimization transfection conditions should also include time course experiments to identify knockdown time points that are able to mediate a biologically relevant knockdown phenotype. Candidate transfection conditions should thus be validated using biochemical analysis to verify efficient target knockdown at a desired time point. The use of positive miRNA controls or siRNAs controls with known targets can be used in conjunction with mRNA or protein expression analyses to this effect (Fig. 5.5). As a general rule of thumb, efficient lipid-mediated siRNA knockdown of mRNA is generally detectable 24–48 hours post transfection, while efficient protein knockdown is detectable 48–96 hours post transfection. It is important to note that these values describe a range of mRNA and protein targets with highly variable half-lives. The choice of an experimental time point for the transfection of miRNA molecules should be selected with cell density considerations and the total time scale of the experimental assay in mind.

Infection

Optimization of the HIV infection assay for HCS analysis is an important parameter of overall experimental design. The choice of viral strain, multiplicity of infection (MOI) and the time point of infection used can impact the range of potential 'HITs' identified. The choice of a viral strain that is able to mediate a strong induction of the reporter system under natural conditions would be preferred for HCS screen analysis. A strong and visually apparent infection phenotype is important for image analysis to be able to readily differentiate between infected and uninfected cells within a mixed population but is also relevant in terms of the potential phenotypic range afforded by the reporter system. The use of pseudotyped HIV strains that do not employ canonical CD4 receptor-mediated entry (e.g. VSVG pseudotyped viruses) will preclude miRNAs that are relevant to receptor-mediated entry and its associated processes from the HIT analysis. This can be a major concern when using a TAT-sensitive reporter systems for example as the potential HIT pool will be further diminished in addition to the restrictions inherent to the reporter system. The use of single cycle viral strains that are still dependent on CD4-mediated entry may thus provide a more comprehensive reflection of miRNA

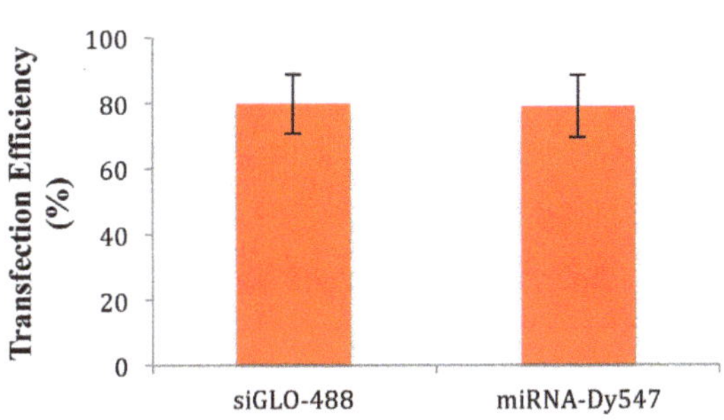

Figure 5.4 Average transfection efficiency for miRNA (25nM) and siRNA (25nM) transfection reactions in GHOST(3) cells using a bespoke image analysis algorithm. Bars represent standard deviation (n=10).

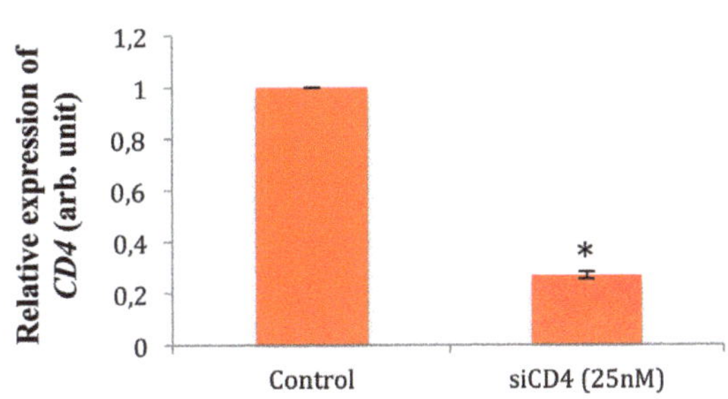

Figure 5.5 qPCR analysis of *CD4* knockdown using siCD4 smart pool at 25 nM. Results show 74% down-regulation of *CD4* transcript in siCD4 (25 nM) treated Ghost(3) cells as compared to untreated control cells. $P<0.05$. $n=3$. Bars represent standard error of the mean (±0.014).

functional roles in HIV infection. Wild type, replication competent viruses would ideally be favoured in order to obtain a more physiologically relevant reflection of host pathogen interactions events but the use of wild-type viruses is not without its drawbacks.

The poor and variable rates of infection efficiency associated with using wild type HIV strains can have broad implications in screening approaches (Rusert *et al.*, 2004; Platt *et al.*, 2010). Asynchronous infection events and poor infection rates can result in a distorted representation of HIV replication phenotypes between wells and may confound downstream image analysis approaches. The addition of polycationic reagents such as dextran calcium-chloride (DEAE) and polybrene to infection reactions have been shown to increase the efficiency of HIV-1 infection by enhancing viral adhesion or adsorption (Callahan, 1994; Platt *et al.*, 2010). The use of these infection-enhancing reagents may thus contribute to the standardization infection events within experimental wells and may improve the accuracy of downstream analyses. The optimal dextran DEAE and polybrene concentration ranges should be optimized to maximize infection efficiency and minimize cytotoxic effects within the reporter cell line used (Rusert *et al.*, 2004; Platt *et al.*, 2010).

Infection time course is an important parameter that must be carefully considered. The optimal infection time point for HCS should be determined in terms of the time point which yields maximal reporter signal output in a given infection model. The GHOST(3) cell line for example, when infected with HIV was shown to illicit a maximal signal at 48–60 hours with signal intensity decreasing at longer time points (Cecilia *et al.*, 1998; Morner *et al.*, 1999). The use of longer infection time points (>72 h) in combination with replication-competent viral strains can also result in asynchronous infection phenotypes due multiple rounds of infections (Platt *et al.*, 2010).

Controls

In addition to the standard biologically relevant experimental controls that should be included on each imaging test plate, high content image analysis requires specific controls, each with discrete functional relevance in terms of downstream image analysis. These include controls required for the characterization of biologically relevant phenotypes in singles cells and cell populations, the calibration of reporter signal thresholds and the normalization of variations between large numbers of experimental plates.

Phenotypic controls

Automated HCS image analysis algorithms require strong visually distinct controls as reference or 'training' tools to be able to accurately categorize experimental wells and thus identify phenotypes of interest (e.g. enhanced or suppressed HIV replication). The use of siRNAs targeting the human CD4 receptor for example represents a visually striking positive control for the inhibition of HIV infection (Fig. 5.6). This is an example of

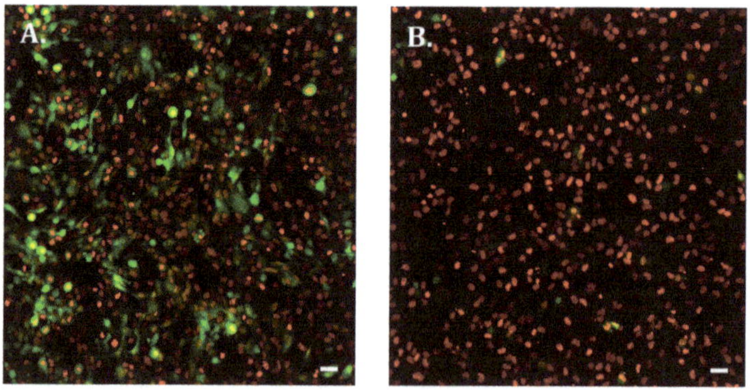

Figure 5.6 GHOST(3) cells 48 hours post infection with HIV-1. MOI = 0.5. Positive infection is denoted by GFP expression (green) and nuclei are stained with DAPI (pseudo-coloured red). (A) Mock transfected and infected control cells. (B) Cells transfected with CD4 siRNA smart pool at 25 nM final concentration 48 hours prior to infection. Scale bars = 20 μM.

a control that can be used to train image analysis to identify phenotypes in which HIV entry and subsequent replication are inhibited. It may be sufficient to use a single entry-inhibition control to identify miRNAs that are potent inhibitors of HIV replication; however additional controls related to different stages of viral replication may be employed to provide greater resolution within the general inhibition phenotype. A control compound that specifically inhibits HIV integration for example can be used in parallel to the siCD4 control to allow image analysis to differentiate between miRNA HITs that inhibit viral entry versus those that inhibit viral integration. The same principles are applicable to the selection of suitable positive controls representative of an enhanced HIV replication phenotype and the selection of phenotypic controls should be cognizant of image analysis requirements.

Standard experimental controls

Routine experimental controls such as mock-transfected, infected 'natural infection' controls and mock-transfected uninfected controls are also valuable in terms of downstream image analysis. The natural infection controls can be used to define the multiparametric space that is characteristic of non-perturbed HIV replication. MiRNA mimic or inhibitor wells, which are classified by this phenotype, will be characterized as having no effect on HIV infection and will thus not be a identified as HITs. Uninfected control cells are essential to determine the average background signal inherent in the reporter system and can be used to subtract background signal or 'noise' from experimental wells. When evaluated in comparison with uninfected, non-targeting miRNA mimic and inhibitor transfection control wells, uninfected mock-transfected controls can also be used to validate that the miRNA transfection process itself has no effect on the cellular reporter system.

Non-targeting miRNA controls

An important assumption with regard to the miRNA mimic and inhibitor screen approach is that the introduction of miRNA mimic and inhibitor molecules alone has no general effects on the reporter readout assay in the presence of HIV. The inclusion of non-targeting miRNA mimic and inhibitor controls are therefore required to verify that the introduction of mimic or inhibitor molecules do not have non-specific effects on HIV replication within a given experimental plate (Fig. 5.7). The inclusion of fluorescent dye-labelled non-targeting miRNA mimic and inhibitor transfection controls in experimental plates may also serve a secondary function by allowing for the normalization of inter-plate variations in transfection efficiency rates.

Sample preparation

As a point of reference, the average time required to image 60 wells of a 96-well plate at 20× magnification, acquiring 4 images per well in two

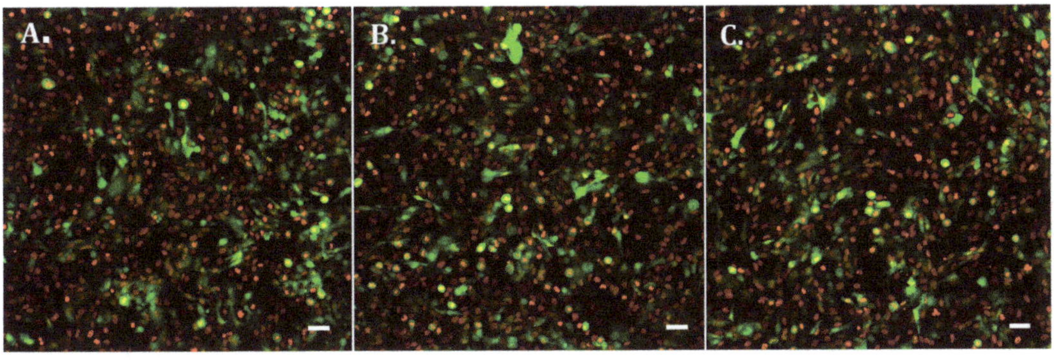

Figure 5.7 GHOST(3) cells 48 hours post infection with HIV-1. Positive infection is denoted by GFP expression (green) and nuclei are stained with DAPI (pseudo coloured red). (A) Untransfected control, (B) non-targeting miRNA mimic transfection control, (C) non-targeting miRNA inhibitor transfection control. Non-targeting transfection controls were shown to have no significant effect on HIV-1 infection as per image analysis. Scale bars = 40 μM.

channels is approximately 20–30 minutes. This estimation is based on using either an automated point scanning fluorescent confocal microscope (Image® Xpress ULTRA, Molecular Devices, USA) or a screen-adapted spinning disk fluorescent confocal microscope (Revolution XD, Andor Technology, UK). Cells in screen plates are therefore generally 'fixed' using cross-linking reagents to ensure that all plates are imaged at a standardized infection time point and to facilitate the downstream sample preparation required for HC analysis of cellular features. A nuclear marker such as DAPI or Hoechst should always be included in HCS sample preparation as nuclear markers are used to correlate cell numbers and are important components of cell segmentation approaches required for HC analysis. More rigorous cellular feature-based analysis may require the use of additional fluorescent markers specific to components of interest. In the case of immunofluorescence-based readout systems, this will also require many additional treatments steps. It should be noted however that additional sample preparation steps might dramatically increase the quantities of consumables required (e.g. micropipette tips) as well the time needed for sample preparation.

Imaging

Advances in integrating novel imaging, miniaturization of experiment and image processing techniques have led to the development of HT microscopy regimes that are able to extract relevant information at the cellular level (Carpenter, 2007). Automated fluorescence microscopy has become an indispensable tool for large studies in cell biology, enabling the emergence of HT techniques and methodologies (Wollman and Stuurman, 2007). Three important advances have facilitated the development of these methodologies, namely (i) novel microscopy methods, (ii) novel fluorescent labelling protocols, and (iii) significant improvements in informatics, image analysis and algorithm development (Carpenter, 2007; Hamilton and Teasdale, 2008; Wollman and Stuurman, 2007). This has allowed large datasets to be interpreted and handled in a routine manner (Hamilton and Teasdale, 2008).

Imaging systems

The automation of fluorescent microscopes has aided reducing the time needed for image acquisition in HT experiments. Initial image-based screens utilized standard microscopes to which automation was applied to the stage, filters and light source. Today many automated imaging systems are available that are able to integrate with automated plate handling systems as well as providing controlled chambers for imaging. A few of the commercially available automated imaging systems are listed in Table 5.3.

Imaging parameters

There are several factors to consider when selecting an imaging platform for use with the HCS experiment; sensitivity of detection, wavelengths available on the imaging platform, speed of acquisition and specimen viability. In wide-field microscopy systems a mercury or xenon light

Table 5.3 Examples of currently available automated imaging systems (Zanella *et al.*, 2010)

Company	Product	Type	Website
BD Biosciences	BD Pathway™ 435	WF/CF	www.bdbiosciences.com
Leica Microsystems	TCS SP5	CF	www.leica-microsystems.com
Molecular Devices	ImageXpress® ULTRA	CF	www.moleculardevices.com
	ImageXpress® MICRO	WF	www.moleculardevices.com
Perkin Elmer	Opera™	CF	www.perkinelmer.com
	Operetta™	WF/CF	www.perkinelmer.com
Thermo Scientific™	Cellomics ArrayScan™ VTI	WF/CF	www.cellomics.com
	CellWoRx™	WF	www.tekontech.com
TTP LabTech	Acumen eX3®	WF	www.ttplabtech.com

WF, widefield; CF, confocal.

source is used and the entire field-of-view is illuminated and all fluorescence emitted from the sample is captured on the detector. In confocal microscopy systems a laser light source is used and out-of-focus light is blocked from the detector using a confocal pinhole. Another important factor in determining the imaging system that is required is the resolution required to differentiate the features of interest in the sample. Often to improve the speed of acquisition objectives with lower magnification are used and this affects the length of exposure as well as the focus plane. One of the most important features in a high-throughput imaging system is having an autofocus module; this ensures the correct focal plane is maintained throughout the imaging experiment. The sensitivity of the detector also influences the length of exposure and as a result the speed of acquisition.

Data management

The effective integration of data storage, data analysis and data accessibility represents a major bottleneck in HCS (Zanella *et al.*, 2010). HCS approaches typically generate large volumes of image-based data. A HCS in which 4 images (2 megabytes each) are acquired per well in two channels across 60 wells of a microwell plate would produce ~1 gigabyte of data per plate. Genome or even miRNome-wide screens utilize hundreds of micro-well-plates and therefore require high fidelity data storage that should be backed up on multiple redundant drives and fail-safe storage solutions. HCS image data should also ideally be hierarchically structured and annotated so that images are readily searchable by treatment (e.g. MiRNA), plate (e.g. mimic plate 1) and screen (e.g. HIV screen replicate 1). Logical indexing and labelling of HCS-derived images is also important in terms of downstream image analysis and the ease with which these analyses can be accessed. The use of integrated data management and analysis systems can dramatically streamline HCS result analysis by improving the overall user experience; thus resulting in increased productivity. There are a number of commercially available data management solutions such as MDCStore (Molecular Devices, USA) and Activity Base (IDBS) as well open source alternatives such as Open Microscopy Environment Repository Objects (OMERO). The Open Microscopy Environment (OME) has developed an open-source software package that uses client-server-based integration for microscopy data storage and annotation management (Allan *et al.*, 2012). This client-server-based software, OMERO, is specifically designed to aid the analysis of large datasets, as those generated in HT or HC experiments. The OMERO configuration is built on open-source technologies allowing the import of data from a large number of acquisition sources, making it versatile for use with many existing imaging systems (already highlighted in Table 5.3). OMERO provides users with data management features via a web browser or its java-based desktop applications which combine the required analysis and storage resources in a user-friendly, interactive interface (Allan *et al.*, 2012). In terms of integrated image analysis, OMERO has an Application Programming Interface (API) to several existing image analysis packages. A few of these include Matlab, Python-based applications (CellProfiler) or Java-based applications (e.g. FIJI, ICY). ICY is an image analysis software package that, in addition to its image modification and analysis modules, integrates the microscope hardware control platform, Micro-Manager (de Chaumont *et al.*, 2012). Data management solutions therefore provide a valuable interface that allows the biologist to efficiently engage with HCS data and its subsequent analysis.

Data analysis

HCS analyses rely on the discrimination of multiple biological descriptors (e.g. GFP intensity, cell size/shape) between images derived from experimental and control wells for HIT identification. Image-based data are susceptible to systematic errors and biological 'noise' and rigorous quality control measures are thus required in order to confidently identify wells/treatments of interest. An example of a bespoke Support Vector Machine (SVM) based pattern recognition image analysis pipeline will be discussed in greater detail in this section.

Initial requirements

As discussed in the previous section on data management, the ability to index and label screen images is crucial for efficient data analysis. The image repository should have a standard system of grouping images for a well, plate and the entire screen. The repository should also have links to a relational database that stores image metadata and labels. In order to aid downstream image processing, the microscope imaging settings need to be fixed throughout the entire screen. HCS is characterized by the aggregation of measurements across a number of plates to determine a threshold for HIT identification. Imaging instrument calibration is thus pivotal to ensure a minimized by potential noise propagation in measurement taking across multiple plates that constitutes a screen. For example, an instrument without fixed calibration for a unit length will have low fidelity in length-based measurements across multiple plates. The efficient extraction of multiple measurements from the images that constitute experimental wells (feature extraction) requires the computing power of a multiple core system as utilizing high performance computing setups and distributed computing.

Feature extraction

HCS analyses extract quantitative subcellular measurements (features) from the acquired images and then identify subsets of these features that can accurately and reproducibly classify specific biological phenotypes within experimental wells. Essentially, any cellular morphological feature can be used as long it permits the extraction of descriptive measurements from the image data (e.g. average size of nuclei and fluorescence intensity). The selection of features should however be carefully considered and be representative of specific phenotypes, which are relevant to the biological model being investigated. The average distances between individual cells and the average length of cell boundaries for example could be used to quantitatively assess syncytia formation in HIV-infected wells (Genevesio *et al.*, 2011).

The use of invariant features improves the performance of classification approaches. Objects in the view field of an imaging instrument are typically not ordered in any manner. The measurements extracted from each image/well should thus not be affected by the relative position of objects in a well and should be independent of common geometric changes like scaling, translation and orientation. The length of the major axis of a cell for example, is dependent on factors such as magnification and pixel size but the ratio of the major and minor axes (eccentricity) is not. Eccentricity would thus be a more robust cell shape feature to use in comparison to major axis length. The feature extraction step involves the utilization of an algorithmic code to extract a set of quantitative measurements related to a specific set of features from all images within a given data set. The algorithm is thus able to generate an average measurement for each feature on a per well basis.

The phenotypic control (e.g. siCD4) and experimental control (positive infection control and uninfected control) data sets from across the HCS are normalized and scaled to ensure that the feature-based measurements across the screen have similar data ranges as well as to normalize the data (Xia and Wong, 2012; Jackson *et al.*, 2010). By performing standard normalization to the data obtained from a HCS an unbiased evaluation of all potential measurements per well is promoted while minimizing the impact of outliers on subsequent analyses (Theodoridis and Koutroumbas, 2003).

Feature set reduction

Feature subset selection involves the selection of the best or most descriptive subset of features. The normalized feature measurement values can be used in the feature reduction step that seeks to exclude features that do not accurately describe the range of phenotypes defined by the control wells, as well as those features that do not add any additional value to the classification of the test images or objects. There are many different approaches that can be used to facilitate feature reduction, e.g. population-based incremental learning (PBIL), branch and bound (B&B) analysis and principal component analysis (PCA).

PBIL is based on evaluating the accuracy of classification for a feature by splitting data into 'training' and 'test' sets and observing how many test objects are classified correctly. This process should be replicated with different data sets and the accuracy of the classification process can be evaluated using the the average result. Greene (2001) proposes the use of Thornton's

Separability Index (SI) to evaluate the merit of specific feature subsets that yields results similar to accuracy of a nearest neighbour classifier obtained with multiple data splits. SI can be used to perform an exhaustive search of the feature space and come up with the best feature subset that describe phenotypes of interest. However, if the number of features is large, an exhaustive search can be computationally expensive. For N features 2^N subsets will be evaluated. PBIL may be used to evaluate the merit of a feature subset based on SI where it is impractical to evaluate every possible subset. Feature subsets are selected probabilistically, resulting in the progressive selection of subsets that yield a higher SI score by using a weighting system for specific features (Baluja, 1994).

The B&B analysis approach uses a monotonic evaluation function to select the best subset of features from the total assessed features (Wolsey, 1980). It models a tree where the root is the set of all features and the leaves are subsets of features. The evaluation function is used to follow the path with the highest evaluation score leading to the identification of the best subset. The disadvantage of the B&B approach is that it generally requires longer processing times however Somol *et al.* (2004) describe a B&B algorithm that does not compute all evaluations but simply predicts selected evaluations to reduce the processing time required.

PCA can be used to reduce the dimensionality of phenotypic control wells that are used to train a classifier. A drawback of PCA is that it is limited to the analysis of square matrices. Conversely, singular value decomposition (SVD) can be performed on matrices of any size. While PCA yields true eigenvectors of a square input matrix, SVD returns singular values. SVD translates true features/measurements to a new feature space, where features are orthogonal to each other. Features corresponding to the largest eigenvalues capture maximum variance across the set of features. In the case of the HIV-miRNA screen being discussed, SVD feature space reduction would facilitate the separation between phenotypic controls in multidimensional space, e.g. the separation of the siCD4 inhibition phenotype from the positive infection control phenotype.

Support vector machines and phenotypic controls

A key consideration when employing SVM analysis is that the classifier is limited by the quality of the training data. The use of non-discriminate features will lead to underperformance of the algorithm, resulting in ill-defined decision boundaries. It is therefore essential that the features selected for use with the SVM algorithm can readily be distinguished by computational analyses and that the phenotypic control images used to train the SVM have distinct readouts with regard to the features being extracted. Using a two-class (two feature) linearly separable classification task, the SVM algorithm establishes a hyper-plane/boundary that differentiates the feature space into classes (Theodoridis and Koutroumbas, 2003), as seen in Fig. 5.8. In cases where the feature values from each control dataset are not linearly separable, a hyper plane is generated that produces a minimum number of misclassified objects from each class/dataset. This constrained optimization problem is solved using Lagrange multipliers. By using a dual Langrangian formulation, the training objects (positive and negative controls) appear as pairs when computing the inner product of the optimization (Theodoridis and Koutroumbas, 2003). The inner product of the two vectors (measurements for each dataset for each feature used) in multi-dimensional space can be expressed as a function of the inner product of the corresponding vectors in the original space (Theodoridis and Koutroumbas, 2003). An example showing the decision boundary of a SVM between the positive (infection control) and negative (siCD4) phenotypic controls for HIV infection is presented in Fig. 5.8 in two-dimensional space.

HIT identification

In terms of the HIT identification using the above-mentioned SVM algorithm, images from the experimental wells are compared to the control phenotype hyper sphere planes as described above based on their proximity to a specific control space. In the example of the HIV-miRNA screen, experimental miRNAs will be classified as inhibitory HITs if they are classified within the siCD4 hyper sphere whereas miRNAs that are classified within the positive infection control

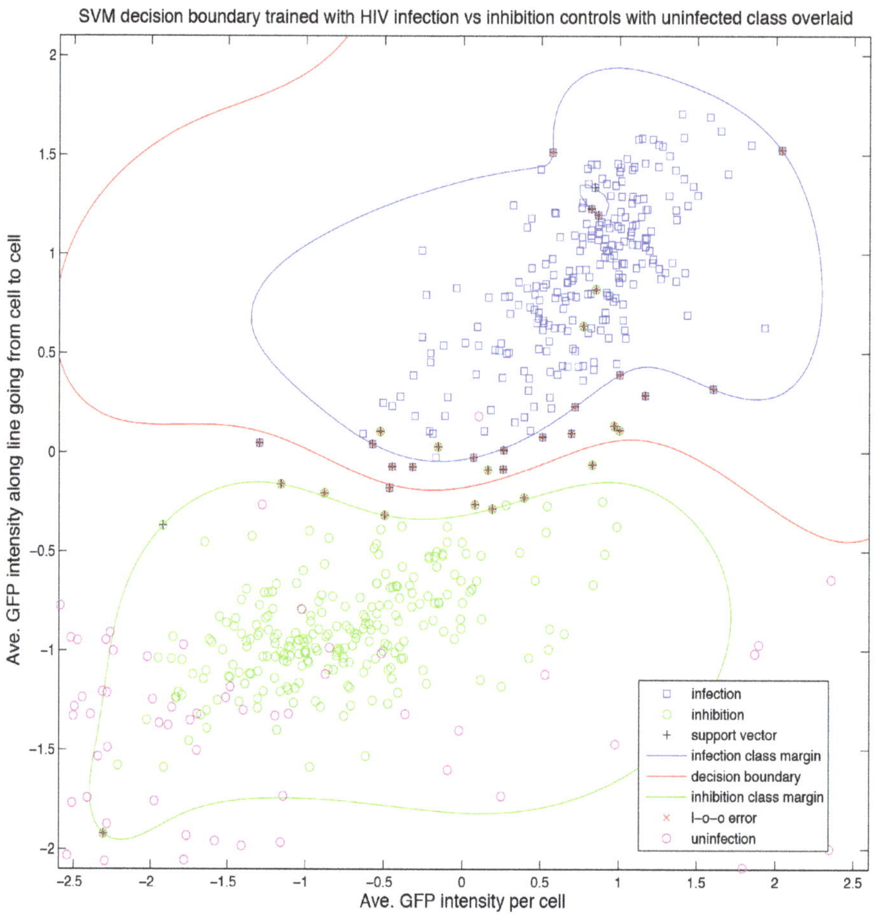

Figure 5.8 Graphic indicating SVM decision boundary for phenotypic controls. HIV positive infection control (blue) and suppressed infection via siCD4 (green).

hyper sphere will be deemed to have no effect on HIV replication (Fig. 5.9). Unlike in HTS, where a HIT is identified by a predetermined cut-off value (e.g. 3 std. deviations above/below the control value); the HCS threshold for HIT identification is derived from biologically relevant phenotypes as defined by classifiers derived from control images. SVM outputs the Euclidean distance of a miRNA HIT from the hyper sphere area, the higher the distance, the more confidence can be derived that the miRNA HIT is correctly classified. The Euclidean distance output can thus also be used to rank HITs in order of their phenotypic similarity to a particular control dataset (e.g. degree of inhibition of HIV infection relative to the siCD4 control). The concordance of HITs between replicate rounds of screening is an important factor in determining the degree of confidence in HIT prediction. A minimum of three screening replicates is required to evaluate the statistical significance of HIT predictions between screen replicates and, such as, the greater the number of replicates, the greater the degree of confidence in HIT prediction.

Quality controls

HT screening approaches, including HCS, require relevant quality control measures. Z'-factor, signal to background ratio and the per cent coefficient of variation are examples of typical HT statistical inter-plate quality controls. The robust per cent coefficient of variation is derived on a per plate basis from the median value and the Median Absolute Deviation (MAD) for each biological

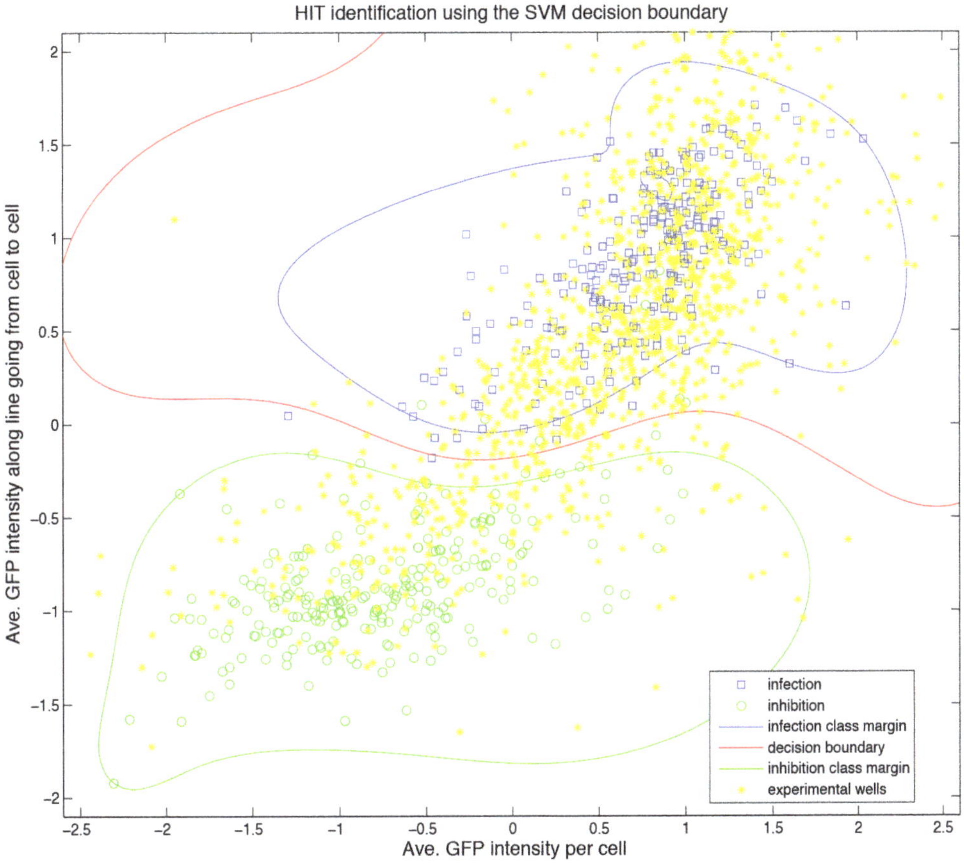

Figure 5.9 HIT identification using SVM decision boundary. HIV positive infection control (blue) and suppressed infection via siCD4 (green). MiRNAs/experimental wells to be classified (yellow).

parameter of interest e.g. GFP signal intensity or transfection efficiency. The MAD of the normalized GFP intensity value, for example, is derived for each well by fitting the GFP intensities to a 2-way median polish algorithm (Gunter *et al.*, 2003). The MAD GFP intensity values can then be used to monitor trends within and between experimental plates. The MAD GFP intensity values can reveal consistently higher readings in the outer wells of experimental plates as compared to the innermost wells or a trend of diminishing GFP signal across the columns of an experimental microwell plate. By normalizing all descriptors utilized in a HCS, plate-specific biases for features of interest can be minimized. The utilization of appropriate quality/normalization controls is thus essential for the accurate comparison and classification of experimental wells across a multi-plate screen.

MiRNA 'HITS' to target identification

Experimental target identification approaches

There are currently a number of experimental methods used for the target identification of small subsets of or individual miRNA HITs (reviewed by Kuhn *et al.*, 2008; Sanchez-Martinez and Murphy, 2013). RNA expression profiling techniques, such as qRT-PCR arrays and Transcriptome sequencing, can be used to assess differential mRNA expression levels in response to the transfection of specific miRNA mimics or inhibitors (Gupta *et al.*, 2011). Given that the mechanism of miRNA-mediated repression does not always involve degradation of target mRNAs however, mRNA expression profiling techniques may not be the most comprehensive approach to miRNA target identification (Kuhn *et al.*, 2008).

Protein expression profiling techniques such protein microarrays (ProtoArray®, Life Technologies; Panorama® Protein Array, Sigma-Aldrich) and Western blot analysis on the other hand represent a more comprehensive approach to infer miRNA targets. Both qRT-PCR and protein technologies are however currently very expensive and represent indirect methods of miRNA target identification as they are not able show a direct interaction between miRNA and target. The use of cloned 3′ UTR reporter systems, tagged-miRNAs and RISC pull down experiments represent direct methods for miRNA target identification as reviewed by Sanchez-Martinez and Murphy (2013). While these approaches represent the 'gold standard' in miRNA target validation the application of these methods to large-scale miRNA-target validations is not always feasible.

Prediction tools/*in silico* analysis

There are a number of open source, web-based bioinformatics prediction tools that can be used to retrieve miRNA targets (Table 5.4). The use of computational analyses represents a non-labour intensive and economical approach for rapidly identifying numerous putative miRNA targets (reviewed by Huang *et al.*, 2010). Computational approaches have been used as a starting point for miRNA target identification with many studies going on to use experimental approaches to verify targets of interest that were computationally predicted. There are a number of available computational algorithms to aid in the prediction of the miRNA–mRNA interaction. These algorithms identify candidate mRNAs by taking into account the seed sequence of the miRNA, the available binding sites in the 3′ UTR of the mRNA sequence and various binding paremeters. While current prediction algorithms are being constantly refined to improve their prediction accuracy, some validated miRNA–mRNA interactions still cannot be predicted using computational methods (Rigoutsos, 2009; Nelson *et al.*, 2010). Additionally, given the promiscuous nature of miRNA binding and variable secondary structure conformations of miRNA molecules, the ability of computational prediction tools to accurately rank the strength of miRNA–mRNA interactions has been an issue of contention within the field (Nelson *et al.*, 2010).

Predicted miRNA–mRNA interactions alone, while useful, are by no means a 'gold standard' and the poor concordance between target prediction and target ranking across different prediction tools highlights the need for more robust algorithms that can accurately characterize the complexity of miRNA–target interactions. An alternative approach to miRNA-target identification is to use published, experimentally validated miRNA–target interactions. Online databases such as TarBase (http://diana.imis.athena-innovation.gr/DianaTools/index.php?r=tarbase/index) (Sethupathy *et al.*, 2006) and miRecords (http://mirecords.biolead.org/) (Xiao *et al.*, 2009) have been established which provide a searchable and user-friendly interface to identify experimentally validated miRNA target interactions. These databases hold great potential but are limited by poor coverage of known miRNAs, with each database only able to provide validated targets for less than 30% of currently identified human miRNAs (miRBase v.19.0). A simple internet search engine approach to miRNA target identification

Table 5.4 Online miRNA target identification resources

Name	Type	Website
Target Scan	Prediction tool	http://www.targetscan.org
miRanda	Prediction tool	http://www.microrna.org/microrna/home.do
miRDB	Prediction tool	http://mirdb.org/miRDB/
DIANA-microT	Prediction tool	http://diana.imis.athena-innovation.gr/DianaTools/index.php?r=microT_CDS/index
Tarbase	Validated database	http://diana.imis.athena-innovation.gr/DianaTools/index.php?r=tarbase/index
miRecords	Validated database	http://mirecords.biolead.org/

highlights the need for these databases to be updated more frequently in order to accurately reflect the current range of experimentally validated miRNA-interactions. As the experimental validation of large numbers of miRNA targets remains expensive and current *in silico* analyses require revision; the importance of updated online databases that catalogue all published miRNA–target interactions will be undoubtedly become increasingly relevant as the demand for large-scale miRNA-target identification becomes more prevalent in the current literature.

Pathway analysis

The enrichment of miRNA targets within specific or related functional pathways has highlighted the importance of individual miRNAs and miRNA networks in the endogenous regulation of specific functional pathways (Gennarino *et al.*, 2012). The use of pathway analysis to directly interrogate miRNA screening 'HITs' thus provides an important additional avenue for the interpretation of miRNA screen data. Current pathway analysis tools such Ingenuity Pathway Analysis (http://www.ingenuity.com/products/ipa), PathVisio (http://www.pathvisio.org, open source) and DIANA miRPath (http://diana.imis.athena-innovation.gr/DianaTools/, open source) allow for the direct evaluation of specific pathways involved in miRNA function, combining bioinformatics-based target prediction and KEGG pathway analyses in a succinct, user friendly interface. The DIANA miRPath application has a very useful feature that allows for pathway analyses to be completed using either prediction analysis or the TarBase experimentally validated database. An excellent example of the use of pathway analysis to complement miRNome-wide data is presented in a study by Gupta *et al.* (2011). Using the Ingenuity Pathway Analysis application, the authors were able to uncover specific miRNA-mediated functional pathways that were preferentially regulated in response to HIV infection. Direct pathway analysis using miRNA 'HITs' may thus represent an extremely powerful tool for the extrapolation of screen data but it should be noted that many of these approaches are still dependent on bioinformatics-based miRNA target prediction tools. The use of pathway analysis following the experimental validation of miRNA 'HITs' represents a more dependable method that can readily complement traditional visual-based miRNA screen analyses.

Validation experiments

Secondary screens are commonly required in HT screening approaches in order to identify and exclude cytotoxic experimental treatments so that they are not falsely categorized by HIT identification analyses (Santhakumar *et al.*, 2010). The confounding effect of cytotoxicity on HIT identification is not an issue of concern for HC visual-based screens. HCS image analyses can directly identify and exclude cytotoxic treatments by defining a minimum threshold or limit relative to the average number of cells present in control wells. Secondary screens can however still be relevant in HCS approaches and can be used to further interrogate specific cellular phenotypes/processes of interest identified by primary screens. These types of secondary screens can be conducted at a smaller experimental scale, focusing on a custom set of miRNAs of interest (e.g. on all HITs with inhibition phenotypes). By employing additional descriptive markers that were not used in the primary screen, which provide novel information on cellular interactions; secondary screens can be exploited to provide greater insight into the molecular mechanisms by which miRNAs and pathways identified in the primary screens are able to modulate HIV infection.

In terms of functional screens using siRNAs, miRNAs or compound libraries, the general practice has been to automatically exclude treatments that are cytotoxic from the subsequent analyses (Brass *et al.*, 2008; Santhakumar *et al.*, 2010; Genovisio *et al.*, 2011). A potentially useful secondary screen approach would be to re-evaluate cytotoxic treatments identified in the primary screen in a secondary screen in which cells are transfected but not infected. This approach holds the potential to identify a significant phenotypic class of miRNAs when considered in comparison with the primary screen data. This phenotypic class would refer to miRNAs that are potentially able to discretely render HIV infected cells susceptible to cytotoxicity and may thus hold tremendous therapeutic potential. The use of this type of secondary screen can thus uncover phenotypes that may have been overlooked by traditional validation experiments and analyses and can also be combined with functional fluorescent markers in order validate the relevance of miRNAs to specific pathways/processes of interest under endogenous conditions.

The high degree of variability in miRNA expression profiles between cell lines necessitates that screen HITs recovered from genetically modified or carcinoma-derived cell lines be validated in physiologically relevant models (Zanella *et al.*, 2010; Witwer *et al.*, 2012;). The efficient introduction of miRNA molecules into primary cell lines (e.g. $CD4^+$ T-cells) may require the use of more cumbersome approaches (e.g. lentiviral-mediated miRNA delivery), changes to HIV replication readout assays and changes to experimental design in general.

Conclusions and future prospects

The interrogation of the host miRNome in models of infection is now readily accessible due the transfer of many advanced gene expression technologies to miRNA platforms. Many studies have evaluated the role of host miRNAs in response to infection by pathogens, including HIV (Gupta *et al.*, 2011; Eulalio *et al.*, 2012; Witwer *et al.*, 2012). These studies have provided invaluable insight into the relationship between infection and the host miRNome.

The advent of automated microscopy techniques has unlocked the full potential of HT screens allowing for the accumulation of large amounts of image-derived biological data in a high throughput manner. In order to fully capitalize on this technology there are three basic requirements, namely, a cell based assay, an automated imaging system and a data management and analysis solution. By exploiting HCS methodologies it is now possible to readily address functional studies at the scale of the entire genome or miRNome. Numerous functional siRNA/shRNA based screens have showcased the high impact of these approaches in the identification of host dependency factors (Reviewed by Pache *et al.*, 2011). These studies have also highlighted the potential role of the host non-coding complement in models of infection. Despite the success of similar siRNA-based approaches in the elucidation of HIV dependency factors, a miRNA homologue of these screens is conspicuously absent from the current literature. In this chapter we have discussed some of the basic framework associated with visual-based miRNA screening with regard to HIV infection.

It is important to note however that while functional screening approaches are powerful tools; they are not without their shortcomings. A major obstacle in genome-scale screening is experimental background and noise. Increasing the number of replicates per screen can reduce the variance between identical samples; however, for practical considerations such as workload and cost, this is not always feasible. Unlike siRNA/shRNA that target specific RNA sequences, each miRNA is predicted/known to have numerous targets, making off targets effects, even if existent, much more difficult to track in miRNA screens. An additional complexity in miRNA-based approaches lies in the interpretation of global miRNA expression and functional phenotypes in context of highly convoluted models of RNA-mediated regulation as proposed by Salmena *et al.* (2011). In infectious models this can be further complicated by the impact of viral encoded miRNAs on the host transcriptional landscape (Umbach and Cullen, 2009; Pfeffer *et al.*, 2009).

miRNome-wide approaches in general and miRNA screens in particular represent a powerful primary experimental approach to address research questions on a global scale. The data obtained from screening approaches must be subject to rigorous validation steps and can present many different potential avenues for further research. Santhakumar *et al.* (2010) compared functional data with miRNA expression analysis, in order to find individual miRNAs involved in viral infection. They found that those miRNAs which exhibited the most significant antiviral properties from the screen data, were also among a small subset of miRNAs, which were down regulated in both murine and human cytomegalovirus infections (Santhakumar *et al.*, 2010). This supports the idea that the investigation of miRNA function in animal models is relevant to human disease (Santhakumar *et al.*, 2010) and implies that we can readily apply screening approaches to identify host miRNAs that are relevant to viral infection and which can be exploited for therapeutic or diagnostic application.

Standardization of methodologies, improved imaging systems and advances in computational technologies will improve the reliability of the HCS approach to genome-scale experimentation (Pache *et al.*, 2011). These advances will also undoubtedly reconcile contradicting information derived from transcriptomic and microarray based approaches (Hogenesch *et al.*, 2001; Culhane *et al.*, 2003; Sontrop *et al.*, 2009; Warnat *et al.*, 2005). Additionally, HCS screening approaches hold the potential to definitively critique the functional significance of miRNA expression profiling data by relating fold changes miRNA expression levels to functional phenotypes. Continuous advances in HCS methodology suggest great promise in miRome-wide screens as a tool to uncover host dependency factors involved in a variety of cellular processes and infectious models. A recent study by Chen *et al.* (2010) has described the use of functionalized nanoparticles for cell type specific delivery of miRNAs *in vivo* with efficiencies high enough to mediate biological effects. Studies of this nature further allude to the potential of miRNA-based research in the identification of novel therapeutic approaches relevant to various models of disease.

In summation, functional miRNA screens hold great promise, not only in context of infection and disease biology but also in obtaining greater insight into the inherent complexity of the mechanisms which govern mammalian cell physiology.

References

Ahlquist, P., Noueiry, A.O., Lee, W., Kushner D.B., and Dye, B.T. (2003). Host factors in positive-strand RNA virus genome replication. J. Virol. 77, 8181–8186.

Aza-Blanc, P., Cooper, C.L., Wagner, K., Batalov, S., Deveraux, Q.L, and Cooke, M.P. (2003). Identification of modulators of TRAIL-induced apoptosis via RNAi-based phenotypic screening. Mol, Cell 12, 627–637.

Baluja S. (1994) Population-Based Incremental Learning: A Method for Integrating Genetic Search Based Function Optimization And Competitive Learning. Technical Report CMU-SC-94-163 (Carnegie-Mellon University, Pittsburgh, PA).

Barad, O., Meiri, E., Avniel, A., Aharonov, R., Barzilai, A., Bentwich, I., Einav, U., Gilad, S., Hurban, P., Karov, Y., *et al.* (2004). MicroRNA expression detected by oligonucleotide microarrays : system establishment and expression profiling in human tissues. Genome Res. *14*, 2486–2494.

Bennasser, Y., Le, S.-Y., Yeung, M.L., and Jeang, K.-T. (2004). HIV-1 encoded candidate micro-RNAs and their cellular targets. Retrovirology *1*, 43.

Betel, D., Wilson, M., Gabow, A., Marks, D.S., and Sander, C. (2008). The microRNA.org resource: targets and expression. Nucleic Acids Res. *36*, D149–53.

Blankson, J.N. (2010). Effector mechanisms in HIV-1 infected elite controllers: highly active immune responses? Antiviral Res. *85*, 295–302.

Brass, A.L., Dykxhoorn, D.M., Benita, Y., Yan, N., Engelman, A., Xavier, R.J., Lieberman, J., and Elledge, S.J. (2008). Identification of host proteins required for HIV infection through a functional genomic screen. Science *319*, 921–926.

Buermans, H.P.J., Ariyurek, Y., Van Ommen, G., Den Dunnen, J.T., and 't Hoen, P.A. (2010). New methods for next generation sequencing based microRNA expression profiling. BMC Genomics *11*, 716.

Callahan, L. (1994). HIV-1 virion–cell interactions: an electrostatic model of pathogenicity and syncytium formation. AIDS Res. Hum. Retroviruses *10*, 231–233.

Carpenter, A.E. (2007). Image-based chemical screening. Nat. Chem. Biol. 3, 461–465.

Cecilia, D., KewalRamani, V.N, O'Leary, J., Volsky, B., Nyambi, P., Burda, S. Xu, S., Littman, D.R., and Zolla-Pazner, S. (1998). Neutralization profiles of primary human immunodeficiency virus type 1 isolates in the context of coreceptor usage. J. Virol. 72, 6988–6996.

Chang, S.T., Sova, P., Peng, X., Weiss, J., Law, G.L., Palermo, R.E., and Katze, M.G. (2011). Next-generation sequencing reveals HIV-1-mediated suppression of T cell activation and RNA processing and regulation of noncoding RNA expression in a CD4+ T Cell Line. MBio. *2*, 1–9.

Chang, S.T., Thomas, M.J., Sova, P., Green, R.R., Palermo, R.E., and Katze, G. (2013). Next-generation sequencing of small RNAs from HIV-infected cells identifies phased microRNA expression patterns and candidate novel microRNAs differentially expressed upon infection. MBio. *4*, 1–10.

Chen, Y., Zhu, X., Zhang, X., Liu, B., and Huang, L. (2010). Nanoparticles modified With tumor-targeting scFv deliver siRNA and miRNA for cancer therapy. Mol. Ther. *18*, 1650–1656.

Chertova, E., Chertov, O., Coren, L.V., Roser, J.D., Trubey, C.M., Bess, J.W., Sowder, R.C., Barsov, E., Hood, B.L., Fisher, R.J., *et al.* (2006). Proteomic and biochemical analysis of purified human immunodeficiency virus type 1 produced from infected monocyte-derived macrophages. J. Virol. *80*, 9039–9052.

Chisari, F.V. (2005). Unscrambling hepatitis C virus–host interactions. Nature *436*, 930–932.

Clavel, F., and Hance, A.J. (2004). HIV Drug Resistance. N. Engl. J. Med. *350*, 1023–1035.

Creighton, C.J., Reid, J.G., and Gunaratne, P.H. (2009). Expression profiling of microRNAs by deep sequencing. Brief. Bioinform. *10*, 490–497.

Culhane, A.C., Perriere, G., and Higgins, D.G. (2003). Cross-platform comparison and visualization of gene expression data using co-inertia analysis. BMC Bioinformatics *4*, 59.

Elbashir, S.M., Harborth, J., Lendeckel, W., Yalcin, A., Weber, K., and Tuschl, T. (2001) Duplexes of 21-nucleotide RNAs mediate RNA interference in cultured mammalian cells. Nature *411*, 494–498.

Eulalio, A., Schulte, L.N., and Vogel, J. (2012). The mammalian microRNA response to bacterial infections. RNA Biol. *9*, 742–750.

Fellay, J., Shianna, K.V., Ge, D., Colombo, S., Ledergerber, B., Weale, M., Zhang, K., Gumbs, C., Castagna, A., Cossarizza, A., *et al.* (2007) A whole genome association study of major determinants for host control of HIV-1. Science *317*, 944–947.

Frankel, D., and Young, J.A. (1998). HIV-1: fifteen proteins and an RNA. Annu. Rev. Biochem. *67*, 1.

Fukunaga, K. (1990). Introduction to Statistical Pattern Recognition. (Academic Press, San Diego).

Gennarino, V.A., D'Angelo, G., and Dharmalingam, G. (2012). Identification of microRNA-regulated gene networks by expression analysis of target genes. Genome Res. *22*, 1163–1172.

Genovesio, A., Kwon, Y.J., Windisch, M.P., Kim, N.Y., Choi, S.Y., Kim, H.C., Jung, S., Mammano, F., Perrin, V., Boese, A.S., *et al.* (2011). Automated genome-wide visual profiling of cellular proteins involved in HIV infection. Journal of Biomolecular Screening. *16*, 945–958.

Gentleman, R.C., Carey, V.J., Bates, D.M., Bolstad, B., Dettling, M., Dudoit, S., Ellis, B., Gautier, L., Ge, Y., Gentry, J., *et al.* (2004). Bioconductor: open software development for computational biology and bioinformatics. Genome Biol. *5*, R80.

Git, A., Dvinge, H., Salmon-Divon, M., Osborne, M., Kutter, C., Hadfield, J., Bertone, P., and Caldas, C. (2010). Systematic comparison of microarray profiling, real-time PCR, and next-generation sequencing technologies for measuring differential microRNA expression. RNA *16*, 991–1006.

Goff, S. (2007). Host factors exploited by retroviruses. Nat. Rev. MicroBiol. *5*, 253.

Greene, J. (2001). Feature subset selection using thornton's separability index and its applicability to a number of sparse proximity-based classifiers. Paper presented at: Twelfth Annual Symposium of the South African Pattern Recognition Association. (Franschoek, South Africa).

Griffiths-Jones, S., Saini, H.K., Van Dongen, S., and Enright, A.J. (2008). miRBase: tools for microRNA genomics. Nucleic Acids Res. *36*, D154–8.

Gunter, B., Brideau, C., Pikounis, B., and Liaw, A. (2003). Statistical and graphical methods for quality control determination of high-throughput screening data. J. Biomol. Screen. *8*, 624–633.

Gupta, A., Nagilla, P., Le, H.-S., Bunney, C., Zych, C., Thalamuthu, A., Bar-Joseph, Z., Mathavan, S., and Ayyavoo, V. (2011). Comparative expression profile of miRNA and mRNA in primary peripheral blood mononuclear cells infected with human immunodeficiency virus (HIV-1). PloS One *6*, e22730.

Hamilton, N. A., and Teasdale, R.D. (2008). Visualizing and clustering high throughput sub-cellular localization imaging. BMC Bioinformatics *9*, 81.

Hao L, Sakurai A, Watanabe T, Sorensen E, Nidom CA, *et al.* (2008). Drosophila RNAi screen identifies host genes important for influenza virus replication. Nature *454*, 890–893.

Hayes, A.M., Qian, S., Yu, L., and Boris-Lawrie*1*, K. (2011). Tat RNA silencing suppressor activity contributes to perturbation of lymphocyte miRNA by HIV-1. Retrovirology *8*, 36.

Hogenesch J.B., Ching K.A., Batalov S., Su A.I, Walker J.R., Zhou Y., Kay S.A., Schultz P.G., and Cooke, M.P. (2001). A comparison of the Celera and Ensembl predicted gene sets reveals little overlap in novel genes. Cell *106*, 413–415.

Hopkins, A.L. and Groom, C.R. (2002). The druggable genome. Nat. Rev. Drug. Discov. *1*, 727–730.

Houzet, L., and Jeang, K. (2011). MicroRNAs and human retroviruses. Biochim. Biophys. Acta. (BBA)-Gene. *1809*, 686–693.

Houzet, L., Yeung, M.L., De Lame, V., Desai, D., Smith, S.M., and Jeang, K.T. (2008). MicroRNA profile changes in human immunodeficiency virus type 1 (HIV-1) seropositive individuals. Retrovirology *5*, 118.

Huang, J., Wang, F., Argyris, E., Chen, K., Liang, Z., Tian, H., Huang, W., Squires, K., Verlinghieri, G., and Zhang, H. (2007). Cellular microRNAs contribute to HIV-1 latency in resting primary CD4+ T lymphocytes. Nat. Med. *13*, 1241–1247.

Jackson, A.L., Burchard, J., Leake, D., Reynolds, A., Schelter, J., Guo, J., Johnson, J.M., Lim, L., Karpilow, J., Nichols, K., *et al.* (2006). Position-specific chemical modification of siRNAs reduces "off-target" transcript silencing. RNA *12*, 1197–1205.

Jackson, D., Lenard M., Zelensky, A., Shaikh, M., Scharpf, J., Shaginaw, R., Nawade, M., Agler, M., Cloutier, N., Fennell, M., *et al.* (2010). HCS road: an enterprise system for integrated HCS data management and analysis. J. Biomol. Screen. *15*, 882–891.

Jones, T.R., Carpenter, A.E, Golland, P., and Sabatini, D.M. (2008) Methods for high-content, high throughput image-based cell screening. In Microscopic Image Analysis for Life Science Applications, Rittscher J, Machiraju R, Wong STC, eds. (Artech House Publishers, Norwood, MA), pp. 209–221.

Kadri, S., Hinman, V.F., and Benos, P. V (2011). RNA deep sequencing reveals differential microRNA expression during development of sea urchin and sea star. PloS One *6*, e29217.

Kar-roy, A., Dong, W., Michael, N., and Li, Y. (2000). Green fluorescence protein as a transcriptional reporter for the long terminal repeats of the human immunodeficiency virus type 1. J. Virol. Methods. *84*, 127–138.

Kim, Y.G., Raunser, S., Munger, C., Wagner, J., Song, Y.L., Cygler, M., Walz, T., Oh, B.H., Sacher, M. (2006). The architecture of the multisubunit TRAPP I complex suggests a model for vesicle tethering. Cell. *127*, 817.

Kong, W., Zhao, J.-J., He, L., and Cheng, J.Q. (2009). Strategies for profiling microRNA expression. J. Cell. Physiol. *218*, 22–25.

Konig, R., Zhou, Y., Elleder, D., Diamond, T.L., Bonamy, G.M., *et al.* (2008). Global analysis of host–pathogen interactions that regulate early-stage HIV-1 replication. Cell. *135*, 49–60.

Kornberg R.D. (2005). Mediator and the mechanism of transcriptional activation. Trends Biochem. Sci. *30*, 235.

Krishnan, M.N., Ng, A., Sukumaran, B., Gilfoy, F.D., and Uchil, P.D. (2008). RNA interference screen for human genes associated with west Nile virus infection. Nature *455*, 242–245.

Linsen, S.E., de Wit, E., Janssens, G., Heater, S., Chapman, L., Parkin, R.K., Fritz, B., Wyman, S.K., de Bruijn, E., Voest, E.E., Kuersten, S., and Cuppen, E. (2009). Limitations and possibilities of small RNA digital gene expression profiling. Nature Methods. *6*, 474–476.

Malone, J.H., and Oliver, B. (2011). Microarrays, deep sequencing and the true measure of the transcriptome. BMC Biol. *9*, 34.

Mardis, E.R. (2008). The impact of next-generation sequencing technology on genetics. Trend Gene. *24*, 133–141.

Mizuguchi, Y., Mishima, T., Yokomuro, S., Arima, Y., Kawahigashi, Y., Shigehara, K., Kanda, T., Yoshida, H., Uchida, E., Tajiri, T., *et al.* (2011). Sequencing and bioinformatics-based analyses of the microRNA transcriptome in hepatitis B-related hepatocellular carcinoma. PloS One *6*, e15304.

Morner, A., Bjorndal, A., Albert, Kewalramani, V.N, Littman, D.R., Inoue, R., Thorstensson, R., Fenyo, E.M., and Bjorling, E. (1999). Primary human immunodeficiency virus type 2 (HIV-2) isolates, like HIV-1 isolates, frequently use CCR5 but show promiscuity in coreceptor usage. J. Virol. *73*, 2343–2349.

Motameny, S., Wolters, S., Nürnberg, P., and Schumacher, B. (2010). Next generation sequencing of miRNAs – strategies, resources and methods. Genes *1*, 70–84.

Narayanan, A., Kehn-Hall, K., Bailey, C., and Kashanchi, F. (2011). Analysis of the roles of HIV-derived microRNAs. Expert Opin. Biol. Ther. *11*, 17–29.

Nelson, P.T., Kiriakidou, M., Mourelatos, Z., Tan, G.S, Jennings, M., Xie, K., Wang, W. (2010) High-throughput experimental studies to identify miRNA targets directly, with special focus on the mammalian brain. Brain Res. *1338*, 122–130.

Orth, A.P., Batalov, S., Perrone, M., Chanda, S.K. (2004) The promise of genomics to identify novel therapeutic targets. Expert Opin. Ther. Targets *8*, 587–596.

Pache, L., König, R., and Chanda, S.K. (2011) Identifying HIV-1 host cell factors by genome-scale RNAi screening. Methods. *53*, 3–12.

Paddison, P.J., Caudy, A.A., Bernstein, E., Hannon, G.J., Conklin, D.S. (2002). Short hairpin RNAs (shRNAs) induce sequence-specific silencing in mammalian cells. Genes Dev. *16*, 948–958.

Parameswaran, P., Sklan, E., Wilkins, C., Burgon, T., Samuel, M.A. (2010) Six RNA viruses and forty-one hosts: Viral small RNAs and modulation of small RNA repertoires in vertebrate and invertebrate systems. PLoS Pathogens. *6*, e1000764.

Pfeffer, S., Zavolan, M., Grässer, F.A., Chien, M., Russo, J.J., Ju, J., John, B., Enright, A.J., Marks, D., Sander, C., and Tuschl, T. (2004) Identification of virus-encoded microRNAs. Science *304*, 734–736.

Platt, E.J., Kozak, S.L., Durnin, J.P, Hope, TJ, Kahat, D. (2010) Rapid dissociation of HIV-1 from cultured cells severely limits infectivity assays, causes the inactivation ascribed to entry inhibitors, and masks the inherently high level of infectivity of virions. J. Virol. *84*, 3106–3110.

Rajewsky, N. (2006). MicroRNA target predictions in animals. Nature Gene. *38*, S8–13.

Rigoutsos, I (2009). New tricks for animal microRNAS: targeting of amino acid coding regions at conserved and nonconserved sites. Cancer Res. *69*, 3245–3248.

Rines, D.R., Tu, B., Miraglia, L., Welch, G.L. and Zhang, J., Hull, M.V., Orth, A.P. and Chanda, S.K., (2006). High-content screening of functional genomic libraries. Methods in Enzymology. *414*, 530–565.

Rusert, P., Fischer, M., Joos, B., Leemann, C., Kuster, H., Flepp, M., Bonhoeffer, S., Gunthard, H. F., and Trkola, A. (2004). Quantification of infectious HIV-1 plasma viral load using a boosted *in vitro* infection protocol. Virology. *326*, 113–129.

Salmena, L., Poliseno, L., Tay, Y., Kats, L., and Pandolfi, P.P. (2011) A ceRNA hypothesis: the rosetta stone of a hidden RNA language? Cell, *146*, 353–358.

Santhakumar, D., Forstern, T., Lagtom, N.N., Fragkoudis, R, Dickinson, P., Abreu-Goodger, C., Manakov, S.A., Choudhury, R., Griffiths, S.J., Vermuelen, A., *et al.* (2010). Combined agonist–antagonist genome-wide functional screening identifies broadly active antiviral microRNAs. PNAS. *107*, 13830–13835.

Sethupathy, P., Corda, B., and Hatzigeorgiou, A.G. (2006). TarBase: A comprehensive database of experimentally supported animal microRNA targets. RNA *12*, 192–197.

Shingara, J., Keiger, K., Shelton, J., Laosinchai-wolf, W., Powers, P., Conrad, R., Brown, D., and Labourier, E. (2005). An optimized isolation and labeling platform for accurate microRNA expression profiling. *11*, 1461–1470.

Shun, T., Lazo, J., Sharlow, E., Johnston, P., (2010). Identifying actives from HTS data sets: practical approached for the selection of an appropriate HTS data-processing method and quality control review. J. of Biomol. Screen. *16*, 1–14.

Sisk, J.M., Clements, J.E., and Witwer, K.W. (2012). MiRNA profiles of monocyte-lineage cells are consistent with complicated roles in HIV-1 restriction. Viruses. *4*, 1844–1864.

Somol, P., Pudil, P., and Kittler, J., (2004). fast branch and bound algorithms for optimal feature selection. IEEE transactions on pattern analysis and machine intelligence. *26*, 900–912.

Sontrop, H.M., Moerland, P.D., van den Ham, R., Reinders, M.J., and Verhaegh, W.F., (2009). A comprehensive sensitivity analysis of microarray breast cancer classification under feature variability. BMC Bioinformatics. *10*, 389.

Studamire, B., and Goff, S.P. (2008) Host proteins interacting with the moloney murine leukaemia virus integrase: multiple transcriptional regulators and chromatin binding factors. Retrovirology. *5*, 48.

Sun, G., and Rossi, J.J. (2011) MicroRNAs and their potential involvement in HIV. Trends Pharm. Sci. *32*, 675–681.

Sun, G., Li, H., Wu, X., Covarrubias, M., Scherer, L., Meinking, K., Luk, B., Chomchan, P., Alluin, J., Gombart, A.F., *et al.*, (2012). Interplay between HIV-1 infection and host microRNAs. Nucleic Acids Res. *40*, 2181–2196.

Swaminathan, S., Zaunders, J., Wilkinson, J., Suzuki, K., and Kelleher, A.D. (2009). Does the presence of anti-HIV miRNAs in monocytes explain their resistance to HIV-1 infection? Blood. *113*, 5029–5030; author reply 5030–5031.

Telenti, A. (2009). HIV-1 host interactions: integration of large-scale datasets. Biology Reports. *1*, 71.

Theodoridis, S., and Koutroumbas, K. (2003) Pattern Recognition, 2nd edn (Academic Press, Elsevier, San Diego).

Triboulet, R., Mari, B., Lin, Y., Chable-bessia, C., Bennasser, Y., Lebrigand, K., Cardinaud, B., Maurin, T., Barbry, P., Baillat, V., *et al.* (2007). Suppression of microRNA-silencing pathway by HIV-1 during virus replication. Science *315*, 1579–1582.

Umbach, J.L., Cullen, B.R. (2009) The role of RNAi and microRNAs in animal virus replication and antiviral immunity. Genes and Development. *23*, 1151–1164.

Ungar, D., Oka, T., Krieger, M., and Hughson, F.M. (2006). Retrograde transport on the COG railway. Trend Cell. Biol. *16*, 113.

Vapnik, V., (1998) Statistical Learning Theory (Springer-Verlag, New York).

Vermeulen A., Robertson, B., Dalby, A.B., Marshall, W.S., Karpilow, J., Leake, D., Khvorova, A., and Baskerville, S. (2007) Double-stranded regions are essential design components. RNA *13*, 723–730.

Wang, B., Howel, P., Bruheim, S., Ju, J., Owen, L.B., Fodstad, O., and Xi, Y. (2011). Systematic evaluation of three microRNA profiling platforms: microarray, beads array, and quantitative real-time PCR array. PloS One *6*, 17167.

Wang, W.-C., Lin, F.-M., Chang, W.-C., Lin, K.-Y., Huang, H.-D., and Lin, N.-S. (2009a). miRExpress: analyzing high-throughput sequencing data for profiling microRNA expression. BMC Bioinformatics *10*, 328.

Wang, X., Ye, L., Hou, W., Zhou, Y., Wang, Y.-J., Metzger, D.S., and Ho, W.-Z. (2009b). Cellular microRNA expression correlates with susceptibility of monocytes/macrophages to HIV-1 infection. Blood. *113*, 671–674.

Warnat, P., Eils, R., and Brors, B. (2005). Cross-platform analysis of cancer microarray data improves gene expression based classification of phenotypes. BMC Bioinformatics. *6*, 265.

Wilhelm, B.T., and Landry, J.R. (2009). RNA-Seq-quantitative measurement of expression through massively parallel RNA-sequencing. Methods. *48*, 249–257.

Witwer, K.W., Watson, A.K., Blankson, J.N., and Clements, J.E. (2012). Relationships of PBMC microRNA expression, plasma viral load, and CD4+ T-cell count in HIV-1-infected elite suppressors and viremic patients. Retrovirology. *9*, 5.

Wollman, R., and Stuurman, N. (2007). High throughput microscopy: from raw images to discoveries. J. Cell Sci. *120*, 3715–3722.

Wolsey, L.A. (1980). Heuristic analysis, linear programming and branch and bound. In Combinatorial Optimization II, Rayward-Smith, V.J., ed. (Springer, Berlin Heidelberg), pp. 121–134.

Wu, Q., Lu, Z., Li, H., Lu, J., Guo, L., and Ge, Q. (2011). Next-generation sequencing of microRNAs for breast cancer detection. J. Biomed. Biotechnol. 597145.

Xia, X., and Wong, S.T. (2012) Concise Review: A High-Content Screening Approach to Stem Cell Research and Drug Discovery. Stem Cells. *30*, 1800–1807.

Xiao, F., Zuo, Z., Cai, G., Kang, S., Gao, X., and Li, T. (2009). miRecords: an integrated resource for microRNA–target interactions. Nucleic Acids Res. *37*, 105–110.

Yang, J.H., Shao, P., Zhou, H., Chen, Y.Q., and Qu, L.H. (2010). deepBase: a database for deeply annotating and mining deep sequencing data. Nucleic Acids Res. *38*, 123–130.

Yeung, M.L., Bennasser, Y., Myers, T.G., Jiang, G., Benkirane, M., and Jeang, K.-T. (2005). Changes in microRNA expression profiles in HIV-1-transfected human cells. Retrovirology. *2*, 81.

Yeung, M.L., Houzet, L., Yedavalli, V.S., and Jeang, K.T, (2009) J. Biol. Chem. *284*, 19463–19473.

Yin, J.Q., Zhao, R.C., and Morris, K. V (2008). Profiling microRNA expression with microarrays. Trends Biotechnol. *26*, 70–76.

Zanella, F., Lorens, J.B. and Link, W. (2010). High content screening: seeing is believing. Trends Biotechnol. *28*, 237–245.

Zhang, H., Wu, T.C., Sang, W., Ruan, Z. (2012). miR-217 is involved in Tat-induced HIV-1 long terminal repeat (LTR) transactivation by down-regulation of SIRT1. Biochim. Biophysic. Acta. *1823*(5), 1017–1023.

Zhou H., Xu M., Huang Q., Gates A.T., Zhang X.D., Castle, J.C., Stec, E., Ferrer, M., Strulovici, B., Hazuda, D.J, and Espeseth, A.S. (2008). Genome-scale RNAi screen for host factors required for HIV replication. Cell Host Microbe. *4*, 495–504.

Zhu, E., Zhao, F., Xu, G., Hou, H., Zhou, L., Li, X., Sun, Z., and Wu, J. (2010). mirTools: microRNA profiling and discovery based on high-throughput sequencing. Nucleic Acids Res. *38*, 392–397.

Synthetic MicroRNA Blocking Agents

6

Peter Järver, Adrian G. Torres and Michael J. Gait

Abstract

This chapter describes how synthetic miRNA blocking agents may be used to inhibit miRNA action in cells and *in vivo*. Particular emphasis is given to chemically synthesized oligonucleotide anti-miRs, types of synthetic analogues, mechanisms of action, and methods used to validate their activities. Also included are brief descriptions of miRNA sponges that are also used to inhibit miRNA activity, as well as other techniques, including use of mRNA target protectors and synthetic miRNA mimics.

Introduction

It is perhaps unsurprising that not long after the discovery in plant and animal cells of hundreds of microRNAs (miRNAs) capable of modulating the activity of mRNAs, attempts would be made to block or mimic miRNA activity by exogenous introduction of reagents into cells. This was first shown by use of antisense DNA oligonucleotides (ODN) complementary to various miRNAs injected into *Drosophila* embryos to inhibit function (Boutla *et al.*, 2003). Soon afterwards, synthetic ODN analogues were shown to be more convenient miRNA blocking agents in cultured cells and *in vivo* (Hutvagner *et al.*, 2004; Meister *et al.*, 2004) and became known as anti-miRs. Since these early demonstrations, a large range of ODN analogue types have been utilized as anti-miRs (for a recent review see Stenvang *et al.*, 2012). The chemistries included in these modified ODN types are compared in this chapter and how particular types of modification have been found to be useful in establishing anti-miR mechanism or therapeutic applications. Note that the word 'antagomiR' is commonly used to describe all anti-miRs but in reality refers to one type of chemically modified anti-miR only.

In addition to anti-miRs, another type of synthetic reagent has been used for miRNA blockage (namely miRNA sponges). Further, synthetic miRNA mimics have also been described for increasing the amount of a miRNA in cells in order to give additional repression to mRNA expression. Although the main purpose of this chapter is to highlight reagents that block miRNA function (shown pictorially in Fig. 6.1), a short section is included that points to where synthetic miRNA mimics have found utility.

Because miRNAs regulate many (sometimes hundreds) of mRNAs, the effects of anti-miRs or similar reagents may have consequences beyond that of the phenotype resultant from enhancement of expression of a single mRNA. Indeed the very concept of up-regulation of multiple mRNAs has been exploited to great effect by design of short ODNs that are fully complementary only to the highly conserved seed region of a miRNA, to obtain deregulation of an entire miRNA family or cluster and subsequent up-regulation of a number of related genes, where this is advantageous [e.g. tiny LNAs Obad *et al.* (2011)].

More normally an anti-miR is designed to be fully complementary to a single RNA strand of a mature miRNA that is known to be responsible for repression of a specific mRNA or group of mRNAs. This is in contrast to the interaction of the miRNA with its mRNA targets where the miRNA is usually only partially complementary, except for the seed region. To what extent partial

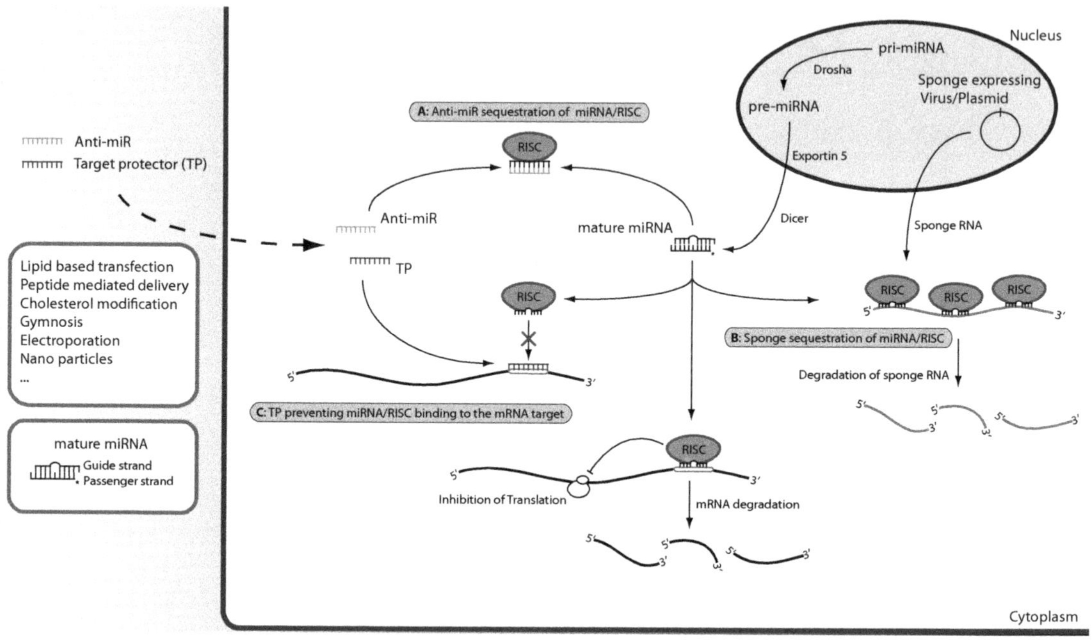

Figure 6.1 Scheme depicting various methods to inhibit the effect of microRNA activity in cells. (A) Anti-miR sequestration of mature miRNA/RISC, (B) sponge sequestration of miRNA/RISC, and (C) target protector preventing miRNA/RISC binding to the mRNA.

complementarity of the designed anti-miR to additional different miRNAs or to other mRNA sites within the transcriptome might lead to off-target effects has rarely been studied, for example through use of a microarray, except for some therapeutically targeted miRNAs. By contrast, the targeting of miRNA precursors by anti-miRs, which are commonly found in the nucleus of cells and then exported to the cytosol (Fig. 6.1), seems to present more difficulties than for mature miRNAs. There are only rare examples that have been published. Touched upon in this chapter is where in a cell an anti-miR targeted mature miRNA might be found, although the trafficking of miRNAs through, and in some cases out of a cell, is not altogether clear and may be dependent both on the miRNA identity and the cell type.

How ODN anti-miRs function once bound to their target miRNAs has been a source of controversy for some years. This has been due to difficulties in obtaining reliable assays (see sections on mechanism and validation of miRNA activity). The evidence is now mostly in favour of sequestration of the miRNA for most anti-miR types, rather than enzyme-directed cleavage. Cleavage of miR-21 in cells by a synthetic hammerhead ribozyme has been claimed (Suryawanshi *et al.*, 2010). This was extended to use of an LNA/DNA-zyme to target various pre-miRNAs in zebrafish embryos (Suryawanshi *et al.*, 2012). Furthermore, targeting of a miRNA precursor at the loop region by a designed siRNA has been published (Li *et al.*, 2009; Lee *et al.*, 2005), but not of a mature miRNA.

Finally, it is also possible to target a miRNA binding site, usually located within the 3′-untranslated region of a specific mRNA, to prevent the repression of expression by a specific miRNA. Such 'target protectors' have been exploited in certain instances, but the delivery of these into cells for mRNA protection may not be identical to that of an anti-miR, since it will be dependent on a mRNA location rather than that of its targeting miRNA.

Anti-miR chemistries

ODNs have been used for over 30 years to suppress expression of mRNA (Zamecnik and Stephenson, 1978) but suppression of miRNA activity by synthetic anti-miR ODNs is much more recent (Hutvagner *et al.*, 2004; Meister *et al.*, 2004). Since unmodified ODNs are unstable in cells and *in vivo*, numerous modification chemistries have been developed to enhance nuclease resistance, to increase binding affinity, and/or to aid cellular uptake. Most modifications have been developed for targeting ODNs to single-stranded mRNA (Kore and Ford, 2008) or for double-stranded siRNAs (Behlke, 2008). Such chemical modification strategies can also be employed to improve function of synthetic anti-miRs. The most widely utilized modifications are shown in Fig. 6.2 and the different anti-miR strategies for such ODN analogues are presented in Fig. 6.3.

Inhibition of a target miRNA function by a modified ODN was first demonstrated using a 31-mer 2′-*O*-methyl (2′-OMe) modified RNA ODN complementary to the let-7 miRNA transcript in both *Drosophila* embryo lysates and HeLa cell S100 extracts or by direct microinjection into *C. elegans* larvae (Hutvagner *et al.*, 2004). Not long after, a 31-mer 2′-OMe ODN was used to inhibit miR-122 in liver cells (Jopling *et al.*, 2005) and ODNs of the same chemistry but complementary to a range of mature miRNAs were also published (Cheng *et al.*, 2005). Later 2′-modifications used include 2′-*O*-methoxyethyl (2′-MOE) and 2′-fluoro (2′-F) (Davis *et al.*, 2006; Davis *et al.*, 2009).

All such negatively charged ODN analogues give rise to significant enhancement of nuclease resistance and binding affinity to their target miRNAs. To further increase stability, the nucleosides may be connected through phosphorothioate (PS) linkages, where a sulphur atom

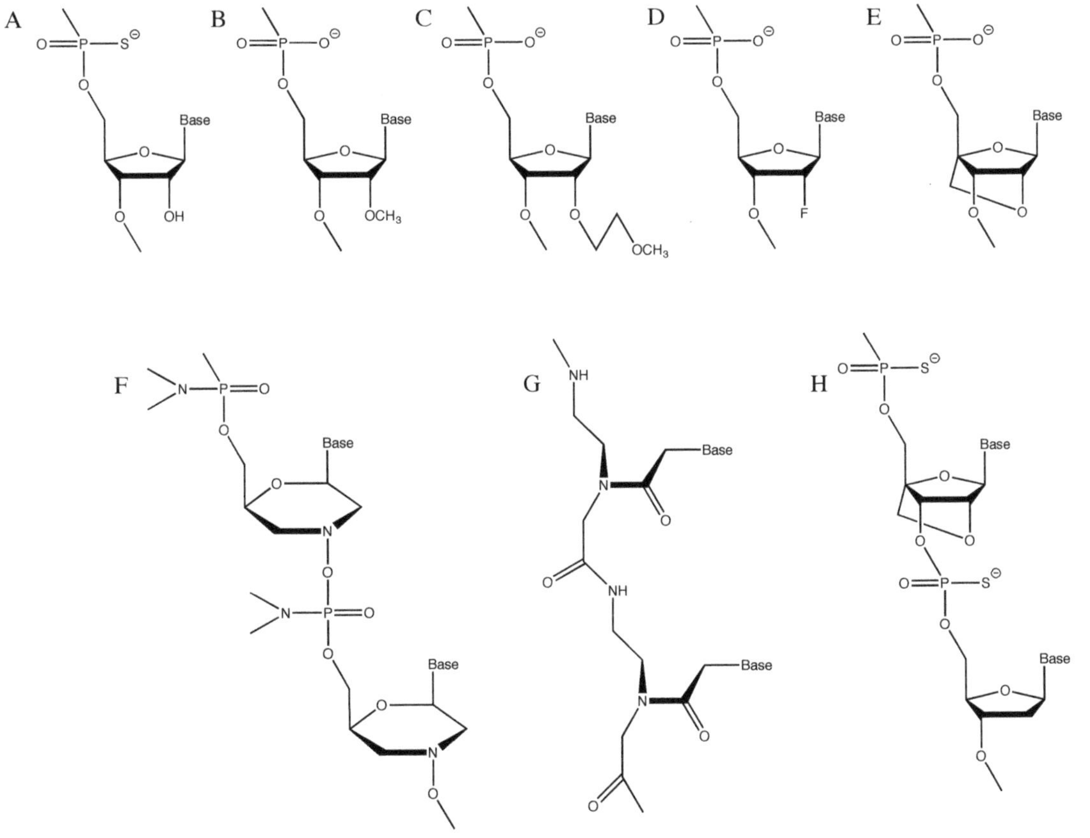

Figure 6.2 A selection of common oligonucleotide analogues used as anti-miRs. (A) RNA with phosphorothioate (PS) linkage, (B) 2′-*O*-methyl (2′-OMe), (C) 2′-*O*-methoxyethyl (2′-MOE) (D) 2′-fluoro (2′-F), (E) locked nucleic acid (LNA), (F) phosphorodiamidate morpholino oligonucleotide (PMO), (G) peptide nucleic acid (PNA), and (H) example of an LNA/DNA mixmer with PS linkage.

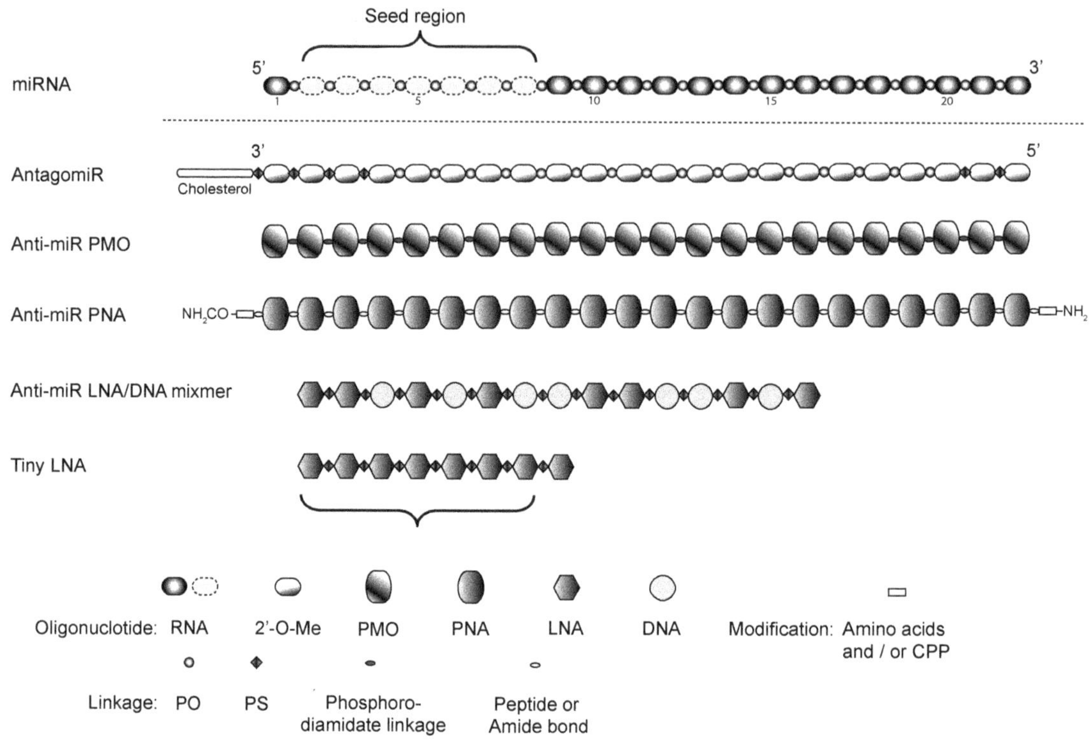

Figure 6.3 Types of anti-miR used for inhibition of miRNA activity in cells.

replaces one of the non-bridging oxygens in the phosphate backbone. It is common to use full PS linkages in anti-miRs but they can be included in any internucleoside linkage within the ODN. PS linkages increase resistance to nucleases but also reduce RNA-target affinity slightly. Further, they enhance interaction with plasma proteins, leading to increases in *in vivo* half-life. Interestingly, the first anti-miR type to show good *in vivo* activity in a range of tissue types involved partial PS linkages (located at only 6 end positions), a 2′-OMe backbone and a cholesterol attachment (Krützfeldt *et al.*, 2005) (Fig. 6.3). These were named 'antagomiRs'. Note that the favourable *in vivo* properties of PS linkages has resulted in their inclusion in all clinically approved ODNs to date.

Since the 7 base-pair seed sequence in the 5′-region is the most important for effective miRNA action (Fig. 6.3), chemical modifications that strongly increase binding affinity to the seed region allow for use of anti-miRs shorter than the full 21 to 24 nt length of the mature miRNA. Shorter ODNs are cheaper to produce and show higher tissue penetration, but full-length anti-miRs usually have advantages in target specificity. Locked nucleic acid (LNA) units, in which the furanose ring in the sugar-phosphate backbone is chemically locked in an RNA-like conformation by the introduction of a 2′-O, 4′-C methylene bridge, are particularly valuable to allow anti-miR shortening without loss of anti-miR activity (Figs. 6.2 and 6.3). This bicyclic LNA modification gives a T_m per residue higher than for all other analogues (T_m of +2°C to +8°C per introduced LNA unit).

However, very high affinity can be a disadvantage in an anti-miR when it comes to selectivity, and it is common to mix LNA units with DNA (or OMe) residues in order to modulate binding strength. LNA/DNA mixmers were first used for detection of miRNAs on Northern blots (Valoczi *et al.*, 2004) but they have now become the most commercially used anti-miRs for diagnostic (Exiqon) and therapeutic (Santaris) applications. The greater binding strength for LNA-containing ODNs increases the risk of off target effects, especially when the ODNs are long. To circumvent this problem, shorter ODNs of 6 or 7 residues that target only the seed sequence and which are composed entirely of LNA residues, referred to as 'tiny-LNAs', have been developed with great success (Obad *et al.*, 2011). Mixmers that include 2′-MOE or 2′-F residues have also been described (Davis *et al.*, 2006, 2009).

The design of mixmers is important. The location of particular analogues at particular nucleotide positions within an anti-miR has a huge impact on target binding selectivity and/or its inhibitory properties. Since several miRNA family members share the same seed sequence, the design becomes particularly crucial when it comes to inclusion of high affinity analogues, such as LNA, into the seed-targeting region of an anti-miR. Functional outcomes, for example, of different mismatches in this region are remarkably variable (Robertson *et al.*, 2010).

Another type of analogue used for steric blocking of miRNA activity is a phosphorodiamidate morpholino oligonucleotide (PMO) in which the backbone deoxyribose sugar rings are replaced by morpholine units and the internucleoside phosphodiesters by phosphorodiamidates (Fig. 6.2). Phosphorodiamidates carry no negative charge, such that PMOs are charge-neutral analogues in the usual physiological pH range. PMOs have found particular utility as anti-miRs following microinjection into zebrafish embryos by targeting either mature miRNA or pri-miRNA (Kloosterman *et al.*, 2007).

A second charge-neutral ODN analogue used as anti-miRs is peptide nucleic acid (PNA), where the backbone is composed of repeating *N*-(2-aminoethyl)-glycine units linked by amide bonds (Fig. 6.2). Fabani and Gait were the first to describe use of PNA as anti-miRs (Fabani and Gait, 2008), but PNAs have now been utilized by others (Brown and Yin, 2012; Oh *et al.*, 2009, 2010). Due to the lack of electrostatic repulsion, the binding between charge-neutral PNA (or PMO) and RNA is stronger than that occurring between two RNA strands. However PNAs are more hydrophobic than negatively charged analogues and therefore they are usually synthesized (in a similar way to peptide synthesis) with attached positively charged Lysine or Arginine residues (Fabani *et al.*, 2010; Torres *et al.*, 2011b), in order to maintain water solubility. Such amino acids also enhance the cell delivery properties of a PNA significantly (Abes *et al.*, 2006).

The cellular potencies of a variety of chemically modified anti-miRs have been studied in miRNA inhibition and in general a correlation between affinity and anti-miR activity is observed (Davis *et al.*, 2006). Amongst the anti-miRs investigated, those with the highest melting temperatures (T_m), for example a uniform 2′-F oligonucleotide and a LNA/2′-MOE mixmer, were the most potent inhibitors of miR-21 in a luciferase based reporter assay. Uniform 2′-OMe and 2′-MOE oligonucleotides included in the study both showed a similarly reduced T_m and were also less inhibitory (Davis *et al.*, 2006). In a study based on the same luciferase reporter system, LNA/2′-OMe mixmers with a PS end linkage, or with a complete PS-substituted backbone, showed approximately 10 times higher potency than a uniform 2′-OMe modified anti-miR (Lennox and Behlke, 2010). Further, inhibition of miR-122 in liver cells by different LNA/DNA mixmers was shown to be affinity dependent (Elmén *et al.*, 2008a). Affinity dependency for inhibition of the same miRNA by anti-miRs of different chemistries was also found by Torres *et al.* (2011b).

Truncation from either the 5′- or 3′-end of a uniformly 2′-MOE-modified anti-miR or of an antagomiR by three or more nucleotides gave rise to substantial loss of efficacy in cultured cells or, in the case of the antagomiR, *in vivo* (Davis *et al.*, 2006; Krützfeldt *et al.*, 2007). Truncation of the 5′-end is less well tolerated as regards to activity (Davis *et al.*, 2006), since this section of anti-miR is complementary to the seed sequence. Similarly for a Cys-K-PNA-K3 anti-miR targeting the liver-specific miR-122, truncations at the 3′-end of the anti-miR were more detrimental than those at the 5′-end (Torres *et al.*, 2012). MiRNA loading into RNA induced silencing complex (RISC) pre-organizes the crucial seed sequence for target recognition of mRNA (Lambert *et al.*, 2011) and this is likely also to be the case for anti-miR–miRNA recognition.

The amount of truncation an anti-miR can tolerate at the 3′-end correlates well with the binding strength of the ODN chemistry type used (LNA > LNA/DNA > PNA > MOE). Thus, truncation at the 3′-end of a 2′-MOE/PS anti-miR by 3 nts (19-mer) resulted in a substantial reduction in anti-miR activity against miR-21 in HeLa cells (Davis *et al.*, 2006). By contrast, similar truncation of a PNA anti-miR against miR-122 led to retention of strong anti-miR potency and only for a 6-nucleotide truncation (17-mer anti-miR) was a significant reduction of anti-miR122 activity observed (Torres *et al.*, 2012). 15-mer LNA/DNA/PS anti-miRs are very strong miR-122 inhibitors (Elmén *et al.*, 2008a,b; Lanford *et al.*, 2010). Further truncation at the 3′-end is only achievable by use of very high affinity all LNA anti-miRs (Obad *et al.*, 2011).

Mechanisms of action and delivery of anti-miRs into cells

Targeting of precursor miRNAs and mature miRNAs

The biosynthesis of mature miRNAs involves processing of RNA precursors (pri-miRNA and pre-miRNA) (Fig. 6.1). Both precursors and mature miRNAs contain the same target RNA sequences. Anti-miRs have been designed to target miRNA precursors alone, rather than a mature miRNA, but this is very challenging. Kloosterman *et al.* (2007) used PMO anti-miRs that target the sites for Drosha or Dicer cleavage on the precursor forms of several miRNAs. In some of these cases, upon direct injection of PMO anti-miRs into zebrafish embryos, a precursor-targeted anti-miR was found to be capable of interfering with miRNA maturation. Moreover, anti-miR-treated zebrafish displayed similar phenotypic effects irrespective of whether mature miR-375 or miR-375 precursors where targeted (Kloosterman *et al.*, 2007).

PNA anti-miRs conjugated to carrier peptides have been proposed as anti-pre-miR agents. The authors showed reduction in the levels of mature miR-210 in K562 cells as seen by quantitative reverse transcriptase polymerase chain reaction (qRT-PCR), but no data were presented demonstrating inhibition of miR-210 function with this approach (Avitabile *et al.*, 2012). Note that in C2C12 cells, an LNA/DNA anti-miR targeting mature miR-125b was suggested to also partially reduce the levels of pre-miR-125b, but this may have been due to an artefact of the Northern blotting technique used (Naguibneva *et al.*, 2006).

However, stem–loop structures of precursor miRNAs are difficult to strand-invade by anti-miRs. By contrast, a mature miRNA lacks the terminal loop and extended stem and is very quickly loaded into RISC, such that the guide strand forms a duplex with its mRNA target, and this is readily denatured upon anti-miR interaction (Kim, 2005). Secondly, when targeting miRNA precursors, anti-miRs may have to compete with RNA-binding proteins necessary for miRNA precursor processing (Davis-Dusenbery and Hata, 2010). Thirdly, a miRNA precursor may be located in a subcellular compartment that is less accessible for the anti-miR, e.g. pri-miRNAs are found in the cell nucleus (Krol *et al.*, 2010).

Kloosterman *et al.* (2007) suggested that by targeting miRNA precursors potential off-target effects may be reduced, since they are longer and there is more opportunity for alternative anti-miR design. By contrast, mature miRNAs are too short for redesign to avoid off-target effects. Even a strongly binding 8-mer all LNA anti-miR that does not target the seed sequence of a miRNA of interest showed a dramatic loss in anti-miR activity as compared to an 8-mer tiny LNA that did target the seed sequence, despite its being specific (Obad *et al.*, 2011). Thus all anti-miRs should at least include a section complementary to the seed sequence of the miRNA.

A second advantage of targeting miRNA precursors is the possibility to inhibit a complete family of miRNAs that share the same seed sequence by targeting Drosha cleavage sites, since Drosha cleavage is required for the maturation and function of all of the individual miRNA species in a family. Nevertheless, a simpler strategy to inhibit a miRNA family is by use of a seed-targeting tiny LNA (Obad *et al.*, 2011). Alternatively, multiple-target anti-miR oligonucleotides (MTg-AMOs), formed of single oligonucleotides containing mostly DNA but with LNA-modified ends, have been designed where several anti-miR sequences are interspersed by 8-nucleotide linkers between each anti-miR (Lu *et al.*, 2009).

Very early on, targeting of mature miRNAs by anti-miRs was shown to be likely to occur once the miRNA is already loaded into RISC (Hutvagner *et al.*, 2004; Meister *et al.*, 2004). Thus, Hutvagner *et al.* (2004) showed that a 31-mer anti-miR targeting Let-7 miRNA was bound by RISC in a stoichiometric 1:1 ratio and had a higher affinity for the miRNA than for the targeting of mRNA. This suggested that other factors beyond the binding affinity of the anti-miR for the miRNA are involved in miRNA/anti-miR recognition. Further, Davis *et al.* (2006) found that a 1-nt truncation at the 3′-end of an anti-miR complementary to the complete 5′-end of a mature miRNA actually improved anti-miR potency. This observation is consistent with the first nucleotide at the 5′-end of a miRNA not being available for binding by an anti-miR when loaded in RISC, as is apparent from a crystal structure of an Argonaute-like PIWI domain-siRNA guide complex (Parker *et al.*, 2005). Indeed in modern design of potent anti-miRs, the nucleotide at the 3′-end of the ODN (complementary to position 1 of the miRNA, Fig. 6.3) is commonly omitted (Elmén *et al.*, 2008a; Lanford *et al.*, 2010; miRCURY LNA inhibitors, Exiqon).

Subcellular localization of anti-miRs targeting mature miRNAs

The precise subcellular compartment remains elusive as to where anti-miRs target mature miRNAs and very few studies have addressed this issue. MiRNAs have been found in cytoplasmic foci known as P-bodies that also contain mRNAs targeted by miRNAs and key components of the RISC complex such as Argonaute proteins (Ago) and trinucleotide-repeat-containing-gene-6 (TNRC6; also known as GW182) (Kulkarni *et al.*, 2010). Likewise, miRNAs and RISC components have also been found in GW-bodies (cytoplasmic foci enriched in GW182), which were found to be independent of P-bodies and that often associate with Multivesicular Body (MVB)-like membranes, an intermediate component of the endosomal pathway (Gibbings *et al.*, 2009). A number of miRNAs have also been found in exosomes or other extracellular bodies (Gibbings *et al.*, 2009; Pegtel *et al.*, 2010; Valadi *et al.*, 2007). A recent model for the subcellular localization of miRNA-mediated gene silencing suggests that the miRNA-loaded RISC complex drives the repressed mRNA to P-bodies where the Cap and poly-A tails of the mRNA are removed. The complex is then thought to move to GW-bodies, where probably RISC disassembly occurs. Both, P-bodies and GW-bodies, may associate with MVBs leading to exosomes for cell-to-cell trafficking or to lysosomes for cargo degradation (Gibbings and Voinnet, 2010).

Krützfeldt *et al.* (2007) showed that in liver from mice treated with Q570-labelled antagomiRs targeting miR-122, both the antagomiR and miR-122 were detected in the same fractions following cellular fractionation on sucrose gradients. Further, laser-scanning microscopy revealed a cytosolic punctate signal for the antagomiR that did not co-localize to GW182-positive compartments (P-bodies/GW-bodies) (Krützfeldt *et al.*, 2007). In Huh7 cells treated in the absence of transfection agents with a PNA anti-miR targeting miR-122 that contained an intrinsically fluorescent base, the PNA anti-miR did not co-localize with P-bodies or GW-bodies as seen by confocal microscopy (Torres *et al.*, 2012). Instead, the cytoplasmic punctate PNA fluorescence was seen in early endosomes (at early time points) and in late endosomes and lysosomes (at later stages), suggesting that the PNA anti-miR was capable of trafficking through the whole endocytotic pathway. Sub-cellular fractionation and immunoprecipitation of endosomal compartments revealed enrichment of PNA anti-miRs and miR-122 in Syntaxin13-positive membrane-bound compartments, likely to be early, recycling and/or late endosomes. Importantly, cell treatment with chloroquine (a lysosomotropic agent used to help release of ODNs from endosomes into the cytosol) did not enhance anti-miR activity, suggesting that endosomal trapping does not limit the anti-miR activity of PNA anti-miRs in Huh7 cells (Torres *et al.*, 2012). Similarly, FAM-labelled tiny LNA anti-miRs, also delivered in the absence of transfection agents, were observed as punctate signals in the cytosol of HEK293 cells and in some cases co-localized with FLAG-tagged Ago2, suggesting that some of these vesicle-like cytosolic compartments contained the active fraction of LNA anti-miRs (Obad *et al.*, 2011). It thus seems likely that anti-miRs may target mature miRNAs within endosomal compartments. It is unclear at present how an anti-miR encounters a miRNA and whether this occurs by fusion of anti-miR-containing endosomal vesicles with a miRNA-containing endosomal compartment or by leakage of anti-miR into a compartment where a miRNA resides (Torres *et al.*, 2012).

MiRNA fate upon anti-miR binding

Although for several years the fate of a miRNA following anti-miR treatment remained controversial, improvements to miRNA isolation techniques in the presence of anti-miRs (see section on Validation of anti-miR activity), led to the current consensus that strongly binding anti-miRs sequester the miRNA without inducing miRNA degradation, whilst lower affinity anti-miRs seem to promote some miRNA degradation (Ameres *et al.*, 2010; Davis *et al.*, 2009; Krützfeldt *et al.*, 2005, 2007; Torres *et al.*, 2011a). Davis *et al.* (2009) reported that high-affinity 2′-F/2′-MOE anti-miRs inhibited miR-122 in mice without inducing miRNA degradation. Similarly, LNA-containing anti-miRs such as LNA/DNA and LNA/OMe, were shown to sequester and form strong stable duplexes with miR-122 in cells and *in vivo* without inducing miRNA decay (Elmén *et al.*, 2008a; Lanford *et al.*, 2010; Obad *et al.*, 2011; Torres *et al.*, 2011a). A cationic PNA anti-miR that efficiently inhibited miR-122 also did not degrade the miRNA in cells or in mice (Torres *et al.*, 2011a). However, lower affinity 2′-MOE/PS anti-miRs (Davis *et al.*, 2009; Esau *et al.*, 2006), 2′-OMe-based anti-miRs such as antagomiRs (Krützfeldt *et al.*, 2005, 2007) and fully 2′-OMe PO anti-miRs (Torres *et al.*, 2011a) all give rise to some miRNA losses following targeting. AntagomiRs targeting miR-16 or miR-21 were also shown to degrade the miRNA in HeLa cells (Ameres *et al.*, 2010).

Delivery of anti-miRs in cell culture

Many techniques have been developed aimed at increasing the cellular uptake of anti-miRs, some of which have also resulted in enhanced delivery *in vivo* (see section on *in vivo* delivery). In cell culture, negatively charged anti-miRs such as those modified at the 2′-ribose position (e.g. 2′-OMe, 2′-MOE/2′-F or LNA) can usually be transfected by means of a standard cationic lipofection agent (Lennox and Behlke, 2010; Torres *et al.*, 2011b). By contrast, this approach cannot be used effectively for charge neutral anti-miRs such as PMO or PNA-based anti-miRs. Instead, PNA anti-miRs have generally been delivered as conjugates with cell penetrating peptides (CPPs) (Avitabile *et al.*, 2012; Fabani *et al.*, 2010; Fabani and Gait, 2008; Oh *et al.*, 2009, 2010).

However, PNA anti-miRs do not need to be conjugated to CPPs in many cases. Only a few attached Lys residues are sufficient for effective miRNA inhibition following incubation with cells in the absence of a transfection agent, as observed for Huh7 liver cells, and for splenic B-cells, as well as in mice (Fabani *et al.*, 2010; Fabani and Gait, 2008; Torres *et al.*, 2011b, 2012). Further, addition of a terminal thiol-reactive group, such as a Cys residue or 3-mercaptopropionic acid, to the PNA anti-miR dramatically enhanced a PNA anti-miR uptake in Huh7 cells, HEK293ET cells (Torres *et al.*, 2012) or in muscle myotubes or myoblasts (Järver *et al.*, unpublished results). Lys-modifed PNA anti-miR carrying a Cys attachment was internalized mainly by clathrin-mediated endocytosis, but also by a cholesterol-dependent clathrin-independent endocytotic pathway. The addition of serum to the media did not affect anti-miR activity (Torres *et al.*, 2012). The presence of thiol-reactive groups may enhance cellular uptake for a number of other synthetic compounds (Torres and Gait, 2012) and it remains to be determined whether other anti-miR chemistries can also benefit from this strategy.

Efficient antisense activity for ODNs of different chemistries have been obtained when these are delivered in the absence of transfection agents (gymnosis or gymnotic delivery) (Fabani *et al.*, 2010; Fabani and Gait, 2008; Koller *et al.*, 2011; Stein *et al.*, 2010; Straarup *et al.*, 2010; Torres *et al.*, 2011b, 2012; Zhang *et al.*, 2012). For mRNA targeting, the antisense activities of LNA/DNA PS gapmers delivered into cells by gymnosis were found to correlate better with *in vivo* ODN activities, compared with cell delivery by use of Lipofectamine 2000 (Stein *et al.*, 2010; Straarup *et al.*, 2010). Anti-miRs of different chemistries, including cationic PNAs (see above), 2′-OMe ODNs with PO or PS backbone and LNA/OMe mixmers with PO or PS backbone, have also been shown to be capable of inhibiting miR-122 function when delivered gymnotically into Huh7 cells (Torres *et al.*, 2011b). Similarly, a tiny-LNA was able to inhibit miR-21 in HEK293 cells without uptake assistance (Obad *et al.*, 2011).

Interestingly, whilst for mRNA targeting efficient gymnotic delivery of ONs required incubation times of several days and high (micromolar) ON concentration (Koller *et al.*, 2011; Stein *et al.*, 2010; Straarup *et al.*, 2010), gymnotic delivery of anti-miRs needed only a few hours of incubation times and concentrations in the nanomolar range for full anti-miR activity, as observed in Huh7 liver cells (Torres *et al.*, 2011b). Since ODNs are known to be trapped in endosomal compartments unless released into the cytosol by a delivery agent, it has been suggested that miRNA targeting by anti-miRs may occur within or in some way associated with endosomal compartments and thus may not need release into the cytosol to access miRNA (Torres *et al.*, 2012). To what extent this finding holds generally true for miRNAs has not yet been established and additional examples of anti-miRs of different chemistries targeting different miRNAs in different cell lines will be needed to clarify this.

Use of peptide coated lipid or polymeric vectors has enhanced cell delivery of anti-miRs in some cases. For example, R_9-coated nanoparticles based on poly(lactic-co-glycolic acid (PLGA) was used to deliver unfunctionalized PNA or PMO anti-miRs and to inhibit miR-155 as measured by a luciferase reporter assay (Cheng and Saltzman, 2012). This and other polymeric delivery systems are discussed further in connection with *in vivo* delivery.

Validation of anti-miR activity in cells and assays

A key factor in development of efficient anti-miR reagents and therapeutics is the correct assessment of anti-miR function in cells and *in vivo*. This is by no means a trivial task. In this section are discussed the different techniques and assays available for screening anti-miR potency, highlighting their advantages and disadvantages. Emphasis is given to the care needed when interpreting results.

Assessment of mature miRNA levels

Initial studies to assess anti-miR potency relied upon detection and quantification of mature miRNAs following anti-miR treatments. Most techniques for miRNA detection are based on the hybridization properties between the target miRNA and a complementary nucleic acid probe. The nature of an anti-miR implies that it will have a sufficiently strong binding affinity for its miRNA target to cause inhibition. Because of this strong binding, miRNA detection in the presence of an anti-miR can be compromised by the presence of an anti-miR, which may mask the miRNA of interest. As discussed below, the use of a hybridization approach to quantify a miRNA is prone to artefacts and is therefore not recommended for assessment of anti-miR potency. However under the right conditions, it can be useful to gain insights into anti-miR mode of action (Davis *et al.*, 2009; Torres *et al.*, 2011a).

It is now well established that mature miRNAs might not be completely recovered from an RNA extract when total RNA extraction is carried out from cells and tissues in the presence of anti-miRs (Davis *et al.*, 2009; Fabani and Gait, 2008; Torres *et al.*, 2011a). The most commonly used methods for miRNA isolation involve total RNA extraction based on use of guanidinium thiocyanate–phenol–chloroform (Chomczynski and Sacchi, 2006), reagents commercially known as TRI-Reagent or TRIzol. These reagents have been designed to remove single-stranded RNA, double-stranded DNA and proteins selectively.

When cells are treated with anti-miRs, strong miRNA: anti-miR duplexes are formed. Under standard RNA extraction procedures, duplexes are not well recovered, resulting in an under-estimate of the amount of a particular miRNA. This occurs especially when high-affinity anti-miRs are used (Davis *et al.*, 2009; Fabani and Gait, 2008; Torres *et al.*, 2011a). For example during the first step of a standard RNA extraction protocol with a TRI-reagent, a 2′-MOE/LNA mixmer anti-miR targeting miR-122 was shown to partition the miRNA partly into the organic phase, instead of wholly into the aqueous RNA-containing phase (Davis *et al.*, 2009). Particularly for 2′-OMe, LNA/OMe and LNA/DNA anti-miRs, proper precipitation of miR-122 from the aqueous phase was prevented during standard iso-propanol addition (Torres *et al.*, 2011a). Instead, full precipitation of miRNA: anti-miR duplexes was obtained by use of a larger volume of iso-propanol, which was followed by removal of other precipitated contaminants through ethanol RNA re-precipitation (Torres *et al.*, 2011a).

Once the RNA has been extracted, Northern blotting and qRT-PCR have been the methods of choice to detect and quantify a specific miRNA. Artefacts seen in Northern blotting due to the presence of anti-miRs have been well studied and some solutions proposed. A key issue is that during hybridization, the probe used for miRNA detection may not be able to displace an anti-miR already bound to a miRNA. Thus, only uncomplexed miRNA would be detected whilst miRNA: anti-miR duplexes would not. For moderate affinity anti-miRs, such as 2′-OMe or antagomiRs, the use of very highly denaturing polyacrylamide gels (e.g. containing 20% formamide and 8 M urea) for RNA electrophoresis is sufficient to break apart miRNA: anti-miR duplexes (Krützfeldt *et al.*, 2005; Torres *et al.*, 2011a), but for very high affinity anti-miRs such as those containing LNA, the binding is so strong to the miRNA, that miRNA: anti-miR duplexes remain unbroken, even after cell lysis and in very high denaturing solutions, such as in Trizol/TRI-Reagent (Davis *et al.*, 2009; Fabani and Gait, 2008; Lu *et al.*, 2009; Torres *et al.*, 2011a).

An alternative strategy is prior to polyacrylamide gel loading to add to the RNA extract small amounts of a PNA oligonucleotide having the same sequence as the miRNA of interest. This PNA acts as a 'decoy' to bind to the anti-miR and to release the miRNA from a miRNA: anti-miR duplex. A Northern blot hybridization probe can recognize both an anti-miR and a decoy PNA, but, since PNAs are charge neutral, an anti-miR: PNA duplex migrates differently under gel electrophoresis and this aids miRNA quantification (Davis *et al.*, 2009).

All these detection artefacts lead to a lack of correlation between miRNA measured amounts and anti-miR inhibition of miRNA in living cells, and hence to unreliability of miRNA quantification as a measure of anti-miR potency. However, through use of protocol modifications, it has now become clear that high affinity anti-miRs (such as 2′-F/MOE, cationic PNA, LNA/DNA and LNA/OMe anti-miRs) do not result in miRNA degradation whilst lower affinity anti-miRs (such as 2′-OMe or 2′-MOE anti-miRs) permit some miRNA decay (Davis *et al.*, 2009; Torres *et al.*, 2011a).

Other techniques to detect and quantify miRNAs, such as qRT-PCR and *in situ* hybridization, have also proved misleading in the presence of anti-miRs (Davis *et al.*, 2009; Elmén *et al.*, 2008b; Lu *et al.*, 2009), but detection artefacts have not been studied in depth. For example, miR-122 could not be detected by *in situ* hybridization in the presence of an LNA/DNA anti-miR even when an LNA-modified probe was used for miRNA detection. By contrast, the LNA/DNA anti-miR itself could be detected with an equivalent LNA detection probe complementary to the anti-miR (Elmén *et al.*, 2008b). Since such LNA/DNA anti-miRs inhibit but do not induce miRNA degradation (Davis *et al.*, 2009; Torres *et al.*, 2011a), this provides further evidence that LNA-modified detection probes are unable to displace an LNA/DNA anti-miR from a miRNA: anti-miR duplex, but can displace a miRNA from such a duplex during *in situ* hybridization.

In the case of qRT-PCR, a dramatic reduction in the level of the targeted miRNA is often observed, even when an anti-miR is added after cell lysis (i.e. when it cannot be biological active) (Davis *et al.*, 2009; Lu *et al.*, 2009). This could be caused by (i) poor recovery of the miRNA into the RNA extract as discussed above, (ii) inhibition of the PCR reaction due to titration of the PCR primers by the anti-miR, or (iii) inhibition of the PCR reaction because of inability of the polymerase enzyme to displace the anti-miR bound to the miRNA

Assessment of mRNA and protein levels of well-validated miRNA targets

In place of quantification of mature miRNA level, a better approach to measure anti-miR potency is to determine the amount by qRT-PCR of an mRNA or its corresponding protein by Western blotting for a known target of a miRNA. Target mRNA degradation is the commonly observed outcome of miRNA-mediated gene inhibition (Guo *et al.*, 2010; Lim *et al.*, 2005). Thus, a rise in the amount of a particular mRNA that is a known target of the miRNA is usually seen following anti-miR treatment. qRT-PCR assay of mRNA is a fast, quantitative and inexpensive technique in 96-well format that allows for processing of multiple sets of samples in single experiments (mid- to high-throughput assays). The readout does not require use of miRNA sensor systems and is suitable for extracts from both cells and tissues.

However, the effects on mRNA up-regulation are frequently very small and the technique may not be always sensitive enough to determine accurately dose-dependent anti-miR effects. An additional drawback is that a miRNA can sometimes repress translation of an mRNA target without induction of mRNA degradation (Baek *et al.*, 2008; Chekulaeva and Filipowicz, 2009; Selbach *et al.*, 2008). In this case, no changes in the levels of those mRNAs will be detected upon anti-miR treatment even if the anti-miR is efficiently inhibiting the miRNA of interest. Of course, it is necessary to first choose or find a validated mRNA target for the miRNA of interest, which may not always be available, especially for a novel miRNA.

Measurement of the amount of protein by Western blotting in cells or tissues is a very reliable approach, and this is a key endpoint for inhibition of a known miRNA target by an anti-miR. In contrast to the mRNA level, analysis at the protein level is independent of the mode of miRNA-induced mRNA inhibition. Like qRT-PCR, Western blotting is relatively inexpensive. However, it is strongly dependent on the availability of suitable antibodies for detection of the protein of interest, it is only semi-quantitative and it is less convenient if the number of samples is high. Interestingly, only a few studies have been able to show changes of gene expression at the protein level following anti-miR treatment and the observed changes were not dramatic (Bhattacharyya *et al.*, 2006; Elmén *et al.*, 2008b; Esau *et al.*, 2006; Lu *et al.*, 2009; Obad *et al.*, 2011; Worm *et al.*, 2009). This may be due to the time needed for protein turnover in cells, the influence of other miRNAs involved in the regulation of the same protein, or alternative regulatory pathways for protein synthesis beyond miRNA control. Thus it is generally accepted that protein up-regulation is not required for validation of anti-miR activity, whereas demonstration of mRNA up-regulation of a miRNA target by qRT-PCR is usually considered essential to verify efficient anti-miR targeting.

Transcriptional and proteomic profiling

An additional approach to study anti-miR potency is to carry out genome-wide assays such as mRNA microarrays or mass spectrometry-based proteomic profiling (Davis *et al.*, 2009; Elmén *et al.*, 2008a,b; Esau *et al.*, 2006; Fabani *et al.*, 2010; Krützfeldt *et al.*, 2005; Lanford *et al.*, 2010; Obad *et al.*, 2011; Selbach *et al.*, 2008). These techniques allow for high-throughput screening of several genes simultaneously and can be carried out without knowledge *a priori* of which genes are validated targets of the miRNA of interest. Moreover, mRNA microarray data can give insights into the function of a miRNA as seen by the specific gene expression changes following miRNA inhibition by an anti-miR (Krützfeldt *et al.*, 2005). Similarly, changes in protein synthesis mediated by miRNAs have been studied by mass spectrometry-based proteomic profiling, for example after targeting Let-7b miRNA by an LNA/DNA anti-miR in cell culture (Selbach *et al.*, 2008).

Genome-wide analysis is advantageous in that the data set may be useful also to address anti-miR off-target effects. For example, Fabani *et al.* (2010) used mRNA microarrays to compare the gene expression of a miR-155 knockout mouse line to that of a wild type mouse that received (or not) treatment with a cationic PNA anti-miR. The PNA anti-miR-treated mice largely recapitulated the gene expression of miR-155 knockout mice, but interestingly a number of additional genes were either up-regulated or down-regulated only in the PNA treated group, pointing to some anti-miR-derived off-target effects. In a different study, Obad *et al.* (2011) used transcriptional and proteomic profiling to rule out off-target effects for tiny LNA anti-miRs targeting miR-122 in mice by comparison of mice treated with different anti-miRs to miR-122, a scrambled anti-miR control or a saline solution. Unfortunately such approaches can prove to be expensive, technically demanding, and do not allow for high-throughput screening of samples simultaneously. Further, a large amount of raw data is obtained that requires special bioinformatics knowledge for processing.

Use of miRNA sensor luciferase expression systems

Sensor luciferase reporter assays are commonly used to assess anti-miR activity (Lennox and Behlke, 2010; Torres *et al.*, 2011b). Such systems have been developed to provide a readout that is fully quantitative and that has a much wider dynamic range than quantification of an endogenous mRNA or protein. Initial reporter systems involved the insertion of the 3′-UTR of an mRNA known to be under the regulation of the miRNA of interest, or instead merely a sequence fully complementary to the full length of the miRNA of interest, within the 3′-UTR of a luciferase gene encoded in a commercial plasmid. A second plasmid coding for another reporter gene was needed as an internal reference (e.g. a different luciferase gene) that was not under the regulation of the miRNA of interest. Co-lipofection of the two plasmids was required followed by analysis of the gene expressions from both reporter plasmids (Davis *et al.*, 2006).

Vermeulen *et al.* (2007) improved such a reporter system by insertion of a DNA sequence recognized by the miRNA of interest at the 3′-UTR of a *Renilla* luciferase gene (RLuc) whilst the same plasmid also contained an expressed Firefly luciferase gene that is not under miRNA regulation (FLuc). This eliminates the need for plasmid co-transfection and improves the quantification and reproducibility of anti-miR effects. It has become the most commonly used miRNA sensor system (Lennox and Behlke, 2010; Robertson *et al.*, 2010; Torres *et al.*, 2011b, 2012; Vermeulen *et al.*, 2007). Cells express low levels of RLuc due to repression by the miRNA of interest present within such cells, but upon anti-miR treatment, the endogenous miRNA is prevented from binding to the recognition site in the RLuc gene, which results in de-repression. Anti-miR potency is shown usually as the ratio between RLuc and the internal control FLuc normalized to control cells.

Luciferase assays are rapid, reliable and can be carried out in 96-well high-throughput format. They do not require large quantities of reagents. Further, sequences inserted at 3′-UTR of RLuc, to bring it under the regulation of the miRNA, can be modified to adjust the sensitivity of the assay. For example, Vermeulen *et al.* (2007) inserted one or more sequences that were not fully complementary to the miRNA of interest (attenuated sites) or a sequence that was fully complementary to the miRNA of interest (cleavage-based site). Assays based on cleavage were more sensitive and had greater dynamic range than assays based on attenuated sites. Finally, such luciferase reporter assays are also useful to uncover mechanistic aspects of anti-miR biology, such as delivery, cellular internalization and sequence optimization for effective miRNA inhibition (Davis *et al.*, 2006; Lennox and Behlke, 2010; Robertson *et al.*, 2010; Torres *et al.*, 2012; Vermeulen *et al.*, 2007).

Limitations of luciferase reporter assays are that they cannot be used efficiently to validate anti-miR activity *in vivo*. Secondly, due to the high sensitivity, anti-miR activity may be overestimated and additional assays may be required to generate comparisons to mRNA/protein levels of endogenous targets. Nevertheless, luciferase reporter assays are useful as an initial screen to find potent anti-miR agents active in particular cell types.

In vivo phenotypes

For development of an anti-miR therapeutic it is helpful also to have a clear biological phenotype *in vivo* linked to miRNA inhibition. One good example is miR-122 where inhibition has resulted in effects on the cholesterol levels in plasma of mice and primates, as well as a reduction in HCV viraemia in primates (Esau *et al.*, 2006; Jopling *et al.*, 2005; Krützfeldt *et al.*, 2005; Lanford *et al.*, 2010). However, for most miRNAs a direct linkage of a single miRNA to a specific phenotypic function, and hence to inhibition by an anti-miR, remains elusive. Cases where efficient *in vivo* anti-miR activity have been demonstrated and linked directly to specific phenotypic changes are discussed in the section below.

Therapeutic anti-miRs and *in vivo* delivery

Oral delivery is usually not an option for ODNs. Therefore, therapeutic anti-miRs need to be able to reach the correct tissue following injection into the circulatory system. Since the size of an anti-miR is about 3–10 kDa and since kidneys excrete molecules smaller than 50 kDa, the half-life of an anti-miR in circulation needs to be lengthened by some means to avoid its rapid elimination. Further, the hydrophilic nature of ODNs makes it challenging for them to cross the vascular endothelium and enter the target tissue from the blood stream. Unless local injection is possible to increase bioavailability or the ODN is highly protein bound, the majority of a therapeutic ODN is likely to be filtered out rapidly by the kidneys. PS ODNs, like most ODN types, are also strongly absorbed by the liver and thus liver-acting anti-miRs have proved to be the most successful therapeutic anti-miRs to date. Once in an organ of interest, the anti-miR must enter the desired cell types and here the eukaryotic cell membrane acts as a further barrier to delivery.

Examples of anti-miRs used *in vivo* are shown in Table 6.1. For antagomiRs, the attached cholesterol functionality helps such 2′-OMe/PS ODNs to enter liver and to show activity, but at only relatively high doses in mice (Krützfeldt *et al.*, 2005, 2007). Three daily mice tail vein injections of antagomiR-16 at 80 mg/kg resulted in silencing of miR-16 in the liver, kidney, lung, heart, skeletal muscle, colon, fat, skin, ovaries, adrenal glands and bone marrow. No activity of the antagomiR was found in brain (Krützfeldt *et al.*, 2005). However antagomiR-16 did decrease miR-16 levels in the

Table 6.1 *In vivo* applications of synthetic anti-miRs

Anti-miR type	miRNA target	Organ target	Reference
AntagomiR	miR-10b	Tumour metastasis	Ma *et al.* (2010)
	miR-17-5p	Tumour xenograft	Fontana *et al.* (2008)
	miR-21	Heart	Thum *et al.* (2008)
	miR-103/107	Liver	Patrick *et al.* (2010)
	miR-122	Liver	Trajkovski *et al.* (2011)
	miR-199b	Heart	Krützfeldt *et al.* (2005)
	miR-328	Heart	Krützfeldt *et al.* (2007), da Costa Martins *et al.* (2010), Lu *et al.* (2010)
2′-OMe	miR-29c	Kidney	Long *et al.* (2011)
	miR-132	Tumour xenograft	Anand *et al.* (2010)
2′-MOE	miR-122	Liver	Esau *et al.* (2006)
2′-F/2′-MOE mixmer	miR-33a	Liver	Rayner *et al.* (2011)
LNA/DNA mixmer	miR-33	Liver	Najafi-Shoushtari *et al.* (2010)
	miR-122	Liver	Elmén *et al.* (2008b)
	miR-155	B-cells/spleen	Elmén *et al.* (2008a)
	miR-380-5p	Tumour xenograft	Lanford *et al.* (2010), Worm *et al.* (2009), Swarbrick *et al.* (2010)
PNA	miR-155	B-cells/spleen	Fabani *et al.* (2010)
		B-cell lymphoma	Babar *et al.* (2012)
Tiny LNA	miR-21	Breast adenocarcinoma	Obad *et al.* (2011)
	miR-34	Heart	Bernardo *et al.* (2012)
	miR-122	Liver	Obad *et al.* (2011)
	miR-155	B-cell lymphoma	Zhang *et al.* (2012)
		B- and T-cells/spleen	Garchow *et al.* (2011)

brain when injected directly into the cortex (Krützfeldt *et al.*, 2007). Furthermore, injections of antagomiR-122 at the same doses led to efficient inhibition of the liver-expressed miR-122 and resulted in de-repression of liver miR-122 target mRNAs (Krützfeldt *et al.*, 2005). Since miR-122 regulates genes involved in fatty-acid metabolism, expression of genes involved in cholesterol biosynthesis was affected (some decreased up to 4.2-fold and some increased up to 3.8-fold) and a 40% decrease in serum cholesterol levels was found to occur in the treated mice.

Since miR-122 is expressed in the developing liver and at high levels in the adult liver, where it makes up 70% of all miRNA (Chang *et al.*, 2004), miR-122 is an attractive and accessible target for therapeutics development. Unconjugated fully PS 2′-MOE, 2′-MOE/2′-F mixmer and LNA/DNA mixmer ODN types have all shown particularly high activities in liver, with significantly lower effective doses than for antagomiRs (Davis *et al.*, 2009; Elmén *et al.*, 2008b; Esau *et al.*, 2006). For example, 75 mg/kg 2′-MOE/PS anti-miR was more effective than the corresponding 2′-OMe ODN, when injected in mice intraperitoneally twice weekly for 4 weeks, in inhibition of miR-122 and up-regulation of four out of five miR-122 target mRNAs was observed (Esau *et al.*, 2006). Interestingly, only three of the four target mRNAs were dose-dependently up-regulated (by 1.5- to 3.5-fold), implying that parts of the observed effect might have been due to a second mechanism. The total plasma cholesterol level was reduced after anti-miR-122 treatment, and this reduction was even more pronounced in high-fat fed mice (Esau *et al.*, 2006).

LNA-based anti-miRs have proved to be the most promising type of anti-miR for therapeutic use to date. A 16-mer LNA/DNA/PS mixmer (31% LNA) anti-miR-122 ODN, when injected daily for 3 days in mouse tail vein at up to 25 mg/kg, resulted in dose-dependent de-repression of four selected target mRNAs all containing the exactly complementary seed sequence to miR-122 (Elmén *et al.*, 2008b). The miR-122 repression was maintained for all genes studied 3 weeks after final treatment, showing that the LNA/DNA/PS anti-miR has a long half-life *in vivo*. Similarly, a 15-mer LNA/DNA/PS mixmer (53% LNA) targeting miR-122 and intravenously injected once daily on alternate days over a five-day period (three injections in total), increased miR122 target mRNA levels and decreased cholesterol levels in mice. Injections varied from 1–200 mg/kg with a median effective dose at 10 mg/kg. Interestingly, it was shown that the PS backbone was essential for activity, since fully PO or partial PO/PS mixmers completely lost anti-miR activity. The 15-mer anti-miR also caused a decrease in cholesterol levels when administered to non-human primates (African green monkeys) (Elmén *et al.*, 2008a).

MiR-122 plays an important positive role in the regulation of hepatitis C (HCV) virus-replication, where the miRNA binds to two closely spaced target sites in the 5′ non-coding region of the HCV genome, resulting in up-regulation of viral RNA levels (Jopling *et al.*, 2005). Two chimpanzees treated with 5 mg/kg 15-mer anti-miR (53% LNA) targeting miR-122, when injected intravenously on a weekly basis for 12 weeks, showed in qRT-PCR of liver-biopsies a decrease in free miR-122 levels by a factor of more than 300, even 8 weeks following the final dose (Lanford *et al.*, 2010). A substantial decline in HCV RNA in serum was detected three weeks after the onset of treatment, with a maximum decrease of 2.6 orders of magnitude two weeks after the last injection. In the liver, HCV RNA was decreased by 2.3 orders of magnitude. Moreover, transcriptional profiling and histopathology of liver-biopsies showed de-repression of target mRNAs containing miR-122 seed sites and improvement of HCV-induced liver pathology. No evidence for viral resistance or side-effects was detected in the treated animals (Lanford *et al.*, 2010). Chronic HCV infection can lead to scarring and cirrhosis of the liver, which will cause liver failure and in some cases liver cancer. This compound has now reached Phase II clinical trials and is looking highly promising (Janssen *et al.*, 2011, http://www.santaris.com).

Encapsulation of anti-miRs in polymer-based nanoparticle or lipid-based delivery systems can offer additional benefits in some cases, such as enhanced stability, high loading densities, and decreased renal clearance. One interesting system was described (Akinc *et al.*, 2008), where a combinatorial library of lipidoids was synthesized and, in complex with cholesterol and a PEG-lipid, screened for their ability to deliver RNAi therapeutics. The best lipidoid, denoted $98N_{12}$-5, was more efficient in complexation and delivery of a 2′-OMe anti-miR having partial (end) PS linkages and in blocking liver-specific miR-122 in a mouse model (5 mg/kg on 3 alternate days), compared to an uncomplexed antagomiR-122 (80 mg/kg on 3 sequential days).

miRNA-155 plays an important role in various physiological and pathological processes. For example, reduction in miR-155 expression may inhibit malignant growth and viral infections, and attenuate the progression of cardiovascular diseases (Mattiske *et al.*, 2012). A 23-mer PNA attached to four Lys residues was shown to block miR-155 activity in mice pre-treated with lipopolysaccharide (in order to induce miR-155 expression) following systemic injection with 50 mg PNA/kg/day for 2 days, and analysis 24 hours post the final injection (Fabani *et al.*, 2010). A comparison of gene expression profiles of B cells from PNA-treated and miR-155 deficient (miR-$155^{-/-}$) mice showed that PNA inhibition of miR-155 recapitulates closely the effect of genetic deletion of miR-155.

The therapeutic value of miR-155 down-regulation by PNA has been shown recently in mice with established B-cell lymphoma. Injections of unconjugated miR-155-targeting PNA encapsulated in nanoparticles, consisting of cell-penetrating peptide (CPP) coated PLGA, resulted in rapid regression of lymphadenopathy (Babar *et al.*, 2012). Growth repression of the malignant lymphocytes was shown to be due to apoptosis as a result of decreased miR-155 activity. A single local injection of anti-miR-155 PNA at 1.5 mg/kg loaded in the CPP-PLGA nanoparticles significantly diminished growth of pre-B-cell tumours over the course of 4–6 days. Not surprisingly, the anti-tumour effect was decreased when the same amount of anti-miR-155 PNA was injected systemically in the absence of nanoparticles. The very low dose (1.5 mg/kg) used suggested that CPP-PLGA might be a powerful method for delivery of PNA and PMO anti-miR therapeutics (Babar *et al.*, 2012).

Silencing of miR-155 was shown in the splenocytes of mice in a mouse inflammation model on treatment with a 15-mer LNA/DNA/PS ODN, with a corresponding up-regulation of granulocyte colony-stimulating factor (Worm *et al.*, 2009). MiR-155 has also been targeted in a mouse xenograft model of B-cell lymphoma. Mice were injected in the tail-vein with an initial 25 mg/kg dose of a fully PS backbone tiny LNA 8-mer targeting the seed sequence of miR-155, followed by weekly maintenance doses of 5 mg/kg until the mice were sacrificed at day 14 or 19. Image analysis post treatment showed that there was a significant reduction in tumour burden at days 14 and 19 in the anti-miR-155–treated mice compared to mice treated with a scrambled control. Furthermore, mRNA levels of six miR-155 targets analysed by qRT-PCR showed that all six targets were de-repressed after treatment with the anti–miR-155 (Zhang *et al.*, 2012).

Another example of anti-tumour anti-miR action involves use of a cyclic RGD peptide targeting integrin $\alpha_v\beta3$ incorporated into a lipid-based nanoparticle to deliver a commercially available 22-mer 2′-OMe anti–miR to miR-132 into endothelium and to suppress angiogenesis and decrease tumour size by restoring endothelial p120RasGAP expression in an orthotopic xenograft mouse model of human breast carcinoma (Anand *et al.*, 2010). Further examples include use of an antagomiR to miR-10b to inhibit metastasis in a mouse mammary tumour model (Ma *et al.*, 2010) and a PNA anti-miR (chemistry not stated but probably the PNA was attached to a Tat peptide) targeting miR-21 that was used to treat MCF-7 cells *ex vivo* and then injected subcutaneously into nude mice to show a reduction in subsequent tumour growth in the mouse xenograft (Yan *et al.*, 2011).

Another organ that has shown particular promise for therapeutic anti-miRs is the heart. Following initial demonstration that an antagomiR to miR-21 reduces MAP kinase signalling in fibroblasts and showed significant effects in treatment of animals with myocardial disease (Thum *et al.*, 2008), further separate studies indicated that use of such an antagomiR or an LNA/DNA PS anti-miR results in stress-dependent cardiac remodelling of mouse heart resultant from inhibition of miR-21 (Patrick *et al.*, 2010). A tiny LNA 8-mer anti-miR to miR-34 has also been shown to result in cardiac remodelling in heart of a mouse model of pressure overload-induced pathological hypertrophy and cardiac dysfunction, and that targeting of the miR-34 family by the tiny LNA was more effective than targeting of miR-34a alone by a 15-mer LNA/DNA/PS ODN (Bernardo *et al.*, 2012).

All these studies of successful anti-miR activity *in vivo* demonstrate the exciting future potential of anti-miR therapeutics. Examples of potential therapeutic anti-miRs are given in a recent review (Stenvang *et al.*, 2012).

MicroRNA sponges for suppression of microRNA activity

The use of miRNA sponges was introduced in 2007 as a method to establish permanent miRNA loss of function in both cell lines and in transgenic animals (Ebert *et al.*, 2007). The sponge RNAs are transcribed from plasmid DNA or introduced through viral delivery (such as lentivirus or retrovirus). The resultant RNA strand typically contains 4–10 miRNA binding sites separated by a few nucleotides, and each binding site contains the sequence complementary to the seed region of a specific miRNA family. When the sponge RNA encounters its target miRNA/RISC complex, the sponge RNA is either degraded, or it sequesters the miRNA/RISC complex and prevents it from acting on its endogenous target mRNA (Fig. 6.1). Binding sites for a particular miRNA seed family are usually perfectly complementary to the seed region, whilst mismatches are introduced at positions 9–12 in order to minimize RNA interference–type cleavage and degradation of the sponge RNA. Introduction of more than 10 miRNA binding sites in a sponge construct does not enhance its effect, since the increased number of binding sites increases the chance of such RNA degradation.

A miRNA sponge, when expressed at high levels, provides a dominant negative method of gene expression modulation where all miRNAs that share the same seed sequence are inhibited simultaneously (Ebert and Sharp, 2010). This may be hard to achieve by construction of knockout cell lines or transgenics, since many miRNAs have seed family members encoded at multiple, separated loci and some are transcribed in clusters, where miRNA family knockout may affect other neighbouring gene expression. A further advantage is that a miRNA sponge responds well in a concentration-dependent manner to exogenously added miRNA.

Compared to chemically modified ODN inhibitors, miRNA sponges are long lasting and are very efficient. For example, they were more potent that 20 nM of transfected anti-miRs (2′-OMe or LNA-based) in cultured cells (Ebert *et al.*, 2007). The ability to utilize alternative promoter sequences opens up the possibility to obtain miRNA sponge expression in selected cell types of tissues, which would be difficult to achieve using synthetic anti-miR inhibitors. Thus, miRNA sponges are particularly good research tools, such as for phenotypic studies. However, anti-miRs are much easier to obtain than sponges and anti-miRs are normally recommended for targeting of single miRNA targeting or for therapeutic applications.

Target protectors for inhibition of microRNA-induced mRNA degradation

Since a miRNA usually has the ability to regulate several mRNA targets, it can prove a major challenge to determine the physiological role of the interaction between that miRNA and a single mRNA target, or to be clear which mRNA target is responsible for an observed phenotype when an anti-miR is added. An alternative technique, namely a target protector (TP), has been developed whereby an exactly complementary ODN is designed to bind to a specific miRNA responsive element in the 3′-UTR of a particular mRNA to interfere with interaction of a single miRNA–mRNA pair (Staton and Giraldez, 2011). Binding of the TP masks the miRNA responsive element and thus inhibits binding of a RISC complex and subsequent mRNA degradation or translational repression (Fig. 6.1).

Initially TPs were introduced as PMO ODNs designed to match perfectly and cover a miRNA responsive element on a desired mRNA (Choi *et al.*, 2007). However, TPs can be any ODN type intended to protect a specific mRNA from miRNA-induced degradation. Commercial TPs, such as miScript (Qiagen), are available, although the chemistry behind these ODNs is not disclosed.

In zebrafish embryos, miR-427 could be specifically suppressed either against the expression of the mRNA of an agonist (*squint*) or an antagonist (*lefty*) of the NODAL signalling pathway by individually designed TPs, to show the physiological balance between pathways, and this also demonstrates the selectivity of this method (Choi *et al.*, 2007). Later, species-specific miR-430/427/302 family control of the mesendodermal cell fate could be established in Xenopus oocytes and in human embryonic stem cells via miRNA target selection amongst ligands of the NODAL family. This study involved use of a single TP that covered both *LeftyA* and *LeftyB* 3′-UTRs to block Lefty expression fully whilst that of the NODAL ligand *xnr5* remained unaffected (Rosa *et al.*, 2009). This feature gives TPs an advantage over an anti-miR strategy when studying specific miRNA–mRNA interactions, since off-target effects are better controlled.

MicroRNA mimics

MiRNA and siRNA duplexes share many common features, but there are important differences also. MiRNAs generally show incomplete base pairing to their mRNA targets, which are often numerous, and binding results in inhibition of translation. By contrast, siRNAs normally pair exactly with a single RNA target and induce mRNA cleavage of that target. Overlap between the two mechanisms can sometimes occur, but in this section, only artificial duplex RNAs that mimic previously known endogenous miRNAs will be discussed.

MiRNA mimic duplexes (Fig. 6.4) usually consist of a guide strand that is identical to the mature miRNA sequence the mimic should replace, and a passenger strand that is partially or fully complementary to the guide strand. The passenger strand does not have to be identical in sequence to that of the natural miRNA, and indeed it is common in design to synthesize a passenger strand that is an exact complement of the guide strand to give a full duplex (Henry *et al.*, 2011). Various alterations are usually included in a miRNA mimic to ensure that it functions in the desired way. Thus the guide strand is usually phosphorylated at the 5′-end, whereas the 5′-end of the passenger strand is chemically protected to prevent phosphorylation. Further, the guide strand usually contains a 3′-end that remains single-stranded when annealed to the passenger strand, whereas the 5′-end is commonly found at a blunt-ended duplex. These features aid loading of the correct guide strand into the RISC complex, and thus avoid off-target effects resultant from loading of the passenger strand.

Other features common to both synthetic miRNAs and siRNAs are the use of chemical modifications to increase stability and reduce susceptibility to nuclease degradation. The passenger strand can be heavily modified (e.g. with about 50% 2′-OMe nucleotides). Such modifications also protect the duplex from immune activation. It is not recommended to place too many very highly stabilizing analogues within the passenger strand (such as LNA), since the duplex may become too resistant to strand dissociation and a completely nuclease resistant passenger strand would have a long life time and could result in off-target effects. By contrast, the guide strand is very sensitive to modifications, which can often result in loss of activity. Usually only the three nucleotides at the 3′-end of the guide strand are modified to increase nuclease resistance (e.g. with 2′-OMe nucleotides), two of which, upon annealing to the passenger strand, become located in the 3′-overhang that is also seen in naturally

5′-•AGAACGCUGCUAACGGUCUa-3′ Passenger strand
3′-AUUCUUGCGACGAUUGCCAGAU*p*-5′ Guide strand

GAU: RNA　a: DNA　*p*: 5′-Phosphate
AUU: 2′OMe　•: Chemical Cap

Figure 6.4 A model design for a microRNA mimic.

occurring miRNAs. Synthetic miRNA mimics are commercially available from several companies (e.g. Dharmacon, Ambion, Sigma Aldrich), but unfortunately the design and chemical modifications are not revealed, as is the case with most commercial RNA interference reagents.

As an example of potential therapeutic use of miRNA mimics, miRNAs are down regulated in several types of cancer. 1 mg/kg lipid-formulated let-7 or miR-34a miRNA mimic injected via tail vein into lung cancer model mice on alternate days for a total of eight injections, showed 48 hours after the last treatment a decreased tumour burden with a maximum of 60% reduction in the tumour area (Trang *et al.*, 2011). Further, a miR-203 mimic (amount not stated) complexed with a commercial Polyethylenimine transfection agent, after subcutaneous injection into a mouse basal cell carcinoma model every 48 h, revealed a smaller number and decreased area of cancer lesions as well as decreased miR-203 target mRNA expression (c-JUN and p63) (Sonkoly *et al.*, 2012). Therapeutic applications of synthetic miRNA mimics, as for siRNA, are dependent on a suitable *in vivo* delivery system.

Even though single-stranded RNA is not normally a good substrate for incorporation by RISC into a complex, 5′-phosphorylated single-stranded miRNA mimics consisting of 2′-F nucleotides and a high pyrimidine content have shown recent potential as inducers of miRNA-induced translation repression and mRNA degradation (Chorn *et al.*, 2012). Such mimics are resistant to nuclease degradation and immune activation, but miRNA activity was increased when a passenger strand was annealed.

Prospects

Many hundreds of miRNAs are now known and characterized, and it is relatively easy to purchase an anti-miR reagent for any individual miRNA. Anti-miR diagnostic probes or knock down ODNs are commercially available, as well as mRNA target protectors and miRNA mimics. Those studying miRNA function need to remember that commercial reagents utilize proprietary chemistry that is rarely disclosed. This chapter has aimed at highlighting the design features and chemical compositions of successful anti-miR types in cells and *in vivo*, their mechanisms of action and some applications. An ideal anti-miR research study should contain an initial screening for anti-miR potency by side-by-side comparison of anti-miRs of different chemistry and nucleotide composition in cell assays using very sensitive miRNA sensor systems, such as those that involve double luciferase reporters. Scrambled and mismatched anti-miR controls should also be tested to evaluate specificity. Once the most promising anti-miR candidates have been selected, it is necessary to confirm the activity of an anti-miR by measuring by qRT-PCR the levels of known (or predicted) mRNA targets of the miRNA. If possible, evaluation of the protein levels by Western blotting for such genes is also desirable. Importantly, off-target effects need to be considered. Evaluation of genome-wide proteomic profiling is currently uncommon but gives very strong evidence for specific anti-miR activity.

A clear biological phenotype in a suitable animal model that is directly linked to an anti-miR is often hard to achieve. However, therapeutic anti-miR agents are now reaching clinical use in favourable cases, particularly for liver diseases. As for all types of ODN, it is to be hoped that advances in the use of liposomal, polymeric and other delivery agents will expand the therapeutic horizon for anti-miRs to other organs and to cancers. Many nucleotide analogue types have been investigated for antisense ODN usage *in vivo*, but the range of analogues used as anti-miRs is currently more limited. As more analogue types and delivery methods become accessible to biologists on a scale suitable for *in vivo* studies, there is considerable scope for discovery of more selective anti-miRs that may more readily penetrate a tissue of interest without generation of unwanted off-target effects or toxicities. Such advances are likely to be mirrored by other recently established ODN analogue applications, such as redirection of nuclear splicing. Furthermore, miRNAs represent only a fraction of the non-coding RNA types now known in a cell, and the discovery of the functions of these non-coding RNAs is opening up amazing opportunities for use of ODNs as steric blocking inhibitors in a similar way to that of anti-miRs.

The development of anti-miR reagents and technology paves the way for such future advances.

Acknowledgements

This work was supported by the Medical Research Council (MRC programme no. U105178803). A.G.T was supported by a Marie Curie Action (COFUND) within the European Union Seventh Framework Programme.

References

Abes, S., Moulton, H.M., Clair, P., Prevot, P., Youngblood, D.S., Wu, R.P., Iversen, P.L., and Lebleu, B. (2006). Vectorization of morpholino oligomers by the (R-Ahx-R)4 peptide allows efficient splicing correction in the absence of endosomolytic agents. J. Control. Release *116*, 304–313.

Akinc, A., Zumbuehl, A., Goldberg, M., Leshchiner, E.S., Busini, V., Hossain, N., Bacallado, S.A., Nguyen, D.N., Fuller, J., Alvarez, R., *et al.* (2008). A combinatorial library of lipid-like materials for delivery of RNAi therapeutics. Nature Biotechnol. *26*, 561–569.

Ameres, S., Horwich, M., Hung, J., Xu, J., Ghildiyal, M., Weng, Z., and Zamore, P. (2010). Target RNA-directed trimming and tailing of small silencing RNAs. Science *328*, 1534–1539.

Anand, S., Majeti, B., Acevedo, L., Murphy, E., Mukthavaram, R., Scheppke, L., Huang, M., Shields, D., Lindquist, J., Lapinski, P., *et al.* (2010). MicroRNA-132-mediated loss of p120RasGAP activates the endothelium to facilitate pathological angiogenesis. Nat. Med. *16*, 909–914.

Avitabile, C., Saviano, M., D'Andrea, L., Bianchi, N., Fabbri, E., Brognara, E., Gambari, R., and Romanelli, A. (2012). Targeting pre-miRNA by peptide nucleic acids: a new strategy to interfere in the miRNA maturation. Artif. DNA, PNA XNA *3*, 88–96.

Babar, I.A., Cheng, C.J., Booth, C.J., Liang, X., Weidhaas, J.B., Saltzman, W.M., and Slack, F.J. (2012). Nanoparticle-based therapy in an *in vivo* microRNA-155 (miR-155)-dependent mouse model of lymphoma. Proc. Natl. Acad. Sci. U.S.A. *109*, E1695–1704.

Baek, D., Villen, J., Shin, C., Camargo, F., Gygi, S., and Bartel, D. (2008). The impact of microRNAs on protein output. Nature *455*, 64–71.

Behlke, M.A. (2008). Chemical modification of siRNAs for *in vivo* use. Oligonucleotides *18*, 305–319.

Bernardo, B.C., Gao, X.M., Winbanks, C.E., Boey, E.J., Tham, Y.K., Kiriazis, H., Gregorevic, P., Obad, S., Kauppinen, S., Du, X.J., *et al.* (2012). Therapeutic inhibition of the miR-34 family attenuates pathological cardiac remodeling and improves heart function. Proc. Natl. Acad. Sci. U.S.A. *109*, 17615–17620.

Bhattacharyya, S.N., Habermacher, R., Martine, U., Closs, E.I., and Filipowicz, W. (2006). Relief of microRNA-mediated translational repression in human cells subjected to stress. Cell *125*, 1111–1124.

Boutla, A., Delidakis, C., and Tabler, M. (2003). Developmental defects by antisense-mediated inactivation of micro-RNAs 2 and 13 in Drosophila and the identification of putative target genes. Nucleic Acids Res. *31*, 4973–4980.

Brown, P.N., and Yin, H. (2012). PNA-based microRNA inhibitors elicit anti-inflammatory effects in microglia cells. Chem. Commun. *49*, 4415–4417.

Chang, J., Nicolas, E., Marks, D., Sander, C., Lerro, A., Buendia, M.A., Xu, C., Mason, W.S., Moloshok, T., Bort, R., *et al.* (2004). miR-*122*, a mammalian liver-specific microRNA, is processed from hcr mRNA and may down-regulate the high affinity cationic amino acid transporter CAT-1. RNA Biol. *1*, 106–113.

Chekulaeva, M., and Filipowicz, W. (2009). Mechanisms of miRNA-mediated post-transcriptional regulation in animal cells. Curr. Opin. Cell Biol. *21*, 452–460.

Cheng, A.M., Byrom, M.W., Shelton, J., and Ford, L.P. (2005). Antisense inhibition of human miRNAs and indications for an involvement of miRNA in cell growth and apoptosis. Nucleic Acids Res. *33*, 1290–1297.

Cheng, C.J., and Saltzman, W.M. (2012). Polymer nanoparticle-mediated delivery of microRNA inhibition and alternative splicing. Mol. Pharm. *9*, 1481–1488.

Choi, W., Giraldez, A., and Schier, A. (2007). Target protectors reveal dampening and balancing of Nodal agonist and antagonist by miR-430. Science *318*, 271–274.

Chomczynski, P., and Sacchi, N. (2006). The single-step method of RNA isolation by acid guanidinium thiocyanate-phenol-chloroform extraction: twenty-something years on. Nat. Protocol. *1*, 581–585.

Chorn, G., Klein-McDowell, M., Zhao, L., Saunders, M.A., Flanagan, W.M., Willingham, A.T., and Lim, L.P. (2012). Single-stranded microRNA mimics. RNA *18*, 1796–1804.

da Costa Martins, P., Salic, K., Gladka, M., Armand, A., Leptidis, S., el Azzouzi, H., Hansen, A., Coenen-de Roo, C., Bierhuizen, M., van der Nagel, R., *et al.* (2010). MicroRNA-199b targets the nuclear kinase Dyrk1a in an auto-amplification loop promoting calcineurin/NFAT signalling. Nat. Cell. Biol. *12*, 1220 – 1227.

Davis, S., Lollo, B., Freier, S., and Esau, C. (2006). Improved targeting of miRNA with antisense oligonucleotides. Nucleic Acids Res. *34*, 2294–2304.

Davis, S., Propp, S., Freier, S., Jones, L., Serra, M., Kinberger, G., Bhat, B., Swayze, E., Bennett, C., and Esau, C. (2009). Potent inhibition of microRNA *in vivo* without degradation. Nucleic Acids Res. *37*, 70–77.

Davis-Dusenbery, B.N., and Hata, A. (2010). Mechanisms of control of microRNA biogenesis. J. Biochem. *148*, 381–392.

Ebert, M., and Sharp, P. (2010). MicroRNA sponges: progress and possibilities. RNA *16*, 2043–2050.

Ebert, M., Neilson, J., and Sharp, P. (2007). MicroRNA sponges: competitive inhibitors of small RNAs in mammalian cells. Nat. Methods *4*, 721–726.

Elmén, J., Lindow, M., Schutz, S., Lawrence, M., Petri, A., Obad, S., Lindholm, M., Hedtjarn, M., Hansen, H., Berger, U., *et al.* (2008a). LNA-mediated microRNA silencing in non-human primates. Nature *452*, 896–899.

Elmén, J., Lindow, M., Silahtaroglu, A., Bak, M., Christensen, M., Lind-Thomsen, A., Hedtjarn, M., Hansen, J., Hansen, H., Straarup, E., *et al.* (2008b). Antagonism of microRNA-122 in mice by systemically administered LNA-antimiR leads to up-regulation of a large set of predicted target mRNAs in the liver. Nucleic Acids Res. *36*, 1153–1162.

Esau, C., Davis, S., Murray, S., Yu, X., Pandey, S., Pear, M., Watts, L., Booten, S., Graham, M., McKay, R., *et al.* (2006). miR-122 regulation of lipid metabolism revealed by *in vivo* antisense targeting. Cell Metab. *3*, 87–98.

Fabani, M.M., and Gait, M.J. (2008). miR-122 targeting with LNA/2′-O-methyl oligonucleotide mixmers, peptide nucleic acids (PNA), and PNA–peptide conjugates. RNA *14*, 336–346.

Fabani, M.M., Abreu-Goodger, C., Williams, D., Lyons, P.A., Torres, A.G., Smith, K.G., Enright, A.J., Gait, M.J., and Vigorito, E. (2010). Efficient inhibition of miR-155 function *in vivo* by peptide nucleic acids. Nucleic Acids Res. *38*, 4466–4475.

Fontana, L., Fiori, M., Albini, S., Cifaldi, L., Giovinazzi, S., Forloni, M., Boldrini, R., Donfrancesco, A., Federici, V., Giacomini, P., *et al.* (2008). Antagomir-17-5p abolishes the growth of therapy-resistant neuroblastoma through p21 and BIM. PloS One *3*, e2236.

Garchow, B.G., Bartulos Encinas, O., Leung, Y.T., Tsao, P.Y., Eisenberg, R.A., Caricchio, R., Obad, S., Petri, A., Kauppinen, S., and Kiriakidou, M. (2011). Silencing of microRNA-21 *in vivo* ameliorates autoimmune splenomegaly in lupus mice. EMBO Mol. Med. *3*, 605–615.

Gibbings, D., and Voinnet, O. (2010). Control of RNA silencing and localization by endolysosomes. Trends Cell Biol. *20*, 491–501.

Gibbings, D.J., Ciaudo, C., Erhardt, M., and Voinnet, O. (2009). Multivesicular bodies associate with components of miRNA effector complexes and modulate miRNA activity. Nature Cell Biol. *11*, 1143–1149.

Guo, H., Ingolia, N., Weissman, J., and Bartel, D. (2010). Mammalian microRNAs predominantly act to decrease target mRNA levels. Nature *466*, 835–840.

Henry, J., Azevedo-Pouly, A., and Schmittgen, T. (2011). MicroRNA replacement therapy for cancer. Pharm. Res. *28*, 3030–3042.

Hutvagner, G., Simard, M., Mello, C., and Zamore, P. (2004). Sequence-specific inhibition of small RNA function. PLoS Biol. *2*, E98.

Janssen, H., Reesink, H., Zeuzem, S., Lawitz, E., Rodriguez-Torres, M., Chen, A., Davis, C., King, B., Levin, A., and Hodges, M. (2011). A randomized, double-blind, placebo (plb) controlled safety and antiviral proof of concept study of miravirsen (MIR), an oligonucleotide targeting miR-*122*, in treatment naive patients with genotype 1 (gt1) chronic HCV infection [abstract]. Hepatology *54*, 1430A.

Jopling, C., Yi, M., Lancaster, A., Lemon, S., and Sarnow, P. (2005). Modulation of hepatitis C virus RNA abundance by a liver-specific microRNA. Science *309*, 1577–1581.

Kim, V.N. (2005). MicroRNA biogenesis: coordinated cropping and dicing. Nature Rev Mol. Cell Biol. *6*, 376–385.

Kloosterman, W., Lagendijk, A., Ketting, R., Moulton, J., and Plasterk, R. (2007). Targeted inhibition of miRNA maturation with morpholinos reveals a role for miR-375 in pancreatic islet development. PLoS Biol. *5*, e203.

Koller, E., Vincent, T.M., Chappell, A., De, S., Manoharan, M., and Bennett, C.F. (2011). Mechanisms of single-stranded phosphorothioate modified antisense oligonucleotide accumulation in hepatocytes. Nucleic Acids Res. *39*, 4795–4807.

Kore, A., and Ford, L. (2008). Chemical modifications of synthetic RNAi agents and *in vivo* delivery techniques. Curr. Bioacti. Compd. *4*, 6–14.

Krol, J., Loedige, I., and Filipowicz, W. (2010). The widespread regulation of microRNA biogenesis, function and decay. Nat. Rev. Genet. *11*, 597–610.

Krützfeldt, J., Rajewsky, N., Braich, R., Rajeev, K., Tuschl, T., Manoharan, M., and Stoffel, M. (2005). Silencing of microRNAs *in vivo* with 'antagomirs'. Nature *438*, 685–689.

Krützfeldt, J., Kuwajima, S., Braich, R., Rajeev, K., Pena, J., Tuschl, T., Manoharan, M., and Stoffel, M. (2007). Specificity, duplex degradation and subcellular localization of antagomirs. Nucleic Acids Res. *35*, 2885–2892.

Kulkarni, M., Ozgur, S., and Stoecklin, G. (2010). On track with P-bodies. Biochem. Soc. Trans. *38*, 242–251.

Lambert, N.J., Gu, S.G., and Zahler, A.M. (2011). The conformation of microRNA seed regions in native microRNPs is prearranged for presentation to mRNA targets. Nucleic Acids Res. *39*, 4827–4835.

Lanford, R., Hildebrandt-Eriksen, E., Petri, A., Persson, R., Lindow, M., Munk, M., Kauppinen, S., and Ørum, H. (2010). Therapeutic silencing of microRNA-122 in primates with chronic hepatitis C virus infection. Science *327*, 198–201.

Lee, Y.S., Kim, H.K., Chung, S., Kim, K.S., and Dutta, A. (2005). Depletion of human micro-RNA miR-125b reveals that it is critical for the proliferation of differentiated cells but not for the down-regulation of putative targets during differentiation. J. Biol. Chem. *280*, 16635–16641.

Lennox, K., and Behlke, M. (2010). A direct comparison of anti-microRNA oligonucleotide potency. Pharm. Res. *27*, 1788–1799.

Li, Y.G., Zhang, P.P., Jiao, K.L., and Zou, Y.Z. (2009). Knockdown of microRNA-181 by lentivirus mediated siRNA expression vector decreases the arrhythmogenic effect of skeletal myoblast transplantation in rat with myocardial infarction. Microvascular Res. *78*, 393–404.

Lim, L., Lau, N., Garrett-Engele, P., Grimson, A., Schelter, J., Castle, J., Bartel, D., Linsley, P., and Johnson, J. (2005). Microarray analysis shows that some microRNAs down-regulate large numbers of target mRNAs. Nature *433*, 769–773.

Long, J., Wang, Y., Wang, W., Chang, B., and Danesh, F. (2011). MicroRNA-29c is a signature microRNA under high glucose conditions that targets Sprouty homolog *1*, and its *in vivo* knockdown prevents progression of diabetic nephropathy. J. Biol. Chem. *286*, 11837–11848.

Lu, Y., Xiao, J., Lin, H., Bai, Y., Luo, X., Wang, Z., and Yang, B. (2009). A single anti-microRNA antisense oligodeoxyribonucleotide (AMO) targeting multiple microRNAs offers an improved approach for microRNA interference. Nucleic Acids Res. *37*, e24.

Lu, Y., Zhang, Y., Wang, N., Pan, Z., Gao, X., Zhang, F., Zhang, Y., Shan, H., Luo, X., Bai, Y., *et al.* (2010). MicroRNA-328 contributes to adverse electrical remodeling in atrial fibrillation. Circulation *122*, 2378–2387.

Ma, L., Reinhardt, F., Pan, E., Soutschek, J., Bhat, B., Marcusson, E., Teruya-Feldstein, J., Bell, G., and Weinberg, R. (2010). Therapeutic silencing of miR-10b inhibits metastasis in a mouse mammary tumor model. Nature Biotechnol. *28*, 341–347.

Mattiske, S., Suetani, R.J., Neilsen, P.M., and Callen, D.F. (2012). The oncogenic role of miR-155 in breast cancer. Cancer Epidemiol. Biomarkers Prev. *21*, 1236–1243.

Meister, G., Landthaler, M., Dorsett, Y., and Tuschl, T. (2004). Sequence-specific inhibition of microRNA- and siRNA-induced RNA silencing. RNA *10*, 544–550.

Naguibneva, I., Ameyar-Zazoua, M., Nonne, N., Polesskaya, A., Ait-Si-Ali, S., Groisman, R., Souidi, M., Pritchard, L.L., and Harel-Bellan, A. (2006). An LNA-based loss-of-function assay for micro-RNAs. Biomed. and Pharmacotherapy *60*, 633–638.

Najafi-Shoushtari, S., Kristo, F., Li, Y., Shioda, T., Cohen, D., Gerszten, R., and Naar, A. (2010). MicroRNA-33 and the SREBP host genes cooperate to control cholesterol homeostasis. Science *328*, 1566–1569.

Obad, S., dos Santos, C., Petri, A., Heidenblad, M., Broom, O., Ruse, C., Fu, C., Lindow, M., Stenvang, J., Straarup, E., *et al.* (2011). Silencing of microRNA families by seed-targeting tiny LNAs. Nat. Genet. *43*, 371–378.

Oh, S.Y., Ju, Y., and Park, H. (2009). A highly effective and long-lasting inhibition of miRNAs with PNA-based antisense oligonucleotides. Molecules and Cells *28*, 341–345.

Oh, S.Y., Ju, Y., Kim, S., and Park, H. (2010). PNA-based antisense oligonucleotides for micrornas inhibition in the absence of a transfection reagent. Oligonucleotides *20*, 225–230.

Parker, J.S., Roe, S.M., and Barford, D. (2005). Structural insights into mRNA recognition from a PIWI domain-siRNA guide complex. Nature *434*, 663–666.

Patrick, D., Montgomery, R., Qi, X., Obad, S., Kauppinen, S., Hill, J., van Rooij, E., and Olson, E. (2010). Stress-dependent cardiac remodeling occurs in the absence of microRNA-21 in mice. J. Clin. Invest. *120*, 3912–3916.

Pegtel, D.M., Cosmopoulos, K., Thorley-Lawson, D.A., van Eijndhoven, M.A., Hopmans, E.S., Lindenberg, J.L., de Gruijl, T.D., Wurdinger, T., and Middeldorp, J.M. (2010). Functional delivery of viral miRNAs via exosomes. Proc. Natl. Acad. Sci. U.S.A. *107*, 6328–6333.

Rayner, K., Sheedy, F., Esau, C., Hussain, F., Temel, R., Parathath, S., van Gils, J., Rayner, A., Chang, A., Suarez, Y., *et al.* (2011). Antagonism of miR-33 in mice promotes reverse cholesterol transport and regression of atherosclerosis. J. Clin. Invest. *121*, 2921–2931.

Robertson, B., Dalby, A.B., Karpilow, J., Khvorova, A., Leake, D., and Vermeulen, A. (2010). Specificity and functionality of microRNA inhibitors. Silence *1*, 10.

Rosa, A., Spagnoli, F.M., and Brivanlou, A.H. (2009). The miR-430/427/302 family controls mesendodermal fate specification via species-specific target selection. Developmental Cell *16*, 517–527.

Selbach, M., Schwanhausser, B., Thierfelder, N., Fang, Z., Khanin, R., and Rajewsky, N. (2008). Widespread changes in protein synthesis induced by microRNAs. Nature *455*, 58–63.

Sonkoly, E., Lovén, J., Xu, N., Meisgen, F., Wei, T., Brodin, P., Jaks, V., Kasper, M., Shimokawa, T., Harada, M., *et al.* (2012). MicroRNA-203 functions as a tumor suppressor in basal cell carcinoma. Oncogenesis *1*, e3.

Staton, A.A., and Giraldez, A.J. (2011). Use of target protector morpholinos to analyze the physiological roles of specific miRNA–mRNA pairs *in vivo*. Nat. Protocol. *6*, 2035–2049.

Stein, C.A., Hansen, J.B., Lai, J., Wu, S., Voskresenskiy, A., Hog, A., Worm, J., Hedtjarn, M., Souleimanian, N., Miller, P., *et al.* (2010). Efficient gene silencing by delivery of locked nucleic acid antisense oligonucleotides, unassisted by transfection reagents. Nuc. Acids Res. *38*, e3.

Stenvang, J., Petri, A., Lindow, M., Obad, S., and Kauppinen, S. (2012). Inhibition of microRNA function by antimiR oligonucleotides. Silence 3, 1.

Straarup, E.M., Fisker, N., Hedtjarn, M., Lindholm, M.W., Rosenbohm, C., Aarup, V., Hansen, H.F., Ørum, H., Hansen, J.B., and Koch, T. (2010). Short locked nucleic acid antisense oligonucleotides potently reduce apolipoprotein B mRNA and serum cholesterol in mice and non-human primates. Nucleic Acids Res. *38*, 7100–7111.

Suryawanshi, H., Scaria, V., and Maiti, S. (2010). Modulation of microRNA function by synthetic ribozymes. Mol. Biosys. *6*, 1807–1809.

Suryawanshi, H., Lalwani, M.K., Ramasamy, S., Rana, R., Scaria, V., Sivasubbu, S., and Maiti, S. (2012). Antagonism of microRNA function in zebrafish embryos by using locked nucleic acid enzymes (LNAzymes). Chem Bio Chem *13*, 584–589.

Swarbrick, A., Woods, S., Shaw, A., Balakrishnan, A., Phua, Y., Nguyen, A., Chanthery, Y., Lim, L., Ashton, L., Judson, R., *et al.* (2010). miR-380-5p represses p53 to control cellular survival and is associated with poor outcome in MYCN-amplified neuroblastoma. Nat. Med. *16*, 1134–1140.

Thum, T., Gross, C., Fiedler, J., Fischer, T., Kissler, S., Bussen, M., Galuppo, P., Just, S., Rottbauer, W., Frantz, S., *et al.* (2008). MicroRNA-21 contributes to myocardial disease by stimulating MAP kinase signalling in fibroblasts. Nature *456*, 980–984.

Torres, A.G., and Gait, M.J. (2012). Exploiting cell surface thiols to enhance cellular uptake. Trends Biotechnol. *30*, 185–190.

Torres, A.G., Fabani, M.M., Vigorito, E., and Gait, M.J. (2011a). MicroRNA fate upon targeting with anti-miRNA oligonucleotides as revealed by an improved Northern-blot-based method for miRNA detection. RNA *17*, 933–943.

Torres, A.G., Threlfall, R.N., and Gait, M.J. (2011b). Potent and sustained cellular inhibition of miR-122 by lysine-derivatized peptide nucleic acids (PNA) and phosphorothioate locked nucleic acid (LNA)/2′-O-methyl (OMe) mixmer anti-miRs in the absence of transfection agents. Artificial DNA, PNA and XNA 2, 71–78.

Torres, A.G., Fabani, M.M., Vigorito, E., Williams, D., Al-Obaidi, N., Wojciechowski, F., Hudson, R.H., Seitz, O., and Gait, M.J. (2012). Chemical structure requirements and cellular targeting of microRNA-122 by peptide nucleic acids anti-miRs. Nucleic Acids Res. *40*, 2152–2167.

Trajkovski, M., Hausser, J., Soutschek, J., Bhat, B., Akin, A., Zavolan, M., Heim, M.H., and Stoffel, M. (2011). MicroRNAs 103 and 107 regulate insulin sensitivity. Nature *474*, 649–653.

Trang, P., Wiggins, J., Daige, C., Cho, C., Omotola, M., Brown, D., Weidhaas, J., Bader, A., and Slack, F. (2011). Systemic delivery of tumor suppressor microRNA mimics using a neutral lipid emulsion inhibits lung tumors in mice. Mol. Ther. *19*, 1116–1122.

Valadi, H., Ekstrom, K., Bossios, A., Sjostrand, M., Lee, J.J., and Lotvall, J.O. (2007). Exosome-mediated transfer of mRNAs and microRNAs is a novel mechanism of genetic exchange between cells. Nature Cell Biol. *9*, 654–659.

Valoczi, A., Hornyik, C., Varga, N., Burgyan, J., Kauppinen, S., and Havelda, Z. (2004). Sensitive and specific detection of microRNAs by northern blot analysis using LNA-modified oligonucleotide probes. Nucleic Acids Res. *32*, e175.

Vermeulen, A., Robertson, B., Dalby, A.B., Marshall, W.S., Karpilow, J., Leake, D., Khvorova, A., and Baskerville, S. (2007). Double-stranded regions are essential design components of potent inhibitors of RISC function. RNA *13*, 723–730.

Worm, J., Stenvang, J., Petri, A., Frederiksen, K., Obad, S., Elmen, J., Hedtjarn, M., Straarup, E., Hansen, J., and Kauppinen, S. (2009). Silencing of microRNA-155 in mice during acute inflammatory response leads to derepression of c/ebp Beta and down-regulation of G-CSF. Nucleic Acids Res. *37*, 5784–5792.

Yan, L.X., Wu, Q.N., Zhang, Y., Li, Y.Y., Liao, D.Z., Hou, J.H., Fu, J., Zeng, M.S., Yun, J.P., Wu, Q.L., *et al.* (2011). Knockdown of miR-21 in human breast cancer cell lines inhibits proliferation, *in vitro* migration and *in vivo* tumor growth. Breast Cancer Res. *13*, R2.

Zamecnik, P.C., and Stephenson, M.L. (1978). Inhibition of Rous sarcoma virus replication and cell transformation by a specific oligodeoxynucleotide. Proc. Natl. Acad. Sci. U.S.A. *75*, 280–284.

Zhang, Y., Roccaro, A.M., Rombaoa, C., Flores, L., Obad, S., Fernandes, S.M., Sacco, A., Liu, Y., Ngo, H., Quang, P., *et al.* (2012). LNA-mediated anti-miR-155 silencing in low-grade B-cell lymphomas. Blood *120*, 1678–1686.

Exploiting MicroRNAs to Regulate Transgene Expression

7

Virginie Pichard, Dejana Ivacik and Nicolas Ferry

Abstract

RNA interference (RNAi) is as highly conserved gene regulatory mechanism which is triggered by double-stranded RNA. The discovery of this naturally functioning gene silencing mechanism has contributed to a deeper understanding of the function and regulation of eukaryotic genes. In mammalian cells, RNAi functions in regulating gene expression via small, non-coding RNA molecules, known as microRNAs (miRNAs). Over the past decade, numerous studies have emphasized the role of miRNAs as powerful transcriptome regulators highlighting the potential usefulness of these small RNA effecters to regulate transgene expression. Precise tissue-specific control of transgene expression is a prerequisite for many investigational and therapeutic applications involving gene transfer. Consequently, miRNAs have been exploited by gene therapists to improve transcriptional regulation of gene transfer vectors and accomplish tight spatial and temporal regulation of transgene expression in different therapeutic contexts. In this chapter, we will emphasize the versatility of miRNAs to improve the efficacy of many gene therapy applications, with a particular focus on their potential for improving specificity and safety of gene therapy as well as their role in averting immune-mediated clearance of gene-modified cells.

Introduction

The field of gene therapy is evolving constantly and a number of clinical successes of this therapeutic strategy have been demonstrated recently (Cartier *et al.*, 2009; Maguire *et al.*, 2008; Nathwani *et al.*, 2011). Viral vectors continue to be one of the most effective tools for delivery of therapeutic transgenes to target cells *in vivo*. Nevertheless, the development of specific as well as non-specific immune and inflammatory responses towards transgene-expressing cells is one of the major obstacles hampering the clinical advancement of this potentially effective therapeutic strategy (Ciesielska *et al.*, 2012; Di Domenico *et al.*, 2005; Gilchrist *et al.*, 2002). Indeed, unintended expression of transgenes in non-target tissues may induce an immune response to the viral capsid, the transgene product or the transduced cells, ultimately resulting in immune-mediated clearance of transgene-expressing cells and diminished expression of the protein encoded by the therapeutic transgene. Low transgene expression is undesirable for many gene therapy applications where high systemic levels of the therapeutic protein are required to attain a therapeutic effect.

Various approaches have been employed by gene therapists to improve low transgene expression and target delivery. An additional layer of regulation can be accomplished by surface and transcriptional targeting of viral vectors. Modification of vector tropism, by engineering viral capsids via pseudotyping or conjugation of capsid proteins with molecular adaptors (specific antibodies, small peptide ligand), has proven to be a useful strategy in overcoming transduction of non-target cells (Anliker *et al.*, 2010; Markusic *et al.*, 2009; Sellner *et al.*, 2008). However, this strategy has shown some limitation *in vivo* as it does not completely abolish vector uptake by non-targeted cell populations. In addition, natural or artificial tissue-specific promoters can be exploited to achieve tight spatial control of transgene expression. Although valuable results were obtained using this approach (Di Domenico *et al.*, 2006; Pastore *et al.*, 1999), tissue-specific promoters are usually weak when compared to more ubiquitously active promoters often resulting in low levels of transgene expression that are inadequate for most therapeutic applications. Moreover, in some instances the length of the specific regulatory sequence can outsize the packaging limitation of most viral vectors causing inefficient packaging of recombinant viral particles. Finally, leakiness of transcriptional regulatory elements can result in transgene expression in non-targeted cells which may initiate pathological immune responses or unintended side-effects (Bell *et al.*, 2011; Wu *et al.*, 2009). As a result, the use of tissue-specific promoters alone is often not sufficient to enable tight and cell-type specific regulation of transgene expression.

Recently, a class of small non-coding RNA molecules, known as microRNAs (miRNA) have emerged as powerful transcriptome regulators and heralded the development of an exciting new field of transcriptional regulation. Gene therapists have exploited this novel class of gene regulatory molecules to add an additional layer of transcriptional regulation when engineering therapeutic viral vectors and accomplish tight spatial and temporal regulation of transgene expression (Brown *et al.*, 2007a,b; Card *et al.*, 2012; Cawood *et al.*, 2009, 2011; Geisler *et al.*, 2011; Karali *et al.*, 2011; Qiao *et al.*, 2011; Suzuki *et al.*, 2008; Wu *et al.*, 2009; Xie *et al.*, 2011). In the following chapter, we will emphasize the versatility of this novel technology to improve the efficacy of many gene therapy applications, with a particular focus on improving the specificity and safety of gene therapy as well as promoting immune tolerance of gene-modified cells.

Using microRNAs avoid immune response to transgene product after gene transfer

In vivo direct delivery of a recombinant viral vector may trigger cellular immune responses to the transgene product leading to tissue inflammation and destruction of transduced cells (Aubert *et al.*, 2002; Ciesielska *et al.*, 2012; Di Domenico *et al.*, 2005). In some instances, such deleterious effects result from expression of the transgene in antigen-presenting cells Indeed, many studies have demonstrated that recombinant retroviral, adenoviral and adeno-associated vectors (AAV) can transduce dendritic cells and generate immune responses to transgene products (Hartigan-O'Connor *et al.*, 2001; Naldini *et al.*, 1996; VandenDriessche *et al.*, 2002; Veron *et al.*, 2007; Xin *et al.*, 2006). The route of administration, the dose and the choice of the promoter can modulate the tissue specificity of transgene expression, and thus have an impact on the generation of immune response towards the encoded transgene (Mays and Wilson, 2011). Indeed, the use of specific promoters to restrict transgene expression to targeted cells while avoiding transgene expression in professional antigen-presenting cells (APCs) has been widely employed to limit transgene-specific immune responses (Di Domenico *et al.*, 2006; Franco *et al.*, 2005; Lee *et al.*, 2011). However, the use of tissue specific promoters alone will not necessarily circumvent the development of transgene-specific neutralizing antibodies and immune-mediated elimination of cells expressing the transgene (Osborn *et al.*, 2011; Schmitt *et al.*, 2010b; Yuasa *et al.*, 2002). Osborn *et al.* (2011) evaluated the efficiency of lentiviral vectors carrying a liver-specific promoter (human apolipoprotein E hepatic locus control region and the human α-1-antitrypsin promoter (sApoE.HCR. hAATp) driving human IDUA (lysosomal hydrolase α-L-iduronidase) cDNA for the treatment of mucopolysaccharidosis type I. They showed that only short-term expression of IDUA was achievable because of immune response against IDUA protein. Additionally, our group has shown that in adult Gunn rats, a model for Crigler–Najjar type I (CN-I), characterized by a high serum level of unconjugated bilirubin, we could not achieve permanent correction of serum bilirubin level using a liver-specific promoter driving expression of the deficient enzyme, uridine diphosphate glucoronyltransferase (UGT1A1) (Schmitt *et al.*, 2010b). In our study lentiviral vectors carrying the human UGT1A1 cDNA under control of a synthetic liver-specific murine transthyretin (mTTR) promoter were delivered systemically to correct serum bilirubin level in adult Gunn rats. We observed a transient correction of bilirubinaemia, followed by clearance of transgene-expressing cells due to priming of effector T cells and the subsequent development of a cytotoxic immune response directed against UGT1A1.

Therefore, precluding transgene expression in APCs would be an attractive strategy to escape the initiation of a deleterious immune response against the transgene product and promote immune tolerance following systemic introduction of therapeutic transgenes. The natural silencing activity of miRNAs can be exploited to add an additional layer of specificity in the regulation of gene expression when engineering gene transfer vectors and expression cassettes (Fig. 7.1). Brown *et al.* (2006) were the first to describe the use of endogenous miRNA activity to accomplish transcriptional tissue-specific regulation of transgenes. They proposed a clinically relevant strategy to negatively regulate transgene expression in APC and evade an unwanted immune response after gene transfer in liver. The strategy is based on the ability of the guide strands of natural miRs to hybridize to the 3′UTRs of a gene and suppress translation. Addition of an artificially engineered miRNA target downstream of a transgene would result in degradation of the transgene mRNA following transcription provided that the guide and target sequences are perfectly complementary. Importantly, miRNAs have a tissue-, lineage- and differentiation-specific activity profiles (Landgraf *et al.*, 2007; Petriv *et al.*, 2010). Therefore it is theoretically possible to silence any transgene in different cell types and at a state-specific level by adding to the transgene transcript sequences that are targets to the corresponding miRNA. Brown *et al.* (2006) demonstrated a proof of principle for the potential utility of this novel technology. They designed a miRNA-regulated lentiviral vector by inserting four tandem copies of a sequence that is perfectly

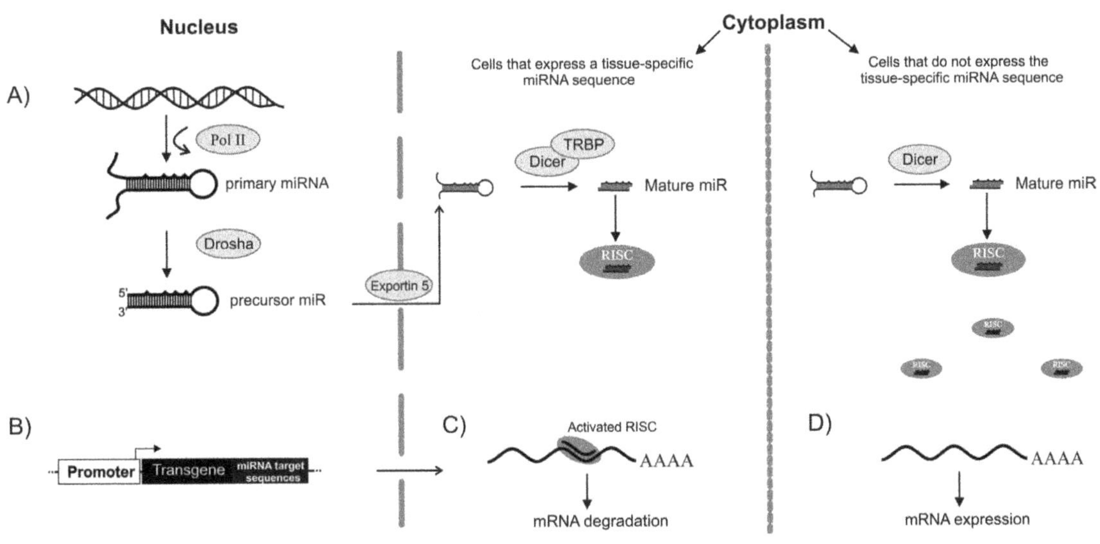

Figure 7.1 Exploiting miRNAs to accomplish safe and specific transgene expression. (A) Schematic representation of miRNA biogenesis. MiRNAs are typically transcribed from RNA polymerase (Pol) II promoters as long, hairpin-like primary miRNA transcripts (pri-miRNA), which are processed to precursor miRNA (pre-miRNA) intermediates by the RNase III endonuclease Drosha. Pre-miRNAs are then exported to the cytoplasm by Exportin-5 and processed to a miRNA duplex by the cytoplasmic RNase III endonuclease Dicer. Dicer together with the TAR RNA binding protein (TRBP) loads the duplex into the RNA-induced silencing complex (RISC). One of the strands in the duplex is selected as the mature miRNA guide strand and mediates sequence-specific gene silencing by binding to the 3′ untranslated regions (UTR) of target mRNAs. (B) Depiction of an expression cassette that encodes a transgene of interest with miRNA target sequences included in its 3′ UTR. (C) Following transcription of the transgene, endogenous miRNAs present in the cell will bind to the miRNA target sequences contained within the 3′ UTR resulting in mRNA silencing. (D) Transgene mRNA is not silenced in the absence of cell-type specific miRNAs targeted to the 3′UTR.

complementary to miRNA-142-3p, an endogenous miR specifically expressed in cells from haematopoietic lineage. The recombinant lentiviral vector encoded a green fluorescent protein (GFP) driven by a ubiquitous phosphoglycerate kinase 1 promoter (LV.PGK-GFP). Inclusion of the miR-142-3p target sequences resulted in a substantial reduction (up to a 100-fold) of GFP expression in haematopoietic lineages while GFP expression in non-haematopoietic cells was retained. In addition, following systemic injection of the recombinant LV.PGK-GFP lentivirus in mice, a widespread pattern of cell expression within the liver was observed, including Kupffer cells, hepatocytes and endothelial cells. In contrast mice injected with vectors containing the miR 142-3p target repeats had undetectable expression of GFP in dendritic and Kupffer cells. Finally, they showed that in mice treated with lentiviral vectors carrying GFP without miRNA-142-3p target repeats, GFP positive hepatocytes were detected at day 5, followed by a rapid decline to almost undetectable levels at day 14 due to the development of an immune response against the GFP protein. In contrast, the proportion of GFP transduced hepatocytes remained stable for more than 4 months in the livers of mice that received miRNA-regulated recombinant lentiviruses. The authors successfully demonstrated that the use of miR 142-targeted vectors is a relevant strategy to de-target transgene expression from APC and circumvent unwanted immune responses following liver gene transfer. The same team went on to demonstrate a potential clinical utility of the novel technology by showing stable correction of haemophilia B in FIX-deficient mice following systemic delivery of lentiviral vectors that encoded miRNA-142-regulated FIX under the control of a hepatocyte-specific promoter (Brown *et al.*, 2007a). Interestingly, inclusion of miRNA-142 targets resulted in superior suppression of transgene-specific immune responses when compared to individual use of an albumin promoter driving liver-specific expression of the transgene.

This novel approach constituted a new paradigm in the design of regulated gene transfer vectors and therapeutic transgenes as it allows for a more stringent regulation of transgene expression. Moreover, we have successfully applied this strategy to achieve long-term correction of hyperbilirubinaemia in Gunn rats (Schmitt *et al.*, 2010a). By using lentiviral vectors that encoded the human UGT1A1 cDNA under control of a liver specific promoter together with four copies of the miR-142-3p target sequence we were able to obtain a complete and permanent normalization of serum bilirubin level after systemic injection in adult Gunn rats. Using this strategy made it possible to avoid the initiation of a UGT1A1-specific immune response that usually takes place after transduction and transgene expression in APCs (Fig. 7.2). Several groups carried out further investigations to get a deeper insight into the mechanisms underlying the absence of transgene-specific immune responses. Investigations carried out by Anoni *et al.* (2009) indicated that combining the use of liver-specific promoters to achieve hepatocyte specific expression with miRNA-142-mediated posttranscriptional silencing of the transgene in APC results in the induction of active transgene tolerance, not immune ignorance as originally thought. Matrai and colleagues further demonstrated that active tolerance is maintained by the induction of antigen specific FoxP3-expressing, regulatory T cells (Matrai *et al.*, 2011). A possible mechanism underlying the induction of tolerance following hepatocyte-targeted transgene expression by miRNA-142-regulated gene-transfer vectors has been proposed recently by Goudy and colleagues (Goudy *et al.*, 2011).

Nevertheless, it is important to note that the induction of transgene tolerance using this approach is dependent on the maintenance of transgene expression in hepatocytes and the absence of transgene expression in APC (Annoni *et al.*, 2009). Indeed, when liver expression was detargeted as well by using target sequences for the miR-122 that is specifically express in hepatocytes, anti GFP response was observed event though the target sequences for miR-142-3p were still present. It should be noted, that in some studies, shut off of transgene expression in APC after gene transfer in liver did not guarantee transgene persistence (Matsui *et al.*, 2011). Matsui and colleagues injected VSV-G pseudotyped lentiviral that encoded a liver-specific promoter driving the expression of a B-domain–deleted FVIII transgene cassette into the circulation of

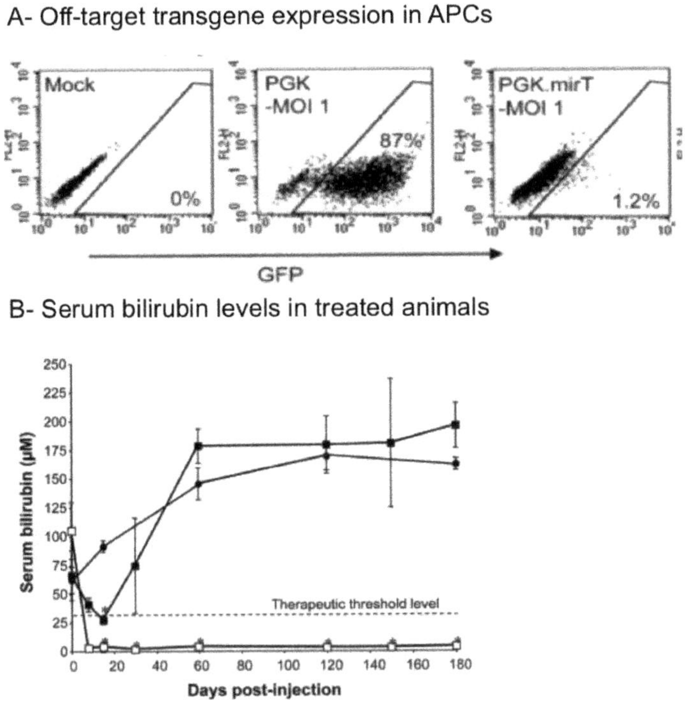

Figure 7.2 Exploiting microRNA-regulated lentiviral vectors to accomplish long-term correction of hyperbilirubinaemia in Gunn rats. (A) Transgene expression was analysed in rat antigen presenting cells (APCs) transduced at a multiplicity of infection (MOI) of 1 with a lentiviral vector encoding a phosphoglycerate kinase 1 promoter expressed green fluorescent protein (GFP) with (PGK-mirT) or without (PGK) microRNA (miRNA)-142-3p target sequences. The percentage of GFP-positive cells is indicated in the lower right corner of each plot. (B) Serum billirubin levels were analysed in adult Gunn rats that had received lentiviral vectors encoding the human UGT1A1 cDNA under the control of the mTTR liver specific promoter together with 4 copies of the miR-142-3p target sequence (mTTR.hUGT1A1.142T) as well as control lentiviral vectors (mTTR.hUGT1A1) which did not contain miR-142-3p target sequences. Each value plotted represents the mean ± SD for uninjected rat controls (n=4; closed circles), mTTR.hUGT1A1-injected rats (n=3; closed squares), and mTTR.hUGT1A1.142T rats (n=5, open squares). Control rats developed severe hyperbilirubinaemia. *P<0.05 injected versus control animals.

haemophilia A mice. Although the transgene cassette comprised four miRNA-142 targets in the 3′ UTR, the authors observed that they were not able to prevent an anti-FVIII immune response *in vivo* using this delivery system. However, they were able to obtain long-term FVIII expression when a GP64 envelope glycoprotein was used to pseudotype the lentivirus. Whether or not this strategy may apply to all vector types is still unclear and some results using non viral as well as viral vectors were less conclusive (Osborn *et al.*, 2011; Kron *et al.*, 2011; Qiao *et al.*, 2011). Osborn *et al* (2011) showed that the use of a liver-specific promoter in conjunction with miRNA-142 target sequences did not result in long-term stable expression of the deficient α-l-iduronidase (IDUA) protein in a mucopolysaccharidosis type I mouse model following a hydrodynamic injection of minicircle DNA carrying a miRNA-142 regulated IDUA transgene. In the same manner, Kron *et al.* (2011) showed that in contrast to lentiviral vectors, recombinant adenovirus (Ad) vectors carrying miR-142-3p target sequences to eliminate the expression of hepatitis B virus surface antigen (HBsAg) in APC, were not capable of inducing immunological tolerance. The authors suggested several hypothesis to explain why the incorporation of miR-142 target sequences did not promote immunological tolerance towards the transgene product. Cross priming mechanisms in which the protein is captured by APC and cross presented to cytotoxic T cells may be involved. Alternatively other cells types, like endothelial cells, which are highly susceptible to adenoviral infection, might be able to present the antigen and prime CD8 T cells.

In conclusion, hepatocyte-targeted gene therapy using miRNA-regulated vectors is currently the best available strategy to overcome short-term transgene expression following gene transfer due to pathological immune responses. Nevertheless, despite encouraging results the immunological outcome of *in vivo* gene transfer is still a complicated process which is highly dependent on many variables such the gene transfer vector system as well as the therapeutic transgene. In addition the precise mechanisms leading to liver-induced transgene tolerance using this strategy will have to be deciphered and the effectiveness, safety of the strategy will have to be determined in clinically relevant animal models before this exciting new technology can be transferred to the clinic.

Using microRNA to improve cell specificity and escape side-effect of gene transfer

Using miRNAs to accomplish tissue-specific transgene expression

Naturally, miRNAs are expressed in a tissue-, lineage- and differentiation specific manner making it possible to exploit miRNA activity to achieve tight spatial and temporal control of transgene expression (Geisler *et al.*, 2011; Karali *et al.*, 2011; Qiao *et al.*, 2011; Xie *et al.*, 2011). Systemic administration of AAV9 vectors results in transduction of several tissues, including the liver, heart, and skeletal muscle. Adding an additional layer of regulation using tissue-specific miRNAs has been used to restrict AAV9 vector transduction to selected tissues. Intravascularly administered AAV9 vectors can cross the blood–brain barrier, making these vectors a potentially effective vehicle for gene therapy of central nervous system (CNS) diseases. In some applications, CNS-directed gene transfer may cause toxicity if the transgene is expressed outside the CNS. Tissue-specific CNS promoters are often too large to be packaged into the recombinant AAV genome. Xi *et al.* (2011) have shown that combination of hepatocyte- and muscle-specific miRNA target sequences can improve central nervous system targeting by AAV9 vectors upon systemic injection. Following an injection of their therapeutic AAV9 miRNA-regulated vector, Xi *et al.* observed reduced transgene expression in the liver (20-fold), heart (100-fold), and muscle (50-fold) while stable transgene expression in the brain was maintained.

In a similar manner, Geisler *et al.* (2011) have shown that hepatocyte de-targeting can enhance the specificity of cardiac gene transfer. They constructed recombinant AAV9 vectors to encode target sequences of the liver specific MiRNA-122 into the 3′ UTR of an enhanced GFP reporter gene expression cassette that was placed under the control of a cytomegalovirus (CMV) or a cardiac specific promoter. Vectors with or lacking miRNA-122 target sequences were injected via the tail vein in mice and GFP expression was quantified in the heart and the liver 4 weeks following gene transfer. Geisler *et al.* (2011) showed that inclusion of miRNA-122 target sequences resulted in suppression of GFP expression in hepatic cells. Nevertheless, they demonstrated that a combination of a tissue-specific promoter together with a miRNA target sequence increased vector specificity, as in the absence of miRNA regulation leaky transgene expression was observed.

MiRNA-mediated regulation of transgene expression has also been studied in the retina, an immune-privileged site. In inherited retinal degenerations, gene expression is typically targeted to the retinal pigment epithelium (RPE) or the photoreceptors (PR). Endogenous RPE- or PR-specific promoter elements have been used to express therapeutic transgene, often resulting in insufficient levels of transgene expression. In an attempt to improve the specificity of transgene expression Karali *et al.* (2011) used RPE specific miR-204, or PR specific miR-124 to restrict expression to either PR or the RPE. They administered subretinally in mice and pigs rAAV2/5 vectors expressing enhanced GFP (EGFP) and containing four tandem copies of miR 124 or miR 204 complementary sequences in the 3′UTR of the transgene expression cassette. Ubiquitous promoters were used to drive expression of the transgene cassettes. The authors demonstrated that they were capable of accomplishing RPE- and PR-specific transgene expression using the miR-124 and miR-204 regulated AAV vectors.

In summary, miRNA-mediated regulation of transgene expression can be applied in different therapeutic contexts both to restrict expression to a specific cell type when using ubiquitous promoters or to provide an additional layer of gene expression regulation when using cell-specific promoters.

Using miRNA regulation to escape side-effect of gene transfer

The use of replication-competent oncolytic viruses to deliver suicide genes or genes encoding enzymes that turn prodrugs into toxic compounds to cancer cells has recently gained popularity as it is proving to be a potentially effective strategy to treat cancer. Nevertheless unintended delivery of these viruses to healthy cells can lead to considerable toxicity and unwanted side-effects. As a result, several teams have evaluated the potential of exploiting miRNA regulation to reduce off-target toxicity arising from the systemic administration of replication-competent oncolytic viruses (Card *et al.*, 2012; Cawood *et al.*, 2009, 2011). Cawood *et al.* (2009) showed that inclusion liver specific miR-122 target sequences had a dramatic effect on reducing the hepatic activity and toxicity of the highly hepatotropic adenovirus serotype 5 (Ad 5) oncolytic virus. They observed a 30-fold reduction in luciferase reporter gene expression in primary hepatocytes infected with an Ad5 vector that encoded an EF1alpha promoter-driven luciferase expression cassette containing four miRNA-122 target sites in the 3′UTR compared to cells infected with wild type Ad 5. Furthermore, to assess the efficiency of their strategy *in vivo* they performed several intravenous injections with PBS, wild type Ad5 or the miRNA regulated Ad5 oncolytic virus in mice bearing HepG2 xenografts. The HepG2 human hepatocellular carcinoma cell line has been reported not to express miRNA-122. Significant hepatic toxicity was observed following administration of the wild type Ad5 oncolytic virus. In contrary, mice injected with the miRNA regulated Ad5 oncolytic virus showed a lower level of hepatic toxicity that was similar to PBS controls. In addition, imaging of the animals confirmed that was no detectable reporter transgene expression in the liver and the tumour volume was significantly reduced. These results demonstrated the usefulness of using miRNA-122-regulated oncolytic adenovirus to control toxicity while maintaining a therapeutic effect.

MiRNA-122-regulated oncolytic viruses have also been exploited to improve the safety and toxicity profile of suicide gene therapy targeting liver tumours (Suzuki *et al.*, 2008; Wu *et al.*, 2009). Cancer suicide gene therapies use either toxic genes or genes encoding enzymes that turn prodrugs into toxic compounds. Gene expression must be tightly restricted to tumour cells to reduce side-effects. Use of tissue-specific promoters in suicide gene therapy does not allow to restrict therapeutic gene expression solely to tumour cells. Suzuki *et al.* (2008) have studied the effects of miR-122 regulation in suicide cancer gene therapy involving herpes simplex virus thymidine kinase (HSVtk)-expressing adenovirus vectors. The principle of the HSVtk/ganciclovir (GCV) system is based on expression of the HSV-tk gene which sensitizes tumour cells to the toxic effect of ganciclovir by converting the prodrug (ganciclovir) into toxic, phosphorylated derivatives. Leakage of transgene expression is commonly observed following intratumoral adenoviral injection, giving rise to HSVtk expression in healthy liver cells with subsequent hepatic damage. Suzuki *et al.* (2008) incorporated four tandem copies of the miR-122 target sequence into the HSV-tk transgene expression cassette in an attempt to de-target hepatic HSV-tk expression. They demonstrated that intratumoral injection of the Ad vector containing the miR-122 target sequences into tumour-bearing mice resulted in a dramatic reduction in hepatic HSV-tk expression and consequently in HSVtk/GCV-induced hepatotoxicity. A single intratumoral injection of Ad-HSVtk-miR-122 or Ad-HSVtk-control into B16 tumour-bearing mice resulted in significant tumour regression. However, a few days after the Ad-HSVtk-control injection was administered, severe hepatotoxicity with inflammatory cell infiltration was observed in the liver, resulting in a significant reduction in body weight and high mortality in the population that received the injection.

In contrast, animals that received an intratumoral injection of the MiRNA-122-regulated Ad-HSVtk vector had low levels of inflammatory cell infiltration in their livers and no mortality was recorded in this group. In an attempt to develop an effective suicide gene therapy treatment for glioma, Wu *et al.* (2009) constructed a baculoviral vector that harboured a glial fibrillary acidic protein (GFAP) gene promoter driving expression of a HSVtk transgene cassette with or without miRNA target sequences. The miRNA-regulated vector contained nine copies of target sequences of three miRNAs target sequences, namely miR-31, miR-127 and miR-143, that have been reported to be down-regulated in glioma cells. Using the respective miRNA target sequences allowed for robust expression of the HSVtk suicide transgene in cells of glial origin while sparing other types of cells like normal astrocytes from cellular toxicity mediated by HSVtk/GCV treatment.

Other groups have exploited cell-specific miRNA activity to re-direct virus tropism and improve the tumour specificity of Coxsackie A21 oncolytic virus (Kelly *et al.*, 2008), measles virus (Kelly *et al.*, 2008), vaccinia virus (Hikichi *et al.*, 2011) as well as vesicular stomatitis virus (Kelly *et al.*, 2010).

The results obtained so far indicate that exploiting miRNA regulation may be an attractive tool to achieve desired tumour selectivity and redirect the tropism of oncolytic viruses. However the possibility of loss of miRNA-mediated regulation due to mutation of the target sequences comprised within the oncolytic virus or unforeseen saturation of the endogenous miRNA machinery has be examined carefully before this promising therapeutic strategy is tested in the clinic.

Novel type of expression system using microRNAs

To improve the versatility of transgene expression control, our group designed an original new system which combines the use of endogenous miRNA activity and the tetR-KRAB artificial transgene-repressor (Pichard *et al.*, 2012). The Kruppel-associated box protein (KRAB) has been demonstrated to be capable of mediating transcriptional repression over a range of tens of kilobases by triggering formation of heterochromatin in the vicinity of its binding site (Groner *et al.*, 2010). A chimeric transrepressor TetR-KRAB protein was engineered by fusing the KRAB domain of human Kox1 to the Tet repressor protein (TetR). The resulting TetR-KRAB was able to bind to the Tet operator DNA sequence (*tetO*) and repress Pol II and III promoters in the absence but not in the presence of Tet (Deuschle *et al.*, 1995; Laurenti *et al.*, 2010; Szulc *et al.*, 2006).

Our system comprises two different expression cassettes. The reporting cassette contains a marker or a transgene under control of a ubiquitously active or a liver specific promoter that has *tetO* binding sites. The second cassette is regulatory and encodes a TetR-KRAB–fusion protein with tissue-specific miRNA targets placed in the 3′ UTR of the transcript. Expression of the TetR-KRAB component is under control of miRNA target sequences that were selected according to the miRNA expression profile of specific cell types. In cells which do not express the tissue-specific miRNA and in the absence of doxycycline, the transgene is not expressed. In contrast, translation of the TetR-KRAB repressor is shut off in permissive cells that express the miRNAs, which in turn results in de-repression of the transgene (Fig. 7.3). Initially we evaluated the efficiency of the system in liver-derived Huh7 cells. We designed a regulatory cassette in which four copies of the miR-122 target sequence were inserted immediately downstream of the TetR-KRAB coding region. In the reporting cassette, GFP was inserted downstream of a CMV promoter or a liver-specific murine trnasthyretin (mTTR) promoter. The cassettes were inserted in a lentiviral backbone to generate lentiviral vectors that are suitable for *in vitro* characterization. We used Huh7 cells, known to express miR-122 as well as 293T cells which do not express MiRNA-122. Huh7 and 293T cells were transduced with the different vectors and we compared GFP expression in the presence or absence of doxycycline in each cell line. Using our TetR-KRAB regulation system without miRNA target sequences, in 293T cells we observed an 82% decrease in GFP mean fluorescence intensity (MFI) when doxycycline was withdrawn. This demonstrated the expected decrease in GFP

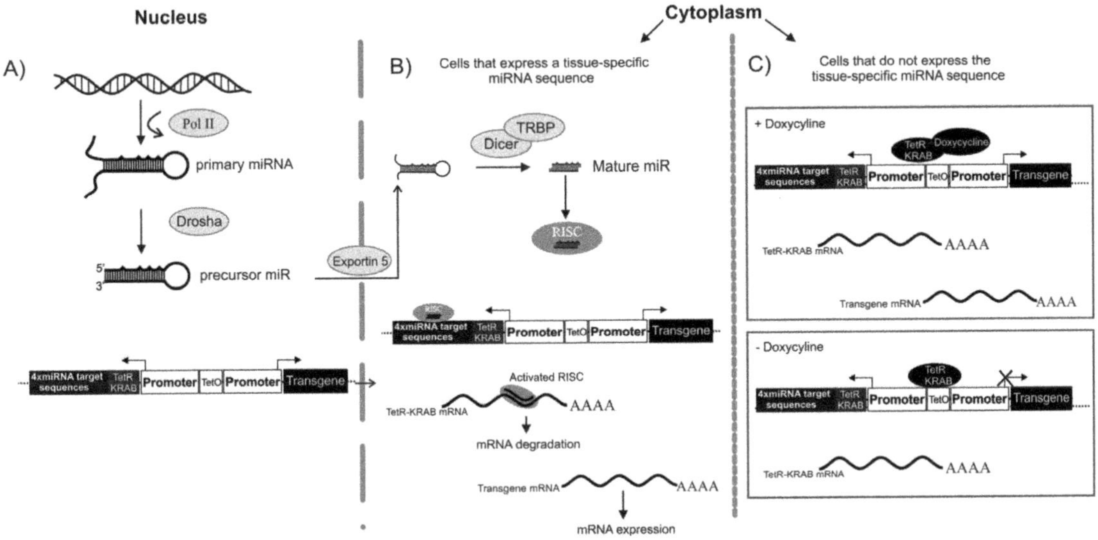

Figure 7.3 MicroRNA-regulated TetR-KRAB transcriptional control of transgene expression (A) Schematic representation of the microRNA (miRNA)-TetR-KRAB expression system. (B) In targeted cells the transgene of interest is expressed as the tissue-specific miRNA binds to its complementary target resulting in degradation of TetR-KRAB mRNA. (C) In the absence of tissue-specific miRNA, the repressor is translated and is free to bind the tetO operator, thus preventing transgene expression. However the transgene is expressed in the presence of a tetracyclin analogue, such as doxycycline, when binding of the TetR-KRAB to the tetO sequence is prevented.

expression effected by the TetR-KRAB protein in the absence of doxycycline. When we used constructs that contained ubiquitous promoters driving GFP we observed a decrease in GFP MFI of about 50% in 293T cells. In Huh7 cells, we observed a decrease in MFI when lentiviral constructs without the miR-122 target sequences were used (76% to 90%) but there was no variation in GFP expression when doxycycline was withdrawn after infection with miR-122 target sequence-containing vectors. Precise control of transgene expression was also demonstrated in macrophage- and muscle-derived cells using 4 copies of perfectly complementary miR-142-3p or miR-133 target sequences respectively.

To test the effectiveness of our system *in vivo*, the cassettes were inserted into recombinant AAV 2/8 vectors. We designed two DNA cassettes containing the TetR-KRAB coding sequence with four copies of miR-133 in the 3'UTR or without miRNA target sequences. The GFP- and TetR-KRAB-coding sequences were under the control of a CMV promoter. These vectors were injected into the two tibialis anterior muscles of subject mice. In mice injected with AAV constructs without miRNA target sequences, we observed a strong fluorescence signal in muscle when animals received doxycycline. In contrast, in the absence of doxycycline, there was a sharp (about 70%) decrease in fluorescence. However, in mice injected with AAV carrying miRNA-133 target sequences, there was no variation of GFP expression in the absence or presence of doxycycline.

In conclusion our multicomponent and modular regulatory system provides a robust and easily adaptable method for achieving regulated transgene expression in different tissue types. The efficiency of regulated transgene expression was influenced by the promoter used and was cell-type specific. Nevertheless the system allows for improvement of tissue-specific expression as it significantly reduces background expression. Moreover, our approach may also address the issue of immune-mediated clearance of transgene-expressing cells as a potential immune response to the TetR-KRAB chimeric protein would be directed to non-targeted cells that express the protein while targeted cells should be spared. This novel system can be exploited to achieve specific transgene expression that is critical for the development of effective gene therapies.

Conclusion

In summary, exploiting miRNA activity to regulate transgene expression has proven to be a versatile and potent technology that may be used to overcome many of the challenges hampering the broad development of gene therapy applications. Indeed, as shown by numerous studies during the past few years, combining this novel technology with existing targeting strategies has allowed gene therapists to accomplish precise, spatio-temporal regulation of transgene expression and improve the specificity and safety of many therapeutic strategies involving gene transfer. The next step will be to test the efficacy and safety of miRNA-mediated transgene regulation in large-animal models before clinical application of this exciting new technology becomes a reality.

References

Anliker, B., Abel, T., Kneissl, S., Hlavaty, J., Caputi, A., Brynza, J., Schneider, I.C., Munch, R.C., Petznek, H., Kontermann, R.E., *et al.* (2010). Specific gene transfer to neurons, endothelial cells and hematopoietic progenitors with lentiviral vectors. Nat. Methods *7*, 929–935.

Annoni, A., Brown, B.D., Cantore, A., Sergi, L.S., Naldini, L., and Roncarolo, M.G. (2009). *In vivo* delivery of a microRNA-regulated transgene induces antigen-specific regulatory T cells and promotes immunologic tolerance. Blood *114*, 5152–5161.

Aubert, D., Menoret, S., Chiari, E., Pichard, V., Durand, S., Tesson, L., Moullier, P., Anegon, I., and Ferry, N. (2002). Cytotoxic immune response blunts long-term transgene expression after efficient retroviral-mediated hepatic gene transfer in rat. Mol. Ther. *5*, 388–396.

Bell, P., Gao, G., Haskins, M.E., Wang, L., Sleeper, M., Wang, H., Calcedo, R., Vandenberghe, L.H., Chen, S.J., Weisse, C., *et al.* (2011). Evaluation of adeno-associated viral vectors for liver-directed gene transfer in dogs. Hum. Gene. Ther. *22*, 985–997.

Brown, B.D., Venneri, M.A., Zingale, A., Sergi Sergi, L., and Naldini, L. (2006). Endogenous microRNA regulation suppresses transgene expression in hematopoietic lineages and enables stable gene transfer. Nat. Med. *12*, 585–591.

Brown, B.D., Cantore, A., Annoni, A., Sergi, L.S., Lombardo, A., Della Valle, P., D'Angelo, A., and Naldini, L. (2007a). A microRNA-regulated lentiviral vector mediates stable correction of hemophilia B mice. Blood *110*, 4144–4152.

Brown, B.D., Gentner, B., Cantore, A., Colleoni, S., Amendola, M., Zingale, A., Baccarini, A., Lazzari, G., Galli, C., and Naldini, L. (2007b). Endogenous microRNA can be broadly exploited to regulate transgene expression according to tissue, lineage and differentiation state. Nat. Biotechnol. *25*, 1457–1467.

Card, P.B., Hogg, R.T., Gil Del Alcazar, C.R., and Gerard, R.D. (2012). MicroRNA silencing improves the tumor specificity of adenoviral transgene expression. Cancer Gene. Ther. *19*, 451–459.

Cartier, N., Hacein-Bey-Abina, S., Bartholomae, C.C., Veres, G., Schmidt, M., Kutschera, I., Vidaud, M., Abel, U., Dal-Cortivo, L., Caccavelli, L., *et al.* (2009). Hematopoietic stem cell gene therapy with a lentiviral vector in X-linked adrenoleukodystrophy. Science *326*, 818–823.

Cawood, R., Chen, H.H., Carroll, F., Bazan-Peregrino, M., van Rooijen, N., and Seymour, L.W. (2009). Use of tissue-specific microRNA to control pathology of wild-type adenovirus without attenuation of its ability to kill cancer cells. PLoS Pathog. *5*, e1000440.

Cawood, R., Wong, S.L., Di, Y., Baban, D.F., and Seymour, L.W. (2011). MicroRNA controlled adenovirus mediates anti-cancer efficacy without affecting endogenous microRNA activity. PLoS One *6*, e16152.

Ciesielska, A., Hadaczek, P., Mittermeyer, G., Zhou, S., Wright, J.F., Bankiewicz, K.S., and Forsayeth, J. (2013). Cerebral infusion of AAV9 vector-encoding non-self proteins can elicit cell-mediated immune responses. Mol. Ther. *21*, 158–166.

Deuschle, U., Meyer, W.K., and Thiesen, H.J. (1995). Tetracycline-reversible silencing of eukaryotic promoters. Mol. Cell. Biol *15*, 1907–1914.

Di Domenico, C., Villani, G.R., Di Napoli, D., Reyero, E.G. Lombardo, A., Naldini, L., and Di Natale, P. (2005). Gene therapy for a mucopolysaccharidosis type I murine model with lentiviral-IDUA vector. Hum. Gene. Ther. *16*, 81–90.

Di Domenico, C., Di Napoli, D., Gonzalez, Y.R.E., Lombardo, A., Naldini, L., and Di Natale, P. (2006). Limited transgene immune response and long-term expression of human alpha-L-iduronidase in young adult mice with mucopolysaccharidosis type I by liver-directed gene therapy. Hum. Gene. Ther. *17*, 1112–1121.

Franco, L.M., Sun, B., Yang, X., Bird, A., Zhang, H., Schneider, A., Brown, T., Young, S.P., Clay, T.M., Amalfitano, A., *et al.* (2005). Evasion of immune responses to introduced human acid alpha-glucosidase by liver-restricted expression in glycogen storage disease type II. Mol. Ther. *12*, 876–884.

Geisler, A., Jungmann, A., Kurreck, J., Poller, W., Katus, H.A., Vetter, R., Fechner, H., and Muller, O.J. (2011). MicroRNA122-regulated transgene expression increases specificity of cardiac gene transfer upon intravenous delivery of AAV9 vectors. Gene. Ther. *18*, 199–209.

Gilchrist, S.C., Ontell, M.P., Kochanek, S., and Clemens, P.R. (2002). Immune response to full-length dystrophin delivered to Dmd muscle by a high-capacity adenoviral vector. Mol. Ther. *6*, 359–368.

Goudy, K.S., Annoni, A., Naldini, L., and Roncarolo, M.G. (2011). Manipulating Immune Tolerance with MicroRNA Regulated Gene Therapy. Front. Microbiol. *2*, 221.

Groner, A.C., Meylan, S., Ciuffi, A., Zangger, N., Ambrosini, G., Denervaud, N., Bucher, P., and Trono, D. (2010). KRAB-zinc finger proteins and KAP1 can mediate long-range transcriptional repression through heterochromatin spreading. PLoS Genet. *6*, e1000869.

Hartigan-O'Connor, D., Kirk, C.J., Crawford, R., Mule, J.J., and Chamberlain, J.S. (2001). Immune evasion by muscle-specific gene expression in dystrophic muscle. Mol. Ther. *4*, 525–533.

Hikichi, M., Kidokoro, M., Haraguchi, T., Iba, H., Shida, H., Tahara, H., and Nakamura, T. (2011). MicroRNA regulation of glycoprotein B5R in oncolytic vaccinia virus reduces viral pathogenicity without impairing its antitumor efficacy. Mol. Ther. *19*, 1107–1115.

Karali, M., Manfredi, A., Puppo, A., Marrocco, E., Gargiulo, A., Allocca, M., Corte, M.D., Rossi, S., Giunti, M., Bacci, M.L., *et al.* (2011). MicroRNA-restricted transgene expression in the retina. PLoS One *6*, e22166.

Kelly, E.J., Hadac, E.M., Greiner, S., and Russell, S.J. (2008). Engineering microRNA responsiveness to decrease virus pathogenicity. Nat. Med. *14*, 1278–1283.

Kelly, E.J., Nace, R., Barber, G.N., and Russell, S.J. (2010). Attenuation of vesicular stomatitis virus encephalitis through microRNA targeting. J. Virol. *84*, 1550–1562.

Kron, M.W., Espenlaub, S., Engler, T., Schirmbeck, R., Kochanek, S., and Kreppel, F. (2011). MiRNA-mediated silencing in hepatocytes can increase adaptive immune responses to adenovirus vector-delivered transgenic antigens. Mol. Ther. *19*, 1547–1557.

Landgraf, P., Rusu, M., Sheridan, R., Sewer, A., Iovino, N., Aravin, A., Pfeffer, S., Rice, A., Kamphorst, A.O., Landthaler, M., *et al.* (2007). A mammalian microRNA expression atlas based on small RNA library sequencing. Cell *129*, 1401–1414.

Laurenti, E., Barde, I., Verp, S., Offner, S., Wilson, A., Quenneville, S., Wiznerowicz, M., Macdonald, H.R., Trono, D., and Trumpp, A. (2010). Inducible gene and shRNA expression in resident hematopoietic stem cells *in vivo*. Stem Cells *28*, 1390–1398.

Lee, C.J., Fan, X., Guo, X., and Medin, J.A. (2011). Promoter-specific lentivectors for long-term, cardiac-directed therapy of Fabry disease. J. Cardiol. *57*, 115–122.

Maguire, A.M., Simonelli, F., Pierce, E.A., Pugh, E.N., Jr., Mingozzi, F., Bennicelli, J., Banfi, S., Marshall, K.A., Testa, F., Surace, E.M., *et al.* (2008). Safety and efficacy of gene transfer for Leber's congenital amaurosis. N. Engl. J. Med. *358*, 2240–2248.

Markusic, D.M., van Til, N.P., Hiralall, J.K., Elferink, R.P., and Seppen, J. (2009). Reduction of liver macrophage transduction by pseudotyping lentiviral vectors with a fusion envelope from Autographa californica GP64 and Sendai virus F2 domain. BMC Biotechnol. *9*, 85.

Matrai, J., Cantore, A., Bartholomae, C.C., Annoni, A., Wang, W., Acosta-Sanchez, A., Samara-Kuko, E., De Waele, L., Ma, L., Genovese, P., *et al.* (2011). Hepatocyte-targeted expression by integrase-defective lentiviral vectors induces antigen-specific tolerance in mice with low genotoxic risk. Hepatology *53*, 1696–1707.

Matsui, H., Hegadorn, C., Ozelo, M., Burnett, E., Tuttle, A., Labelle, A., McCray, P.B., Jr., Naldini, L., Brown, B., Hough, C., *et al.* (2011). A microRNA-regulated and GP64-pseudotyped lentiviral vector mediates stable expression of FVIII in a murine model of Hemophilia A. Mol. Ther. *19*, 723–730.

Mays, L.E., and Wilson, J.M. (2011). The complex and evolving story of T cell activation to AAV vector-encoded transgene products. Mol. Ther. *19*, 16–27.

Naldini, L., Blomer, U., Gallay, P., Ory, D., Mulligan, R., Gage, F.H., Verma, I.M., and Trono, D. (1996). *In vivo* gene delivery and stable transduction of nondividing cells by a lentiviral vector. Science *272*, 263–267.

Nathwani, A.C., Tuddenham, E.G. Rangarajan, S., Rosales, C., McIntosh, J., Linch, D.C., Chowdary, P., Riddell, A., Pie, A.J., Harrington, C., *et al.* (2011). Adenovirus-associated virus vector-mediated gene transfer in hemophilia B. N. Engl. J. Med. *365*, 2357–2365.

Osborn, M.J., McElmurry, R.T., Lees, C.J., DeFeo, A.P., Chen, Z.Y., Kay, M.A., Naldini, L., Freeman, G., Tolar, J., and Blazar, B.R. (2011). Minicircle DNA-based gene therapy coupled with immune modulation permits long-term expression of alpha-L-iduronidase in mice with mucopolysaccharidosis type I. Mol. Ther. *19*, 450–460.

Pastore, L., Morral, N., Zhou, H., Garcia, R., Parks, R.J., Kochanek, S., Graham, F.L., Lee, B., and Beaudet, A.L. (1999). Use of a liver-specific promoter reduces immune response to the transgene in adenoviral vectors. Hum. Gene. Ther. *10*, 1773–1781.

Petriv, O.I., Kuchenbauer, F., Delaney, A.D., Lecault, V., White, A., Kent, D., Marmolejo, L., Heuser, M., Berg, T., Copley, M., *et al.* (2010). Comprehensive microRNA expression profiling of the hematopoietic hierarchy. Proc. Natl. Acad. Sci. U.S.A. *107*, 15443–15448.

Pichard, V., Aubert, D., Boni, S., Battaglia, S., Ivacik, D., Nguyen, T.H., Arbuthnot, P., and Ferry, N. (2012). Specific microRNA-regulated TetR-KRAB transcriptional control of transgene expression in viral vector-transduced cells. PLoS One 7, e51952.

Qiao, C., Yuan, Z., Li, J., He, B., Zheng, H., Mayer, C., and Xiao, X. (2011). Liver-specific microRNA-122 target sequences incorporated in AAV vectors efficiently inhibits transgene expression in the liver. Gene. Ther. *18*, 403–410.

Schmitt, F., Flageul, M., Dariel, A., Pichard, V., Pontes, C.A., Boni, S., Podevin, G., Myara, A., Ferry, N., and Nguyen, T.H. (2010a). Transient increase in intrahepatic pressure mediates successful treatment of the Gunn rat with reduced doses of lentiviral vector. Hum. Gene. Ther. *21*, 1349–1356.

Schmitt, F., Remy, S., Dariel, A., Flageul, M., Pichard, V., Boni, S., Usal, C., Myara, A., Laplanche, S., Anegon, I., *et al.* (2010b). Lentiviral vectors that express UGT1A1 in liver and contain miR-142 target sequences normalize hyperbilirubinemia in Gunn rats. Gastroenterology *139*, 999–1007, 1007 e1001–1002.

Sellner, L., Stiefelhagen, M., Kleinschmidt, J.A., Laufs, S., Wenz, F., Fruehauf, S., Zeller, W.J., and Veldwijk, M.R. (2008). Generation of efficient human blood progenitor-targeted recombinant adeno-associated viral vectors (AAV) by applying an AAV random peptide library on primary human hematopoietic progenitor cells. Exp Hematol *36*, 957–964.

Suzuki, T., Sakurai, F., Nakamura, S., Kouyama, E., Kawabata, K., Kondoh, M., Yagi, K., and Mizuguchi, H. (2008). miR-122a-regulated expression of a suicide gene prevents hepatotoxicity without altering antitumor effects in suicide gene therapy. Mol. Ther. *16*, 1719–1726.

Szulc, J., Wiznerowicz, M., Sauvain, M.O., Trono, D., and Aebischer, P. (2006). A versatile tool for conditional gene expression and knockdown. Nat. Methods *3*, 109–116.

VandenDriessche, T., Thorrez, L., Naldini, L., Follenzi, A., Moons, L., Berneman, Z., Collen, D., and Chuah, M.K. (2002). Lentiviral vectors containing the human immunodeficiency virus type-1 central polypurine tract can efficiently transduce nondividing hepatocytes and antigen-presenting cells *in vivo*. Blood *100*, 813–822.

Veron, P., Allo, V., Riviere, C., Bernard, J., Douar, A.M., and Masurier, C. (2007). Major subsets of human dendritic cells are efficiently transduced by self-complementary adeno-associated virus vectors 1 and 2. J. Virol. *81*, 5385–5394.

Wu, C., Lin, J., Hong, M., Choudhury, Y., Balani, P., Leung, D., Dang, L.H., Zhao, Y., Zeng, J., and Wang, S. (2009). Combinatorial control of suicide gene expression by tissue-specific promoter and microRNA regulation for cancer therapy. Mol. Ther. *17*, 2058–2066.

Xie, J., Xie, Q., Zhang, H., Ameres, S.L., Hung, J.H., Su, Q., He, R., Mu, X., Seher Ahmed, S., Park, S., *et al.* (2011). MicroRNA-regulated, systemically delivered rAAV9, a step closer to CNS-restricted transgene expression. Mol. Ther. *19*, 526–535.

Xin, K.Q., Mizukami, H., Urabe, M., Toda, Y., Shinoda, K., Yoshida, A., Oomura, K., Kojima, Y., Ichino, M., Klinman, D., *et al.* (2006). Induction of robust immune responses against human immunodeficiency virus is supported by the inherent tropism of adeno-associated virus type 5 for dendritic cells. J. Virol. *80*, 11899–11910.

Yuasa, K., Sakamoto, M., Miyagoe-Suzuki, Y., Tanouchi, A., Yamamoto, H., Li, J., Chamberlain, J.S., Xiao, X., and Takeda, S. (2002). Adeno-associated virus vector-mediated gene transfer into dystrophin-deficient skeletal muscles evokes enhanced immune response against the transgene product. Gene. Ther. *9*, 1576–1588.

Use of Artificial MicroRNAs for Gene Silencing

8

Mohube B. Mowa and Abdullah Ely

Abstract

The discovery of the RNA interference (RNAi) pathway in 1998 resulted in intense excitement and interest in elucidating the mechanisms of this gene regulatory pathway. These studies laid the groundwork for manipulating the RNAi pathway to silence genes of interest in functional genomics studies and for therapeutic purposes. Mammalian RNAi is mainly activated by non-coding double stranded RNA molecules called microRNAs (miRNAs). Induction of RNAi can be achieved with exogenous RNA molecules that mimic the various intermediates of the pathway. This review discusses application of RNAi intermediate mimics for therapeutic gene silencing in mammalian cells. Like any other drug development strategy, using gene silencing as a therapeutic approach has not been without challenges. The most daunting of which include off-targeting effects, toxicity and identifying effective delivery vehicles. Intense dedication to overcome these problems has led to the approval of antisense based drugs for clinical use and progression of many more gene silencing lead candidates to clinical trials. The same dedication will see the use of artificial miRNAs in clinical settings.

Introduction

Extensive studies to elucidate the mechanisms of gene silencing through RNA interference (RNAi) have added another layer of complexity to gene regulation in biological systems. This has led to an intense interest in manipulating RNAi to mediate selective post-transcriptional down-regulation of target genes. RNAi has potential as both a valuable tool in studying gene function in biological research and development of therapeutics against human genetic disease (Walder *et al.*, 2011) and viral infections (Ely *et al.*, 2009; Pu *et al.*, 2011). The major class of small non-coding RNAs which play a fundamental role in sequence specific RNAi-based gene silencing in mammals are the microRNAs (miRNAs). These exist as RNA sequences of 21–22 nts which act by Watson–Crick base-pairing to target mRNA sequences and inhibit gene expression (Zeng *et al.*, 2003).

First discovered in *Caenorhabiditis elegans*, miRNAs are small endogenous RNA duplexes, encoded in both eukaryotic (reviewed in Bartel, 2009) and viral (Kincaid *et al.*, 2012) genomes. miRNAs commonly base pair with partial complementarity to the 3′ UTR of the target mRNA and inhibit its translation (Baek *et al.*, 2008). The fact that >800 miRNAs (http://www.mirbase.org/) have been identified in humans and regulate >60% of human genes (Friedman *et al.*, 2009) suggests an essential role for these small non-coding RNAs in eukaryotic systems. There is clear evidence that miRNAs play vital roles in a wide range of cellular functions e.g. cell development and regulation of immune responses (Liston *et al.*, 2008; Saito *et al.*, 2009; Xu *et al.*, 2009). Several studies, mainly from high throughput analyses assessing global gene expression in clinical samples, have also shown a clear link between aberrant miRNA expression patterns and development of human disease (Wang *et al.*, 2012; Xiao *et al.*, 2011). These observations indicate that the RNAi pathway may not only be exploited for potential therapeutic purposes, but may also be the target for disease intervention, by either antagonizing or restoring miRNA function (reviewed in Bader *et al.*, 2010; Soifer *et al.*, 2007). The discussion in this review will be limited the use of miRNAs as gene silencing agents.

In mammals, most miRNAs are first transcribed as a single transcript by RNA polymerase (Pol) II from coding or non-coding regions of the genome. These long transcripts (thousands of bp) called primary microRNAs (pri-miRNAs), fold back on themselves to form monocistronic or polycistronic hairpin structures, which are normally capped at the 5′ end and polyadenylated at the 3′ end (Cai *et al.*, 2004). Intronic miRNAs are expressed from the same promoter and co-regulated with their host transcripts, whereas miRNAs from non-coding regions are regulated and transcribed independently (Cai *et al.*, 2004; Saito *et al.*, 2009). The pri-miRNA transcripts are later processed in two steps by members of the RNase III family of endonucleases. In the nucleus pri-miRNAs are cleaved by a nuclear microprocessor complex consisting of the enzyme Drosha and its partner DGCR8 to produce stem–loop structures of about 60–70 nt with a 2 nt 3′ overhang called precursor miRNA (pre-miRNA) (Han *et al.*, 2006, 2009; Zeng *et al.*, 2005). A third subset of miRNAs, called mirtrons, are processed independently of Drosha and DGCR8. Studies in *C. elegans* and *Drosophila melanogaster* suggest that mirtrons do not require Drosha processing but are cleaved by pre-mRNA splicing machinery to produce pre-miRNAs (Ruby *et al.*, 2007).

Unlike plant pre-miRNAs, which are processed into mature ~22 nt duplexes in the nucleus before export in to the cytoplasm (Papp *et al.*, 2003) mammalian pre-miRNAs are exported to the cytoplasm in a mechanism dependent on exportin-5 and its cofactor Ran-GTP (Yi *et al.*, 2003). In the cytoplasm the pre-miRNAs are cleaved by another endoribonuclease Dicer to produce a mature miRNA duplex of ~22 bp. The duplex is loaded into the RNA-induced silencing complex (RISC) made up of proteins from the Argonaute family with RNase activity. In the RISC–miRNA complex, one strand is commonly selected to serve as a guide and interact with the target mRNA by base pairing at nucleotide position 2–8 of the miRNA 5′ end (seed region), subsequently silencing the message. The other strand (passenger) is released and degraded (reviewed in Liu and Paroo, 2010).

Activation of the mammalian RNAi pathway to silence genes of interest can be achieved by either delivering chemically synthesized or expressed RNA sequences that mimic miRNA intermediates (Table 8.1). Chemically synthesized activators are typically 19–21 nt duplexes with 2 nt 3′ overhangs called short interfering RNAs (siRNAs) and resemble miRNA duplexes. Expressed RNAi activators are typically designed as Pol III- or Pol II-driven DNA expression cassettes capable of transcribing RNA sequences that resemble naturally occurring pri- and pre-miRNAs. These expressed or synthetic sequences are incorporated into the endogenous RNAi pathway to confer robust silencing of target gene expression.

Evolution of artificial miRNA as therapeutic agents

Broadly, artificial miRNA (amiRNA) systems can be divided into two groups, artificial pre-miRNAs (pre-miRNA mimics) and artificial pri-miRNAs (pri-miRNA mimics). The discovery and exploitation of RNAi through the use of exogenous activators of the pathway preceded the elucidation of the mechanisms underlying miRNA biogenesis (Lee *et al.*, 2002b). As a consequence, initial efforts aimed at reprogramming the RNAi pathway through the introduction of exogenous sequences focussed on the use of synthetic siRNAs (Elbashir *et al.*, 2001), expressed siRNAs (Lee *et al.*, 2002a; Miyagishi and Taira, 2002) and expressed short hairpin RNAs (shRNAs) (Brummelkamp *et al.*, 2002; Paddison *et al.*, 2002; Paul *et al.*, 2002) (Fig. 8.1). To date most applications exploiting the RNAi pathway to silence genes of interest have either employed chemically synthesized siRNAs or Pol III promoter-driven shRNA expression cassettes. Both technologies have their inherent advantages and disadvantages (Table 8.1). Addressing weaknesses in synthetic RNAi activators involves introduction of chemical modifications within the RNA sequences to confer desirable properties such as improved stability and efficacy, lowered immunostimulation and reduced non-specific activity (reviewed

Table 8.1 Properties of RNAi activators

Property	siRNAs	shRNAs	amiRNAs
Propagation method	Synthetic	Expressed	Expressed
miRNA pathway intermediate equivalent	miRNA duplex	pre-miRNA	pre-miRNA or pri-miRNA
RNAi processing machinery used	RISC	Exportin-5, Dicer and RISC	Drosha-DGCR8, Exportin-5, Dicer and RISC
Point of entry into RNAi pathway	Cytoplasm	Nucleus	Nucleus
Commonly induced gene silencing mechanism	mRNA degradation	mRNA degradation	mRNA degradation
Advantages	Ease of large scale production	Compatible with viral vector delivery systems	Compatible with inducible systems and tissue specific promoter systems
	Easy chemical modification to confer desired properties e.g. stability and reduction of off-target effects	Long term and stable knockdown can be achieved	Long term and stable knockdown
	siRNA dose can be controlled easily	Large scale production is cost effective	Simulate RNAi more closely by entering at early stages of the pathway
	Do not require nuclear delivery to be effective		Efficient expression from Pol II promoters enables dose regulation and tissue specific expression
			Compatible with viral vector delivery systems
			Large-scale production is cost effective
Disadvantages	Large-scale production is expensive	Large scale production of virally encoded sequences is labour intensive	Require nuclear delivery to be functional
	Results in transient knockdown	Dosing is limited to expression regulation	Large-scale production of virally encoded sequences is labour intensive
	Not compatible with viral vector delivery systems	Not efficiently expressed from Pol II promoters	Dosing is limited to expression regulation
	Not compatible with inducible or tissue specific promoter systems	Expression from Pol III results in toxicity	Require nuclear delivery to be functional

in Bramsen and Kjems, 2011; Gaglione and Messere, 2010).

Improving expressed RNAi activators on the other hand is limited to regulating expression of the RNA sequence and optimization of the expressed sequence itself. The first examples of expressed RNAi activators employed Pol III promoters, such as the U6 or H1 promoters, to drive expression of each individual strand of an siRNA (Lee *et al.*, 2002a; Miyagishi and Taira, 2002) or a shRNA sequence with [U6+27 shRNA (Paul *et al.*, 2002)] or without [U6/H1 shRNA (Brummelkamp *et al.*, 2002; Paddison *et al.*, 2002)] additional endogenous small nuclear RNA sequences (Fig. 8.1). The reason for using Pol III promoters was as a result of the seminal work by Elbashir and colleagues who demonstrated that chemically synthesized siRNAs are capable of

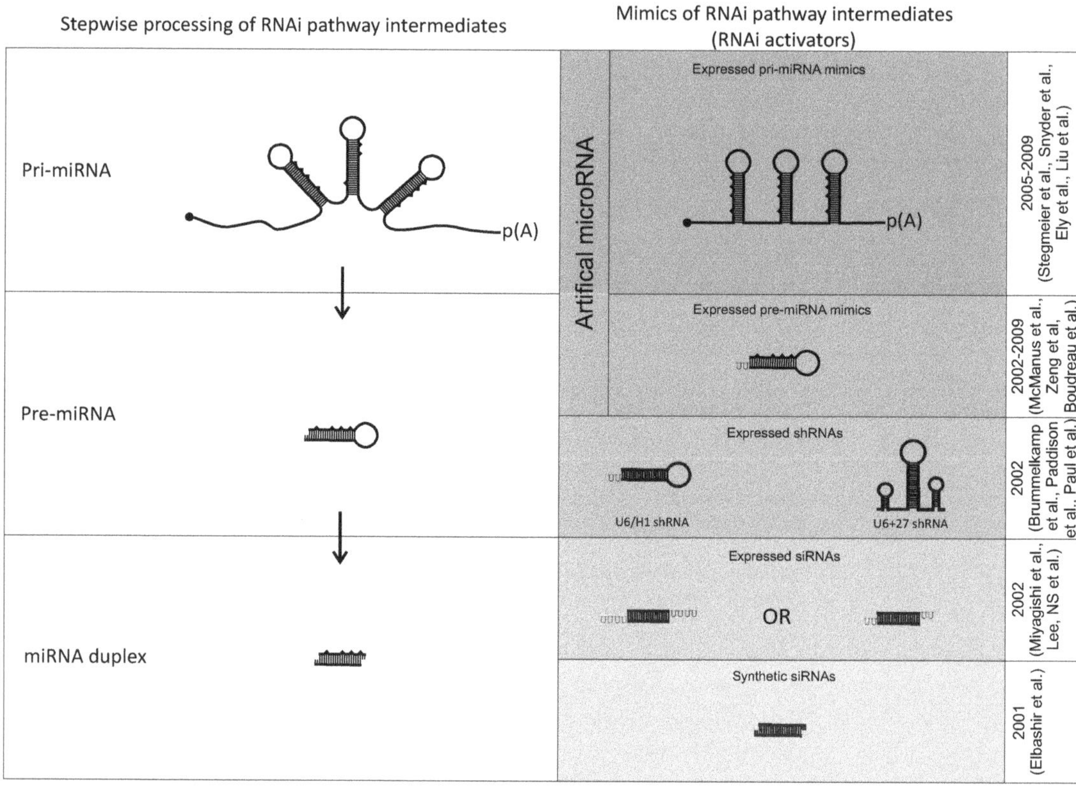

Figure 8.1 The evolution of artificial microRNAs for mammalian gene silencing. Schematic representing the three RNA intermediates of the microRNA biogenesis pathway (pri-miRNA, pre-miRNA and miRNA duplexes, left panel) and some commonly used mimics of these intermediates (right panel). Synthetic siRNAs were the first RNA sequences unequivocally shown to be capable of activating RNAi in mammalian cells (Elbashir *et al.*, 2001). Elucidation of the microRNA biogenesis pathway paved the way for the development of activators that more closely resemble RNAi intermediates, i.e. artificial microRNA (pre-miRNA mimics and pri-miRNA mimics).

triggering RNAi in mammalian cells (Elbashir *et al.*, 2001). Prior to this, long double stranded RNA was typically used to induce RNAi-mediated gene silencing in invertebrate systems. In mammalian cells, however long double-stranded RNAs induce a non-specific interferon response (Gantier and Williams, 2007). The ability of synthetic siRNAs to induce RNAi without non-specific immune stimulation in mammalian cells led researchers to develop expression systems capable of producing sequences that resemble siRNAs as much as possible. Pol III promoters were a convenient choice for these expression systems as they contain all *cis* regulatory elements necessary for transcription located upstream of the transcription start site and consequently transcribe precisely defined sequences (Brummelkamp *et al.*, 2002; Hannon *et al.*, 1991; Lobo *et al.*, 1990). Since the first reports describing the use of Pol III-driven siRNA and shRNA expression cassettes numerous groups have exploited the utility of these systems, in particular U6 shRNA expression, for studies in functional genomics and as a potential gene therapy. Expressed RNAi activators are particularly well-suited to gene therapy against chronic diseases and a major drive of researchers world-wide has been to make the potential of this exciting technology a reality.

A major limitation of Pol III promoter-driven shRNA expression cassettes, especially the U6 shRNA system, was highlighted by Grimm *et al.* (2006). Delivery of anti-hepatitis B U6 shRNA expression cassettes within a recombinant adeno-associated viral vector caused severe toxicity and lethality in mice. The observed toxicity was as a result of oversaturation of the RNAi machinery by the U6 shRNA sequences. Specifically saturation of exportin-5 by the shRNAs prevented export of pre-miRNAs and consequently maturation of endogenous miRNAs. The resultant loss of miRNAs led to loss of liver function and ultimately death of the mice. Saturation of the miRNA biogenesis pathway was as a direct result of overexpression of shRNA sequences from a U6 promoter and emphasized the need for regulatable expressed RNAi activators if these systems are ever to be feasible as potential therapeutic modalities. On hindsight it became clear that shRNA are able to trigger RNAi-mediated gene silencing as they mimic pre-miRNAs and enter the pathway at the point of nuclear export by exportin-5. This led a number of researchers to postulate whether expressed sequences that more closely resemble intermediates of the RNAi pathway would be more effective at reprogramming the pathway to achieve a therapeutic effect (Boudreau *et al.*, 2009a; Ely *et al.*, 2008, 2009; Liu *et al.*, 2008; McBride *et al.*, 2008; Snyder *et al.*, 2008) (Fig. 8.1). shRNAs are typically designed to have an arbitrary loop sequence and the two strands of the stem perfectly complementary to each other. miRNAs on the other hand exist in highly specialized secondary structures and contain very defined loop sequences. An early report demonstrated that incorporating the loop sequence of miR-23 into a shRNA sequence significantly improved its nuclear export (Kawasaki and Taira, 2003). It is therefore logical to assume that incorporating features of endogenous miRNAs into RNAi activators could significantly improve the function of these expressed sequences. Prior work already demonstrated that amiRNAs made using miR-30 as a template were able to effect gene silencing (McManus *et al.*, 2002; Zeng *et al.*, 2002). Engineering amiRNAs follows a simple formula; the guide sequence of the naturally occurring miRNA is replaced with a guide sequence against the target message and the passenger strand is designed to complement the new guide sequence. Depending on the extent of miRNA sequences included, the amiRNA would either be a pre-miRNA mimic or a pri-miRNA mimic. Although these initial studies did not explore whether amiRNAs represented an improved therapeutic modality it was demonstrated that engineered sequences were functional. Subsequently research was focussed on further developing amiRNAs to counter disease phenotypes.

Application of amiRNA

One of the first studies employing amiRNAs to combat disease was described by McBride *et al.* (2008). While initially developing U6 promoter-driven shRNA sequences against mutant *huntingtin* (*HTT*) gene as a means of combating Huntington's disease, it was observed that delivery of two of the therapeutic sequences caused severe neurotoxicity in mice. The observed toxicity did not correlate to silencing of the mutant *HTT* gene as a third U6 shRNA did not induce any toxicity. Additionally, an unrelated control shRNA sequence also induced neurotoxicity in mice. Toxicity was correlated to amount of guide strand produced from the various U6 shRNA sequences. Northern blot analysis revealed a significantly higher amount of guide strand processed from shRNA sequences that caused cytotoxicity and was in line with the results of Grimm *et al.* (2006). To overcome the observed toxicity, the stem sequences of the anti-huntingtin shRNAs was used to create amiRNAs based on previously described designs exploiting the miR-30 backbone (McManus *et al.*, 2002; Zeng *et al.*, 2002). In addition to achieving nearly identical levels of gene suppression with the U6 promoter-driven amiRNAs as with their shRNA counterparts; there was no indication of neurotoxicity from the amiRNAs *in vivo*. Further analysis by this group conclusively demonstrated that amiRNAs exhibit an improved safety profile as compared to equivalent shRNA sequences (Boudreau *et al.*, 2009a). It was also shown that the observed toxicity induced by U6 promoter-driven shRNAs is likely as a consequence of the introduced sequences outcompeting endogenous miRNAs for entry into the RNAi pathway. An important feature of amiRNAs highlighted by these and other studies is that equivalent levels of gene silencing can be achieved at much lower concentrations of guide strand thereby improving safety even when expressed from the constitutive U6 promoter. Typically, endogenous miRNAs are expressed as Pol II transcripts, which raised the interesting possibility of expressing amiRNAs from Pol II promoters.

As opposed to the commonly used Pol III promoters, Pol II promoters exhibit spatial and/or temporal expression of their transcripts. In addition, natural and engineered Pol II promoters may also be induced upon exposure to a predefined stimulus. These are convenient features to exploit for the expression of a therapeutic gene and have been assessed by a number of groups for its utility in expressing amiRNAs. Stegmeier *et al.* (2005) developed amiRNAs based on the previously described miR-30 designs that are driven from a Pol II promoter responsive to doxycycline. To explore spatial regulation of RNAi activators, Snyder *et al.* (2008) assessed expression of a miR-30-derived amiRNA from a liver specific promoter. Although the study employed a reporter system to evaluate silencing in different cell lines it was demonstrated that target knockdown was limited to a liver-derived cell line. In contrast, attempts to express shRNAs from Pol II promoters were met with limited success or yielded constructs with functionality limited to specific sequences (Giering *et al.*, 2008; Peng *et al.*, 2007; Yuan *et al.*, 2006). Early reports clearly demonstrate that amiRNAs provide a much more reliable scaffold for the development of effective Pol II-driven RNAi activators (Ely *et al.*, 2008; Snyder *et al.*, 2008; Stegmeier *et al.*, 2005). In addition to enabling expression from Pol II promoters, amiRNAs also allow concatemerization of multiple sequences to produce a polycistronic sequence expressed from a single promoter. Silencing multiple targets simultaneously using RNAi is especially important in the field of antiviral therapy as viruses often develop resistance against the therapeutic intervention. Ely *et al.* (2009) developed polycistronic amiRNAs based on the naturally occurring miR-31, targeted against the hepatitis B virus (HBV). These polycistronic amiRNAs, which mimicked pri-miR-31, efficiently silenced HBV replication. In addition, these constructs generated guide sequences against three distinct targets on the viral genome and was able to silence each target independently. Similarly, Liu *et al.* (2008) developed multi-targeting amiRNAs against HIV-1 based on an existing miRNA polycistron (miR-17-92). Various artificial polycistrons were generated targeting two, three or four regions within the HIV-1 genome. The polycistrons exhibited superior silencing activity and were able to suppress various HIV-derived sequences. These studies demonstrate the utility of amiRNAs as potential therapy to silence disease-causing phenotypes.

More recently, amiRNAs have been used to silence cancer-causing genes or genes that sensitizes cancer cells to radiotherapy for therapeutic benefit (McLaughlin *et al.*, 2007; Tseng *et al.*, 2012; Zheng *et al.*, 2012, 2013). In addition, amiRNAs exhibit an improved safety profile as compared to their first generation Pol III-driven shRNA counterparts. Furthermore, amiRNAs allow expression from Pol II promoters to achieve tissue-specific and regulatable expression of RNAi activators. miR-30 has been extensively studied and as a consequence has often been used for the generation of amiRNAs (Keck *et al.*, 2009; Silva *et al.*, 2005; Snyder *et al.*, 2008; Stegmeier *et al.*, 2005; Zhang *et al.*, 2012b). amiRNAs however are by no means limited to the use of the miR-30 backbone. In addition to miR-30-based constructs amiRNAs based on miR-155 (Chung *et al.*, 2006; Pu *et al.*, 2011; Zhang *et al.*, 2012b), miR-122 (Ely *et al.*, 2008), miR-31 (Ely *et al.*, 2009; Ely *et al.*, 2008) and the polycistronic miR-17-92 (Liu *et al.*, 2008) have also been developed and demonstrates the versatility of this approach to designing RNAi activators.

Challenges using amiRNAs as therapeutic agents

Insights gained from studies investigating application of ribozymes, antisense oligonucleotides and synthetic siRNAs as therapeutics (reviewed in Burnett and Rossi, 2012) have had a significant impact on the development of amiRNA-based therapies. Despite the differences in all these RNA therapies, the main challenges are common and include stability, safety, target escape by mutations and delivery to target tissue (Barros and Gollob, 2012; Elsabahy *et al.*, 2011; Fedorov *et al.*, 2006; Grimm *et al.*, 2006). However, currently there is intense interest in trying to develop different solutions and several strategies have been described

(Boudreau *et al.*, 2011; Suhy *et al.*, 2012; Xie *et al.*, 2011).

Generating efficient amiRNA expression constructs can pose an obstacle to the therapeutic application of this technology. To ensure efficient gene silencing, several factors have to be considered when designing amiRNAs. The RNA molecules designed have to be efficiently recognized by the endogenous RNAi machinery (Gu *et al.*, 2012; Han *et al.*, 2006; Zeng and Cullen, 2004). It is also crucial to ensure that the target selected is accessible by RISC (Ameres *et al.*, 2007), off target effects are avoided (Fedorov *et al.*, 2006) and chances of escape by target mutations is reduced (Gitlin *et al.*, 2005). While most of these requirements are met by basing amiRNAs on naturally occurring pri-miRNAs, the necessity of long oligonucleotides (~96 nt) for construction of amiRNAs may hamper development of amiRNAs. Hu and colleagues have attempted to overcome this obstacle by designing a simple method of generating both single and multi-amiRNAs using four < 60 nt primers in a one-step PCR (Hu *et al.*, 2009). Targeting conserved regions and simultaneous silencing of multiple targets may also improve silencing effect and minimize RNAi escape through mutations (Boudreau *et al.*, 2013; Friedman *et al.*, 2009; Guirey *et al.*, 2010).

Insights into RNAi therapeutics is mostly derived from studies using synthetic siRNAs and expressed shRNAs. This has led to numerous attempts to directly apply the data obtained from siRNA and shRNAs studies to amiRNA design, mainly by incorporating guide sequences from siRNA or shRNAs into amiRNAs (Boudreau *et al.*, 2009a; Ely *et al.*, 2008; McBride *et al.*, 2008). However this approach has proven to be more challenging than has been anticipated. The paucity of information regarding the exact miRNA processing pattern of shRNAs/siRNA sequences incorporated into pri-miRNA scaffolds poses a challenge. Maczuga and colleagues (2013) showed that siRNA sequences incorporated into pri-miRNA mimics can be differentially processed, resulting in guide sequences of different lengths, 5′ and 3′ cleavage sites and abundance of mature guides. Interestingly, authors observed reduced potency in some of the siRNA-derived amiRNAs as compared to the original siRNA sequences. This might make it difficult to predict off-targeting effects and have serious implications in clinical application of amiRNAs.

The short length of the hexameric 'seed' sequence required for miRNA function in mammalian cells results in a high number of possible targets/miRNA (estimated to be > 200/miRNA) (Friedman *et al.*, 2009). Consequently, RNAi activators designed to target pathology-causing genes may exhibit off-target silencing. Finding effective delivery vehicles which can target amiRNAs to diseased cells may alleviate this problem. A tremendous amount of effort is currently dedicated to alleviating the challenge of safe, efficient and target specific delivery and a comprehensive review on this topic can be found in previous publications (Mowa *et al.*, 2010; Sliva and Schnierle, 2010; Zhang *et al.*, 2012a).

Most criteria for RNAi effector selection are biased for gene silencing efficiency over safety. This is supported by a recent literature survey, which classified most potent published siRNA sequences as not safe. The study analysed the frequency of all possible hexamers (4096) across the 3′ UTR transcriptome. Subsequent classification of siRNAs according to seed complement frequencies, allowing selection of siRNAs with enhanced specificity, revealed that siRNAs with low seed complement frequencies had fewer off-target phenotypes (assessed by cell viability). Only 7 out of 80 assessed siRNAs were classified as safe, defined as those with a scarcity of seed complements within known human 3′ UTRs or exons (Anderson *et al.*, 2008).

Using microarray studies, it was shown that using 'safe seeds' minimizes off-targeting (Boudreau *et al.*, 2011). A recent report using > 50 available microarray data to design a more efficient bioinformatics tool that can predict off-targeting potential of siRNAs will make using 'safe seed' selection criteria much easier (Boudreau *et al.*, 2013). Therefore, these selection criteria should enrich for sequences with less off-targeting potential and serve as an essential strategy for preclinical development of amiRNA-based therapies and improve the chances of selecting lead amiRNA candidates for clinical trials. These studies have been done in rodents (Boudreau *et al.*, 2011) and primates (McBride *et al.*, 2011) and still need to be confirmed in human sequences to account for differences in 3′UTRs between species.

amiRNA progress to clinic

The lack of good therapeutic strategies to counter complex human diseases has led scientists to devise innovative solutions. Gene silencing using exogenous nucleic acid sequences holds great promise for development of molecular therapeutics. Approval of the first antisense oligonucleotide-based drug vitravene in 1998 for treatment of cytomegalovirus retinitis in immunocompromised patients (Mulamba *et al.*, 1998), garnered significant enthusiasm for the field of gene silencing therapeutics. Since then, several antisense oligonucleotides entered clinical trials.

Recently, a locked nucleic acid (LNA)-modified oligonucleotide against the liver specific miR-122 (LNA-miR-122), designed for hepatitis C virus (HCV) infection treatment showed promising outcomes in primates. This molecule results in prolonged suppression of HCV viraemia with a good safety profile, no evidence of viral resistance and no rebound during a 12 week treatment period in chronically infected chimpanzees (Elmen *et al.*, 2008; Lanford *et al.*, 2010; Thibault *et al.*, 2013). A phase I clinical trial conducted by Santaris Pharmaceuticals in 2008 for an anti-miR-122 oligonucleotide to treat HCV called miravirsen showed no evidence of toxicity. Recent data from the phase II clinical trials of miravirsen in chronically infected patients showed significant reduction in viral loads following four weeks of treatment, with no evidence of viral resistance (http://www.santaris.com).

In January 2013, mipomersen, a 2′-*O*-methoxyethyl-modified antisense oligonucleotide developed by ISIS Pharmaceuticals was approved by the US Food and Drug Administration. Mipomersen inhibits synthesis of apolipoprotein B-100 and is used for treatment of familial hypercholesterolaemia, a genetic disorder that causes over-production of the low-density lipoprotein cholesterol. Currently, ISIS Pharmaceuticals and its partners have over 30 antisense oligonucleotides tested in clinical trials, with ~3 oligonucleotides in phase III trials (http://www.isispharm.com).

Despite the fact that there is very little information on the safety of amiRNAs *in vivo*, the overwhelming information from studies shows the high therapeutic efficacy (Boudreau *et al.*, 2009c; Ely *et al.*, 2008; Hu *et al.*, 2009; Pu *et al.*, 2011; Schwab *et al.*, 2006; Ye *et al.*, 2011), lack of saturation of the endogeneous RNAi machinery, lack of toxicity in animal models of disease and indicate that these sequences have a better safety profile as compared to shRNA counterparts (Boudreau *et al.*, 2009a). Currently, the potential for using amiRNAs in a clinical setting have been well demonstrated in culture, mouse and primate models (Ely *et al.*, 2008, 2009; McBride *et al.*, 2008, 2011; Pu *et al.*, 2011). Recently, promising outcomes were observed when miR-30 scaffolds expressed from adeno-associated viral vectors were tested in a non-human primate model of Huntington's disease. In this study, partial suppression of *HTT* by miR-30 mimics prevented neuropathological abnormalities and was well tolerated (McBride *et al.*, 2011).

Polygenic diseases involve deregulation of multiple genes and are difficult to target with classical gene therapy approaches. The natural ability of polycistronic pri-miRNAs to target multiple genes simultaneously (reviewed in Olive *et al.*, 2010), makes pri-miRNA mimics potential candidates for treatment of polygenic disorders like cancer. Following the approval of Gendicine, a recombinant adenovirus expressing a tumour-suppressor protein p53 (Adp53), for cancer treatment in 2003 in China (Guan *et al.*, 2005; Peng, 2005), a promising Adp53 derivative amiRNA-based therapy was designed. This expresses both tumour-suppressor p53 and a polycistronic amiRNA against a p53 antagonist p21. Co-expression of p53 and the anti-p21 amiRNA augmented anti-p53 cancer effects in nude mice (Idogawa *et al.*, 2009).

The possibility of using amiRNA-based strategies in combinatorial disease prevention was recently demonstrated by the progression of the FANG vaccine (bishRNAi^{furin}/GMCSF DNA) from phase I to phase II clinical trials. The FANG vaccine is a potential cancer vaccine made of a plasmid encoding a granulocyte–macrophage colony stimulating factor (GMCSF, important for maturation of DCs and MHC II expression) and a shRNA against furin convertase (inhibitor of GMCSF and activator of oncogenic and immunosuppressive TGF-β1 and -β2). The vaccine showed a safe profile and elicited prolonged survival in patients with advanced cancer (Senzer *et al.*, 2012). A phase II clinical trial with the FANG vaccine is currently under way.

Conclusions

RNA-based gene silencing therapeutics has evolved significantly over the years. Application of antisense oligonucleotides has set the benchmark for the development of gene silencing therapeutics. Since the discovery of the RNAi pathway, there has been a significant progress made in investigating the mechanism and function of RNAi-mediated gene silencing in a wide range of living organisms. These have contributed dramatically to the application of this phenomenon in the development of novel nucleic acid based therapeutic agents. At first, synthetic siRNAs became the preferred molecules to artificially induce this pathway and silence disease related genes. Recently, the more advanced gene silencing approach by amiRNAs has been developed and garnered greater attention. Along this path of gene silencing transformation, lessons were learned and used to improve the new generation of RNA therapeutics. Innovative strategies were developed to overcome hurdles encountered along the way and these have given rise to more advanced and ground breaking approaches to therapeutic gene silencing. Impressive data from recent studies also indicate that RNAi-based therapies may be well suited for combinatorial therapies to augment the currently available disease treatments. With the growing list of antisense oligonucleotides entering the clinic and performing well in clinical trials, the future for amiRNA-based drugs seems bright.

References

Ameres, S.L., Martinez, J., and Schroeder, R. (2007). Molecular basis for target RNA recognition and cleavage by human RISC. Cell *130*, 101–112.

Anderson, E.M., Birmingham, A., Baskerville, S., Reynolds, A., Maksimova, E., Leake, D., Fedorov, Y., Karpilow, J., and Khvorova, A. (2008). Experimental validation of the importance of seed complement frequency to siRNA specificity. RNA *14*, 853–861.

Bader, A.G., Brown, D., and Winkler, M. (2010). The promise of microRNA replacement therapy. Cancer Res. *70*, 7027–7030.

Baek, D., Villen, J., Shin, C., Camargo, F.D., Gygi, S.P., and Bartel, D.P. (2008). The impact of microRNAs on protein output. Nature *455*, 64–71.

Barros, S.A., and Gollob, J.A. (2012). Safety profile of RNAi nanomedicines. Adv. Drug Deliv. Rev. *64*, 1730–1737.

Bartel, D.P. (2009). MicroRNAs: target recognition and regulatory functions. Cell *136*, 215–233.

Boudreau, R.L., Martins, I., and Davidson, B.L. (2009a). Artificial microRNAs as siRNA shuttles: improved safety as compared to shRNAs *in vitro* and *in vivo*. Mol. Ther. *17*, 169–175.

Boudreau, R.L., McBride, J.L., Martins, I., Shen, S., Xing, Y., Carter, B.J., and Davidson, B.L. (2009b). Nonallele-specific silencing of mutant and wild-type huntingtin demonstrates therapeutic efficacy in Huntington's disease mice. Mol. Ther. *17*, 1053–1063.

Boudreau, R.L., Spengler, R.M., and Davidson, B.L. (2011). Rational design of therapeutic siRNAs: minimizing off-targeting potential to improve the safety of RNAi therapy for Huntington's disease. Mol. Ther. *19*, 2169–2177.

Boudreau, R.L., Spengler, R.M., Hylock, R.H., Kusenda, B.J., Davis, H.A., Eichmann, D.A., and Davidson, B.L. (2013). siSPOTR: a tool for designing highly specific and potent siRNAs for human and mouse. Nucleic Acids Res. *41*, e9.

Bramsen, J.B., and Kjems, J. (2011). Chemical modification of small interfering RNA. Methods Mol. Biol. *721*, 77–103.

Brummelkamp, T.R., Bernards, R., and Agami, R. (2002). A system for stable expression of short interfering RNAs in mammalian cells. Science *296*, 550–553.

Burnett, J.C., and Rossi, J.J. (2012). RNA-based therapeutics: current progress and future prospects. Chem Biol *19*, 60–71.

Cai, X., Hagedorn, C.H., and Cullen, B.R. (2004). Human microRNAs are processed from capped, polyadenylated transcripts that can also function as mRNAs. RNA *10*, 1957–1966.

Chung, K.H., Hart, C.C., Al-Bassam, S., Avery, A., Taylor, J., Patel, P.D., Vojtek, A.B., and Turner, D.L. (2006). Polycistronic RNA polymerase II expression vectors for RNA interference based on BIC/miR-155. Nucleic Acids Res. *34*, e53.

Elbashir, S.M., Harborth, J., Lendeckel, W., Yalcin, A., Weber, K., and Tuschl, T. (2001). Duplexes of 21-nucleotide RNAs mediate RNA interference in cultured mammalian cells. Nature *411*, 494–498.

Elmen, J., Lindow, M., Schutz, S., Lawrence, M., Petri, A., Obad, S., Lindholm, M., Hedtjarn, M., Hansen, H.F., Berger, U., *et al.* (2008). LNA-mediated microRNA silencing in non-human primates. Nature *452*, 896–899.

Elsabahy, M., Nazarali, A., and Foldvari, M. (2011). Non-viral nucleic acid delivery: key challenges and future directions. Curr. Drug Deliv. *8*, 235–244.

Ely, A., Naidoo, T., Mufamadi, S., Crowther, C., and Arbuthnot, P. (2008). Expressed anti-HBV primary microRNA shuttles inhibit viral replication efficiently *in vitro* and *in vivo*. Mol. Ther. *16*, 1105–1112.

Ely, A., Naidoo, T., and Arbuthnot, P. (2009). Efficient silencing of gene expression with modular trimeric Pol II expression cassettes comprising microRNA shuttles. Nucleic Acids Res. *37*, e91.

Fedorov, Y., Anderson, E.M., Birmingham, A., Reynolds, A., Karpilow, J., Robinson, K., Leake, D., Marshall, W.S., and Khvorova, A. (2006). Off-target effects by siRNA can induce toxic phenotype. RNA *12*, 1188–1196.

Friedman, R.C., Farh, K.K., Burge, C.B., and Bartel, D.P. (2009). Most mammalian mRNAs are conserved targets of microRNAs. Genome Res. *19*, 92–105.

Gaglione, M., and Messere, A. (2010). Recent progress in chemically modified siRNAs. Mini. Rev. Med. Chem. *10*, 578–595.

Gantier, M.P., and Williams, B.R. (2007). The response of mammalian cells to double-stranded RNA. Cytokine Growth Factor Rev *18*, 363–371.

Giering, J.C., Grimm, D., Storm, T.A., and Kay, M.A. (2008). Expression of shRNA from a tissue-specific pol II promoter is an effective and safe RNAi therapeutic. Mol. Ther. *16*, 1630–1636.

Gitlin, L., Stone, J.K., and Andino, R. (2005). Poliovirus escape from RNA interference: short interfering RNA-target recognition and implications for therapeutic approaches. J. Virol. *79*, 1027–1035.

Grimm, D., Streetz, K.L., Jopling, C.L., Storm, T.A., Pandey, K., Davis, C.R., Marion, P., Salazar, F., and Kay, M.A. (2006). Fatality in mice due to oversaturation of cellular microRNA/short hairpin RNA pathways. Nature *441*, 537–541.

Gu, S., Jin, L., Zhang, Y., Huang, Y., Zhang, F., Valdmanis, P.N., and Kay, M.A. (2012). The loop position of shRNAs and pre-miRNAs is critical for the accuracy of dicer processing *in vivo*. Cell *151*, 900–911.

Guan, Y.S., Liu, Y., Zhou, X.P., Li, X., He, Q., and Sun, L. (2005). p53 gene (Gendicine) and embolisation overcame recurrent hepatocellular carcinoma. Gut *54*, 1318–1319.

Guirey, E., Bees, M., Martin, A., and Srokosz, M. (2010). Persistence of cluster synchronization under the influence of advection. Phys. Rev. E. Stat. Nonlin. Soft Matter Phys. *81*, 051902.

Han, J., Lee, Y., Yeom, K.H., Nam, J.W., Heo, I., Rhee, J.K., Sohn, S.Y., Cho, Y., Zhang, B.T., and Kim, V.N. (2006). Molecular basis for the recognition of primary microRNAs by the Drosha–DGCR8 complex. Cell *125*, 887–901.

Han, J., Pedersen, J.S., Kwon, S.C., Belair, C.D., Kim, Y.K., Yeom, K.H., Yang, W.Y., Haussler, D., Blelloch, R., and Kim, V.N. (2009). Posttranscriptional crossregulation between Drosha and DGCR8. Cell *136*, 75–84.

Hannon, G.J., Chubb, A., Maroney, P.A., Hannon, G., Altman, S., and Nilsen, T.W. (1991). Multiple cis-acting elements are required for RNA polymerase III transcription of the gene encoding H1 RNA, the RNA component of human RNase P. J. Biol. Chem. *266*, 22796–22799.

Hu, T., Fu, Q., Chen, P., Ma, L., Sin, O., and Guo, D. (2009). Construction of an artificial MicroRNA expression vector for simultaneous inhibition of multiple genes in mammalian cells. International journal of molecular sciences *10*, 2158–2168.

Idogawa, M., Sasaki, Y., Suzuki, H., Mita, H., Imai, K., Shinomura, Y., and Tokino, T. (2009). A single recombinant adenovirus expressing p53 and p21-targeting artificial microRNAs efficiently induces apoptosis in human cancer cells. Clin. Cancer Res. *15*, 3725–3732.

Kawasaki, H., and Taira, K. (2003). Short hairpin type of dsRNAs that are controlled by tRNA(Val) promoter significantly induce RNAi-mediated gene silencing in the cytoplasm of human cells. Nucleic Acids Res. *31*, 700–707.

Keck, K., Volper, E.M., Spengler, R.M., Long, D.D., Chan, C.Y., Ding, Y., and McCaffrey, A.P. (2009). Rational design leads to more potent RNA interference against hepatitis B virus: factors effecting silencing efficiency. Mol. Ther. *17*, 538–547.

Kincaid, R.P., Burke, J.M., and Sullivan, C.S. (2012). RNA virus microRNA that mimics a B-cell oncomiR. Proc. Natl. Acad. Sci. U.S.A. *109*, 3077–3082.

Lanford, R.E., Hildebrandt-Eriksen, E.S., Petri, A., Persson, R., Lindow, M., Munk, M.E., Kauppinen, S., and Orum, H. (2010). Therapeutic silencing of microRNA-122 in primates with chronic hepatitis C virus infection. Science *327*, 198–201.

Lee, N.S., Dohjima, T., Bauer, G., Li, H., Li, M.J., Ehsani, A., Salvaterra, P., and Rossi, J. (2002a). Expression of small interfering RNAs targeted against HIV-1 rev transcripts in human cells. Nat. Biotechnol. *20*, 500–505.

Lee, Y., Jeon, K., Lee, J.T., Kim, S., and Kim, V.N. (2002b). MicroRNA maturation: stepwise processing and subcellular localization. EMBO J. *21*, 4663–4670.

Liston, A., Lu, L.F., O'Carroll, D., Tarakhovsky, A., and Rudensky, A.Y. (2008). Dicer-dependent microRNA pathway safeguards regulatory T cell function. J Exp Med *205*, 1993–2004.

Liu, Q., and Paroo, Z. (2010). Biochemical principles of small RNA pathways. Annu Rev Biochem *79*, 295–319.

Liu, Y.P., Haasnoot, J., ter Brake, O., Berkhout, B., and Konstantinova, P. (2008). Inhibition of HIV-1 by multiple siRNAs expressed from a single microRNA polycistron. Nucleic Acids Res. *36*, 2811–2824.

Lobo, S.M., Ifill, S., and Hernandez, N. (1990). cis-acting elements required for RNA polymerase II and III transcription in the human U2 and U6 snRNA promoters. Nucleic Acids Res. *18*, 2891–2899.

McBride, J.L., Boudreau, R.L., Harper, S.Q., Staber, P.D., Monteys, A.M., Martins, I., Gilmore, B.L., Burstein, H., Peluso, R.W., Polisky, B., *et al.* (2008). Artificial miRNAs mitigate shRNA-mediated toxicity in the brain: implications for the therapeutic development of RNAi. Proc. Natl. Acad. Sci. U.S.A. *105*, 5868–5873.

McBride, J.L., Pitzer, M.R., Boudreau, R.L., Dufour, B., Hobbs, T., Ojeda, S.R., and Davidson, B.L. (2011). Preclinical safety of RNAi-mediated HTT suppression in the rhesus macaque as a potential therapy for Huntington's disease. Mol. Ther. *19*, 2152–2162.

McLaughlin, J., Cheng, D., Singer, O., Lukacs, R.U., Radu, C.G., Verma, I.M., and Witte, O.N. (2007). Sustained suppression of Bcr-Abl-driven lymphoid leukaemia by microRNA mimics. Proc. Natl. Acad. Sci. U.S.A. *104*, 20501–20506.

McManus, M.T., Petersen, C.P., Haines, B.B., Chen, J., and Sharp, P.A. (2002). Gene silencing using micro-RNA designed hairpins. RNA *8*, 842–850.

Maczuga, P., Lubelski, J., van Logtenstein, R., Borel, F., Blits, B., Fakkert, E., Costessi, A., Butler, D., van Deventer, S., Petry, H., *et al.* (2013). Embedding siRNA sequences targeting apolipoprotein B100 in shRNA and miRNA scaffolds results in differential processing and *in vivo* efficacy. Mol. Ther. *21*, 217–227.

Miyagishi, M., and Taira, K. (2002). U6 promoter-driven siRNAs with four uridine 3′ overhangs efficiently suppress targeted gene expression in mammalian cells. Nat. Biotechnol. *20*, 497–500.

Mowa, M.B., Crowther, C., and Arbuthnot, P. (2010). Therapeutic potential of adenoviral vectors for delivery of expressed RNAi activators. Expert Opin. Drug Deliv. *7*, 1373–1385.

Mulamba, G.B., Hu, A., Azad, R.F., Anderson, K.P., and Coen, D.M. (1998). Human cytomegalovirus mutant with sequence-dependent resistance to the phosphorothioate oligonucleotide fomivirsen (ISIS 2922). Antimicrob. Agents Chemother. *42*, 971–973.

Olive, V., Jiang, I., and He, L. (2010). mir-17–92, a cluster of miRNAs in the midst of the cancer network. Int. J. Biochem. Cell Biol. *42*, 1348–1354.

Paddison, P.J., Caudy, A.A., Bernstein, E., Hannon, G.J., and Conklin, D.S. (2002). Short hairpin RNAs (shRNAs) induce sequence-specific silencing in mammalian cells. Genes Dev. *16*, 948–958.

Papp, I., Mette, M.F., Aufsatz, W., Daxinger, L., Schauer, S.E., Ray, A., van der Winden, J., Matzke, M., and Matzke, A.J. (2003). Evidence for nuclear processing of plant microRNA and short interfering RNA precursors. Plant Physiol. *132*, 1382–1390.

Paul, C.P., Good, P.D., Winer, I., and Engelke, D.R. (2002). Effective expression of small interfering RNA in human cells. Nat. Biotechnol. *20*, 505–508.

Peng, Y., Lu, J.X., and Shen, X.F. (2007). shRNA driven by Pol II/T7 dual-promoter system effectively induce cell-specific RNA interference in mammalian cells. Biochem. Biophys. Res. Commun. *360*, 496–500.

Peng, Z. (2005). Current status of gendicine in China: recombinant human Ad-p53 agent for treatment of cancers. Hum. Gene. Ther. *16*, 1016–1027.

Pu, C., Wang, L., Miao, X., Zhang, Y., Jiang, C., Liu, W., Sun, W., and Gao, Q. (2011). Optimized tandem amiRNA mediates stronger inhibitory effects on hepatitis B virus infection. J. Gastrointestin. Liver Dis. *20*, 271–278.

Ruby, J.G., Jan, C.H., and Bartel, D.P. (2007). Intronic microRNA precursors that bypass Drosha processing. Nature *448*, 83–86.

Saito, Y., Friedman, J.M., Chihara, Y., Egger, G., Chuang, J.C., and Liang, G. (2009). Epigenetic therapy up-regulates the tumor suppressor microRNA-126 and its host gene EGFL7 in human cancer cells. Biochem. Biophys. Res. Commun. *379*, 726–731.

Schwab, R., Ossowski, S., Riester, M., Warthmann, N., and Weigel, D. (2006). Highly specific gene silencing by artificial microRNAs in Arabidopsis. Plant Cell *18*, 1121–1133.

Senzer, N., Barve, M., Kuhn, J., Melnyk, A., Beitsch, P., Lazar, M., Lifshitz, S., Magee, M., Oh, J., Mill, S.W., *et al.* (2012). Phase I trial of 'bi-shRNAi(furin)/GMCSF DNA/autologous tumor cell' vaccine (FANG) in advanced cancer. Mol. Ther. *20*, 679–686.

Silva, J.M., Li, M.Z., Chang, K., Ge, W., Golding, M.C., Rickles, R.J., Siolas, D., Hu, G., Paddison, P.J., Schlabach, M.R., *et al.* (2005). Second-generation shRNA libraries covering the mouse and human genomes. Nat. Genet. *37*, 1281–1288.

Sliva, K., and Schnierle, B.S. (2010). Selective gene silencing by viral delivery of short hairpin RNA. Virol. J. *7*, 248.

Snyder, L.L., Esser, J.M., Pachuk, C.J., and Steel, L.F. (2008). Vector design for liver-specific expression of multiple interfering RNAs that target hepatitis B virus transcripts. Antiviral Res. *80*, 36–44.

Soifer, H.S., Rossi, J.J., and Saetrom, P. (2007). MicroRNAs in disease and potential therapeutic applications. Mol. Ther. *15*, 2070–2079.

Stegmeier, F., Hu, G., Rickles, R.J., Hannon, G.J., and Elledge, S.J. (2005). A lentiviral microRNA-based system for single-copy polymerase II-regulated RNA interference in mammalian cells. Proc. Natl. Acad. Sci. U.S.A. *102*, 13212–13217.

Suhy, D.A., Kao, S.C., Mao, T., Whiteley, L., Denise, H., Souberbielle, B., Burdick, A.D., Hayes, K., Wright, J.F., Lavender, H., *et al.* (2012). Safe, long-term hepatic expression of anti-HCV shRNA in a nonhuman primate model. Mol. Ther. *20*, 1737–1749.

Thibault, P.A., Huys, A., Dhillon, P., and Wilson, J.A. (2013). MicroRNA-122-dependent and -independent replication of Hepatitis C Virus in Hep3B human hepatoma cells. Virology *436*, 179–190.

Tseng, C.W., Huang, H.C., Shih, A.C., Chang, Y.Y., Hsu, C.C., Chang, J.Y., Li, W.H., and Juan, H.F. (2012). Revealing the anti-tumor effect of artificial miRNA p-27-5p on human breast carcinoma cell line T-47D. Int. J. Mol. Sci. *13*, 6352–6369.

Walder, R.Y., Gautam, M., Wilson, S.P., Benson, C.J., and Sluka, K.A. (2011). Selective targeting of ASIC3 using artificial miRNAs inhibits primary and secondary hyperalgesia after muscle inflammation. Pain *152*, 2348–2356.

Wang, L., Zhang, D., Zhang, C., Zhang, S., Wang, Z., Qu, C., and Liu, S. (2012). A microRNA expression signature characterizing the properties of tumor-initiating cells for breast cancer. Oncol. Lett. 3, 119–124.

Xiao, J., Liang, D., Zhang, Y., Liu, Y., Zhang, H., Li, L., Liang, X., Sun, Y., and Chen, Y.H. (2011). MicroRNA expression signature in atrial fibrillation with mitral stenosis. Physiol Genomics 43, 655–664.

Xie, J., Xie, Q., Zhang, H., Ameres, S.L., Hung, J.H., Su, Q., He, R., Mu, X., Seher Ahmed, S., Park, S., *et al.* (2011).

MicroRNA-regulated, systemically delivered rAAV9, a step closer to CNS-restricted transgene expression. Mol. Ther. *19*, 526–535.

Xu, H., Cheung, I.Y., Guo, H.F., and Cheung, N.K. (2009). MicroRNA miR-29 modulates expression of immunoinhibitory molecule B7-H3, potential implications for immune based therapy of human solid tumors. Cancer Res. *69*, 6275–6281.

Ye, X., Liu, Z., Hemida, M.G., and Yang, D. (2011). Targeted delivery of mutant tolerant anti-coxsackievirus artificial microRNAs using folate conjugated bacteriophage Phi29 pRNA. PLoS One *6*, e21215.

Yi, R., Qin, Y., Macara, I.G., and Cullen, B.R. (2003). Exportin-5 mediates the nuclear export of pre-microRNAs and short hairpin RNAs. Genes Dev. *17*, 3011–3016.

Yuan, J., Wang, X., Zhang, Y., Hu, X., Deng, X., Fei, J., and Li, N. (2006). shRNA transcribed by RNA Pol II promoter induce RNA interference in mammalian cell. Mol. Biol. Rep. *33*, 43–49.

Zeng, Y., and Cullen, B.R. (2004). Structural requirements for pre-microRNA binding and nuclear export by Exportin-5. Nucleic Acids Res. *32*, 4776–4785.

Zeng, Y., Wagner, E.J., and Cullen, B.R. (2002). Both natural and designed microRNAs can inhibit the expression of cognate mRNAs when expressed in human cells. Mol. Cell. *9*, 1327–1333.

Zeng, Y., Yi, R., and Cullen, B.R. (2003). MicroRNAs and small interfering RNAs can inhibit mRNA expression by similar mechanisms. Proc. Natl. Acad. Sci. U.S.A. *100*, 9779–9784.

Zeng, Y., Yi, R., and Cullen, B.R. (2005). Recognition and cleavage of primary microRNA precursors by the nuclear processing enzyme Drosha. EMBO J. *24*, 138–148.

Zhang, S., Zhao, Y., and Zhi, D. (2012a). Non-viral vectors for the mediation of RNAi. Bioorg. Chem. *40*, 10–18.

Zhang, T., Cheng, T., Wei, L., Cai, Y., Yeo, A.E., Han, J., Yuan, Y.A., Zhang, J., and Xia, N. (2012b). Efficient inhibition of HIV-1 replication by an artificial polycistronic miRNA construct. Virol. J. *9*, 118.

Zheng, Z., Ng, W.L., Zhang, X., Olson, J.J., Hao, C., Curran, W.J., and Wang, Y. (2012). RNAi-mediated targeting of noncoding and coding sequences in DNA repair gene messages efficiently radiosensitizes human tumor cells. Cancer Res. *72*, 1221–1228.

Zheng, Z., Wang, P., Wang, H., Zhang, X., Wang, M., Cucinotta, F.A., and Wang, Y. (2013). Combining heavy ion radiation and artificial microRNAs to target the homologous recombination repair gene efficiently kills human tumor cells. Int. J. Radiat. Oncol. Biol. Phys. *85*, 466–471.

Harnessing RNAi for the Treatment of Viral Infections

9

Lorea Blázquez and Puri Fortes

Abstract

No treatment currently exists for many devastating viral infections. These include infections with emerging viruses or with viruses such as influenza that change very rapidly. Other viral infections can be treated, but the therapies may exert damaging unwanted side-effects. Moreover, many viruses may evolve easily to drug resistant variants. Therefore, development of novel therapies for the treatment of viral infections is mandatory. RNA interference (RNAi) is a widely used technique to inhibit gene expression with a tremendous potential as an antiviral. Since RNAi degrades RNA in a sequence-specific manner, the only requirement for the development of RNAi inhibitors is to know the sequence of the target gene or the viral genome. Furthermore, host cell factors that are essential for viral infection but dispensable for cell viability, can be easily identified with genome-wide screenings and targeted by RNAi. The combination of different RNAi inhibitors, or the combination of RNAi with alternative therapies, should avoid the emergence of escape mutants resistant to the treatment. Over the last decade, several studies have used synthetic or gene-expressed short-interfering RNAs (siRNAs) to treat viral infections. Despite promising results in preclinical models, translation to clinical trials has been slow. Delivery remains the main challenge for the therapeutic application of RNAi. Several chemical formulations have been developed that increase delivery, stability and specificity of synthetic RNAi inhibitors or that target them to a specific organ. Several viral and non viral vectors have been employed to increase the delivery or the targeting of gene-expressed siRNAs. However, further efforts to improve delivery and decrease unwanted side-effects of RNAi are still required. In the present review we discuss the current status of RNAi as an antiviral for the treatment of viral hepatitis, haemorrhagic fever viruses, respiratory viruses and other viruses.

Introduction

RNA interference (RNAi) is a widely used technique to inhibit gene expression. The inhibitors that mediate RNAi are small double stranded RNA molecules named short-interfering RNAs (siRNAs). One strand of the siRNA is designed to be perfectly complementary to the RNA that will be inhibited and, therefore, the expression of the gene will be compromised. RNAi can be triggered by directly introducing into cells chemically synthesized siRNAs, or DNAs with expression cassettes that transcribe for short hairpin RNAs (shRNAs). The DNA is used to transduce cells that will then produce a shRNA in the nucleus and process it to siRNA using the cellular machinery required to mature cellular shRNA-like molecules named microRNAs (miRNAs). Similar to miRNAs, the shRNAs are first transported to the cytoplasm by Exportin-5. There, shRNAs are processed by the Dicer endonuclease into siRNA duplexes. Dicer transfers the siRNA to the RNA-induced silencing complex (RISC), which can also load chemically synthesized siRNAs. RISC retains one strand, sufficient to guide the RISC complex to complementary RNAs. Perfectly complementary RNAs are cleaved by the Argonaute 2 protein of the RISC complex and thus efficient gene silencing is achieved (Rivas *et al.*, 2005).

The choice between synthetic siRNAs or gene-encoded shRNAs has important implications for therapeutic applications. Both chemically synthesized siRNAs and DNAs expressing shRNAs have been used for the pre-clinical treatment of acute infections (McCaffrey *et al.*, 2002; Hamasaki *et al.*, 2003; Wang *et al.*, 2005; Ying *et al.*, 2007). However, treatment of chronic infections or prophylactic use of RNAi requires sustained inhibitions. This is easier to obtain with gene-encoded shRNAs. To obtain sustained inhibitions with synthetic siRNAs, these have to be administered periodically, even when the stability of synthetic siRNAs has been improved by chemical modifications. The undeniable advantages of synthetic siRNAs versus gene-encoded shRNAs are safety and versatility. Synthetic siRNAs can be easily produced to high yields under controlled GMP conditions and can be chemically modified to have increased efficacy and stability and decreased off target effects including those caused by immunostimulation. Efficacy can be improved with the incorporation of phosphorothioate modifications in the backbone, inclusion of 2′-*O*-methyl RNA and 2′-fluoro ribose moieties or use of locked nucleic acid derivatives (Elmen *et al.*, 2005; Morrissey *et al.*, 2005; Muhonen *et al.*, 2007; Bramsen *et al.*, 2009). *In vitro* and *in vivo* silencing efficacy as well as stability of synthetic siRNAs can also be improved by the incorporation of 2′-O-guanidinopropyl (GP) moieties in the antisense strand (Brzezinska *et al.*, 2012; Marimani *et al.*, 2013). Stability can be enhanced when sugar altritol substitutes the ribose (Hean *et al.*, 2010). Inhibitors with the second nucleotide modified with 2′-*O*-methyl show decreased off-target effects caused by binding to seed sequences in a miRNA fashion (Jackson *et al.*, 2006). When the terminal ribose of the sense strand incorporates a 5′-*O*-methyl or when the end of the guide strand binds weakly to the sense strand, proper RISC loading of the antisense strand is favoured (Khvorova *et al.*, 2003; Chen *et al.*, 2008a). For delivery, synthetic siRNAs can be administered as naked RNAs, but the efficiency of cell-entry can be enhanced by encapsulation in synthetic vehicles. Several compositions have been used such as cationic liposomes (lipoplexes), polymer complexes (polyplexes) and polymer-modified lipoplexes (lipopolyplexes) (Carmona *et al.*, 2009; Hean *et al.*, 2010; Schafer *et al.*, 2010). Besides liposomes and lipidoids, siRNAs have been also coupled to nanotubes, dendrimers and carbohydrates (Yokota *et al.*, 2007; Kesharwani *et al.*, 2012). siRNAs can also be conjugated to antibodies, membrane-translocating peptides or aptamers to target particular cell types or infected cells (Song *et al.*, 2005; Neff *et al.*, 2011; Ren *et al.*, 2012). Efficient carriers should counteract the negative charges of the siRNAs, avoid aggregation and coupling to neutralizing serum host factors, allow transport through target cell membranes, release themselves from endosomes and escape from immune detection, which is essential to permit the repeated administrations of the drug required for silencing in chronic diseases.

Gene-encoded shRNAs can also be coupled to nanoparticles (Ding *et al.*, 2013). However, they are usually delivered using viral vectors. Viral vectors are also useful to restrict infection to target tissues. Examples exist in which the viral vector that delivers shRNAs derives from the target virus (ter Brake and Berkhout, 2007). Thus, both parental and recombinant viruses transduce the same cells. The use of tissue-specific and viral infection-specific promoters can also restrict shRNA expression to therapeutically relevant cells (Strayer *et al.*, 2005; Snyder *et al.*, 2008). Polymerase III promoters have been also used to express shRNAs, because they are simple and generate short RNAs with high efficiency. However, shRNA overexpression from U6 Pol III promoters can induce toxicity in mouse and primate models due to saturation of the cellular miRNA machinery that leads to inhibition of endogenous miRNA function (Grimm *et al.*, 2006; Suhy *et al.*, 2012). To avoid toxicity due to shRNA overexpression, use of Pol II promoters (Ely *et al.*, 2008; Giering *et al.*, 2008) or inclusion of specific regulatory elements into Pol III promoters (Suhy *et al.*, 2012) is recommended.

Inhibition of the expression of damaging factors using RNAi has great therapeutic potential. In fact, RNAi has several applications in biomedical research for the treatment of infections, cardiovascular and cerebrovascular diseases, metabolic diseases, neurodegenerative disorders and cancer. Among infections, the most progress has been made using RNAi as an antiviral. As RNAi degrades RNA in a sequence-specific manner, to apply RNAi, the only requirement is to know the DNA sequence of the target gene whose expression should be inhibited. For most diseases, identification of target genes requires complex and time-consuming molecular biology experiments. In the case of viral infections, the target genes can be the viral genes. Genome sequences from several strains of most pathogenic viruses are publicly available. Also, new emerging pathogenic viruses can be rapidly identified and sequenced. Therefore, these sequences are sufficient to design antiviral RNAi inhibitors against viral transcripts that block viral gene expression. For RNA viruses, RNAi could also target the viral genome or the replicating intermediates (Jacque *et al.*, 2002). Thus, these viruses are susceptible to degradation before transcription of viral mRNAs, during their initial infection and entry into a cell. However antiviral RNAi is not always successful. Targeting viral genome can fail when the viral genome is protected by the viral capsid, specific structures or it is inaccessible to the RNAi machinery in specific cellular compartments (Bitko and Barik, 2001). In the case of RNA viruses, mutation rates during viral genome replication are very high and this endows viruses with impressive sequence diversity (Duffy *et al.*, 2008). These viruses evolve rapidly and become resistant to RNAi (Boden *et al.*, 2003; Wilson and Richardson, 2005). To circumvent viral escape, highly conserved regions of the viral genome should be targeted because they might compromise viral function if they mutate (Chevalier *et al.*, 2007; Ray and Kanda, 2009). Also, viral escape is reduced or avoided when several RNAi inhibitors that target viral sequences are used at the same time or when the inhibitors target cellular genes required for viral replication (Takigawa *et al.*, 2004; Hong *et al.*, 2012). Viruses are intracellular parasites that rely on a number of host factors to mediate their replication. Therefore, when information about the viral life cycle and viral interaction with the host cell is available, non-essential cellular genes whose function is important for virus viability could also be targeted through RNAi. Identification of host factors involved in viral infections can be easily conducted with large-scale RNAi screens (Randall *et al.*, 2007). Finally, increased RNAi success and delayed viral escape can be achieved using combinatorial RNAi approaches that target both highly conserved viral sequences and host factors essential for viral replication (Bian *et al.*, 2012).

RNAi antiviral strategies can be developed to treat already infected patients but also for prophylaxis in individuals with a high risk of infection. This is mandatory, for instance, before liver transplantation of HBV- or HCV-infected patients, where reinfection of the transplanted liver occurs very rapidly (Olivera-Martinez and Gallegos-Orozco, 2007; Roche and Samuel, 2008). Unfortunately, no current methods are able to deliver siRNAs efficiently into the liver to be transplanted. For a standard or prophylactic RNAi therapy to be effective, ideally, all the patient cells sensitive to be infected by a given virus should express the RNAi-based inhibitors. Such an efficient delivery of siRNAs or gene-encoded shRNAs is very difficult to achieve in most tissues. However, delivery to a smaller proportion of cells could still show therapeutic efficacy. This could be the case when an RNAi treatment leads to viral antigen levels sufficiently low to allow reactivation of a blunted immune response, which is ultimately responsible for viral clearance. In tissues able to regenerate, such as the haematopoietic system or the liver, harnessing a percentage of cells with gene-encoded antiviral shRNAs integrated in their genome, may be enough to obtained sustained virological responses. Viral sensitive cells, not expressing the antiviral shRNA, may die because of the cytopathic effects caused by the infection or by the cytotoxic response of the immune system against the cells that express viral antigens. Treated cells resistant to the infection or unable to express viral antigens may have a survival advantage and proliferate to regenerate an entire resistant organ. Note that when resistant cells are invisible for the immune system, they should have a complete inhibition of the expression of viral antigens, as expression of only few peptides allows immune detection.

Besides improving delivery efficiency, limiting off-target effects also remains a challenge. Checking for sequence homology between the target and the host genome is mandatory to minimize the silencing of unintended targets. It is important to consider that both strands of the siRNA can be potentially loaded in RISC and lead to off-target inhibitions. Moreover, siRNAs can behave in a similar way to microRNAs and affect RNA stability or the translation efficiency of off-targets only bound by a 6- to 8 nt-long seed sequence (Saxena *et al.*, 2003; Birmingham *et al.*, 2006). In addition, some RNAi effectors are sensed by the cell leading to activation of the interferon response (Bridge *et al.*, 2003; Sledz *et al.*, 2003). Therefore, the potential effect of pro-inflammatory stimulation on the therapeutic effect should be adequately evaluated. Finally, the impact of viral silencing suppressors on the efficacy of RNAi antiviral therapy needs to be evaluated further.

RNAi-based therapy needs to be tailored for each virus and in the following sections the current status of RNAi therapy for specific viral infections will be reviewed.

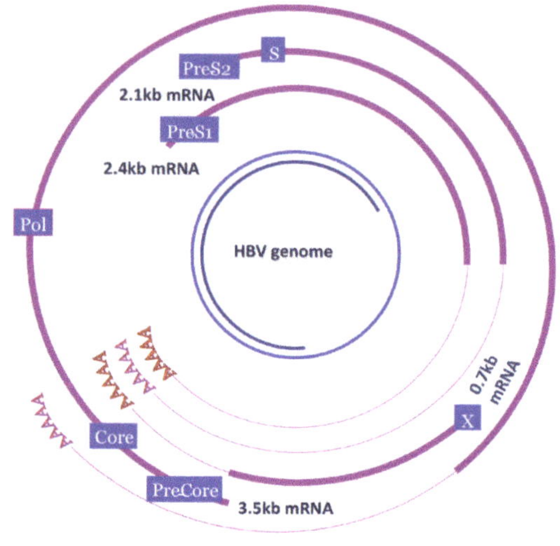

Figure 9.1 HBV genome organization. The diagram shows the partly double-stranded HBV DNA genome (inner blue circles). The genome serves as a template for the four HBV transcripts, which share the same polyadenylation site. The 3.5 kb transcript is the pregenomic RNA that is retrotranscribed to generate HBV DNA and translated to yield the viral core and polymerase proteins. The 2.4 kb and 2.1 kb mRNAs encode for the viral envelope proteins. The 0.7 kb mRNA encodes for the viral X protein.

RNAi to treat viral hepatitis

Hepatitis B virus infection

Hepatitis B virus (HBV) is the sole human pathogen in the family of Hepadnaviridae that predominantly infects the liver. It is a non-cytopathic enveloped virus with a partly double-stranded DNA genome of 3.2 kbps. In the infected hepatocyte, the viral genome is converted to a covalently closed circular DNA (cccDNA) that contains regulatory regions embedded in four overlapping open reading frames that encode for core, polymerase, surface and X proteins (Fig. 9.1). Interestingly, the viral DNA molecule is not replicated directly in the viral cycle. Instead, the viral DNA transcribes for a pregenomic RNA that moves to the cytoplasm and encapsidates. There, the viral polymerase retrotranscribes the pregenomic RNA to generate a negative stranded DNA. Then, the pregenomic RNA is degraded and a positive strand is synthesized. Once the DNA genome is formed, the particle can move to the nucleus to increase the levels of viral cccDNA or can travel to the endoplasmatic reticulum for assembly of envelope proteins and budding out of the cell (Levrero *et al.*, 2009).

Eight HBV genotypes (A–H) have been described which have similar sequences but distinct clinical characteristics and geographical distribution (Schaefer, 2007). Though there is an effective vaccine against HBV, it is only useful for the prevention of viral infection and not for the approximately 2 billion people who are infected with HBV. When infection counteracts the immune response, it leads to chronic HBV infections that can evolve to chronic inflammation and development of liver disease, including liver cirrhosis and hepatocellular carcinoma (Hollinger and Liang, 2001). It is estimated that 400 million people are chronic HBV carriers (Romano *et al.*, 2006). Currently, the two main treatments for antiviral therapy of chronic HBV infection are nucleoside/nucleotide analogues, such as entecavir or lamivudine, and interferon α (Scaglione and Lok, 2012). Nucleoside/nucleotide analogues affect polymerase function decreasing the retrotranscription and the synthesis of the positive-strand DNA. Therefore, the treatment leads to a drastic decrease in viral particles in serum, but does not cure the infection in most patients. Viral

DNA in infected cells is unaffected, as is transcription, viral protein expression or HBV antigen load. Interruption of the treatment normally leads to a rebound in viral particle production. Likewise, interferon-α has limited efficacy due to side-effects and the contraindication profile (Nebbia *et al.*, 2012). Therefore, novel therapies are required for HBV treatment and HBV RNAi-mediated silencing could be a promising option. Indeed, unlike nucleoside/nucleotide analogues, HBV antiviral RNAi can down-regulate surface antigen expression, the major cause of liver damage.

There is no ideal model to test therapies against HBV infections. *In vitro*, RNAi efficiency can be tested in any of the several tissue culture cells that support viral replication, such as HepG2.2.15. Moreover, assays can be performed in cell lines transfected with plasmids that express a little more than the complete HBV genome (normally 1.3 times the genome), which drive viral replication. Cells transfected with plasmids that contain most HBV sequences modified to include reporter genes such as luciferase can also be used. These cells do not show viral replication but allow an easy and quantitative evaluation of the inhibition of the expression obtained when HBV sequences are targeted by RNAi (Ely *et al.*, 2008; Blazquez *et al.*, 2012). *In vivo*, there are several mouse models that can be used, all hampered by the fact that mice support viral replication but not infection by HBV, as they lack the HBV receptor (Yan *et al.*, 2012; Schieck *et al.*, 2013). Several lines of transgenic mice have been established that are able to express HBV proteins and replicate the HBV genome but that are immune tolerant to HBV antigens (Guidotti *et al.*, 1995; Uprichard *et al.*, 2005; Carmona *et al.*, 2006). In addition, viral DNA can be introduced into mouse liver by hydrodynamic injection or viral vector delivery leading to immune competent mice able to support the complete HBV cell cycle with the exception of infection (Huang *et al.*, 2006, 2011). Transgenic animals and models obtained by hydrodynamic injections lack cccDNA, whose efficient elimination should be evaluated in any antiviral treatment to ensure sustained virological responses. cccDNA can be generated in chimeric mouse models where human hepatocytes are engrafted and support all the steps of HBV infection. The drawback of the model is that it requires immunodeficiency to allow liver humanization. Therefore, the immune effects of the antiviral therapy cannot be studied. The generation of transgenic mice expressing the human HBV receptor leading to immune competent models able to recapitulate all the steps of the viral cycle including infection would be highly desirable. In these animals, infection should establish after introduction of purified HBV from patients or after transient expression of HBV pregenomic RNA in hepatocytes. These models should lead to good levels of HBV cccDNA in the liver to test the efficacy of RNAi or any other antiviral treatment. Before these animals are available, a good experimental model is woodchucks naturally infected with woodchuck hepatitis virus, which is very similar to HBV (Summers *et al.*, 1978).

There are several examples of efficient RNAi-mediated inhibition of HBV replication where all four HBV transcripts have been targeted (Shlomai and Shaul, 2003; Yang *et al.*, 2005; Ying *et al.*, 2007) (Table 9.1). In fact, the HBV genome is so compact that single RNAi effectors can silence all viral genes (Ely *et al.*, 2008). Anti-HBV RNAi studies include siRNAs (Hamasaki *et al.*, 2003; Jiao *et al.*, 2007), endoribonuclease-prepared siRNAs (esiRNAs) (Xuan *et al.*, 2006; Tan *et al.*, 2007), gene-expressed shRNAs (Chen *et al.*, 2003; Liu *et al.*, 2004; Ren *et al.*, 2005) or long hairpin RNAs (lhRNAs), which are longer hairpins than shRNAs that can be processed by Dicer to several siRNAs (Weinberg *et al.*, 2007). Anti-HBV sequences have also been expressed from single or polycistronic miRNA shuttles (Ely *et al.*, 2008, 2009; Snyder *et al.*, 2008; Pu *et al.*, 2011; Xiangji *et al.*, 2011) (Table 9.1).

For siRNA delivery, targeting to the liver has been achieved by coupling siRNAs with apolipoprotein A-I (Kim *et al.*, 2007). A single intravenous administration of a low dose (≤ 2 mg/kg) of this compound showed liver specificity and persistent antiviral effect up to 8 days. Liver targeting was also obtained with cholesterol-conjugated siRNAs directed against conserved HBV sequences mixed with N-acetylgalactosamine-conjugated melittin-like peptide (NAG-MLP). A single injection of the mixture in the transgenic mouse model resulted in multilog repression of HBV that lasted

Table 9.1 Examples of RNAi to treat human viral infections

Virus	Target	Mode	Model	Reference
HBV	Core	siRNA	Huh7 and HepG2 cells	Hamasaki *et al.* (2003)
HBV	Core, Pol, X	shRNA	Murine hydrodynamic injection	McCaffrey *et al.* (2003)
HBV	X, Core	shRNA	Huh7 cells	Shlomai and Shaul (2003)
HBV	X	lhRNA	Murine hydrodynamic injection	Weinberg *et al.* (2007)
HBV	X	miRNA shuttles	Murine hydrodynamic injection	Ely *et al.* (2008)
HBV	S Ag	AAV8/PolII-shRNA	HBV transgenic mice	Giering *et al.* (2008)
HBV	Core, S Ag, X	shRNA	HBV-infected patients	Gish *et al.* (2011)
HBV	S Ag, Hsc70	shRNA	HepG2.2.15	Bian *et al.* (2012)
HBV	S Ag, X	shRNA/U1in	Murine hydrodynamic injection	Blazquez *et al.* (2012)
HBV	S Ag, X	NAG-MLP chol-siRNA	Murine hydrodynamic injection/ HBV transgenic mice	Wooddell *et al.* (2013)
HCV	NS5B	siRNA	Murine hydrodynamic injection	McCaffrey *et al.* (2002)
HCV	5′UTR	siRNA and shRNA	HCV replicon system in Huh7	Yokota *et al.* (2003)
HCV	Core	siRNA	Murine hydrodynamic injection	Kim *et al.*, (2006)
HCV	Cellular CD81/ IRES/NS5B	Lentivirus delivered shRNA	HCV replicon system in Huh7	Henry *et al.* (2006)
HCV	5′UTR, 3′UTR	lhRNA	HCV replicon system in Huh7	Watanabe *et al.* (2006)
HCV	Hsp90	siRNA	HCV replicon system in Huh7	Nakagawa *et al.* (2007)
HCV	5′UTR of GBV-B	CL-siRNA	Non-human primate surrogate model for HCV	Yokota *et al.*, (2007)
HCV	Core	DTC-rhApo/siRNA lipoplexes	Murine hydrodynamic injection	Lee *et al.* (2009)
HCV	Cellular MKNK1	siRNA	Huh7 inoculated with HJ3-5 virus or HCVpp	Kim *et al.* (2013)
HCV	5′UTR, Core, NS5B	AAV8 delivered miRNA cluster	Murine hydrodynamic injection	Yang *et al.* (2013)
RSV	P	siRNA	RSV-infected mice	Bitko *et al.* (2005)
RSV	NS1	siRNA	RSV-infected mice	Zhang *et al.* (2005)
RSV	N	siRNA	Humans experimentally infected with RSV	DeVincenzo *et al.*, (2010)
RSV	Cellular Nucleolin	siRNA	RSV-infected mice	Tayyari *et al.* (2011)
MV	Cellular SLAM	shRNA	MV infected B95-8 marmoset cells	Hu *et al.* (2005)
MV	L	siRNA/shRNA	MV-infected Vero/SLAM monkey cells	Otaki *et al.* (2006)
MV	Cellular Rab9 GTPase, viral L and N	shRNA	MV-infected Vero-E6 monkey cells	Shi *et al.* (2012)
SARS-CoV	RNA polymerase	shRNA	Vero monkey cells	Wang *et al.* (2004)

Table 9.1 continued

Virus	Target	Mode	Model	Reference
SARS-CoV	Leader, TRS, 3′-UTR, Spike	siRNA	SARS-CoV infected Vero-E6 monkey cells	Wu *et al.* (2005a)
SARS-CoV	Spike, ORF1b	Intranasally administered siRNAs	Rhesus macaque SARS model	Tang *et al.* (2008)
Influenza	NP/PA/PB1	PEI-siRNA/shRNA	Experimentally infected mice	Ge *et al.* (2004)
Influenza	NP/PA	siRNA	Experimentally infected mice	Tompkins *et al.* (2004)
Influenza	PB1	Immunostimulatory siRNAs	H5N1 infected HD-11 chicken cells	Stewart *et al.* (2011)
EBOV	L	SNALP-siRNAs	EBOV infected Guinea pigs	Geisbert *et al.*, (2006)
EBOV	L/VP24/VP35	SNALP-siRNAs	Non-human primate model of EBOV	Geisbert *et al.*, (2010)
EBOV	TIM-1	siRNA	Vero cells	Kondratowicz *et al.*, (2011)
MARV	NP/VP35/VP30	siRNA	Vero cells	Fowler *et al.* (2005)
DENV	prM	dsRNA	Aedes aegypti mosquitoes	Franz *et al.* (2006)
DENV	5′CS	Silencer *in vivo* Ready siRNA	AG129 mouse model	Stein *et al.*, (2011)
YFV	NS1/E	Intracraneal injection of shRNA plasmids	YFV infected mice	Pacca *et al.*, (2009)
WNV	Envelope	Lentivirally delivered shRNA or lipid-complexed siRNA	WNV inoculated mice	Kumar *et al.* (2006)
HSV	UL27/UL29	Vaginal instillation of siRNA	HSV-2 infected mice	Palliser *et al.* (2006)
HPV	E6/E7	Bifunctional siRNAs	HPV-driven tumours in mice	Khairuddin *et al.* (2012)
Rotavirus	VP4	siRNAs	RRV infected MA104 monkey cell line	Dector *et al.* (2002)
RABV	N/L	Adenovirus delivered shRNAs	RABV infected mice	Gupta *et al.* (2012)

HBV, hepatitis B virus; HCV, hepatitis C virus; RSV, respiratory syncytial virus; MV, measles virus; SARS-CoV, severe acute respiratory syndrome coronavirus; EBOV, Ebola virus; MARV, Marburg virus; DENV, dengue virus; YFV, yellow fever virus; WNV, West Nile virus; HSV, herpes simplex virus; HPV, human papillomavirus; RABV, rabies virus.

for one month (Wooddell *et al.*, 2013). siRNAs targeting HBV have also been delivered as nanoparticles. Cationic liposomes that incorporate an aminoxy cholesteryl lipid for postcoupling of polyethylene glycol (PEG) result in PEGylated siRNA-nanoparticles with uniform small sizes of 80–100 nm in diameter (Carmona *et al.*, 2009). This is the size of the liver sinusoid filter that the particles need to cross to reach hepatocytes. It should be taken into consideration that chronic HBV infections may lead to liver cirrhosis, which courses with a capillarization of the sinusoids and therefore the 100 nm-wide windows to enter the liver are closed. Repeated systemic administration of PEGylated nanoparticles led to a decrease of markers of HBV replication by up to 3-fold over a 28-day period. Lipoplexes were used to deliver altritol-containing synthetic siRNAs to the transgenic mouse model resulting in an inhibition of HBV replication of approximately 50% without evident toxicity (Hean *et al.*, 2010). In addition, successful delivery of antiHBV siRNAs *in vitro* and *in vivo* has been obtained with a PreS1 HBV ligand combined with a synthetic 9-mer arginine peptide (PreS1-9Arg) (Huang *et al.*, 2013).

Although there are examples of HBV inhibition with synthetic siRNAs, most studies have inhibited HBV with shRNAs. It has been clearly demonstrated that overexpression of anti-HBV shRNAs from polymerase III promoters can be fatal in mice due to saturation of the cellular silencing machinery (Grimm *et al.*, 2006). Therefore, several studies have analysed the safety and efficacy of different promoters. Pol III tRNA(Lys3) promoters have proven to be efficient for a controlled expression of shRNAs that target HBV (Dyer *et al.*, 2010). In addition, Pol II promoters with ubiquitous or liver specific expression of shRNAs are also functional (Giering *et al.*, 2008). Pol II promoters are also used for the expression of anti-HBV sequences from miRNA-like transcripts (Ely *et al.*, 2008; Snyder *et al.*, 2008; Pu *et al.*, 2011; Xiangji *et al.*, 2011). When using miRNA-based vectors for therapy, competition between exogenous miRNAs and endogenous miRNAs should also be considered (Keck *et al.*, 2009).

Anti-HBV shRNA therapy has to face the issue of proper delivery. The first demonstration of anti-HBV RNAi *in vivo* involved hydrodynamic co-delivery of an HBV replicon and an anti-HBV shRNA plasmid in mice (McCaffrey *et al.*, 2003). This study demonstrated a significant knockdown of the HBV core antigen in hepatocytes. However, the hydrodynamic procedure has not proven feasible thus far in larger mammals. Several viral vectors that target hepatocytes have been used to deliver gene-expressed RNAi effectors. The vectors of choice have been based on first generation and helper dependent adenovirus (Uprichard *et al.*, 2005; Carmona *et al.*, 2006; Romano *et al.*, 2006) and adeno-associated virus (AAV) (Grimm *et al.*, 2006; Chen *et al.*, 2007; Giering *et al.*, 2008), although others, such as lentivirus vectors (Deng *et al.*, 2009) or prototype foamy virus vectors (PFV) (Moore *et al.*, 2005), have also been tested.

Adenoviruses have a natural tropism for the liver after systemic administration. However, these first generation adenoviral vectors can induce toxicity that results from immune stimulation and have the disadvantage of short-term expression. Therefore, a better option to deliver anti-HBV shRNAs is the use of helper-dependent adenoviral vectors, with all viral protein-encoding genes deleted, with decreased immune stimulation and longer transgene expression (Rauschhuber *et al.*, 2008). PEG-modified adenoviral vectors can also decrease the immune response in HBV transgenic mice (Crowther *et al.*, 2008). However, most *in vivo* studies to date have been performed with AAV vectors. Among several AAV serotypes, AAV8 has an exquisite preference and efficiency to transduce mouse hepatocytes and has resulted in efficient inhibition of HBV replication in transgenic models (Giering *et al.*, 2008). However, a gradual loss of AAV genome in mouse liver can lead to a decrease in the inhibitory effect. Thus, if repeated injections of viral vectors are needed, the administration of immunologically distinct AAV serotypes could be a strategy to prolong transgene expression (Chen *et al.*, 2009). Studies performed with AAV8 delivered shRNAs have also highlighted the importance of optimizing the dose of inhibitor. Vectors expressing 25-mer anti-HBV shRNAs consistently caused toxicity in the treated mice, while low doses of 19-mer inhibitors lead to safe and long-term antiviral activity (Grimm *et al.*, 2006).

Anti-HBV therapy has also been performed with RNAi inhibitors that target cellular factors required for HBV infection. Thus, inhibition of human La protein expression results in a significant decrease in HBV mRNA (Ni *et al.*, 2004) and HBV replication can be reduced modulating HNF-4α expression (Hong *et al.*, 2012). Moreover, combinatorial RNAi targeting the HBV surface open reading frame and human heat stress cognate 70 (Hsc70) is highly effective in suppressing ongoing HBV expression and replication (Bian *et al.*, 2012). Combinatorial strategies may be essential for successful HBV therapy because HBV polymerase lacks proofreading activity resulting in rapid mutagenesis of the HBV genome and emergence of resistant variants. However, viral escape of HBV against single RNAi targets has not been observed in experimental models. Thus, sustained inhibition of HBV has been observed in tissue culture for longer than 30 passages (Ren *et al.*, 2006) and in transgenic mice for 5 months after a single injection of viral vectors expressing the antiviral siRNAs (Uprichard *et al.*, 2005; Carmona *et al.*, 2006). These experiments should be interpreted with caution as the mutants generated in mouse models cannot infect mouse hepatocytes, cannot be amplified and cannot be selected as escape mutants. Therefore, it is still possible that RNAi effectors may force selection pressure on newly generated or preexistent viral mutants (Wu *et al.*, 2005b). Application of allele-specific RNAi could be useful to target these resistant variants, including lamivudine resistance mutations in the HBV genome (Teng *et al.*, 2011). Simultaneous delivery of two or more shRNAs that target different regions or combination of RNAi with pharmacological treatments could also reduce viral escape (Chen *et al.*, 2003; Ely *et al.*, 2009). Inhibition of viral gene expression by RNAi does not require active viral replication, so it can be combined with conventional treatments that target the reverse transcription required to produce the viral genome or with IFN. In fact, the IFN response can be activated directly by modified siRNAs. siRNAs that target multiple HBV transcripts of all genotypes have been coupled to a 5′-triphosphate (3p) to activate retinoic acid-inducible gene I protein (RIG-I) and induce a strong type I IFN response. Indeed, these 3p-modified siRNAs controlled replication of HBV more efficiently and for longer periods of time than 3p-RNAs without silencing capacity or siRNAs that targeted identical sequences but did not contain 3p (Ebert *et al.*, 2011). Finally, anti-HBV RNAi therapy can also be combined with other techniques to inhibit gene expression, such as U1 interference (U1i) (Fortes *et al.*, 2003). Combination of RNAi with U1i was intended to reduce viral escape but also resulted in increased synergistic inhibitions, higher than those obtained using either of the inhibitors on their own. This has great therapeutic interest because using this strategy good inhibition can be obtained with lower doses of U1i or RNAi-based inhibitors, thus minimizing toxicity (Blazquez *et al.*, 2012).

Combination of different RNAi inhibitors or different antiviral strategies may also be interesting to develop a single antiviral strategy against all genotypes of HBV. There are eight main genotypes of HBV (A to H) with over 8% sequence diversity. There are few conserved regions across all genotypes and they are localized to constrained areas of functional relevance. In fact, only a 17-mer contiguous sequence has been described to be conserved among all relevant HBV variants (Sun *et al.*, 2010). The results using less stringent criteria that allowed for certain differences suggest that common sequences account for less than 500 nt of the circulating HBV genomes. A pan-antiviral HBV could also be achieved using esiRNAs which can tolerate limited target sequence variations without losing their inhibitory capacity (Tan *et al.*, 2007). Alternatively, different antiviral strategies may be developed for each of the different HBV genotypes.

Although HBV inhibition with RNAi has been successful in cell culture and small animal models, translation to clinical trials has been slow. To date, only one phase I trial has been conducted (Gish *et al.*, 2011). In this study, a plasmid encoding four distinct shRNAs mapping to different regions of the HBV genome was delivered to patients. The results showed a reasonable safety profile. However, there was only a modest effect against HBV and evidence of an immune response. Thus, the authors conclude that further evaluations are needed before plasmid delivered shRNA can undergo Phase II or III trials. Viral vectors may be required to deliver the RNAi inhibitors to most hepatocytes and decrease viral pregenomic RNA and mRNA in patients. This should lead to reduced viral particles and viral proteins. Theoretically, in this scenario the viral cccDNA genome should still be stable in the nucleus of infected cells. In fact, even if some controversy exists, several studies have shown that RNAi impacts on HBV expression and therefore reduces the formation of newly synthesized cccDNA but does not affect pre-existent cccDNA (Li *et al.*, 2007; Starkey *et al.*, 2009). As the cccDNA is preserved, if the inhibitors are only expressed transiently, as soon as inhibition is lost, the viral RNAs will be transcribed from the stable cccDNA and the infection will rebound. However, a durable expression of RNAi inhibitors will lead to a long term decrease in the expression of surface antigen,

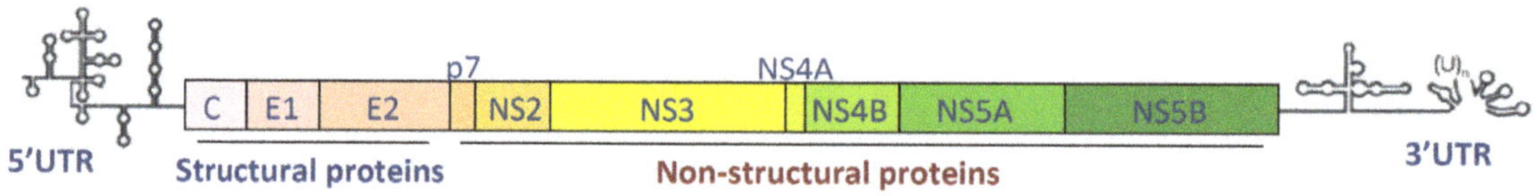

Figure 9.2 HCV genome organization. The HCV genome is a positive-strand RNA of ~9.6 kb, which is translated as a single polypeptide. The polypeptide is cleaved to produce three structural and seven non-structural proteins. Structural proteins are the core protein, known to bind the viral genome and packaged it into the capsid (C), and envelope glycoproteins (E1 and E2) involved in the attachment and entry process of the virus. Non-structural proteins are the ion channel p7, the protease NS2, the serine protease and helicase NS3, the serine protease cofactor NS4A, NS4B, involved in the generation of intracellular membrane rearrangements where replication occurs, NS5A, required for replication and assembly and the RNA-dependent RNA polymerase NS5B.

the major cause of liver damage, and decrease the progression of the disease to liver cirrhosis and hepatocarcinoma (Grimm *et al.*, 2006; Chen *et al.*, 2007, 2009). It could also be possible that a decrease in viral proteins may allow the activation of the immune response against the virus and the cure of the infection (Chisari *et al.*, 2010).

Hepatitis C virus infection

Hepatitis C virus (HCV) is a member of the Flaviridae family with a 9.6-kb single-stranded RNA genome of positive polarity. The genome contains a single open reading frame that encodes a polyprotein cleaved post-transcriptionally into three structural proteins (core, E1 and E2) and seven non-structural proteins (p7, NS2, NS3, NS4A, NS4B, NS5A and NS5B) (Fig. 9.2). The open reading frame is flanked by untranslated regions (UTR) that contain several well-conserved domains important for the HCV life cycle. The 5′UTR includes a highly conserved region important for replication that functions as an internal ribosome entry site (IRES) to initiate polyprotein translation. Replication occurs attached to cytoplasmic membranes and requires NS5B, the RNA-dependent RNA polymerase that lacks proof reading activity and leads to imperfect copies of the viral genome. This results in distinct but related HCV genomes that coexist in the infected individual and are known as quasi-species (Martell *et al.*, 1992).

There are six major and more than 80 subtypes of HCV. This classification is based on nucleotide variation among different HCV isolates. They are found in different proportion in different parts of the world, with genotype 1a and 1b being the most common in the USA and Europe (McOmish *et al.*, 1994). Approximately 170 million people (2.5% of the world population) are infected with HCV and it is estimated that many of them will develop a chronic liver disease, such as liver fibrosis, cirrhosis or hepatocellular carcinoma. There is no vaccine for prevention of HCV infection and current therapy for HCV is the combination of pegylated interferon-α with the nucleoside analogue ribavirin. This therapy is effective in 30–80% of patients, depending on the HCV serotype. However, this therapy is expensive and often associated with side-effects that may lead to discontinuation of therapy (Sherman, 2012). Recently, NS3 protease inhibitors that improve the rate of sustained virological response, boceprevir and telaprevir, have been approved for the treatment of HCV (Poordad and Dieterich, 2012). Additionally, there are other HCV antivirals in the pipeline expected soon to be on the market. However, given the high prevalence of the infection and the ability of HCV to generate escape mutants, there is still a need for the development of new therapies for chronic HCV. RNAi-based HCV antivirals could be an interesting option.

HCV RNA was predicted to be highly susceptible to RNAi because viral replication occurs in the cytoplasm and the viral genome resembles an mRNA. The results have confirmed the prediction. Indeed, RNAi-mediated inhibition is more effective in reducing HCV RNA levels than high doses of IFN (Kapadia *et al.*, 2003). HCV has a positive-sense RNA genome but requires the generation of a small number of negative genomes for replication. Although the less abundant antigenome is an attractive target for RNAi, it is not targetable by RNAi during HCV replication (Lisowski *et al.*, 2013).

There are no ideal models to test the inhibition of HCV by RNAi. *In vitro*, most studies are performed with HCV replicons that persistently replicate in stably transfected human hepatoma HuH7 cell lines (Lohmann *et al.*, 1999; Blight *et al.*, 2000). Replicons can be full-length or subgenomic RNAs, which lack the structural sequences, and express reporter or resistance genes. Most laboratories have used replicons derived from HCV 1b genotype. Once the replicon RNA is introduced into the cell, viral proteins are expressed able to replicate the viral RNA, which produces sufficient amounts of the reporter or the resistance protein to allow selection for replicon-expressing cells. However, the replicon system does not produce viral particles. Complete replication of HCV can be achieved in HuH7 with JFH1, a particular clone of HCV genotype 2a (Lindenbach *et al.*, 2005; Wakita *et al.*, 2005; Zhong *et al.*, 2005). Introduction of RNA from the JFH1 full-length clone leads to viral replication and production of viral particles able to infect fresh HuH7 cells and chimpanzees (Wakita *et al.*, 2005). A similar system with genotype 1a has just been published (Li *et al.*, 2012). Available animal models for the study of HCV are limited, mainly because of the selective host range of the virus, limited to humans and primates. Traditionally, *in vivo* models for HCV infection were limited to chimpanzees, which currently are not available for biomedical research. More recently, several labs have developed immunodeficient mice with chimeric humanized livers after being grafted with human hepatocytes (Mercer *et al.*, 2001; MacArthur *et al.*, 2012). These animals can be infected with HCV and produce viral particles, but lack a functional immune system (Turrini *et al.*, 2006; Ohira *et al.*, 2009; Washburn *et al.*, 2011; Bility *et al.*, 2012). HCV infection of fully immunocompetent inbred mice has been achieved after expression of two human genes, CD81 and occludin, required for HCV entry into hepatocytes (Dorner *et al.*, 2011). Unfortunately, mouse hepatocytes do not support HCV replication, so these animals only serve to study cell entry. Further humanization of these animals with human factors required for replication may lead to the generation of a model to recapitulate all the steps of the viral cycle.

The *in vitro* systems have been used to evaluate the efficacy of synthetic siRNAs or gene-expressed shRNAs targeting the conserved 5'UTR (Yokota *et al.*, 2003; Seo *et al.*, 2003; Takigawa *et al.*, 2004; Kanda *et al.*, 2007; Zekri *et al.*, 2009; Khaliq *et al.*, 2011) and structural genes (Randall *et al.*, 2003; Takigawa *et al.*, 2004; Prabhu *et al.*, 2005; Liu *et al.*, 2006; Ansar *et al.*, 2011). Among non-structural proteins, NS3 (Kapadia *et al.*, 2003; Wilson *et al.*, 2003; Takigawa *et al.*, 2004; Prabhu *et al.*, 2005), NS4 (Kim *et al.*, 2006) and NS5 (Kapadia *et al.*, 2003; Sen *et al.*, 2003; Wilson *et al.*, 2003; Takigawa *et al.*, 2004; Prabhu *et al.*, 2005; Trejo-Avila *et al.*, 2007) have been targeted (Table 9.1). Targeting of all these sequences with RNAi inhibitors leads to a decrease in the HCV genome and viral protein expression, suggesting that many regions of HCV RNA are accessible to the RNAi machinery. Importantly, the effects of RNAi on HCV are sequence specific and not caused by non-specific induction of host-defence pathways (Seo *et al.*, 2003; Wilson *et al.*, 2003). These results are encouraging, because viruses have evolved strategies to overcome host antiviral effects, including RNAi. Indeed, HCV can suppress RNAi by the interaction of HCV core and E2 protein with Dicer or the Ago-2 component of RISC respectively (Wang *et al.*, 2006; Ji *et al.*, 2008) and this was expected to represent a hurdle for antiviral therapy. It is still possible that higher inhibitions of HCV could be obtained when RNAi inhibitors are used in conditions in which viral silencing suppressors are not functional.

Another viral strategy to avoid RNA interference is the generation of escape mutants. Given the high rate of mutation of HCV genome, it is not surprising that after treatment with anti-HCV siRNAs, HCV escape mutants emerge with several point mutations in the siRNA targeted sequence (Wilson and Richardson, 2005). One way to decrease viral escape might be to select target sequences that are conserved because they are essential for viral replication. According to the level of sequence conservation among different genotypes, highly conserved sequences are in the 5'UTR IRES and in the region of the AUG translation initiation codon, where mutations may lead to loss of function (Chevalier *et al.*, 2007; Ray and Kanda, 2009). Other regions such as 3′UTR and NS3 and NS5B protein coding regions are controversial target sites because of the relatively low degree of conservation. Combination of RNAi effectors targeting multiple regions of HCV genome may also prevent viral escape. This can be achieved by expression of multiple shRNAs from independent expression cassettes in a single vector (Takigawa *et al.*, 2004; Henry *et al.*, 2006; Shin *et al.*, 2009) or by expression of lhRNAs that are processed to different siRNAs (Hamazaki *et al.*, 2004; Akashi *et al.*, 2005; Watanabe *et al.*, 2006). To decrease IFN response induced by lhRNAs, multiple point mutations can be inserted within the sense strand (Akashi *et al.*, 2005; Watanabe *et al.*, 2006). Several shRNAs can also be expressed from a single construct design to transcribe a polycistron. This is normally performed by mimicking miRNA polycistrons (Liu *et al.*, 2008; Ely *et al.*, 2009). Indeed, several polycistrons that express 4–5 miRNA-like transcripts complementary to the HCV genome have recently been described (Yang *et al.*, 2010, 2013).

Likewise, RNAi mediated inhibition of host cellular proteins that promote HCV replication should limit the emergence of resistant viruses. Several cellular factors have been identified that are required for HCV entry, replication and translation in the host cell (Evans *et al.*, 2007; Tai *et al.*, 2009; Vaillancourt *et al.*, 2009; Lupberger *et al.*, 2011; Metz *et al.*, 2012). Indeed, a number of laboratories have shown that RNAi against host cellular genes can decrease HCV infection. In human hepatoma cells, HCV infection has been decreased by RNAi-mediated inhibition of proteins required for HCV cell entry such as CD81, scavenger receptor class B type 1 (SR-B1), low density lipoprotein receptor (LDLR), claudin1 (CLDN1), occludin, EGFR and EphA2 (Zhang *et al.*, 2004a; Zeisel *et al.*, 2007; Jahan *et al.*, 2011b). In addition, the MAP kinase interacting serine/threonine kinase 1 (MKNK1) has recently been identified as a host factor involved in HCV entry. Indeed, siRNA-mediated knockdown of MKNK1 significantly impaired entry of HCV in culture cells (Kim *et al.*, 2013). Cellular proteins involved in HCV replication which have been targeted with RNAi include PI4-kinase (PI4K) III alpha, an enzymatic protein involved in phospholipid metabolism (Berger *et al.*, 2009); Polo-like kinase 1 (Plk1), an NS5A phosphokinase (Chen *et al.*, 2010); cytoplasmic cyclophilins A, B and C (Nakagawa *et al.*, 2005; Watashi *et al.*, 2005); proline-serine-threonine phosphatase-interacting protein 2 (PSTPIP2), a protein with membrane deforming activity (Chao *et al.*, 2012); hepcidin, an iron modulator (Bartolomei *et al.*, 2011) and RNA helicase A (RHA), which is recruited to the viral replication complex in the cytoplasm (Isken *et al.*, 2007). Human VAMP-associated protein A (hVAP-A), which binds to HCV NS5A and NS5B (Zhang *et al.*, 2004b), La autoantigen, which binds to HCV 5′UTR and polypyrimidine track-binding protein (PTB), which binds to both the 5′ and 3′ UTRs of HCV have also been targeted with RNAi and their down-regulation reduces HCV replication in cell culture with no impact upon cell viability (Zhang *et al.*, 2004b; Domitrovich *et al.*, 2005; Chang and Luo, 2006; Xue *et al.*, 2007). In hepatoma cells that constitutively produce infectious HCV, HCV production was reduced by siRNA directed against apolipoprotein B (ApoB). This protein is required for VLDL assembly, which forms a complex with secreted HCV in blood (Huang *et al.*, 2007). Another enzyme of the lipid metabolism whose RNAi-mediated knockdown severely impairs infectious virion production is triglyceride-synthesizing enzyme diacylglycerol acyltransferase-1 (DGAT1) (Herker *et al.*, 2010). *In vivo*, inhibition of heat shock protein 90 (Hsp90) suppressed HCV genome replication in HCV-infected chimeric mice with humanized liver (Nakagawa *et al.*, 2007).

Systematic RNAi screens have also identified the host genes implicated in HCV infection (Lupberger *et al.*, 2011; Chao *et al.*, 2012; Coller *et al.*, 2012). In 2007, Randall and colleagues designed siRNAs for 62 host genes that encode proteins that interact with the HCV genome or belong to cellular pathways that may modulate HCV infection. *In vitro*, RNAi-mediated inhibition of 26 of these host genes resulted in a 3-fold reduction in HCV infection. Interestingly, Dicer and other components of the RNAi pathway were among them. This was surprising as it has been proposed that the HCV genome forms dsRNA regions that can be targeted by Dicer, suggesting that Dicer inhibition should improve HCV replication. Moreover, it has been described that HCV core protein is a silencing suppressor that blocks Dicer (Chen *et al.*, 2008b). However, the RNAi machinery is required for production of miR-122, a liver specific miRNA required for HCV replication. In fact, miR-122 blockade with antisense oligonucleotides impairs HCV replication (Randall *et al.*, 2007; Lanford *et al.*, 2010) and could represent a therapeutic option for HCV infection in the future. Indeed, silencing of many host cellular genes that affect HCV viability could also be the basis for therapies to treat chronic HCV, but it should be taken into consideration that long-term silencing of such host genes will be required, which could result in unwanted side-effects.

HCV replication can also be inhibited by combining RNAi effectors that target host cellular genes and the HCV genome (Randall *et al.*, 2003; Korf *et al.*, 2005; Henry *et al.*, 2006; Jahan *et al.*, 2011a). To avoid off-target effects due to the expression of multiple RNAi effectors, RNAi can also be combined with other therapies such as hairpin ribozymes targeting the 3'UTR of the HCV genome (Jarczak *et al.*, 2005) or IFN. RNAi and IFN combination can lead to combined antiviral activity (Pan *et al.*, 2009) or can even potentiate the ability of IFN to inhibit HCV RNA replication (Randall *et al.*, 2006). As USP18 modulates the anti-HCV type I IFN response, silencing of this protease in human cells with RNAi potentiates the antiviral activity of IFN against HCV infection (Murray *et al.*, 2011).

Most of the inhibitors described so far have been assayed in tissue culture cells expressing the HCV replicons or in cells infected with JFH1 clone of HCV. However, some of the inhibitors have been tested in mice. NS5B and IRES have been targeted by RNAi in adult mice (McCaffrey *et al.*, 2002; Wang *et al.*, 2005). A transient HCV mouse model system has been used to test siRNAs targeting HCV core protein (Kim *et al.*, 2006). More recently, efficacy of anti-HCV inhibitors targeting the 5'UTR, core and NS5B expressed from a micro-RNA cluster has been tested in mice expressing luciferase-HCV reporter (Yang *et al.*, 2010, 2013). RNAi therapy has also been examined by using a non-human primate surrogate model for HCV (Yokota *et al.*, 2007). Three shRNAs that simultaneously target multiple conserved regions of the HCV genome have been efficiently delivered in a non-human primate model (Suhy *et al.*, 2012).

For delivery, synthetic siRNAs have been intravenously administered after assembly with cationic liposomes and recombinant Apolipoprotein AI (Apo AI) in a HCV mouse model (Lee *et al.*, 2009). Moreover, siRNAs improve their activity and do not show immunotoxicity when two U sequences on its sense strand are modified by 2'-O-Methyl modification (Kim *et al.*, 2009). In another study, sustained inhibition of HCV replication and infection in hepatocyte-derived cells were obtained with anti HCV siRNAs delivered by cell-degradable multilayered polyelectrolyte films (MPFs) (Dimitrova *et al.*, 2008). Viral vectors have been used for the delivery of gene-encoded shRNAs, although other non-viral carriers have been also explored (Ding *et al.*, 2013). Sustained expression of anti-HCV RNAi has been obtained using retrovirus (Korf *et al.*, 2005; Nakagawa *et al.*, 2005), lentivirus (Takigawa *et al.*, 2004; Henry *et al.*, 2006), adenovirus (Zhang *et al.*, 2004b) and AAV viral vectors (Yang *et al.*, 2010; Suhy *et al.*, 2012). In a recent study, Pan and colleagues have observed that NOD/SCID mice engrafted with human hepatoma cells producing CD81 siRNA, also showed suppression of CD81 expression in mouse hepatocytes (Pan *et al.*, 2012). This result suggests that there may be an RNAi transmission in the liver *in vivo*, which could potentially extend the therapeutic reach of RNAi-based therapies against HCV and other liver diseases. Further experiments are required to determine if this is indeed the case. Also, the promising results obtained so far with RNAi anti-HCV in preclinical experiments should move forward into clinical studies.

RNAi for respiratory viruses

Respiratory syncytial virus infection

Respiratory syncytial virus (RSV) infection is a major cause of seasonal colds, which have a huge economic impact on the health systems. Moreover, RSV can cause bronchiolitis in children and is the leading cause of infant hospitalization. In addition, RSV can produce severe respiratory diseases in patients with compromised immune systems. An RSV vaccine is not available and the only approved antiviral therapy for RSV is the nucleoside analogue ribavirin in an aerosolized formulation. However, the clinical utility of this drug is controversial, specially in the paediatric population. Therefore, a safe and efficacious RSV therapy is long awaited for both paediatric and adult populations.

The RSV genome is a negative stranded RNA of approximately 15.2 kb. It is transcribed into 10 transcripts encoding 11 distinct proteins, which include non-structural NS1 and NS2, the RNA-dependent RNA polymerase (P) and nucleocapsid (N) proteins.

RNAi has been used to target viral RSV genes or host genes required for RSV infection. RSV gene expression and growth has been inhibited in culture with siRNAs targeting, for instance, P (Bitko and Barik, 2001), NS1 (Zhang *et al.*, 2005) or the fusion protein (Vig *et al.*, 2009) (Table 9.1). These siRNAs were also administered intranasally in mice to show that RSV infection can be prevented with anti-P siRNAs (Bitko *et al.*, 2005). Nanoparticles containing anti-NS1 siRNAs administered before or after infection with RSV also lead to decreased virus titres compared to controls (Zhang *et al.*, 2005). Decreased RSV infection was also observed upon RNAi-mediated inhibition of cellular nucleolin both in tissue culture and in mouse lung, confirming that nucleolin is a functional RSV receptor *in vivo* (Tayyari *et al.*, 2011).

The effect of a siRNA that targets a well conserved sequence throughout naturally occurring RSV-A and B genotypes in the viral N transcript has been studied in detail. This compound, named ALN-RSV01, is an unmodified naked siRNA that when delivered intranasally in mice reduced RSV replication prophylactically and therapeutically in a dose-dependent fashion. ALN-RSV01 exerted the antiviral activity without off-target proinflammatory effects and generated RNAi-mediated RNA cleavage products (Alvarez *et al.*, 2009). Therefore ALN-RSV01 was selected for clinical development and was the first antiviral RNAi therapy approved for clinical trial.

In an initial Phase I trial, two safety and tolerability studies with ALN-RSV01 were conducted involving healthy adults. ALN-RSV01 appeared safe and well tolerated when delivered intranasally (DeVincenzo *et al.*, 2008). Therefore, a Phase II trial was performed in adults experimentally infected with wild-type RSV. In this study, ALN-RSV01 showed a significant antiviral activity against human RSV infection. Moreover, the RNAi-based mechanism of action of the drug was validated in humans (DeVincenzo *et al.*, 2010). Another phase II trial assessed the safety and antiviral activity of ALN-RSV01 in lung transplant patients experiencing acute infection with RSV. ALN-RSV01 was safe, well tolerated and associated with a statistically significant improvement in symptoms (Zamora *et al.*, 2011). However, no statistically significant antiviral effect was demonstrated and a larger Phase IIb trial is now being conducted to confirm and extend previous findings.

Other respiratory viruses

RNAi-based therapy has been tested in other respiratory viruses such as measles virus, severe acute respiratory syndrome (SARS) coronavirus, influenza, coxsackievirus and other enteroviruses, human metapneumovirus and parainfluenza viruses (Lu *et al.*, 2004; Tan *et al.*, 2008; Wu *et al.*, 2008; Yao *et al.*, 2012).

Measles is a highly contagious respiratory disease caused by the negative-strand RNA measles virus (MV). MV replication has been inhibited by RNAi using siRNAs targeting viral genes (Otaki *et al.*, 2006), host cellular proteins such as the main receptor of MV, signalling lymphocyte activation molecule (SLAM) (Hu *et al.*, 2005) or by the combinatorial RNAi approach targeting both host and viral genes (Shi *et al.*, 2012) (Table 9.1).

SARS coronavirus (SARS-CoV) has a positive stranded RNA genome and causes a viral respiratory disease. Viral replication has been efficiently suppressed in culture using synthetic (Wu *et al.*, 2005a) or plasmid-derived siRNAs (Wang *et al.*, 2004; Ni *et al.*, 2005) against the viral genome (Table 9.1). Moreover, synthetic siRNAs that were functional *in vitro* (Zheng *et al.*, 2004) were also efficient for prophylactic and therapeutic treatment of SARS infection in a rhesus macaque

SARS model (Li *et al.*, 2005; Tang *et al.*, 2008) (Table 9.1).

Influenza virus causes a highly contagious disease responsible for seasonal epidemics and recurring pandemics. There is a large variety of influenza viral strains that infect human, swine and avian species. Occasionally, two strains can infect the same individual and recombine to generate a new virus. In addition, the high mutation rates of the viral polymerase facilitate the generation of viral escape mutants, rendering vaccines and drugs directed against virus-encoded targets potentially ineffective. The viral genome consists of eight negative-strand RNA segments that code for 11 distinct viral proteins. RNAi against influenza virus should target the viral mRNAs, since the RNA genome is encapsidated. To date, several studies have evaluated RNAi as a therapy against influenza virus. Most have focused on influenza A viral strains (Ge *et al.*, 2003; Sui *et al.*, 2009), including the novel influenza A (H1N1) virus detected in 2009 (Zhiqiang *et al.*, 2010). The results show that RNAi is effective in preventing and treating influenza virus infection in mice (Ge *et al.*, 2004; Tompkins *et al.*, 2004) (Table 9.1). Interestingly, RNAi-induced off-target activation of the immune system has been exploited to potentiate the antiviral effects of siRNAs targeting influenza virus genes (Stewart *et al.*, 2011). The immunostimulatory properties of siRNAs were enhanced by introducing the nucleoside sequence 5′-UGUGU-3′ at the 5′-end of the sense strand of siRNAs. A siRNA against influenza PB1 gene with this sequence induced more type I IFN and faster than controls, and therefore, it should have a stronger antiviral activity. Moreover, human host factors crucial for influenza virus replication have been identified (Karlas *et al.*, 2010), whose RNAi-mediated decrease could also represent an antiviral strategy to inhibit viral infection. Given Influenza virus ability to generate escape mutants, combination of several RNAi-inhibitors or combination of RNAi with other antiviral strategies will be required to develop a durable therapy.

A host component whose inhibition led to a mild increase in viral production and apoptosis of the infected cell was Dicer (Matskevich and Moelling, 2007). This result suggests that RNAi induces an antiviral response, and Dicer is important for protection against influenza A virus infection. Increased susceptibility to influenza virus has also been described in mice double-knockout for the RNAi components Argonaute 1 and 3 (Van Stry *et al.*, 2012). These mice exhibited increased mortality after infection, which was not associated with increased viral replication. Influenza virus may counteract the antiviral RNAi with the silencing suppressor activity of NS1 protein, although this activity is still controversial (Li *et al.*, 2004; Kok and Jin, 2006; de Vries *et al.*, 2009). Even if NS1 or any other viral protein has silencing suppressor activity, the inhibition of influenza virus replication with siRNAs seems strong enough to counteract suppression.

RNAi for haemorrhagic fever viruses

Viral haemorrhagic fevers are severe zoonotic diseases caused by RNA-viruses classified into 4 families: *Arenaviridae, Bunyaviridae, Filoviridae,* and *Flaviviridae*. The most important representatives are Ebola (EBOV), Marburg (MARV), yellow fever (YFV), West Nile virus (WNV) and dengue (DENV). Haemorrhagic fever is characterized by subcutaneous bleeding or bleeding from body cavities and internal organs that may lead to death. Viral haemorrhagic fevers are most common in Africa, Southeast Asia and South America and are sometimes transferred to other areas by tourists, immigrants or natural vectors. There are no effective drugs against these viruses. The current therapy relies on symptomatic treatment. Therefore, RNAi-mediated antiviral silencing could represent a promising therapeutic option to inhibit viral gene expression and replication.

Ebola

Ebola virus (EBOV) is a negative stranded RNA virus that causes severe haemorrhagic fever with mortality rates that might exceed 80%. At present there are no vaccines or effective therapies available for humans. Therefore, RNAi could represent a potential therapy for EBOV infection. Feasibility of RNAi-based therapy for Ebola was first assessed in guinea pigs. siRNAs targeting the Zaire Ebola virus (ZEBOV) RNA polymerase L protein formulated in stable nucleic acid-lipid particles (SNALPs) completely protected guinea pigs when administered shortly after a lethal ZEBOV challenge. Importantly, IFN response was not induced by the siRNA–SNALP complexes (Geisbert *et al.*, 2006) (Table 9.1). In subsequent studies, Geisbert and colleagues demonstrated that siRNAs targeting the polymerase L and VP24 and VP35 viral proteins in the form of SNALPs had antiviral activity in *Rhesus macaque* monkeys after ZEBOV challenge (Geisbert *et al.*, 2010).

Host cellular genes required for infection have been also targeted by RNAi. For instance, Niemann-Pick C1 protein (NPC1) plays a key role in EBOV infection. This protein is an endo/lysosomal cholesterol transporter that mediates the escape of viral particles from lysosomes to the cytoplasm. Cells defective for NPC1 are resistant to infection with EBOV and NPC1 knockout mice survived infection with a lethal dose of Ebola virus. The same results were observed when Marburg virus was used for infection (Carette *et al.*, 2011). Likewise, TIM-1 protein serves as a receptor for Ebola and Marburg viruses and reduction of cell-surface expression of TIM-1 by RNAi decreases EBOV infection in highly permissive cells (Kondratowicz *et al.*, 2011).

More detailed research is needed to determine whether the antiviral effect of RNAi for EBOV infection can be increased by silencing EBOV proteins that act as silencing suppressors by direct interaction with the mammalian RNAi pathway (Fabozzi *et al.*, 2011).

Marburg

Marburg (MARV) is a negative stranded RNA virus whose infection leads to mortality rates ranging from 30% to 90%. The genome contains seven genes arranged in a linear order that express NP, VP35, VP40, GP, VP30, VP24 and L proteins. To date, only one study has evaluated RNAi efficacy against MARV infection. MARV nucleocapsid proteins NP, VP35 and VP30 were efficiently silenced *in vitro* when produced alone via plasmid transfection and in the natural viral-infection environment. In MARV-infected cells, decreasing the cellular levels of these proteins invariably resulted in a decreased release of progeny virions in the supernatant (Fowler *et al.*, 2005) (Table 9.1). Host cellular genes required for infection have been also targeted by RNAi (see 'Ebola', above).

Dengue

Dengue (DENV) is a positive-stranded RNA virus. The genome is translated into a polyprotein that is later processed to 3 structural and seven non-structural proteins. Symptoms upon DENV infection range from mild fever to potentially fatal haemorrhagic fever and hypovolaemic shock. Currently, there is neither a preventative vaccine nor an effective antiviral therapy against dengue disease. Thus, RNAi approaches have been studied for DENV infection. Adenoviral and AAV delivered shRNAs targeting the untranslated regions of DENV genome have efficiently reduced DENV infection in cultured cells (Zhang *et al.*, 2004c; Korrapati *et al.*, 2012). Likewise, synthetic siRNAs targeting highly conserved sequences of the DENV genome have suppressed dengue virus replication in cultured cells (Subramanya *et al.*, 2010) and in a mouse model of Dengue disease (Stein *et al.*, 2011) (Table 9.1). Finally, lentivirus-delivered polycistronic miRNA shuttles have very recently demonstrated DENV virus replication inhibition in a dose-dependent manner (Xie *et al.*, 2013). Other studies have focused on silencing cellular proteins implicated in DENV entry into host cells, with promising results that show attenuated dengue infection (Alhoot *et al.*, 2011, 2012).

Since DENV is transmitted by the *Aedes aegypti* mosquito, an alternative approach against DENV infection involves the use of RNAi to render mosquito cells resistant to infection. Replacement of wild-type mosquitoes with transgenic DENV-resistant vectors would control virus transmission. In an initial study, cultured mosquito cells transfected with a plasmid-expressing dsRNAs derived from the DENV genome, became resistant to DENV challenge (Adelman *et al.*, 2002). Later, transgenic mosquito lines that transcribed the effector dsRNAs from constitutive promoters were developed. Though generation of DENV-specific siRNAs was demonstrated, their production in critical sites for viral replication, such as midgut and salivary glands, was not observed (Travanty *et al.*, 2004). In a subsequent study, dsRNA expression was driven by an alternative promoter and mosquitoes exhibited decreased viral replication. However, silencing of DENV was not complete and viral replication occurred in some of the transgenic insects (Franz *et al.*, 2006). In fact, transgenic flies that express a viral inverted repeat able to generate siRNAs that target dengue virus, showed a strong viral resistance phenotype but only for some generations (Franz *et al.*, 2009). The reason for this is unclear. The transgene sequence is not mutated but expression of the transgene is not detected.

Further improvements may be obtained by the silencing of mosquito proteins required for viral infection, such as some midgut serine proteases (Brackney *et al.*, 2008; Guo *et al.*, 2010; Sim *et al.*, 2012) or required for blocking components of the mosquito antiviral response, such as PIAS, the negative regulator of the JAK-STAT pathway (Souza-Neto *et al.*, 2009). Indeed, one of the pathways required to control viral infection in the mosquito is RNAi. Viral infection generates dsRNAs and production of DENV2-specific siRNAs (Scott *et al.*, 2010). However, virus replication persists, suggesting that the virus expresses silencing suppressors. Accordingly, silencing the expression of Dicer or Argonaute, increases viral replication (Sanchez-Vargas *et al.*, 2009; Mukherjee and Hanley, 2010; Khoo *et al.*, 2013).

Other haemorrhagic fever viruses

RNAi-based approaches have also been employed to target other haemorrhagic fever viruses such as yellow fever virus (YFV), West Nile Virus (WNV), Junín virus (JUNV), Rift Valley fever virus (RVFV) and Crimean Congo fever virus (CCHFV) (Flusin *et al.*, 2011).

YFV contains a positive-stranded RNA genome that encodes three structural and seven nonstructural proteins. Yellow fever is a tropical disease of great importance for public health with no available treatment. In 2009, Pacca and colleagues showed that shRNAs against YFV inhibited viral replication in stable cell lines and protected adult mice against YFV challenge (Pacca *et al.*, 2009) (Table 9.1).

WNV is a single-stranded RNA virus that causes fatal encephalitis in humans, mice, and other vertebrates. Prophylactic use of synthetic siRNAs or lentivirally delivered shRNAs against WNV infection has been successful *in vitro* and in mice (Bai *et al.*, 2005; Kumar *et al.*, 2006) (Table 9.1). In addition, plasmid or retrovirus-mediated delivery of shRNAs has efficiently targeted the WNV genome in cell culture (Ong *et al.*, 2008; Yang *et al.*, 2008; Anthony *et al.*, 2009). Finally, identification of host cellular proteins by genome-wide RNAi screens associated with WNV infection has revealed novel potential antiviral targets (Krishnan *et al.*, 2008). As has been described previously for other viruses, WNV also infects mosquitoes as part of its natural transmission cycle. Therefore, RNAi could be used to develop transgenic mosquitoes resistant to WNV infection. In fact, mosquitoes use the RNAi machinery to control WNV replication and WNV responds with the generation of quasispecies preferentially mutated in the regions that are more intensely targeted by RNAi (Brackney *et al.*, 2009). Moreover, a subgenomic flavivirus RNA (sfRNA) from WNV generated from the 3′-UTR acts as a silencing suppressor of siRNAs and miRNAs in both mammalian and insect cells (Schnettler *et al.*, 2012).

Other viruses targeted by RNAi

Other viruses targeted by RNAi include herpes simplex virus (HSV), human papillomavirus (HPV), rotaviruses, rabies virus (RABV) and lentiviruses such as HIV virus, which deserves a special chapter in this book.

Herpes simplex viruses 1 and 2 (HSV-1 and HSV-2) are two members of the Herpesviridae family that infect humans. They have a double-stranded, linear DNA genome. HSV-1 causes a variety of infections, including oral and skin infections, genital herpes, ocular infections, and the rare but severe HSV encephalitis. HSV-2 is a sexually transmitted virus whose infection causes genital herpes and is an important cofactor for the transmission of HIV infection.

RNAi have been applied to suppress HSV-1 and HSV-2 infections in cultured cells (Bhuyan *et al.*, 2004; Qing *et al.*, 2011). Recently canonical siRNA and longer Dicer substrate siRNAs (DsiRNAs) were designed to block HSV-1 infection in human skin and nervous system-derived cell lines (Romanovskaya *et al.*, 2012). Moreover, RNAi has yielded promising results in an animal model of HSV-2 where vaginal instillation of siRNAs targeting HSV-2 protected mice from lethal infection (Palliser *et al.*, 2006) (Table 9.1).

Human papillomavirus (HPV) is a small DNA virus that infects keratinocytes from skin or mucous membranes. HPV infection is the major risk factor for cervical cancer. Infection of HPV types 16 and 18 is associated with greater risk and causes approximately 50% and 20% of cervical cancer cases respectively (Walboomers *et al.*, 1999). Vaccination provides significant prophylactic protection but effective therapeutic treatments are still required.

The RNAi approach to treating HPV has focused on targeting HPV 16 and 18 types. Though any viral gene could potentially be targeted, studies have focused on E6 and E7 early non-structural genes, since their simultaneous expression is pivotal for transformation and maintenance of malignant phenotypes. The first successful attempt to silence HPV using siRNAs was published in 2002 (Jiang and Milner, 2002). siRNAs targeting either HPV16 E6 or E7 were transfected into cervical cancer cell lines. Knockdown of E6 reduced cell growth while knockdown of E7 induced apoptosis. In addition, siRNAs against the HPV18 E6 oncogene restored dormant tumour-suppressor pathways on HPV-positive cervical cancer cells (Butz *et al.*, 2003; Yamato *et al.*, 2006). Moreover, targeting HPV oncogenes by RNAi increased sensitivity of tumour cells to other therapeutic approaches (Putral *et al.*, 2005; Courtete *et al.*, 2007). *In vivo*, inoculation of cervical cancer cells pre-treated with E6/E7 siRNAs in immune-deficient mice resulted in smaller tumours compared to those obtained after inoculation of cells treated with control siRNAs (Yoshinouchi *et al.*, 2003; Khairuddin *et al.*, 2012) (Table 9.1). All these results favour translation to the clinical setting but there are significant technical hurdles to overcome, particularly in the delivery area, since varying success has been reported when delivering siRNAs to the vaginal tract.

Rotaviruses, the leading cause of severe dehydrating diarrhoea in infants and young children worldwide, are non-enveloped viruses with a double-stranded RNA genome. RNAi targeting VP4 viral protein has been used to inhibit rotaviral infection, as VP4 is essential for receptor binding and cell penetration. Thus, monkey kidney epithelial cells transfected with synthetic siRNAs against VP4 sequences produced fewer viral particles (Dector *et al.*, 2002) (Table 9.1).

Rabies is a viral disease that causes acute encephalitis in warm-blooded animals. It is caused by the rabies virus (RABV), which is transmitted from one species to another, such as from dogs to humans, commonly by a bite from an infected animal. The RABV genome consists of a negative-stranded RNA molecule of about 12 kb that encodes N, P, M, G and L viral proteins. siRNAs targeting the conserved region of the RABV N gene have proven efficient for the inhibition of viral replication *in vitro* (Yang *et al.*, 2012). Moreover, siRNAs protected mice against lethal RABV challenge (Gupta *et al.*, 2012; Yang *et al.*, 2012) (Table 9.1). Anti-RABV RNAi effectors were expressed as shRNAs (Yang *et al.*, 2012), from adenoviruses (Gupta *et al.*, 2012), from AAVs (Wu *et al.*, 2013) or from miRNA-like expressing vectors (Israsena *et al.*, 2009).

RNAi has also been used to avoid replication of many other viruses, which are not described here for reasons of space. In summary, RNAi could theoretically be used to inhibit the replication of any virus, probably with the exception of subviral pathogens such as prions, which do not contain a DNA or an RNA genome. In most cases, viral mRNAs are targeted, even if the viral genome of some RNA viruses has also been successfully targeted. Also, targeting host genes required for viral replication has shown antiviral efficacy. To prevent the emergence of resistant variants, several inhibitors should be combined based on RNAi or other strategies that target viral or non-essential host genes relevant for viral infection.

RNAi for the treatment of non-viral infections

Infections by bacteria and parasites could also be treated with RNAi-mediated silencing. Bacteria can silence genetic material by an RNA-guided system, termed clustered regularly interspaced short palindromic repeat (CRISPR)-interference (Wiedenheft *et al.*, 2012). However, this process is distinct from RNA interference (RNAi) in eukaryotes. Moreover, bacterial genes are expressed outside the host cell. Therefore, bacterial genes cannot be silenced by RNAi. Alternatively, bacterial infections can be treated by RNAi by down-regulation of host genes required for bacterial invasions or for the immune response that leads to damaging side-effects in some bacterial infections (Thornbrough *et al.*, 2012).

Among parasites, malaria remains a leading cause of morbidity and mortality worldwide. RNAi could also be used for the treatment of malaria infections. In malaria parasites, although it has been previously suggested that RNAi is not functional (Baum *et al.*, 2009), a putative, non-canonical RNAi pathway might exist that allows knockdown of plasmodial proteins resulting in lower parasitaemia (Schwentke *et al.*, 2012). In addition, functional RNAi machinery has been described in the principal malaria vector in Africa, *Anopheles gambiae*. Therefore, malaria could also be treated by down-regulation of non-essential mosquito genes required for malaria transmission (Brown *et al.*, 2003). Parasite resistant mosquito strains could then be released to replace natural disease-transmitting populations and achieve malaria eradication.

Conclusions

Since the mechanism of RNAi was first elucidated, several studies have shown its potential as a clinical tool to treat viral infections. RNAi effectors can be easily designed based on simple sequence information and thermodynamic guidelines. Furthermore, genome-wide RNAi screenings have identified a whole range of host cell factors which could also be targeted by RNAi to modulate viral infection. However, all the success in preclinical antiviral RNAi has not yet had any impact in clinical settings. Delivery remains a challenge for the therapeutic application of RNAi and has limited its potential. Future progress in increasing transduction with viral and non-viral particles as well as the positive results of several ongoing clinical trials will progressively alleviate these shortcomings and will help RNAi to occupy a place in clinical practice.

References

Adelman, Z.N., Sanchez-Vargas, I., Travanty, E.A., Carlson, J.O., Beaty, B.J., Blair, C.D., and Olson, K.E. (2002). RNA silencing of dengue virus type 2 replication in transformed C6/36 mosquito cells transcribing an inverted-repeat RNA derived from the virus genome. J. Virol. *76*, 12925–12933.

Akashi, H., Miyagishi, M., Yokota, T., Watanabe, T., Hino, T., Nishina, K., Kohara, M., and Taira, K. (2005). Escape from the interferon response associated with RNA interference using vectors that encode long modified hairpin-RNA. Mol. Biosyst. *1*(5–6), 382–390.

Alhoot, M.A., Wang, S.M., and Sekaran, S.D. (2011). Inhibition of dengue virus entry and multiplication into monocytes using RNA interference. PLoS Negl. Trop. Dis. *5*, e1410.

Alhoot, M.A., Wang, S.M., and Sekaran, S.D. (2012). RNA interference mediated inhibition of dengue virus multiplication and entry in HepG2 cells. PLoS One *7*, e34060.

Alvarez, R., Elbashir, S., Borland, T., Toudjarska, I., Hadwiger, P., John, M., Roehl, I., Morskaya, S.S., Martinello, R., Kahn, J., *et al.* (2009). RNA interference-mediated silencing of the respiratory syncytial virus nucleocapsid defines a potent antiviral strategy. Antimicrob. Agents Chemother. *53*, 3952–3962.

Ansar, M., Ashfaq, U.A., Shahid, I., Sarwar, M.T., Javed, T., Rehman, S., Hassan, S., and Riazuddin, S. (2011). Inhibition of full length hepatitis C virus particles of 1a genotype through small interference RNA. Virol. J. *8*, 203.

Anthony, K.G., Bai, F., Krishnan, M.N., Fikrig, E., and Koski, R.A. (2009). Effective siRNA targeting of the 3′ untranslated region of the west Nile virus genome. Antiviral Res. *82*, 166–168.

Bai, F., Wang, T., Pal, U., Bao, F., Gould, L.H., and Fikrig, E. 2005. Use of RNA interference to prevent lethal murine west Nile virus infection. J. Infect. Dis. *191*, 1148–1154.

Bartolomei, G., Cevik, R.E., and Marcello, A. (2011). Modulation of hepatitis C virus replication by iron and hepcidin in Huh7 hepatocytes. J. Gen. Virol. *92*, 2072–2081.

Baum, J., Papenfuss, A.T., Mair, G.R., Janse, C.J., Vlachou, D., Waters, A.P., Cowman, A.F., Crabb, B.S., and de Koning-Ward, T.F. (2009). Molecular genetics and comparative genomics reveal RNAi is not functional in malaria parasites. Nucleic Acids Res. *37*, 3788–3798.

Berger, K.L., Cooper, J.D., Heaton, N.S., Yoon, R., Oakland, T.E., Jordan, T.X., Mateu, G., Grakoui, A., and Randall, G. (2009). Roles for endocytic trafficking and phosphatidylinositol 4-kinase III alpha in hepatitis C virus replication. Proc. Natl. Acad. Sci. U.S.A. *106*, 7577–7582.

Bhuyan, P.K., Kariko, K., Capodici, J., Lubinski, J., Hook, L.M., Friedman, H.M., and Weissman, D. (2004). Short interfering RNA-mediated inhibition of herpes simplex virus type 1 gene expression and function during infection of human keratinocytes. J. Virol. *78*, 10276–10281.

Bian Z, Xiao A, Cao M, Liu M, Liu S, Jiao Y, Yan W, Qi Z, and Zheng, Z. (2012). Anti-HBV efficacy of combined siRNAs targeting viral gene and heat shock cognate 70. Virol. J. *9*, 275.

Bility MT, Zhang L, Washburn ML, Curtis TA, Kovalev GI, and Su, L. (2012). Generation of a humanized mouse model with both human immune system and liver cells to model hepatitis C virus infection and liver immunopathogenesis. Nat. Protoc. *7*, 1608–1617.

Birmingham, A., Anderson, E.M., Reynolds, A., Ilsley-Tyree, D., Leake, D., Fedorov, Y., Baskerville, S., Maksimova, E., Robinson, K., Karpilow, J., *et al.* (2006). 3′ UTR seed matches, but not overall identity, are associated with RNAi off-targets. Nat. Methods *3*, 199–204.

Bitko, V., and Barik, S. (2001). Phenotypic silencing of cytoplasmic genes using sequence-specific double-stranded short interfering RNA and its application in the reverse genetics of wild type negative-strand RNA viruses. BMC Microbiol. *1*, 34.

Bitko, V., Musiyenko, A., Shulyayeva, O., and Barik, S. (2005). Inhibition of respiratory viruses by nasally administered siRNA. Nat. Med. *11*, 50–55.

Blazquez, L., Gonzalez-Rojas, S.J., Abad, A., Razquin, N., Abad, X., and Fortes, P. (2012). Increased *in vivo* inhibition of gene expression by combining RNA interference and U1 inhibition. Nucleic Acids Res. *40*, e8.

Blight, K.J., Kolykhalov, A.A., and Rice, C.M. (2000). Efficient initiation of HCV RNA replication in cell culture. Science *290*,1972–1974.

Boden, D., Pusch, O., Lee, F., Tucker, L., and Ramratnam, B. (2003). Human immunodeficiency virus type 1 escape from RNA interference. J. Virol. *77*, 11531–11535.

Brackney, D.E., Foy, B.D., and Olson, K.E. (2008). The effects of midgut serine proteases on dengue virus type 2 infectivity of aedes aegypti. Am. J. Trop. Med. Hyg. *79*, 267–274.

Brackney, D.E., Beane, J.E., and Ebel, G.D. (2009). RNAi targeting of west Nile virus in mosquito midguts promotes virus diversification. PLoS Pathog. *5*, e1000502.

ter Brake, O., and Berkhout, B. (2007). Lentiviral vectors that carry anti-HIV shRNAs: Problems and solutions. J. Gene. Med. 9,743–750.

Bramsen, J.B., Laursen, M.B., Nielsen, A.F., Hansen, T.B., Bus, C., Langkjaer, N., Babu, B.R., Hojland, T., Abramov, M., Van Aerschot, A., *et al.* (2009). A large-scale chemical modification screen identifies design rules to generate siRNAs with high activity, high stability and low toxicity. Nucleic Acids Res. *37*, 2867–2881.

Bridge, A.J., Pebernard, S., Ducraux, A., Nicoulaz, A.L., and Iggo, R. (2003). Induction of an interferon response by RNAi vectors in mammalian cells. Nat. Genet. *34*, 263–264.

Brown, A.E., Bugeon, L., Crisanti, A., and Catteruccia, F. (2003). Stable and heritable gene silencing in the malaria vector anopheles stephensi. Nucleic Acids Res. *31*, e85.

Brzezinska, J., D'Onofrio, J., Buff, M.C., Hean, J., Ely, A., Marimani, M., Arbuthnot, P., and Engels, J.W. (2012). Synthesis of 2′-O-guanidinopropyl-modified nucleoside phosphoramidites and their incorporation into siRNAs targeting hepatitis B virus. Bioorg. Med. Chem. *20*, 1594–1606.

Butz, K., Ristriani, T., Hengstermann, A., Denk, C., Scheffner, M., and Hoppe-Seyler, F. (2003). siRNA targeting of the viral E6 oncogene efficiently kills human papillomavirus-positive cancer cells. Oncogene *22*, 5938–5945.

Carette, J.E., Raaben, M., Wong, A.C., Herbert, A.S., Obernosterer, G., Mulherkar, N., Kuehne, A.I., Kranzusch, P.J., Griffin, A.M., Ruthel, G., *et al.* (2011). Ebola virus entry requires the cholesterol transporter niemann-pick C1. Nature *477*, 340–343.

Carmona, S., Ely, A., Crowther, C., Moolla, N., Salazar, F.H., Marion, P.L., Ferry, N., Weinberg, M.S., and Arbuthnot, P. (2006). Effective inhibition of HBV replication *in vivo* by anti-HBx short hairpin RNAs. Mol. Ther. *13*, 411–421.

Carmona, S., Jorgensen, M.R., Kolli, S., Crowther, C., Salazar, F.H., Marion, P.L., Fujino, M., Natori, Y., Thanou, M., Arbuthnot, P., *et al.* (2009). Controlling HBV replication *in vivo* by intravenous administration of triggered PEGylated siRNA-nanoparticles. Mol. Pharm. *6*, 706–717.

Chang, K.S., and Luo, G. (2006). The polypyrimidine tract-binding protein (PTB) is required for efficient replication of hepatitis C virus (HCV) RNA. Virus Res. *115*, 1–8.

Chao, T.C., Su, W.C., Huang, J.Y., Chen, Y.C., Jeng, K.S., Wang, H.D., and Lai, M.M. (2012). Proline-serine-threonine phosphatase-interacting protein 2 (PSTPIP2), a host membrane-deforming protein, is critical for membranous web formation in hepatitis C virus replication. J. Virol. *86*, 1739–1749.

Chen, C.C., Ko, T.M., Ma, H.I., Wu, H.L., Xiao, X., Li, J., Chang, C.M., Wu, P.Y., Chen, C.H., Han, J.M., *et al.* (2007). Long-term inhibition of hepatitis B virus in transgenic mice by double-stranded adeno-associated virus 8-delivered short hairpin RNA. Gene. Ther. *14*, 11–19.

Chen, C.C., Sun, C.P., Ma, H.I., Fang, C.C., Wu, P.Y., Xiao, X., and Tao, M.H. (2009). Comparative study of anti-hepatitis B virus RNA interference by double-stranded adeno-associated virus serotypes 7, 8, and 9. Mol. Ther. *17*, 352–359.

Chen, P.Y., Weinmann, L., Gaidatzis, D., Pei, Y., Zavolan, M., Tuschl, T., and Meister, G. (2008a). Strand-specific 5′-O-methylation of siRNA duplexes controls guide strand selection and targeting specificity. RNA *14*, 263–274.

Chen, W., Zhang, Z., Chen, J., Zhang, J., Zhang, J., Wu, Y., Huang, Y., Cai, X., and Huang, A. (2008b). HCV core protein interacts with dicer to antagonize RNA silencing. Virus Res. *133*, 250–258.

Chen, Y., Du, D., Wu, J., Chan, C.P., Tan, Y., Kung, H.F., and He, M.L. (2003). Inhibition of hepatitis B virus replication by stably expressed shRNA. Biochem. Biophys. Res. Commun. *311*, 398–404.

Chen, Y.C., Su, W.C., Huang, J.Y., Chao, T.C., Jeng, K.S., Machida, K., Lai, M.M. (2010). Polo-like kinase 1 is involved in hepatitis C virus replication by hyperphosphorylating NS5A. J. Virol. *84*, 7983–7993.

Chevalier, C., Saulnier, A., Benureau, Y., Flechet, D., Delgrange, D., Colbere-Garapin, F., Wychowski, C., and Martin, A. (2007). Inhibition of hepatitis C virus infection in cell culture by small interfering RNAs. Mol. Ther. *15*, 1452–1462.

Chisari, F.V., Isogawa, M., and Wieland, S.F. (2010). Pathogenesis of hepatitis B virus infection. Pathol. Biol. (Paris) *58*, 258–266.

Coller, K.E., Heaton, N.S., Berger, K.L., Cooper, J.D., Saunders, J.L., and Randall, G. (2012). Molecular determinants and dynamics of hepatitis C virus secretion. PLoS Pathog. *8*, e1002466.

Courtete, J., Sibler, A.P., Zeder-Lutz, G., Dalkara, D., Oulad-Abdelghani, M., Zuber, G., and Weiss, E. (2007). Suppression of cervical carcinoma cell growth by intracytoplasmic codelivery of anti-oncoprotein E6 antibody and small interfering RNA. Mol. Cancer Ther *6*, 1728–1735.

Crowther, C., Ely, A., Hornby, J., Mufamadi, S., Salazar, F., Marion, P., and Arbuthnot, P. (2008). Efficient inhibition of hepatitis B virus replication *in vivo*, using polyethylene glycol-modified adenovirus vectors. Hum. Gene. Ther. *19*, 1325–1331.

Dector, M.A., Romero, P., Lopez, S., and Arias, C.F. (2002). Rotavirus gene silencing by small interfering RNAs. EMBO Rep. *3*, 1175–1180.

Deng, L., Li, G., Xi, L., Yin, A., Gao, Y., You, W., Wang,. X, and Sun, B. (2009). Hepatitis B virus inhibition in mice by lentiviral vector mediated short hairpin RNA. BMC Gastroenterol. *9*, 73.

DeVincenzo, J., Cehelsky, J.E., Alvarez, R., Elbashir, S., Harborth, J., Toudjarska, I., Nechev, L., Murugaiah, V., Van Vliet, A., Vaishnaw A.K., *et al.* (2008). Evaluation of the safety, tolerability and pharmacokinetics of ALN-RSV01, a novel RNAi antiviral therapeutic directed against respiratory syncytial virus (RSV). Antiviral Res. *77*, 225–231.

DeVincenzo, J., Lambkin-Williams, R., Wilkinson, T., Cehelsky, J., Nochur, S., Walsh, E., Meyers, R., Gollob, J., and Vaishnaw, A. (2010). A randomized, double-blind, placebo-controlled study of an RNAi-based therapy directed against respiratory syncytial virus. Proc. Natl. Acad. Sci. U.S.A. *107*, 8800–8805.

Dimitrova, M., Affolter, C., Meyer, F., Nguyen, I., Richard, D.G., Schuster, C., Bartenschlager, R., Voegel, J.C., Ogier, J., and Baumert, T.F. (2008). Sustained delivery of siRNAs targeting viral infection by cell-degradable multilayered polyelectrolyte films. Proc. Natl. Acad. Sci. U.S.A. *105*, 16320–16325.

Ding, Y., Zhang, H., Li, Y., Wu, D., He, S., Wang, Y., Li, Y., Wang, F., and Niu, J. (2013). Inhibition of HCV 5′-NTR and core expression by a small hairpin RNA delivered by a histone gene carrier, HPhA. Int. J. Med. Sci. *10*, 957–964.

Domitrovich, A.M., Diebel, K.W., Ali, N., Sarker, S., and Siddiqui, A. (2005). Role of la autoantigen and polypyrimidine tract-binding protein in HCV replication. Virology *335*, 72–86.

Dorner, M., Horwitz, J.A., Robbins, J.B., Barry, W.T., Feng, Q., Mu, K., Jones, C.T., Schoggins, J.W., Catanese, M.T., and Burton, D.R., *et al.* (2011). A genetically humanized mouse model for hepatitis C virus infection. Nature *474*, 208–211.

Duffy, S., Shackelton, L.A., and Holmes, E.C. 2008. Rates of evolutionary change in viruses: Patterns and determinants. Nat. Rev. Genet. *9*, 267–276.

Dyer, V., Ely, A., Bloom, K., Weinberg, M., and Arbuthnot, P. (2010). tRNA Lys3 promoter cassettes that efficiently express RNAi-activating antihepatitis B virus short hairpin RNAs. Biochem. Biophys. Res. Commun. *398*, 640–646.

Ebert, G., Poeck, H., Lucifora, J., Baschuk, N., Esser, K., Esposito, I., Hartmann, G., and Protzer, U. (2011). 5′ triphosphorylated small interfering RNAs control replication of hepatitis B virus and induce an interferon response in human liver cells and mice. Gastroenterology *141*, *696–706*, 706.e1–3.

Elmen, J., Thonberg, H., Ljungberg, K., Frieden, M., Westergaard, M., Xu, Y., Wahren, B., Liang, Z., Orum, H., Koch, T., *et al.* (2005). Locked nucleic acid (LNA) mediated improvements in siRNA stability and functionality. Nucleic Acids Res. *33*, 439–447.

Ely, A., Naidoo, T., Mufamadi, S., Crowther, C., and Arbuthnot, P. (2008). Expressed anti-HBV primary microRNA shuttles inhibit viral replication efficiently *in vitro* and *in vivo*. Mol. Ther. *16*, 1105–1112.

Ely, A., Naidoo, T., and Arbuthnot, P. (2009). Efficient silencing of gene expression with modular trimeric pol II expression cassettes comprising microRNA shuttles. Nucleic Acids Res. *37*, e91.

Evans, M.J., von Hahn, T., Tscherne, D.M., Syder, A.J., Panis, M., Wolk, B., Hatziioannou, T., McKeating, J.A., Bieniasz, P.D., and Rice, C.M. (2007). Claudin-1 is a hepatitis C virus co-receptor required for a late step in entry. Nature *446*, 801–805.

Fabozzi, G., Nabel, C.S., Dolan, M.A., Sullivan, N.J. (2011). Ebolavirus proteins suppress the effects of small interfering RNA by direct interaction with the mammalian RNA interference pathway. J. Virol. *85*, 2512–2523.

Flusin, O., Vigne, S., Peyrefitte, C.N., Bouloy, M., Crance, J.M., and Iseni, F. (2011). Inhibition of hazara nairovirus replication by small interfering RNAs and their combination with ribavirin. Virol. J. *8*, 249.

Fortes, P., Cuevas, Y., Guan, F., Liu, P., Pentlicky, S., Jung, S.P., Martinez-Chantar, M.L., Prieto, J., Rowe, D., and Gunderson, S.I. (2003). Inhibiting expression of specific genes in mammalian cells with 5′ end-mutated U1 small nuclear RNAs targeted to terminal exons of pre-mRNA. Proc. Natl. Acad. Sci. U.S.A. *100*, 8264–8269.

Fowler, T., Bamberg, S., Moller, P., Klenk, H.D., Meyer, T.F., Becker, S., and Rudel, T. (2005). Inhibition of marburg virus protein expression and viral release by RNA interference. J. Gen. Virol. *86*, 1181–1188.

Franz, A.W., Sanchez-Vargas, I., Adelman, Z.N., Blair, C.D., Beaty, B.J., James, A.A., and Olson, K.E. (2006). Engineering RNA interference-based resistance to dengue virus type 2 in genetically modified aedes aegypti. Proc. Natl. Acad. Sci. U.S.A. *103*, 4198–4203.

Franz, A.W., Sanchez-Vargas, I., Piper, J., Smith, M.R., Khoo, C.C., James, A.A., and Olson, K.E. (2009). Stability and loss of a virus resistance phenotype over time in transgenic mosquitoes harbouring an antiviral effector gene. Insect Mol. Biol. *18*, 661–672.

Ge, Q., McManus, M.T., Nguyen, T., Shen, C.H., Sharp, P.A., Eisen, H.N., and Chen, J. (2003). RNA interference of influenza virus production by directly targeting mRNA for degradation and indirectly inhibiting all viral RNA transcription. Proc. Natl. Acad. Sci. U.S.A. *100*, 2718–2723.

Ge, Q., Filip, L., Bai, A., Nguyen, T., Eisen, H.N., and Chen, J. (2004). Inhibition of influenza virus production in virus-infected mice by RNA interference. Proc. Natl. Acad. Sci. U.S.A. *101*, 8676–8681.

Geisbert, T.W., Hensley, L.E., Kagan, E., Yu, E.Z., Geisbert, J.B., Daddario-DiCaprio, K., Fritz, E.A., Jahrling, P.B., McClintock, K., Phelps, J.R., *et al.* (2006). Postexposure protection of guinea pigs against a lethal ebola virus challenge is conferred by RNA interference. J. Infect. Dis. *193*, 1650–1657.

Geisbert, T.W., Lee, A.C., Robbins, M., Geisbert, J.B., Honko, A.N., Sood, V., Johnson, J.C., de Jong, S., Tavakoli, I., Judge, A., *et al.* (2010). Postexposure protection of non-human primates against a lethal ebola virus challenge with RNA interference: A proof-of-concept study. Lancet *375*, 1896–1905.

Giering, J.C., Grimm, D., Storm, T.A., and Kay MA. (2008). Expression of shRNA from a tissue-specific pol II promoter is an effective and safe RNAi therapeutic. Mol. Ther. *16*, 1630–1636.

Gish, R.G., Satishchandran, C., Young, M., and Pachuk, C. (2011). RNA interference and its potential applications to chronic HBV treatment: Results of a phase I safety and tolerability study. Antivir. Ther. *16*, 547–554.

Grimm, D., Streetz, K.L., Jopling, C.L., Storm, T.A., Pandey, K., Davis, C.R., Marion, P., Salazar, F., and Kay, M.A.. (2006). Fatality in mice due to oversaturation of cellular microRNA/short hairpin RNA pathways. Nature *441*, 537–541.

Guidotti, L.G., Matzke, B., Schaller, H., and Chisari, F.V. (1995). High-level hepatitis B virus replication in transgenic mice. J. Virol. *69*, 6158–6169.

Guo, X., Xu, Y., Bian, G., Pike, A.D., Xie, Y., and Xi, Z. (2010). Response of the mosquito protein interaction network to dengue infection. BMC Genomics *11*, 380.

Gupta, P.K., Sonwane, A.A., Singh, N.K., Meshram, C.D., Dahiya, S.S., Pawar, S.S., Gupta, S.P., Chaturvedi, V.K., and Saini, M. (2012). Intracerebral delivery of small interfering RNAs (siRNAs) using adenoviral vector protects mice against lethal peripheral rabies challenge. Virus Res. *163*, 11–18.

Hamasaki, K., Nakao, K., Matsumoto, K., Ichikawa, T., Ishikawa, H., and Eguchi, K. (2003). Short interfering RNA-directed inhibition of hepatitis B virus replication. FEBS Lett. *543*, 51–54.

Hamazaki, H., Ujino, S., Abe, E., Miyano-Kurosaki, N., Shimotohno, K., and Takaku, H. (2004). RNAi expression mediated inhibition of HCV replication. Nucleic Acids Symp. Ser. (Oxf) (48), 307–308.

Hean, J., Crowther, C., Ely, A., Ul Islam, R., Barichievy, S., Bloom, K., Weinberg, M.S., van Otterlo, W.A., de Koning, C.B., Salazar, F., *et al.* (2010). Inhibition of hepatitis B virus replication *in vivo* using lipoplexes containing altritol-modified antiviral siRNAs. Artif. DNA PNA XNA *1*, 17–26.

Henry, S.D., van der Wegen, P., Metselaar, H.J., Tilanus, H.W., Scholte, B.J., and van der Laan, L.J. (2006). Simultaneous targeting of HCV replication and viral binding with a single lentiviral vector containing multiple RNA interference expression cassettes. Mol. Ther. *14*, 485–493.

Herker, E., Harris, C., Hernandez, C., Carpentier, A., Kaehlcke, K., Rosenberg, A.R., Farese, R.V.Jr., and Ott, M. (2010). Efficient hepatitis C virus particle formation requires diacylglycerol acyltransferase-1. Nat. Med. *16*, 1295–1298.

Hollinger, F., and Liang, T. (2001). Hepatitis B virus. In: Fields Virology, Knipe D, Howley P, eds. (Lippincott-Raven Publishers, Philadelphia, PA). pp. 2971–3036.

Hong, M.H., Chou, Y.C., Wu, Y.C., Tsai, K.N., Hu, C.P., Jeng, K.S., Chen, M.L., and Chang, C. (2012). Transforming growth factor-beta1 suppresses hepatitis B virus replication by the reduction of hepatocyte nuclear factor-4alpha expression. PLoS One *7*, e30360.

Hu, L., Wang, Z., Hu, C., Liu, X., Yao, L., Li, W., and Qi, Y. (2005). Inhibition of measles virus multiplication in cell culture by RNA interference. Acta Virol *49*, 227–234.

Huang, H., Sun, F., Owen, D.M., Li, W., Chen, Y., Gale, M.Jr., and Ye, J. (2007). Hepatitis C virus production by human hepatocytes dependent on assembly and secretion of very low-density lipoproteins. Proc. Natl. Acad. Sci. U.S.A. *104*, 5848–5853.

Huang, L.R., Wu, H.L., Chen, P.J., and Chen, D.S. (2006). An immunocompetent mouse model for the tolerance of human chronic hepatitis B virus infection. Proc. Natl. Acad. Sci. U.S.A. *103*, 17862–17867.

Huang, W., Li, X., Yi, M., Zhu, S., and Chen, W. (2013). Targeted delivery of siRNA against hepatitis B virus by preS1 peptide molecular ligand. Hepatol. Res. (Epub ahead of print).

Huang, Y.H., Fang, C.C., Tsuneyama, K., Chou, H.Y., Pan, W.Y., Shih, Y.M., Wu, P.Y., Chen, Y., Leung, P.S., Gershwin, M.E., *et al.* (2011). A murine model of hepatitis B-associated hepatocellular carcinoma generated by adeno-associated virus-mediated gene delivery. Int. J. Oncol. *39*, 1511–1519.

Isken, O., Baroth, M., Grassmann, C.W., Weinlich, S., Ostareck, D.H., Ostareck-Lederer, A., and Behrens, S.E. (2007). Nuclear factors are involved in hepatitis C virus RNA replication. RNA *13*, 1675–1692.

Israsena, N., Supavonwong, P., Ratanasetyuth, N., Khawplod, P., and Hemachudha, T. (2009). Inhibition of rabies virus replication by multiple artificial microRNAs. Antiviral Res. *84*, 76–83.

Jackson, A.L., Burchard, J., Leake, D., Reynolds, A., Schelter, J., Guo, J., Johnson, J.M., Lim, L., Karpilow, J., Nichols, K., *et al.* (2006). Position-specific chemical modification of siRNAs reduces 'off-target' transcript silencing. RNA *12*, 1197–1205.

Jacque, J.M., Triques, K., and Stevenson, M. (2002). Modulation of HIV-1 replication by RNA interference. Nature *418*, 435–438.

Jahan, S., Khaliq, S., Samreen, B., Ijaz, B., Khan, M., Ahmad, W., Ashfaq, U.A., and Hassan, S. (2011a). Effect of combined siRNA of HCV E2 gene and HCV receptors against HCV. Virol. J. *8*, 295.

Jahan, S., Samreen, B., Khaliq, S., Ijaz, B., Khan, M., Siddique, M.H., Ahmad, W., and Hassan, S. (2011b). HCV entry receptors as potential targets for siRNA-based inhibition of HCV. Genet. Vaccines Ther. *9*, 15.

Jarczak, D., Korf, M., Beger, C., Manns, M.P., and Kruger, M. (2005). Hairpin ribozymes in combination with siRNAs against highly conserved hepatitis C virus sequence inhibit RNA replication and protein translation from hepatitis C virus subgenomic replicons. FEBS. J. *272*, 5910–5922.

Ji, J., Glaser, A., Wernli, M., Berke, J.M., Moradpour, D., and Erb, P. (2008). Suppression of short interfering RNA-mediated gene silencing by the structural proteins of hepatitis C virus. J. Gen. Virol. *89*, 2761–2766.

Jiang, M., and Milner, J. 2002. Selective silencing of viral gene expression in HPV-positive human cervical carcinoma cells treated with siRNA, a primer of RNA interference. Oncogene *21*, 6041–6048.

Jiao, J., Cao, H., Chen, X.W., Zhou, M.J., Liu, Z.H., and Ding, Z.H. (2007). Down-regulation of HBx mRNA in HepG2.2.15 cells by small interfering RNA. Eur. J. Gastroenterol. Hepatol. *19*, 1114–1118.

Kanda, T., Steele, R., Ray, R., and Ray, R.B. (2007). Small interfering RNA targeted to hepatitis C virus 5′ non-translated region exerts potent antiviral effect. J. Virol. *81*,669–676.

Kapadia, S.B., Brideau-Andersen, A., and Chisari, F.V. (2003). Interference of hepatitis C virus RNA replication by short interfering RNAs. Proc. Natl. Acad. Sci. U.S.A. *100*, 2014–2018.

Karlas, A., Machuy, N., Shin, Y., Pleissner, K.P., Artarini, A., Heuer, D., Becker, D., Khalil, H., Ogilvie, L.A., Hess, S., *et al.* (2010). Genome-wide RNAi screen identifies human host factors crucial for influenza virus replication. Nature *463*, 818–822.

Keck, K., Volper, E.M., Spengler, R.M., Long, D.D., Chan, C.Y., Ding, Y., and McCaffrey, A.P. (2009). Rational design leads to more potent RNA interference against hepatitis B virus: Factors effecting silencing efficiency. Mol. Ther. *17*, 538–547.

Kesharwani, P., Gajbhiye, V., and Jain, N.K. 2012. A review of nanocarriers for the delivery of small interfering RNA. Biomaterials *33*,7138–7150.

Khairuddin, N., Gantier, M.P., Blake, S.J., Wu, S.Y., Behlke, M.A., Williams. B.R., and McMillan, N.A. (2012). siRNA-induced immunostimulation through TLR7 promotes antitumoral activity against HPV-driven tumors *in vivo*. Immunol. Cell Biol. *90*, 187–196.

Khaliq, S., Jahan, S., Pervaiz, A., Ali Ashfaq, U., and Hassan, S. (2011). Down-regulation of IRES containing 5′UTR of HCV genotype 3a using siRNAs. Virol. J. *8*, 221.

Khoo, C.C., Doty, J.B., Heersink, M.S., Olson, K.E., and Franz, A.W. (2013). Transgene-mediated suppression of the RNA interference pathway in aedes aegypti interferes with gene silencing and enhances sindbis virus and dengue virus type 2 replication. Insect Mol. Biol. *22*,104–114.

Khvorova, A., Reynolds, A., and Jayasena, S.D. (2003). Functional siRNAs and miRNAs exhibit strand bias. Cell *115*, 209–216.

Kim, M., Shin, D., Kim, S.I., and Park, M. (2006). Inhibition of hepatitis C virus gene expression by small interfering RNAs using a tri-cistronic full-length viral replicon and a transient mouse model. Virus Res. *122*, 1–10.

Kim, S.I., Shin, D., Choi, T.H., Lee, J.C., Cheon, G.J., Kim, K.Y., Park, M., and Kim, M. (2007). Systemic and specific delivery of small interfering RNAs to the liver mediated by apolipoprotein A-I. Mol. Ther. *15*, 1145–1152.

Kim, S.I., Shin, D., Lee, H., Ahn, B.Y., Yoon, Y., and Kim, M. (2009). Targeted delivery of siRNA against hepatitis C virus by apolipoprotein A-I-bound cationic liposomes. J. Hepatol. *50*, 479–488.

Kim, S., Ishida, H., Yamane, D., Yi, M., Swinney, D.C., Foung, S., and Lemon, S.M. (2013). Contrasting roles of mitogen-activated protein kinases in cellular entry and replication of hepatitis C virus: MKNK1 facilitates cell entry. J. Virol. *87*, 4214–4224.

Kok, K.H., and Jin, D.Y. (2006). Influenza A virus NS1 protein does not suppress RNA interference in mammalian cells. J. Gen. Virol. *87*, 2639–2644.

Kondratowicz, A.S., Lennemann, N.J., Sinn, P.L., Davey, R.A., Hunt, C.L., Moller-Tank, S., Meyerholz, D.K., Rennert, P., Mullins, R.F., Brindley, M., *et al.* (2011). T-cell immunoglobulin and mucin domain 1 (TIM-1) is a receptor for zaire ebolavirus and lake victoria marburgvirus. Proc. Natl. Acad. Sci. U.S.A. *108*, 8426–8431.

Korf, M., Jarczak, D., Beger, C., Manns, M.P., and Kruger, M. (2005). Inhibition of hepatitis C virus translation and subgenomic replication by siRNAs directed against highly conserved HCV sequence and cellular HCV cofactors. J. Hepatol. *43*, 225–234.

Korrapati, A.B., Swaminathan, G., Singh, A., Khanna, N., and Swaminathan, S. (2012). Adenovirus delivered short hairpin RNA targeting a conserved site in the 5′ non-translated region inhibits all four serotypes of dengue viruses. PLoS Negl. Trop. Dis. *6*, e1735.

Krishnan, M.N., Ng, A., Sukumaran, B., Gilfoy, F.D., Uchil, P.D., Sultana, H., Brass, A.L., Adametz, R., Tsui, M., Qian, F., *et al.* (2008). RNA interference screen for human genes associated with west Nile virus infection. Nature *455*, 242–245.

Kumar, P., Lee, S.K., Shankar, P., and Manjunath, N. (2006). A single siRNA suppresses fatal encephalitis induced by two different flaviviruses. PLoS Med. 3, e96.

Lanford, R.E., Hildebrandt-Eriksen, E.S., Petri, A., Persson, R., Lindow, M., Munk, M.E., Kauppinen, S., and Orum, H. (2010). Therapeutic silencing of microRNA-122 in primates with chronic hepatitis C virus infection. Science 327, 198–201.

Lee, H., Kim, S.I., Shin, D., Yoon, Y., Choi, T.H., Cheon, G.J., and Kim, M. (2009). Hepatic siRNA delivery using recombinant human apolipoprotein A-I in mice. Biochem. Biophys. Res. Commun. *378*, 192–196.

Levrero, M., Pollicino, T., Petersen, J., Belloni, L., Raimondo, G., and Dandri, M. (2009). Control of cccDNA function in hepatitis B virus infection. J. Hepatol. *51*, 581–592.

Li, B.J., Tang, Q., Cheng, D., Qin, C., Xie, F.Y., Wei, Q., Xu, J., Liu, Y., Zheng, B.J., Woodle, M.C., *et al.* (2005). Using siRNA in prophylactic and therapeutic regimens against SARS coronavirus in rhesus macaque. Nat. Med. *11*, 944–951.

Li, G.Q., Gu, H.X., Li, D., and Xu, W.Z. (2007). Inhibition of hepatitis B virus cccDNA replication by siRNA. Biochem. Biophys. Res. Commun. *355*, 404–408.

Li, W.X., Li, H., Lu, R., Li, F., Dus, M., Atkinson, P., Brydon, E.W., Johnson, K.L., Garcia-Sastre, A., Ball, L.A., *et al.* (2004). Interferon antagonist proteins of influenza and vaccinia viruses are suppressors of RNA silencing. Proc. Natl. Acad. Sci. U.S.A. *101*, 1350–1355.

Li, Y.P., Ramirez, S., Jensen, S.B., Purcell, R.H., Gottwein, J.M., and Bukh, J. (2012). Highly efficient full-length hepatitis C virus genotype 1 (strain TN) infectious culture system. Proc. Natl. Acad. Sci. U.S.A. *109*, 19757–19762.

Lindenbach, B.D., Evans, M.J., Syder, A.J., Wolk, B., Tellinghuisen, T.L., Liu, C.C., Maruyama, T., Hynes, R.O., Burton, D.R., McKeating, J.A., *et al.* (2005). Complete replication of hepatitis C virus in cell culture. Science *309*, 623–626.

Lisowski, L., Elazar, M., Chu, K., Glenn, J.S., and Kay, M.A. (2013). The anti-genomic (negative) strand of hepatitis C virus is not targetable by shRNA. Nucleic Acids Res. *41*, 3688–3698.

Liu, J., Guo, Y., Xue, C.F., Li, Y.H., Huang, Y.X., Ding, J., Gong, W.D., and Zhao, Y. (2004). Effect of vector-expressed siRNA on HBV replication in hepatoblastoma cells. World J. Gastroenterol. *10*, 1898–1901.

Liu, M., Ding, H., Zhao, P., Qin, Z.L., Gao, J., Cao, M.M., Luan, J., Wu, W.B., and Qi, Z.T.. (2006). RNA interference effectively inhibits mRNA accumulation and protein expression of hepatitis C virus core and E2 genes in human cells. Biosci. Biotechnol. Biochem. *70*, 2049–2055.

Liu, Y.P., Haasnoot, J., ter Brake, O., Berkhout, B., and Konstantinova, P. (2008). Inhibition of HIV-1 by multiple siRNAs expressed from a single microRNA polycistron. Nucleic Acids Res. *36*, 2811–2824.

Lohmann, V., Korner, F., Koch, J., Herian, U., Theilmann, L., and Bartenschlager, R. (1999). Replication of subgenomic hepatitis C virus RNAs in a hepatoma cell line. Science *285*, 110–113.

Lu, W.W., Hsu, Y.Y., Yang, J.Y., and Kung, S.H. (2004). Selective inhibition of enterovirus 71 replication by short hairpin RNAs. Biochem. Biophys. Res. Commun. *325*, 494–499.

Lupberger, J., Zeisel, M.B., Xiao, F., Thumann, C., Fofana, I., Zona, L., Davis, C., Mee, C.J., Turek, M., Gorke, S., *et al.* (2011). EGFR and EphA2 are host factors for hepatitis C virus entry and possible targets for antiviral therapy. Nat. Med. *17*, 589–595.

MacArthur, K.L., Wu, C.H., and Wu, G.Y. (2012). Animal models for the study of hepatitis C virus infection and replication. World J. Gastroenterol. *18*, 2909–2913.

McCaffrey, A.P., Meuse, L., Pham, T.T., Conklin, D.S., Hannon, G.J., and Kay, M.A. (2002). RNA interference in adult mice. Nature *418*, 38–39.

McCaffrey, A.P., Nakai, H., Pandey, K., Huang, Z., Salazar, F.H., Xu, H., Wieland, S.F., Marion, P.L., and Kay, M.A. (2003). Inhibition of hepatitis B virus in mice by RNA interference. Nat. Biotechnol. *21*, 639–644.

McOmish, F., Yap, P.L., Dow, B.C., Follett, E.A., Seed, C., Keller, A.J., Cobain, T.J., Krusius, T., Kolho, E., and Naukkarinen, R. (1994). Geographical distribution of hepatitis C virus genotypes in blood donors: An international collaborative survey. J. Clin. Microbiol. *32*, 884–892.

Marimani, M.D., Ely, A., Buff, M.C., Bernhardt, S., Engels, J.W., and Arbuthnot, P. (2013). Inhibition of hepatitis B virus replication in cultured cells and *in vivo* using 2′-O-guanidinopropyl modified siRNAs. Bioorg. Med. Chem. *15*, 6145–6155.

Martell, M., Esteban, J.I., Quer, J., Genesca, J., Weiner, A., Esteban, R., Guardia, J., and Gomez, J. (1992). Hepatitis C virus (HCV) circulates as a population of different but closely related genomes: Quasispecies nature of HCV genome distribution. J. Virol. *66*, 3225–3229.

Matskevich, A.A., and Moelling, K. 2007. Dicer is involved in protection against influenza A virus infection. J. Gen. Virol. *88*, 2627–2635.

Mercer, D.F., Schiller, D.E., Elliott, J.F., Douglas, D.N., Hao, C., Rinfret, A., Addison, W.R., Fischer, K.P., Churchill, T.A., Lakey, J.R., *et al.* (2001). Hepatitis C virus replication in mice with chimeric human livers. Nat. Med. *7*, 927–933.

Metz, P., Dazert, E., Ruggieri, A., Mazur, J., Kaderali, L., Kaul, A., Zeuge, U., Windisch, M.P., Trippler, M., Lohmann, V., *et al.* (2012). Identification of type I and type II interferon-induced effectors controlling hepatitis C virus replication. Hepatology *56*, 2082–2093.

Moore, M.D., McGarvey, M.J., Russell, R.A., Cullen, B.R., and McClure, M.O. (2005). Stable inhibition of hepatitis B virus proteins by small interfering RNA expressed from viral vectors. J. Gene. Med. *7*, 918–925.

Morrissey, D.V., Lockridge, J.A., Shaw, L., Blanchard, K., Jensen, K., Breen, W., Hartsough, K., Machemer, L., Radka, S., Jadhav, V., *et al.* (2005). Potent and persistent *in vivo* anti-HBV activity of chemically modified siRNAs. Nat. Biotechnol. *23*, 1002–1007.

Muhonen, P., Tennila, T., Azhayeva, E., Parthasarathy, R.N., Janckila, A.J., Vaananen, H.K., Azhayev, A., and Laitala-Leinonen, T. (2007). RNA interference tolerates 2′-fluoro modifications at the Argonaute2 cleavage site. Chem. Biodivers. *4*, 858–873.

Mukherjee, S., and Hanley, K.A. (2010). RNA interference modulates replication of dengue virus in drosophila melanogaster cells. BMC Microbiol. *10*, 127.

Murray, E.J., Burden, F., Horscroft, N., Smith-Burchnell, C., and Westby, M. (2011). Knockdown of USP18 increases alpha 2a interferon signaling and induction of interferon-stimulating genes but does not increase antiviral activity in Huh7 cells. Antimicrob. Agents Chemother. *55*, 4311–4319.

Nakagawa, M., Sakamoto, N., Tanabe, Y., Koyama, T., Itsui, Y., Takeda, Y., Chen, C.H., Kakinuma, S., Oooka, S., Maekawa, S., *et al.* (2005). Suppression of hepatitis C virus replication by cyclosporin a is mediated by blockade of cyclophilins. Gastroenterology *129*, 1031–1041.

Nakagawa, S., Umehara, T., Matsuda, C., Kuge, S., Sudoh, M., and Kohara, M. (2007). Hsp90 inhibitors suppress HCV replication in replicon cells and humanized liver mice. Biochem. Biophys. Res. Commun. *353*, 882–888.

Nebbia, G., Peppa, D., and Maini, M.K. (2012). Hepatitis B infection: Current concepts and future challenges. QJM *105*, 109–113.

Neff, C.P., Zhou, J., Remling, L., Kuruvilla, J., Zhang, J., Li, H., Smith, D.D., Swiderski, P., Rossi, J.J., and Akkina, R. (2011). An aptamer-siRNA chimera suppresses HIV-1 viral loads and protects from helper CD4(+) T cell decline in humanized mice. Sci. Transl. Med. *3*, 66ra6.

Ni, B., Shi, X., Li, Y., Gao, W., Wang, X., and Wu, Y. (2005). Inhibition of replication and infection of severe acute respiratory syndrome-associated coronavirus with plasmid-mediated interference RNA. Antivir. Ther. *10*, 527–533.

Ni, Q., Chen, Z., Yao, H.P., Yang, Z.G., Liu, K.Z., and Wu, L.L. (2004). Inhibition of human la protein by RNA interference down-regulates hepatitis B virus mRNA in 2.2.15 cells. World J. Gastroenterol. *10*, 2050–2054.

Ohira, M., Ishiyama, K., Tanaka, Y., Doskali, M., Igarashi, Y., Tashiro, H., Hiraga, N., Imamura, M., Sakamoto, N., Asahara, T., *et al.* (2009). Adoptive immunotherapy with liver allograft-derived lymphocytes induces anti-HCV activity after liver transplantation in humans and humanized mice. J. Clin. Invest. *119*, 3226–3235.

Olivera-Martinez, M.A., and Gallegos-Orozco, J.F. (2007). Recurrent viral liver disease (hepatitis B and C) after liver transplantation. Arch. Med. Res. *38*, 691–701.

Ong, S.P., Chu, J.J., and Ng, M.L. (2008). Inhibition of west Nile virus replication in cells stably transfected with vector-based shRNA expression system. Virus Res. *135*, 292–297.

Otaki, M., Sada, K., Kadoya, H., Nagano-Fujii, M., and Hotta, H. (2006). Inhibition of measles virus and subacute sclerosing panencephalitis virus by RNA interference. Antiviral Res. *70*, 105–111.

Pacca, C.C., Severino, A.A., Mondini, A., Rahal, P., D'avila, S.G., Cordeiro, J.A., Nogueira, M.C., Bronzoni, R.V., and Nogueira, M.L. (2009). RNA interference inhibits yellow fever virus replication *in vitro* and *in vivo*. Virus Genes *38*, 224–231.

Palliser, D., Chowdhury, D., Wang, Q.Y., Lee, S.J., Bronson, R.T., Knipe, D.M., and Lieberman, J. (2006). An siRNA-based microbicide protects mice from lethal herpes simplex virus 2 infection. Nature *439*, 89–94.

Pan, Q., Henry, S.D., Metselaar, H.J., Scholte, B., Kwekkeboom, J., Tilanus, H.W., Janssen, H.L., and van der Laan, L.J. (2009). Combined antiviral activity of interferon-alpha and RNA interference directed against hepatitis C without affecting vector delivery and gene silencing. J. Mol. Med. (Berl) *87*, 713–722.

Pan, Q., Ramakrishnaiah, V., Henry, S., Fouraschen, S., de Ruiter, P.E., Kwekkeboom, J., Tilanus, H.W., Janssen, H.L., and van der Laan, L.J. (2012). Hepatic cell-to-cell transmission of small silencing RNA can extend the therapeutic reach of RNA interference (RNAi). Gut *61*, 1330–1339.

Poordad, F., and Dieterich, D. (2012). Treating hepatitis C: Current standard of care and emerging direct-acting antiviral agents. J. Viral Hepat. *19*, 449–464.

Prabhu, R., Vittal, P., Yin, Q., Flemington, E., Garry, R., Robichaux, W.H., and Dash, S. (2005). Small interfering RNA effectively inhibits protein expression and negative strand RNA synthesis from a full-length hepatitis C virus clone. J. Med. Virol. *76*, 511–519.

Pu, C., Wang, L., Miao, X., Zhang, Y., Jiang, C., Liu, W., Sun, W., and Gao, Q. (2011). Optimized tandem amiRNA mediates stronger inhibitory effects on hepatitis B virus infection. J. Gastrointestin. Liver Dis. *20*, 271–278.

Putral, L.N., Bywater, M.J., Gu, W., Saunders, N.A., Gabrielli, B.G., Leggatt, G.R., and McMillan, N.A. (2005). RNA interference against human papillomavirus oncogenes in cervical cancer cells results in increased sensitivity to cisplatin. Mol. Pharm. *68*, 1311–1319.

Qing, G., Weili, W., Fanqin, Z., Rongchang, Z., Yijin, L., and Jianqun, D. (2011). Research of UL54-specific siRNA on herpes simplex virus type II replication. Int. J. Dermatol. *50*, 362–366.

Randall, G., Grakoui, A., and Rice, C.M. (2003). Clearance of replicating hepatitis C virus replicon RNAs in cell culture by small interfering RNAs. Proc. Natl. Acad. Sci. U.S.A. *100*, 235–240.

Randall, G., Chen, L., Panis, M., Fischer, A.K., Lindenbach, B.D., Sun, J., Heathcote, J., Rice, C.M., Edwards, A.M., and McGilvray, I.D. (2006). Silencing of USP18 potentiates the antiviral activity of interferon against hepatitis C virus infection. Gastroenterology *131*, 1584–1591.

Randall, G., Panis, M., Cooper, J.D., Tellinghuisen, T.L., Sukhodolets, K.E., Pfeffer, S., Landthaler, M., Landgraf, P., Kan, S., Lindenbach, B.D., *et al.* (2007). Cellular cofactors affecting hepatitis C virus infection and replication. Proc. Natl. Acad. Sci. U.S.A. *104*, 12884–12889.

Rauschhuber, C., Xu, H., Salazar, F.H., Marion, P.L., and Ehrhardt, A. (2008). Exploring gene-deleted adenoviral vectors for delivery of short hairpin RNAs and reduction of hepatitis B virus infection in mice. J. Gene. Med. *10*, 878–889.

Ray, R.B., and Kanda, T. (2009). Inhibition of HCV replication by small interfering RNA. Methods Mol. Biol. *510*, 251–262.

Ren, X.R., Zhou, L.J., Luo, G.B., Lin, B., and Xu, A. (2005). Inhibition of hepatitis B virus replication in 2.2.15 cells by expressed shRNA. J. Viral Hepat. *12*, 236–242.

Ren, X., Luo, G., Xie, Z., Zhou, L., Kong, X., and Xu, A. (2006). Inhibition of multiple gene expression and virus replication of HBV by stable RNA interference in 2.2.15 cells. J. Hepatol. *44*, 663–670.

Ren, Y., Cheung, H.W., von Maltzhan, G., Agrawal, A., Cowley, G.S., Weir, B.A., Boehm, J.S., Tamayo, P., Karst, A.M., Liu, J.F., *et al.* (2012). Targeted tumor-penetrating siRNA nanocomplexes for credentialing the ovarian cancer oncogene ID4. Sci. Transl. Med. *4*, 147ra112.

Rivas, F.V., Tolia, N.H., Song, J.J., Aragon, J.P., Liu, J., Hannon, G.J., and Joshua-Tor, L. (2005). Purified Argonaute2 and an siRNA form recombinant human RISC. Nat. Struct. Mol. Biol. *12*, 340–349.

Roche, B., and Samuel, D. (2008). Liver transplantation in viral hepatitis: Prevention of recurrence. Best Pract. Res. Clin. Gastroenterol. *22*, 1153–1169.

Romano, P.R., McCallus, D.E., and Pachuk, C.J. (2006). RNA interference-mediated prevention and therapy for hepatocellular carcinoma. Oncogene *25*, 3857–3865.

Romanovskaya, A., Paavilainen, H., Nygardas, M., Bamford, D.H., Hukkanen, V., and Poranen, M.M. (2012). Enzymatically produced pools of canonical and dicer-substrate siRNA molecules display comparable gene silencing and antiviral activities against herpes simplex virus. PLoS One *7*, e51019.

Sanchez-Vargas, I., Scott, J.C., Poole-Smith, B.K., Franz, A.W., Barbosa-Solomieu, V., Wilusz, J., Olson, K.E., and Blair, C.D. (2009). Dengue virus type 2 infections of aedes aegypti are modulated by the mosquito's RNA interference pathway. PLoS Pathog. *5*, e1000299.

Saxena, S., Jonsson, Z.O., and Dutta, A. (2003). Small RNAs with imperfect match to endogenous mRNA repress translation. implications for off-target activity of small inhibitory RNA in mammalian cells. J. Biol. Chem. *278*, 44312–44319.

Scaglione, S.J., and Lok, A.S. (2012). Effectiveness of hepatitis B treatment in clinical practice. Gastroenterology *142*, 1360–1368.e1.

Schaefer, S. (2007). Hepatitis B virus taxonomy and hepatitis B virus genotypes. World J. Gastroenterol. *13*, 14–21.

Schafer, J., Hobel, S., Bakowsky, U., and Aigner, A. (2010). Liposome–polyethylenimine complexes for enhanced DNA and siRNA delivery. Biomaterials *31*, 6892–6900.

Schieck, A., Schulze, A., Gahler, C., Muller, T., Haberkorn, U., Alexandrov, A., Urban, S., and Mier, W. (2013). Hepatitis B virus hepatotropism is mediated by specific receptor recognition in the liver and not restricted to susceptible hosts. Hepatology *58*, 43–53.

Schnettler, E., Sterken, M.G., Leung, J.Y., Metz, S.W., Geertsema, C., Goldbach, R.W., Vlak, J.M., Kohl, A., Khromykh, A.A., and Pijlman, G.P. (2012). Non-coding flavivirus RNA displays RNA interference suppressor activity in insect and mammalian cells. J. Virol. *86*, 13486–13500.

Schwentke, A., Krepstakies, M., Mueller, A.K., Hammerschmidt-Kamper, C., Motaal, B.A., Bernhard, T., Hauber, J., and Kaiser, A. (2012). *In vitro* and *in vivo* silencing of plasmodial dhs and eIf-5a genes in a putative, non-canonical RNAi-related pathway. BMC Microbiol. *12*, 107.

Scott, J.C., Brackney, D.E., Campbell, C.L., Bondu-Hawkins, V., Hjelle, B., Ebel, G.D., Olson, K.E., and Blair, C.D. (2010). Comparison of dengue virus type 2-specific small RNAs from RNA interference-competent and -incompetent mosquito cells. PLoS Negl. Trop. Dis. *4*, e848.

Sen, A., Steele, R., Ghosh, A.K., Basu, A., Ray, R., and Ray, R.B. (2003). Inhibition of hepatitis C virus protein expression by RNA interference. Virus Res. *96*, 27–35.

Seo, M.Y., Abrignani, S., Houghton, M., and Han, J.H. (2003). Small interfering RNA-mediated inhibition of hepatitis C virus replication in the human hepatoma cell line huh-7. J. Virol. *77*, 810–812.

Sherman, K.E. (2012). Managing adverse effects and complications in completing treatment for hepatitis C virus infection. Top Antivir. Med. *20*, 125–128.

Shi, J., Wang, M., Wang, J., Wang, S., and Luo, E. (2012). Comparison of inhibitory efficacy of short interfering RNAs targeting different genes on measles virus replication. J. Basic Microbiol. *52*, 332–339.

Shin, D., Lee, H., Kim, S.I., Yoon, Y., and Kim, M. (2009). Optimization of linear double-stranded RNA for the production of multiple siRNAs targeting hepatitis C virus. RNA *15*, 898–910.

Shlomai, A., and Shaul, Y. (2003). Inhibition of hepatitis B virus expression and replication by RNA interference. Hepatology *37*, 764–770.

Sim, S., Ramirez, J.L., and Dimopoulos, G. (2012). Dengue virus infection of the aedes aegypti salivary gland and chemosensory apparatus induces genes that modulate infection and blood-feeding behavior. PLoS Pathog. *8*, e1002631.

Sledz, C.A., Holko, M., de Veer, M.J., Silverman, R.H., and Williams, B.R. (2003). Activation of the interferon system by short-interfering RNAs. Nat. Cell Biol. *5*, 834–839.

Snyder, L.L., Esser, J.M., Pachuk, C.J., and Steel, L.F. (2008). Vector design for liver-specific expression of multiple interfering RNAs that target hepatitis B virus transcripts. Antiviral Res. *80*, 36–44.

Song, E., Zhu, P., Lee, S.K., Chowdhury, D., Kussman, S., Dykxhoorn, D.M., Feng, Y., Palliser, D., Weiner, D.B., Shankar, P., *et al.* (2005). Antibody mediated *in vivo* delivery of small interfering RNAs via cell-surface receptors. Nat. Biotechnol. *23*, 709–717.

Souza-Neto, J.A., Sim, S., and Dimopoulos, G. (2009). An evolutionary conserved function of the JAK-STAT pathway in anti-dengue defense. Proc. Natl. Acad. Sci. U.S.A. *106*, 17841–17846.

Starkey, J.L., Chiari, E.F., and Isom, H.C. (2009). Hepatitis B virus (HBV)-specific short hairpin RNA is capable of reducing the formation of HBV covalently closed circular (CCC) DNA but has no effect on established CCC DNA *in vitro*. J. Gen. Virol. *90*, 115–126.

Stein, D.A., Perry, S.T., Buck, M.D., Oehmen, C.S., Fischer, M.A., Poore, E., Smith, J.L., Lancaster, A.M., Hirsch, A.J., Slifka, M.K., *et al.* (2011). Inhibition of dengue virus infections in cell cultures and in AG129 mice by a small interfering RNA targeting a highly conserved sequence. J. Virol. *85*, 10154–10166.

Stewart, C.R., Karpala, A.J., Lowther, S., Lowenthal, J.W., and Bean, A.G. (2011). Immunostimulatory motifs enhance antiviral siRNAs targeting highly pathogenic avian influenza H5N1. PLoS One *6*, e21552.

Strayer, D.S., Feitelson, M., Sun, B., and Matskevich, A.A. (2005). Paradigms for conditional expression of RNA interference molecules for use against viral targets. Methods Enzymol. *392*, 227–241.

Subramanya, S., Kim, S.S., Abraham, S., Yao, J., Kumar, M., Kumar, P., Haridas, V., Lee, S.K., Shultz, L.D., Greiner D, *et al.* (2010). Targeted delivery of small interfering RNA to human dendritic cells to suppress dengue virus infection and associated proinflammatory cytokine production. J. Virol. *84*, 2490–2501.

Suhy, D.A., Kao, S.C., Mao, T., Whiteley, L., Denise, H., Souberbielle, B., Burdick, A.D., Hayes, K., Wright, J.F., Lavender, H., *et al.* (2012). Safe, long-term hepatic expression of anti-HCV shRNA in a nonhuman primate model. Mol. Ther. *20*, 1737–1749.

Sui, H.Y., Zhao, G.Y., Huang, J.D., Jin, D.Y., Yuen, K.Y., and Zheng, B.J. (2009). Small interfering RNA targeting m2 gene induces effective and long term inhibition of influenza A virus replication. PLoS One *4*, e5671.

Summers, J., Smolec, J.M., and Snyder, R. (1978). A virus similar to human hepatitis B virus associated with hepatitis and hepatoma in woodchucks. Proc. Natl. Acad. Sci. U.S.A. *75*, 4533–4537.

Sun, D., Rosler, C., Kidd-Ljunggren, K., and Nassal, M. (2010). Quantitative assessment of the antiviral potencies of 21 shRNA vectors targeting conserved, including structured, hepatitis B virus sites. J. Hepatol. *52*, 817–826.

Tai, A.W., Benita, Y., Peng, L.F., Kim, S.S., Sakamoto, N., Xavier, R.J., and Chung, R.T. (2009). A functional genomic screen identifies cellular cofactors of hepatitis C virus replication. Cell Host Microbe *5*, 298–307.

Takigawa, Y., Nagano-Fujii, M., Deng, L., Hidajat, R., Tanaka, M., Mizuta, H., and Hotta, H. (2004). Suppression of hepatitis C virus replicon by RNA interference directed against the NS3 and NS5B regions of the viral genome. Microbiol. Immunol.*48*, 591–598.

Tan, C., Xuan, B., Hong, J., Dai, Z., Hao, R., Li, Z., and Huang, W. (2007). RNA interference against hepatitis B virus with endoribonuclease-prepared siRNA despite of the target sequence variations. Virus Res. *126*, 172–178.

Tan, E.L., Marcus, K.F., and Poh, C.L. (2008). Development of RNA interference (RNAi) as potential antiviral strategy against enterovirus 70. J. Med. Virol. *80*, 1025–1032.

Tang, Q., Li, B., Woodle, M., and Lu, P.Y. (2008). Application of siRNA against SARS in the rhesus macaque model. Methods Mol. Biol. *442*, 139–158.

Tayyari, F., Marchant, D., Moraes, T.J., Duan, W., Mastrangelo, P., and Hegele, R.G. 2011. Identification of nucleolin as a cellular receptor for human respiratory syncytial virus. Nat. Med. *17*, 1132–1135.

Teng, X., Liu, J.Y., Li, D., Fang, Y., Wang, X.Y., Ma, Y.X., Chen, S.J., Zhao, Y.X., Xu, W.Z., and Gu, H.X. (2011). Application of allele-specific RNAi in hepatitis B virus lamivudine resistance. J. Viral Hepat. *18*, e491–8.

Thornbrough, J.M., Hundley, T., Valdivia, R., and Worley, M.J. (2012). Human genome-wide RNAi screen for host factors that modulate intracellular salmonella growth. PLoS One 7, e38097.

Tompkins, S.M., Lo, C.Y., Tumpey, T.M., and Epstein, S.L. (2004). Protection against lethal influenza virus challenge by RNA interference *in vivo*. Proc. Natl. Acad. Sci. U.S.A. *101*, 8682–8686.

Travanty, E.A., Adelman, Z.N., Franz, A.W., Keene, K.M., Beaty, B.J., Blair, C.D., James, A.A., and Olson, K.E. (2004). Using RNA interference to develop dengue virus resistance in genetically modified aedes aegypti. Insect Biochem. Mol. Biol. *34*, 607–613.

Trejo-Avila, L., Elizondo-Gonzalez, R., Trujillo-Murillo, Kdel. C., Zapata-Benavides, P., Rodriguez-Padilla, C., and Rivas-Estilla, A.M. (2007). Antiviral therapy: Inhibition of hepatitis C virus expression by RNA interference directed against the NS5B region of the viral genome. Ann. Hepatol. *6*, 174–180.

Turrini, P., Sasso, R., Germoni, S., Marcucci, I., Celluci, A., Di Marco, A., Marra, E., Paonessa, G., Eutropi, A., Laufer, R., *et al.* (2006). Development of humanized mice for the study of hepatitis C virus infection. Transplant. Proc. *38*, 1181–1184.

Uprichard, S.L., Boyd, B., Althage, A., and Chisari, F.V. (2005). Clearance of hepatitis B virus from the liver of transgenic mice by short hairpin RNAs. Proc. Natl. Acad. Sci. U.S.A. *102*, 773–778.

Vaillancourt, F.H., Pilote, L., Cartier, M., Lippens, J., Liuzzi, M., Bethell, R.C., Cordingley, M.G., and Kukolj, G. (2009). Identification of a lipid kinase as a host factor involved in hepatitis C virus RNA replication. Virology *387*, 5–10.

Van Stry, M., Oguin, T.H.3rd., Cheloufi, S., Vogel, P., Watanabe, M., Pillai, M.R., Dash, P., Thomas, P.G., Hannon, G.J., and Bix, M. (2012). Enhanced susceptibility of Ago1/3 double-null mice to influenza A virus infection. J. Virol. *86*, 4151–4157.

Vig, K., Lewis, N., Moore, E.G. Pillai, S., Dennis, V.A., and Singh, S.R. (2009). Secondary RNA structure and its role in RNA interference to silence the respiratory syncytial virus fusion protein gene. Mol. Biotechnol. *43*, 200–211.

de Vries, W., Haasnoot, J., Fouchier, R., de Haan, P., and Berkhout, B. (2009). Differential RNA silencing suppression activity of NS1 proteins from different influenza A virus strains. J. Gen. Virol. *90*, 1916–1922.

Wakita, T., Pietschmann, T., Kato, T., Date, T., Miyamoto, M., Zhao, Z., Murthy, K., Habermann, A., Krausslich, H.G., Mizokami, M., *et al.* (2005). Production of infectious hepatitis C virus in tissue culture from a cloned viral genome. Nat. Med. *11*, 791–796.

Walboomers, J.M., Jacobs, M.V., Manos, M.M., Bosch, F.X., Kummer, J.A., Shah, K.V., Snijders, P.J., Peto, J., Meijer, C.J., and Munoz, N. (1999). Human papillomavirus is a necessary cause of invasive cervical cancer worldwide. J. Pathol. *189*, 12–19.

Wang, Q., Contag, C.H., Ilves, H., Johnston, B.H., and Kaspar, R.L. (2005). Small hairpin RNAs efficiently inhibit hepatitis C IRES-mediated gene expression in human tissue culture cells and a mouse model. Mol. Ther. *12*, 562–568.

Wang, Y., Kato, N., Jazag, A., Dharel, N., Otsuka, M., Taniguchi, H., Kawabe, T., and Omata, M. (2006). Hepatitis C virus core protein is a potent inhibitor of RNA silencing-based antiviral response. Gastroenterology *130*, 883–892.

Wang, Z., Ren, L., Zhao, X., Hung, T., Meng, A., Wang, J., and Chen, Y.G. (2004). Inhibition of severe acute respiratory syndrome virus replication by small interfering RNAs in mammalian cells. J. Virol. *78*, 7523–7527.

Washburn, M..L., Bility, M.T., Zhang, L., Kovalev, G.I., Buntzman, A., Frelinger, J.A., Barry, W., Ploss, A., Rice, C.M., and Su, L. (2011). A humanized mouse model to study hepatitis C virus infection, immune response, and liver disease. Gastroenterology *140*, 1334–1344.

Watanabe, T., Sudoh, M., Miyagishi, M., Akashi, H., Arai, M., Inoue, K., Taira, K., Yoshiba, M., and Kohara, M. (2006). Intracellular-diced dsRNA has enhanced efficacy for silencing HCV RNA and overcomes variation in the viral genotype. Gene. Ther. *13*, 883–892.

Watashi, K., Ishii, N., Hijikata, M., Inoue, D., Murata, T., Miyanari, Y., and Shimotohno, K. (2005). Cyclophilin B is a functional regulator of hepatitis C virus RNA polymerase. Mol. Cell. *19*, 111–122.

Weinberg, M.S., Ely, A., Barichievy, S., Crowther, C., Mufamadi, S., Carmona, S., and Arbuthnot, P. (2007). Specific inhibition of HBV replication *in vitro* and *in vivo* with expressed long hairpin RNA. Mol. Ther. *15*, 534–541.

Wiedenheft, B., Sternberg, S.H., and Doudna, J.A. (2012). RNA-guided genetic silencing systems in bacteria and archaea. Nature *482*, 331–338.

Wilson, J.A, and Richardson, C.D. (2005). Hepatitis C virus replicons escape RNA interference induced by a short interfering RNA directed against the NS5b coding region. J. Virol. *79*, 7050–7058.

Wilson, J.A., Jayasena, S., Khvorova, A., Sabatinos, S., Rodrigue-Gervais, I.G., Arya, S., Sarangi, F., Harris-Brandts, M., Beaulieu, S., and Richardson, C.D. (2003). RNA interference blocks gene expression and RNA synthesis from hepatitis C replicons propagated in human liver cells. Proc. Natl. Acad. Sci. U.S.A. *100*, 2783–2788.

Wooddell, C.I., Rozema, D.B., Hossbach, M., John, M., Hamilton, H.L., Chu, Q., Hegge, J.O., Klein, J.J., Wakefield, D.H., Oropeza, C.E., *et al.* (2013). Hepatocyte-targeted RNAi therapeutics for the treatment of chronic hepatitis B virus infection. Mol. Ther. *21*, 973–985.

Wu, C.J., Huang, H.W., Liu, C.Y., Hong, C.F., and Chan, Y.L. (2005a). Inhibition of SARS-CoV replication by siRNA. Antiviral Res. *65*, 45–48.

Wu, H.L., Huang, L.R., Huang, C.C., Lai, H.L., Liu, C.J., Huang, Y.T., Hsu, Y.W., Lu, C.Y., Chen, D.S., and Chen, P.J. (2005b). RNA interference-mediated control of hepatitis B virus and emergence of resistant mutant. Gastroenterology *128*, 708–716.

Wu, H.X., Wang, H.L., Guo, X.F., Yang, Y.J., Ma, J.Z., Wang, T.C., Gao, Y.W., Zhao, Y.K., Yang, S.T., and Xia, X.Z. (2013). Adeno-associated viruses serotype 2-mediated RNA interference efficiently inhibit rabies virus replication *in vitro* and *vivo*. J. Vet. Med. Sci. *75*, 1355–1361.

Wu, Z., Gao, Y., Sun, L., Tien, P., and Jin, Q. (2008). Quick identification of effective small interfering RNAs that inhibit the replication of coxsackievirus A16. Antiviral Res. *80*, 295–301.

Xiangji, L., Feng, X., Qingbao, C., Weifeng, T., Xiaoqing, J., Baihe, Z., Feng, S., Hongyang, W., and Mengchao, W. (2011). Knockdown of HBV surface antigen gene expression by a lentiviral microRNA-based system inhibits HBV replication and HCC growth. J. Viral Hepat. *18*, 653–660.

Xie, P.W., Xie, Y., Zhang, X.J., Huang, H., He, L.N., Wang, X.J., and Wang, S.Q. (2013). Inhibition of dengue virus 2 replication by artificial MicroRNAs targeting the conserved regions. Nucleic Acid Ther. *23*, 244–252.

Xuan, B., Qian, Z., Hong, J., and Huang, W. (2006). EsiRNAs inhibit hepatitis B virus replication in mice model more efficiently than synthesized siRNAs. Virus Res. *118*, 150–155.

Xue, Q., Ding, H., Liu, M., Zhao, P., Gao, J., Ren, H., Liu, Y., and Qi, Z.T.. (2007). Inhibition of hepatitis C virus replication and expression by small interfering RNA targeting host cellular genes. Arch. Virol. *152*, 955–962.

Yamato, K., Fen, J., Kobuchi, H., Nasu, Y., Yamada, T., Nishihara, T., Ikeda, Y., Kizaki, M., and Yoshinouchi, M. (2006). Induction of cell death in human papillomavirus 18-positive cervical cancer cells by E6 siRNA. Cancer Gene. Ther. *13*, 234–241.

Yan, H., Zhong, G., Xu, G., He, W., Jing, Z., Gao, Z., Huang, Y., Qi, Y., Peng, B., Wang, H., *et al.* (2012). Sodium taurocholate cotransporting polypeptide is a functional receptor for human hepatitis B and D virus. Elife *1*, e00049.

Yang, X., Haurigot, V., Zhou, S., Luo, G., and Couto, L.B. (2010). Inhibition of hepatitis C virus replication using adeno-associated virus vector delivery of an exogenous anti-hepatitis C virus microRNA cluster. Hepatology *52*, 1877–1887.

Yang, X., Marcucci, K., Anguela, X., and Couto, L.B. (2013). Preclinical evaluation of an anti-HCV miRNA cluster for treatment of HCV infection. Mol. Ther. *21*, 588–601.

Yang, Y., Wu, C., Wu, J., Nerurkar, V.R., Yanagihara, R., and Lu, Y. (2008). Inhibition of west Nile virus replication by retrovirus-delivered small interfering RNA in human neuroblastoma cells. J. Med. Virol. *80*, 930–936.

Yang, Y.J., Zhao, P.S., Zhang, T., Wang, H.L., Liang, H.R., Zhao, L.L., Wu, H.X., Wang, T.C., Yang, S.T., Xia, X.Z. (2012). Small interfering RNAs targeting the rabies virus nucleoprotein gene. Virus Res. *169*, 169–174.

Yang, Z.G., Chen, Z., Ni, Q., Xu, N., Shao, J.B., and Yao, H.P. (2005). Inhibition of hepatitis B virus surface antigen expression by small hairpin RNA *in vitro*. World J. Gastroenterol. *11*, 498–502.

Yao, H., Zhang, Y., He, F., Wang, C., Xiao, Z., Zou, J., Wang, F., and Liu, Z. (2012). Short hairpin RNA targeting 2B gene of coxsackievirus B3 exhibits potential antiviral effects both *in vitro* and *in vivo*. BMC Infect. Dis. *12*, 177.

Ying, R.S., Zhu, C., Fan, X.G., Li, N., Tian, X.F., Liu, H.B., and Zhang, B.X. (2007). Hepatitis B virus is inhibited by RNA interference in cell culture and in mice. Antiviral Res. 73, 24–30.

Yokota, T., Sakamoto, N., Enomoto, N., Tanabe, Y., Miyagishi, M., Maekawa, S., Yi, L., Kurosaki, M., Taira, K., Watanabe, M., and Mizusawa, H. (2003). Inhibition of intracellular hepatitis C virus replication by synthetic and vector-derived small interfering RNAs. EMBO Rep. *4*, 602–608.

Yokota, T., Iijima, S., Kubodera, T., Ishii, K., Katakai, Y., Ageyama, N., Chen, Y., Lee, Y.J., Unno, T., Nishina, K., *et al.* (2007). Efficient regulation of viral replication by siRNA in a non-human primate surrogate model for hepatitis C. Biochem. Biophys. Res. Commun. *361*, 294–300.

Yoshinouchi, M., Yamada, T., Kizaki, M., Fen, J., Koseki, T., Ikeda, Y., Nishihara, T., and Yamato, K. (2003). *In vitro* and *in vivo* growth suppression of human papillomavirus 16-positive cervical cancer cells by E6 siRNA. Mol. Ther. *8*, 762–768.

Zamora, M.R., Budev, M., Rolfe, M., Gottlieb, J., Humar, A., Devincenzo, J., Vaishnaw, A., Cehelsky, J., Albert, G., Nochur, S., *et al.* (2011). RNA interference therapy in lung transplant patients infected with respiratory syncytial virus. Am. J. Respir. Crit. Care Med. *183*, 531–538.

Zeisel, M.B., Koutsoudakis, G., Schnober, E.K., Haberstroh, A., Blum, H.E., Cosset, F.L., Wakita, T., Jaeck, D., Doffoel, M., Royer, C., *et al* (2007). Scavenger receptor class B type I is a key host factor for hepatitis C virus infection required for an entry step closely linked to CD81. Hepatology *46*, 1722–1731.

Zekri, A.R., Bahnassy, A.A., El-Din, H.M., and Salama, H.M. (2009). Consensus siRNA for inhibition of HCV genotype-4 replication. Virol. J. *6*, 13.

Zhang, J., Randall, G., Higginbottom, A., Monk, P., Rice, C.M., and McKeating, J.A. (2004a). CD81 is required for hepatitis C virus glycoprotein-mediated viral infection. J. Virol. *78*, 1448–1455.

Zhang, J., Yamada, O., Sakamoto, T., Yoshida, H., Iwai, T., Matsushita, Y., Shimamura, H., Araki, H., and Shimotohno, K. (2004b). Down-regulation of viral replication by adenoviral-mediated expression of siRNA against cellular cofactors for hepatitis C virus. Virology *320*, 135–143.

Zhang, W., Singam, R., Hellermann, G., Kong, X., Juan, H.S., Lockey, R.F., Wu, S.J., Porter, K., and Mohapatra, S.S. (2004c). Attenuation of dengue virus infection by adeno-associated virus-mediated siRNA delivery. Genet. Vaccines Ther. *2*, 8.

Zhang, W., Yang, H., Kong, X., Mohapatra, S., San Juan-Vergara, H., Hellermann, G., Behera, S., Singam, R., Lockey, R.F., and Mohapatra, S.S. (2005). Inhibition of respiratory syncytial virus infection with intranasal siRNA nanoparticles targeting the viral NS1 gene. Nat. Med. *11*, 56–62.

Zheng, B.J., Guan, Y., Tang, Q., Du, C., Xie, F.Y., He, M.L., Chan, K.W., Wong, K.L., Lader, E., Woodle, M.C., *et al.* (2004). Prophylactic and therapeutic effects of small interfering RNA targeting SARS-coronavirus. Antivir. Ther. *9*, 365–374.

Zhiqiang, W., Yaowu, Y., Fan, Y., Jian, Y., Yongfeng, H., Lina, Z., Jianwei, W., and Qi, J. (2010). Effective siRNAs inhibit the replication of novel influenza A (H1N1) virus. Antiviral Res. *85*, 559–561.

Zhong, J., Gastaminza, P., Cheng, G., Kapadia, S., Kato, T., Burton, D.R., Wieland, S.F., Uprichard, S.L., Wakita, T., and Chisari, F.V. (2005). Robust hepatitis C virus infection *in vitro*. Proc. Natl. Acad. Sci. U.S.A. *102*, 9294–9299.

Roles of MicroRNAs in Cancers Associated with Human Tumour Viruses

10

Xianzhi Lin, Deguang Liang and Ke Lan

Abstract

Considerable accumulated lines of evidence have shown that microRNAs (miRNAs) play important roles in regulating gene expression in many critical physiological conditions as well as disease progression. The regulation of gene expression by miRNAs could affect mRNA cleavage, mRNA degradation, and inhibition of mRNA translation. Seven tumour viruses have so far been identified in humans. The infection of the host by these human tumour viruses could result in the development of many severe diseases or even several malignant ones like cancers. These human tumour viruses have taken advantage of the pathways that cellular miRNAs have been involved in by encoding their own viral miRNAs and/or hijacking host miRNAs. In this chapter, we will review recent advances on the roles miRNAs played in cancers associated with human tumour viruses, with the focus on human tumorigenic herpesviruses.

Introduction

MicroRNA (miRNA) was first identified in *Caenorhabditis elegans* with regulatory function in developmental stage transition (Lee *et al.*, 1993; Reinhart *et al.*, 2000). Many other similar molecules were then shown to exist in both invertebrates and vertebrates, suggesting their broad expression and they might potentially play key roles in gene regulations (Lagos-Quintana *et al.*, 2001; Lau *et al.*, 2001; Lee and Ambros, 2001). Later on, several large DNA viruses from herpesvirus family were also shown to encode their own viral miRNAs (Cai *et al.*, 2005; Pfeffer *et al.*, 2005; Samols *et al.*, 2005; Sullivan *et al.*, 2005). Like their cellular counterparts, viral miRNAs have been shown to be involved in many critical processes. Viral miRNAs can be utilized by viruses to repress host cell apoptosis, evade host immune system, control viral gene expression, and regulate their own life cycle (Grey *et al.*, 2008). To date, there are seven tumour viruses identified in humans. They are Epstein–Barr virus (EBV), hepatitis B virus (HBV), human T-cell lymphotropic virus type 1 (HTLV-1), human papillomavirus (HPV), hepatitis C virus (HCV), Kaposi's sarcoma-associated herpesvirus (KSHV), and Merkel cell polyomavirus (MCV). These seven human tumour viruses take advantage of cellular miRNA pathway for their own gene expression either by encoding viral miRNAs or hijacking the host miRNA pathway. In the following parts, we will summarize our current understanding of how miRNAs are involved in the pathogenesis of cancers associated with human tumour viruses.

Biogenesis of microRNAs

Canonical pathway for miRNA production

MiRNAs is one type of non-coding RNAs with approximately 22 nt in length. They have been reported to have regulatory functions and are expressed by all eukaryotic cells and many viruses (Bartel, 2009; Cullen, 2009). In the canonical pathway for miRNAs biogenesis, primary miRNAs (pri-miRNAs) that formed an imperfect stem–loop with a hairpin bulge are first transcribed by RNA Pol III from miRNA genes. Then pri-miRNAs are processed by a microprocessor composed of RNase III Drosha and co-factor DiGeorge syndrome critical region gene 8 (DGCR8). The processed products with a hairpin structure of ~70 nt-long named precursor miRNAs (pre-miRNAs) are exported from nucleus to cytoplasm by Ran GTPase Exportin-5. In the cytoplasm, the stem–loops of pre-miRNAs are further cleaved off by another RNase III Dicer and HIV-1 transactivating response (TAR) RNA-binding protein (TRBP), leaving a ~22 nt RNA duplex. Normally, one strand of the duplex is then preferentially incorporated into the RNA-induced silencing complex (RISC). The other strand named miRNA star (miRNA*) or passenger strand, is subsequently degraded. Mature miRNAs then will guide RISCs to pair with target transcripts at their 3′ untranslated region (3′UTR) and induce post-transcriptional silencing through either mRNA cleavage, mRNA degradation, or translational inhibition (Bartel, 2009).

Non-canonical pathways to produce miRNAs

Besides the canonical pathways, many non-canonical pathways for generating miRNAs independent of Drosha or Dicer were also reported (Table 10.1, details reviewed in Libri *et al.*, 2013; Yang and Lai, 2011). MiRNAs could be generated from precursor tRNAs after processed by RNase P and RNase Z or from direct processing by Dicer (Lee *et al.*, 2009). Viral tRNAs are also used as templates to produce viral miRNAs (Bogerd *et al.*, 2010; Diebel *et al.*, 2010). After transcribed by PolII, either box C/D or H/ACA small nucleolar RNAs (snoRNAs) are able to be cleaved by Dicer to make miRNAs (Saraiya and Wang, 2008; Taft *et al.*, 2009). One viral snoRNA-generated miRNA case from EBV has also been reported (Hutzinger *et al.*, 2009). Additionally, MiRNAs could also be generated from introns, which were named miRtrons (Ladewig *et al.*, 2012; Ruby *et al.*, 2007). The precursors of herpesvirus saimiri (HVS) Sm class U RNAs (HSURs) are first processed by Integrator 11 after their transcription. Pre-miRNA is also generated during this process and is exported to cytoplasm to be further processed by Dicer to make mature miRNA (Cazalla *et al.*, 2011). Endo-shRNA and endo-siRNA substrates have been shown to be excised by Dicer cleavage directly to produce mature miRNA (Babiarz *et al.*, 2008). Cytoplasmic vault RNAs (vRNAs) from conserved organelles vault particles have also been reported to utilize Dicer cleavage to generate mature miRNA, which could target a key enzyme in drug metabolism (Persson *et al.*, 2009). This finding may help to explain the correlation between vault particles and drug resistance. Another exception is hsa-miR-451, which can be sliced by Ago2 but not processed by Dicer during maturation after its precursor was transported from nuclear to cytoplasm (Cheloufi *et al.*, 2010; Cifuentes *et al.*, 2010; Yang *et al.*, 2010).

Table 10.1 Non-canonical pathways for miRNA biogenesis

Characteristics	Examples
Drosha/DGCR8-independent, Dicer-dependent	miRNAs generated from tRNAs (Lee *et al.*, 2009), miRNAs generated from murine gamma herpesvirus 68 (MHV68) tRNAs (Bogerd *et al.*, 2010; Diebel *et al.*, 2010)
	miRNAs generated from snoRNAs (Saraiya and Wang, 2008; Taft *et al.*, 2009), miRNA generated from EBV snoRNA (Hutzinger *et al.*, 2009)
	miRtrons (Ladewig *et al.*, 2012; Ruby *et al.*, 2007)
	miRNAs generated from Herpesvirus saimiri (HVS) Sm class U RNAs (HSURs) (Cazalla *et al.*, 2011)
	miRNAs generated from endo-shRNA/siRNA (Babiarz *et al.*, 2008)
	miRNA generated from Vault RNAs (Persson *et al.*, 2009)
Drosha/DGCR8-dependent, Dicer-independent	hsa-miR-451 (Cheloufi *et al.*, 2010; Cifuentes *et al.*, 2010; Yang *et al.*, 2010)

In general, all these non-canonical pathways partially bypass the canonical pathway for miRNA biogenesis. However, data from recent reports also suggest the existence of RNase III-independent pathway for miRNA biogenesis (Maurin *et al.*, 2012) and cytoplasmic microprocessor to initiate pri-miRNA processing (Shapiro *et al.*, 2010, 2012). It is likely that other unknown pathways for miRNA biogenesis await discovery.

Viruses that encode miRNAs

Viruses from the herpesvirus family were first shown to be able to encode viral own miRNAs (Cai *et al.*, 2005; Pfeffer *et al.*, 2004, 2005; Samols *et al.*, 2005). Later on, many other viruses with large DNA genome, most of which are herpesvirus and polyomavirus, were also reported to produce one or several viral miRNAs (see reviews Cullen, 2006, 2009, 2010; Plaisance-Bonstaff and Renne, 2011 and references therein). The evidence that only DNA viruses were shown to encode viral miRNAs and no viral RNase III exists suggested that only viruses having a nuclear life cycle are able to encounter the first cellular machinery microprocessor to generate viral miRNA (Cullen, 2010). Those viruses that have a nuclear lifecycle are all DNA viruses except vaccinia virus, influenza viruses, and retroviruses. Indeed, the number of DNA viruses shown to encode viral miRNAs is gradually growing in the last few years (Cullen, 2006, 2009, 2010; Plaisance-Bonstaff and Renne, 2011). However, not all the nuclear viruses are actually able to encode viral miRNAs (Cai *et al.*, 2006a; Umbach *et al.*, 2010). Especially for retrovirus human immunodeficiency virus (HIV), whether it encodes viral miRNAs has been debated for several years (Bennasser *et al.*, 2004; Klase *et al.*, 2007; Lin and Cullen, 2007; Omoto and Fujii, 2005; Omoto *et al.*, 2004; Ouellet *et al.*, 2008; Pfeffer *et al.*, 2004). Recent deep sequencing data using RNA extracted from HIV infected T cells demonstrated that HIV does encode viral miRNAs, although at a relative low level (Schopman *et al.*, 2012). Another study from a bovine oncogenic retrovirus clearly showed that oncogenic miRNA could be encoded by this leukaemia virus (Kincaid *et al.*, 2012). Surprising results coming from a cytoplasmic virus called West Nile virus (WNV) indicate that a miRNA encoded by WNV was able to facilitate viral replication in the host cell (Hussain *et al.*, 2012). Taken together, all these data suggest that most of DNA viruses may hijack the canonical pathway for miRNA biogenesis to generate viral miRNAs, while some other viruses may utilize the existing non-canonical pathways or even some unknown mechanism to produce viral miRNAs.

Human tumour viruses and their encoding miRNAs

Among those seven human tumour viruses, four have been reported to encode viral own miRNAs. EBV was the first one demonstrated to be able to produce viral miRNAs (Pfeffer *et al.*, 2004). Altogether there are 25 precursor miRNAs encoded by EBV, which will subsequently be processed into 48 mature miRNAs (Cai *et al.*, 2006b; Cosmopoulos *et al.*, 2009; Grundhoff *et al.*, 2006; Landgraf *et al.*, 2007; Lung *et al.*, 2009; Pfeffer *et al.*, 2004; Zhu *et al.*, 2009). KSHV was shown to encode 12 precursor miRNAs and after processing and editing they could finally generate 25 mature viral miRNAs (Cai *et al.*, 2005; Grundhoff *et al.*, 2006; Landgraf *et al.*, 2007; Lin *et al.*, 2010; Pfeffer *et al.*, 2005; Samols *et al.*, 2005; Umbach and Cullen, 2010). Two viral miRNAs were predicted from the genome of HBV. Only one virally encoded miRNA vmiRNA#2 could be detected by northern blot in HBV-infected patients (Jin *et al.*, 2007). However, the sequence of this vmiRNA#2 spans the stem loop region in the precursor miRNA, which is likely to be cleaved off during miRNA maturation (Lee *et al.*, 2002). In addition, that no other studies have confirmed this miRNA nor have the functions been demonstrated further adds suspicion on its existence. MCV was identified most recently in merkel cell carcinoma (MCC), a rare but aggressive type of human skin cancer that commonly affects elderly and immunosuppressed individuals (Feng *et al.*, 2008). One miRNA precursor found to be encoded by MCV will finally give rise to two mature viral miRNAs (Seo *et al.*, 2009). One type of high-risk human papillomavirus 31 has been tested and no viral miRNAs could be detected during either latent infection or productive virus replication (Cai *et al.*, 2006a). However, since more than 100 types of HPV have been identified and characterized (Munoz *et al.*, 2006), other types of HPV should also be tested for viral miRNA expression before making a conclusion that HPVs do not encode any viral miRNA. Other three human tumour viruses have so far not been demonstrated to encode any viral miRNA. Nevertheless, two retroviruses HIV and bovine leukaemia virus-4 (BLV-4), with similar genomic size to HTLV-1 and HCV, have already been proven to encode viral miRNAs

(Kincaid *et al.*, 2012; Klase *et al.*, 2013; Schopman *et al.*, 2012). Thus, it will be really interesting and worthy testing that whether these two RNA human tumour viruses encode viral miRNA. Table 10.2 summarizes all viral miRNAs encoded by human tumour viruses.

Epstein–Barr virus (EBV)

The Epstein–Barr virus, which is also called human herpesvirus 4 (HHV-4), is one of the oncogenic human herpesviruses. EBV has infected more than 90% of the human population and is aetiologically related to several human diseases, including infectious mononucleosis (IM), nasopharyngeal carcinoma (NPC), Burkitt's lymphoma (BL), Hodgkin's lymphoma (HL), diffuse large B-cell lymphoma (DLBCL), nasal NK/T-cell lymphoma and gastric carcinoma (Delecluse *et al.*, 2007).

EBV is a large enveloped DNA virus, which mainly targets B cells and epithelial cells *in vivo*. *In vitro*, infection of primary B cells leads to B cell transformation, which is a direct proof of its oncogenic properties. Unlike another oncogenic gamma-herpesvirus KSHV, EBV has different latent gene expression programmes in different settings. For example, in EBV transformed lymphoblastoid cell lines and in diffuse large B-cell lymphoma, a subset of viral latent genes are expressed including the latent membrane proteins (LMP) 1, 2A, and 2B, Epstein–Barr nuclear antigens (EBNAs) 1, 2, 3A, 3B, 3C, and LP, the small non-coding EBER RNAs, *BART* miRNAs and *BHRF1* miRNAs, which is called latency III gene expression program. In nasopharyngeal carcinoma and Hodgkin's lymphoma, EBV set up latency II gene expression program, in which only EBNA1, LMP1, LMP2A, LMP2B, EBERs, *BART* miRNAs are expressed. While in Burkitt's lymphoma, only EBNA1, EBERs and *BART* miRNAs are expressed, which is called latency I gene expression programme (reviewed in Kieff and Rickinson, 2007).

Table 10.2 Viral miRNAs encoded by human tumour viruses

Human tumour virus	Pre-miRNA
EBV	*ebv-mir*-BART1–*ebv-mir*-BART-22, *ebv-mir-BHRF1-1*, *ebv-mir-BHRF-2*, *ebv-mir-BHRF-3*
KSHV	kshv-miR-K12-1–kshv-miR-K12-9, kshv-miR-K12-10a, kshv-miR-K12-10b, kshv-miR-K12-11, kshv-miR-K12-12
MCV	mcv-mir-M1
HBV[a]	vmiRNA#2

[a]See details in the text.

EBV was the first human virus reported to encode miRNAs (Pfeffer *et al.*, 2004). Up to now, there are 25 EBV miRNAs precursors identified in *BHRF1* region and in *BART* region, generating as many as 48 mature miRNAs. The *BHRF1* transcript encodes three pre-miRNAs (*ebv-miR-BHRF1-1*, *-2* and *-3*). In the *BART* region, there are two clusters of total 22 pre-miRNAs encoded. Cluster 1, located between *BART* exons I and IB, encodes eight miRNA precursors (*ebv-miR-BART1*, *-BART3–6* and *-BART15–17*). Cluster 2, located between *BART* exons IB and III, encodes 13 miRNA precursors (*ebv-miRBART7–14* and *-BART18–22*). The *ebv-miR-BART2* is singly located between exons IV and V (Cai *et al.*, 2006b; Chen *et al.*, 2010; Grundhoff *et al.*, 2006; Lung *et al.*, 2009; Pfeffer *et al.*, 2004; Zhu *et al.*, 2009). *BHRF1* miRNAs are expressed only in type III latency programme and in lytic replication, while the *BART* miRNAs are expressed in all forms of latency.

EBV not only encodes its own miRNAs, but also targets the cellular miRNA system. Multiple human miRNAs are reported to be dramatically up or down regulated in various EBV associated cancers (reviewed in Barth *et al.*, 2011). The functions of miRNA in EBV oncogenesis have been extensively studied in recent years. We will summarize the function of both viral and cellular miRNAs in EBV associated oncogenesis in different disease settings.

Nasopharyngeal carcinoma (NPC)

Nasopharyngeal carcinoma is a squamous cell carcinoma that usually develops around the ostium of the eustachian tube in the lateral wall of the nasopharynx (Sham *et al.*, 1990). There are three major contributors to the development and progression of NPC, which are genetic susceptibility, environmental factors and Epstein–Barr virus (EBV) infection; however, the molecular mechanism of NPC pathogenesis is still not fully understood. The expression profile of EBV miRNAs in NPC has been extensively studied. Several reports have confirmed abundant expression of the *BART* miRNAs but absence of *BHRF1* miRNAs in NPC tissues (Cosmopoulos *et al.*, 2009; Wong *et al.*, 2012). Moreover, alterations in cellular miRNA expression have been observed in NPC. Some dysregulated miRNAs, including miR-26a, miR-29c, miR-200a, miR-100, and miR-141, have been shown to regulate cell growth, apoptosis and the metastasis of NPC (reviewed in He *et al.*, 2012). These findings suggest that aberrant miRNA expression may involve in the oncogenesis of NPC.

Anti-apoptotic function of the *BART* miRNAs in NPC

BART miRNAs have been demonstrated to target p53 pathway inactivation in NPC. *ebv-miR-BART5* is reported to suppresses the expression of p53 up-regulated modulator of apoptosis (PUMA) (Choy *et al.*, 2008). Depletion of *ebv-miR-BART5* or overexpression of PUMA in EBV-positive epithelial cells increased the rate of apoptosis in response to the DNA-damaging agent Doxorubicin. Inhibition of PUMA by *BART* miRNA may facilitate cancer progression by protecting the cells from apoptosis (Choy *et al.*, 2008). *BART* miRNAs have also been shown to target Bim, another pro-apoptotic protein. Both the *BART* cluster1 and cluster 2 can inhibit apoptosis independently in response to etoposide, a potent inducer of p53-dependent apoptosis in an epithelial cell line. Bioinformatics analysis indicated multiple potential binding sites for several *BART* miRNAs in the Bim 3'UTR (Marquitz *et al.*, 2011). TOM22, a mitochondrial receptor for the proapoptotic protein Bax, as a target of *ebv-miR-BART-16* is also reported (Dolken *et al.*, 2010). Taken together, *BART* miRNAs exert anti-apoptosis function in NPC by targeting pro-apoptotic proteins, which is not only beneficial to viral persistence but also beneficial to neoplastic cell growth (Lo *et al.*, 2012).

Immune evasion function of the *BART* miRNAs in NPC

MHC class I-related chain B (*MICB*), a potent activating receptor on NK cells and a co-stimulatory receptor on γ/δ T cells and CD8+ T cells has been shown to be targeted by *ebv-miR-BART2-5p* (Nachmani *et al.*, 2009). In the presence of *ebv-miR-BART2-5p*, the expression of MICB is suppressed in EBV-infected cells, protecting infected cells from recognition and killing by NK cells. *MICB* appears to be an important and common immune gene targeted by viral miRNAs to attenuate host cell-mediated immune responses, since *MICB* is also a miRNA target of KSHV and HCMV herpesviruses (Nachmani *et al.*, 2009). Importin 7 (IPO7), which is implicated in innate immunity as inhibition of IPO7 in macrophages by siRNA reduced IL-6 secretion upon stimulation with LPS (Yang *et al.*, 2009), was identified as a target of *ebv-miR-BART3* (Dolken *et al.*, 2010).

In addition to host immune regulating genes, other studies show that *BART* miRNAs suppress both *LMP1* and *LMP2A* latent genes, which are highly immunogenic in NPC cells (Lo *et al.*, 2007; Lung *et al.*, 2009). LMP1, which is the major EBV-encoded oncogenic protein and considered to play an important role in NPC pathogenesis, is shown to be targeted by *ebv-miR-BART1-5p, 16* and *17-5p* (Lo *et al.*, 2007). Negative regulation of LMP1 expression by *BART* miRNAs may favour NPC development by 'fine-tuning' the levels of LMP1 and altering the balance between the growth-promoting and pro-apoptotic actions of LMP1 (Lo *et al.*, 2012). As memory T cells specific for LMP1 epitopes are detected in NPC tissues and in the circulating blood of NPC patients and healthy donors (Li *et al.*, 2007; Lin *et al.*, 2008), tight control of LMP1 protein expression may protect NPC cells from immune surveillance (Lo *et al.*, 2012). LMP2A is reported to be targeted by *ebv-miRBART22*, leading to a reduction in LMP2A expression (Lung *et al.*, 2009). Since LMP2A is highly immunogenic relative to other EBV latent proteins expressed in NPC tumours (Hislop *et al.*, 2007), modulation of LMP2A expression by *BART* miRNAs may therefore protect NPC cells from immunological attack (Lo *et al.*, 2012).

Latency maintenance function of the *BART* miRNAs in NPC

ebv-miR-BART6-5p contributes to viral latency maintenance by suppressing the expression of Dicer (Iizasa *et al.*, 2010). The silencing of Dicer by *ebv-miR-BART6-5p* has great impact on the EBV infection programme. *miR-BART6-5p* RNAs suppress the EBNA2 viral oncogene required for transition from immunologically less responsive type I and type II latency to the more immune-reactive type III latency as well as Zta and Rta viral proteins essential for lytic replication, revealing the regulatory function of *miR-BART6* in EBV infection and latency (Iizasa *et al.*, 2010). Another example for *BART* miRNAs in viral latency maintenance is *miR-BART2-5p* inhibits the expression of the viral DNA polymerase BALF5 (Barth *et al.*, 2008). Forced expression of *ebv-miR-BART2-5p* during lytic cycle leads to reduction of both BALF5 and viral maturation (Barth *et al.*, 2008).

vRecently, various attempts have been carried out to discover the human miRNA profiles in clinical NPC biopsies, and even clustered with tumour–node–metastasis (TNM) stages. A bunch of aberrant miRNAs have been identified in NPC specimens and/or NPC cell lines. Among them, some of their targets and corresponding functions in nasopharyngeal carcinogenesis have been further demonstrated by experiments (reviewed in He *et al.*, 2012).

It was demonstrated that miR-200a controls epithelial-mesenchymal transition by targeting ZEB2, and it regulates the stem-like transition differentially and specifically by β-catenin signalling (Xia *et al.*, 2010), which is a key step for carcinogenesis. MiR-26a is commonly down-regulated in NPC specimens and NPC cell lines with important functional consequences. Ectopic expression of miR-26a dramatically suppressed cell proliferation and colony formation by inducing G_1-phase cell-cycle arrest. It was demonstrated that miR-26a strongly reduced the expression of EZH2 oncogene in NPC cells (Lu *et al.*, 2011). Another study has validated that miR-26a, miR-101, and miR-98 as bona fide regulators of EZH2 expression (Alajez *et al.*, 2010). MiR-100 which is underexpressed in NPC leads to Plk1 overexpression, which in turn contributes to NPC progression. Targeting Plk1 will cause mitotic catastrophe, with significant cytotoxicity both *in vitro* and *in vivo*, underscoring the important therapeutic opportunity of Plk1 in NPC (Shi *et al.*, 2010). MiR-218 is frequently down-regulated in in primary NPC tissues and cell lines where it plays a critical role in NPC progression. Suppression of miR-218 was associated with epigenetic silencing of SLIT2 and SLIT3, ligands of ROBO receptors that have been previously implicated in tumour angiogenesis (Alajez *et al.*, 2011). c-Myc, a key factor for cell growth and carcinogenesis, was found to regulate or to be regulated by a bunch of microRNAs, including miR-34b, -34c, -18a, -29a/b, -100, -141, -221 and let-7 family members (reviewed in He *et al.*, 2012). MiR-138 is another down-regulated miRNA in NPC. MiR-138 could suppress the growth of NPC cells through the repression of CCND1 and its downstream signalling (Liu *et al.*, 2012).

Exosomal miRNA and NPC

It was shown that five *BART* miRNAs including cluster1 (*miR-BART-1-5p* and *miR-BART-5*) and cluster2 (*miR-BART-7-3p*, *miR-BART-12* and *miR-BART-13*) were released by NPC cells *in vitro* as well as in plasma samples from NPC xenografted nude mice and NPC patients. This study provides a basis to explore their potential as a source of novel tumour biomarkers and their possible role in communications between malignant and non-malignant cells (Gourzones *et al.*, 2010). Another study not only detected the presence of certain enriched *BART*-miRNAs in exosomes secreted from NPC cell lines but also identified the viral oncoprotein LMP1 as a messenger within the exosomes (Meckes *et al.*, 2010). The function of exosomal miRNAs in NPC is still obscure. However, they may be monitored as markers of NPC progression.

B-cell lymphoma

EBV is associated with various B cell lymphomas, including Burkitt's lymphoma (BL), Hodgkin's lymphoma (HL), diffuse large B-cell lymphoma (DLBCL). EBV transforms B cells through the expression of the latent viral proteins EBNA and latent membrane protein (LMP). The EBV B95-8 laboratory strain that readily transforms primary B cells is deleted for most of the *BART* miRNAs indicates that they are not required for B-cell transformation. However, it has become recently apparent that viral miRNAs also contribute to EBV's oncogenic properties (Feederle *et al.*, 2011). Recombinant EBVs that lack the *BHRF1* miRNA cluster display a reduced ability to transform B lymphocytes *in vitro*. Furthermore, infected cells evince a marked up-regulation of the EBNA genes. It was further shown that all three *BHRF1* miRNAs contribute to B-cell transformation. Recombinants that lacked *miR-BHRF1-2* or *miR-BHRF1-3* displayed enhanced EBNA expression initiated at the Cp and Wp promoters (Feederle *et al.*, 2011).

Cellular miRNAs are also targeted by EBV in EBV associated B cell lymphoma. MiR-155 is an oncogenic miRNA, processed from a primary transcript, known as 'BIC' (B-cell integration cluster). Transgenic mouse studies demonstrated that B-cell targeted expression of BIC leads to the development of B-cell malignancies (Costinean *et al.*, 2006). MiR-155 is one of the miRNAs most frequently implicated in cancers. It was shown that miR-155 expression is much higher in EBV-immortalized B cells than in EBV-negative B cells (Gatto *et al.*, 2008; Rahadiani *et al.*, 2008). In addition, substantial expression of BIC/miR-155 was detected in latency III-, but not in latency I-type cells. It was found that LMP1, but not LMP2, is responsible for the up-regulation of miR-155. With the use of pathway inhibitors, it was shown that LMP1-induced strong BIC expression, primarily through NFkappaB (NFκB) and p38/MAPK pathways. These results suggest that BIC/miR-155 play a role in lymphomagenesis through NFκB and p38/MAPK pathways in response to activation by EBV LMP1 (Gatto *et al.*, 2008; Rahadiani *et al.*, 2008). It was further demonstrated that miR-155 is the most highly expressed miRNA in LCLs and that the selective inhibition of miR-155 function specifically inhibits the growth of both LCLs and the DLBCL cell line IBL-1. Therefore, miR-155 plays a key role in B-cell immortalization by EBV (Linnstaedt *et al.*, 2010). Similarly, miR-146a, a modulator of lymphocyte signalling is up-regulated by LMP1 via NF-κB signalling (Cameron *et al.*, 2008; Motsch *et al.*, 2007). It is possible that the induction of miR-146a plays a role in the induction or maintenance of EBV latency by modulating innate immune responses to the virus infected host cell (Motsch *et al.*, 2007).

One miRNA described as a p53-targeted tumour suppressor, miR-34a, was strongly induced by EBV infection and expressed in many EBV and Kaposi's sarcoma-associated herpesvirus (KSHV)-infected lymphoma cell lines. MiR-34a was also regulated by LMP-1 via NFκB activation but independent of functional p53. Surprisingly, overexpression of miR-34a was not toxic in several B lymphoma cell lines, and inhibition of miR-34a impaired the growth of EBV-transformed cells. This study identifies a progrowth role for a tumour-suppressive miRNA, highlighting the importance of studying miRNA function in different cellular contexts (Forte *et al.*, 2012).

Nasal NK/T cell lymphomas (NKTCL)

Nasal NK/T cell lymphomas are a subset of aggressive Epstein–Barr virus (EBV)-associated non-Hodgkin's lymphomas. NKTCL are mainly found in South-east Asia where they constitute about 3–9% of all malignant lymphoma (Reviewed in Delecluse *et al.*, 2007). Only a little information about the expression and function of miRNAs in NKTCL is known. One study showed that all EBV *BART* miRNAs except BHRF-derived miRNAs were expressed and some of these miRNAs are expressed at higher levels than in nasopharyngeal carcinomas. Modulating the expression of *BART9* with antisense RNAs consistently reduced the proliferation of SNK6 and SNT16. It is indicated that *BART9* is involved in NKTCL proliferation, and one of its mechanisms of action appears to be regulating LMP-1 levels (Ramakrishnan *et al.*, 2011).

Gastric carcinoma

About 6–16% of gastric carcinoma cases worldwide are associated with Epstein–Barr virus (EBV) (Takada, 2000). Marquitz *et al.* (2012) showed that *in vitro* infection of a gastric epithelial cell line with EBV alters growth properties and induces growth in soft agar. The infected cells have high levels of expression of *BART* miRNAs and limited viral protein expression. Microarray data showed that the expression changes are highly enriched for genes involved in cell motility and transformation pathways. Importantly, the transcripts decreased by microarray are significantly enriched in both experimentally and bioinformatically predicted *BART* miRNA targets, which suggest that the *BART* miRNAs are major contributors to the transformed growth properties of the EBV-infected cells (Marquitz *et al.*, 2012).

Hepatitis B virus (HBV)

The hepatitis B virus (HBV) is a partially double stranded DNA virus that is tightly correlated to chronic hepatitis, liver cirrhosis, and hepatocellular carcinoma (HCC) (Lavanchy, 2004).

miRNAs that modulate HBV replication

Although one viral miRNA encoded by HBV has been reported and predicted to be able to target only viral genome itself (Jin *et al.*, 2007), further studies are needed to confirm its existence and to investigate its function in regulation of viral replication. However, there are indeed many cellular miRNAs reported to regulate viral replication (see review in Liu *et al.*, 2011a). MiR-199a-3p, miR-210 and miR-125a-5p were shown to inhibit HBV replication through direct targeting *HBV surface antigen* (HBsAg) transcript (Potenza *et al.*, 2011; Zhang *et al.*, 2010). On the contrary, miR-1 was demonstrated to enhance HBV replication through targeting histone deacetylase 4 (HDAC4) (Zhang *et al.*, 2011). MiR-372/373 was shown to promote HBV replication by targeting nuclear factor I/B (NFIB) (Guo *et al.*, 2011).

miRNAs as biomarkers for HBV infection and HBV-related HCC

In addition to manipulating HBV replication, a set of miRNAs was demonstrated to be elevated with HBV infection, suggesting the potential of these miRNAs as biomarker for diagnosis of HBV infection (see review in Liu *et al.*, 2011a). Among those up-regulated miRNAs miR-17-92 polycistron and miR-21 may contribute to the malignant phenotype of HBV-related HCC (Connolly *et al.*, 2008), providing a substantial target for treating HBV-positive HCC. A group of thirteen miRNAs was shown to be differentially expressed in HBV-positive serum (Li *et al.*, 2010). It was also shown that HBV-positive HCC cases were clearly separated from control cases while exploring only three of these differentially expressed miRNAs as biomarker. And more strikingly, HBV-positive HCC could be accurately predicted by using miR-375 alone as biomarker (Li *et al.*, 2010).

Hepatitis C virus (HCV)

Hepatitis C virus is positive-sense single-stranded RNA virus belonging to the family of *Flaviviridae*. Similar to HBV infection, HCV infection in patients can cause hepatitis C, liver cirrhosis, and HCC (Kiyosawa *et al.*, 1990; Saito *et al.*, 1990). Cellular miRNAs have been demonstrated to play critical role in HCV replication gene expression, and pathogenesis in HCV-positive HCC. Among these miRNAs, miR-199a* was shown to inhibit HCV replication through targeting 5′UTR of HCV (Murakami *et al.*, 2009). miR-196 was reported to repress HCV gene expression in HCV replicon cell line and HCV culture system (Hou *et al.*, 2010). These data suggest that overexpression of miR-199a* and/or miR-196 may hold promise as a novel strategy to control hepatitis C infection and protect against liver injury in chronic HCV infection. On the other hand, liver-specific miR-122 could facilitate viral replication through interacting with the 5′ UTR of HCV RNA genome (Jopling *et al.*, 2005) as well as viral translation and infectious virus production (Henke *et al.*, 2008; Jangra *et al.*, 2010; Randall *et al.*, 2007). All these characteristics of miR-122 on helping HCV replication, translation, and producing infectious viral progeny make miR-122 as a perfect target for antiviral intervention. In the HCV chronically infected primate animal model, the treatment using locked nucleic acid (LNA) complementary to miR-122 leads to long-lasting suppression of HCV viraemia and improvement of HCV induced liver pathology, which really holds promise for new antiviral therapy to HCV-related pathogenesis (Lanford *et al.*, 2010).

Moreover, a set of miRNAs are up-regulated in tumour samples compare with normal tissue controls (Pineau *et al.*, 2010; Wong *et al.*, 2008). Among them, overexpression of miR-221 was shown to contribute liver tumorigenesis through targeting cyclin-dependent kinase (CDK) inhibitor p27 (Pineau *et al.*, 2010). On the other hand, down-regulation of miR-223 was successfully used as biomarker to differentiate HCC from adjacent non-tumoural liver tissue (Wong *et al.*, 2008). Deep sequencing small RNAs extracted from HCC cell lines revealed that miR-1323 expressed quite abundant in HCC and its distinct correlation in tumour arising from liver cirrhosis (Law *et al.*, 2013). Taken together, all the evidence suggests that some miRNAs could be used as biomarkers for HCC diagnosis and may be potential target for HCC treatment.

Human T-cell lymphotropic virus type 1 (HTLV-1)

Human T-cell lymphotropic virus type 1, also called adult T-cell lymphoma virus type 1 or Adult T-cell leukaemia lymphoma (ATLL) is first human retrovirus which causes several diseases including adult T-cell leukaemia (ATL) in human (Matsuoka and Jeang, 2007). A number of cellular miRNAs were shown to be dysregulated in HTLV-1 adult T cell leukaemia or HTLV-1-transformed T cells (Bellon *et al.*, 2009; Pichler *et al.*, 2008; Yeung *et al.*, 2008). MiR-146a was directly stimulated via its promoter by Tax encoded by HTLV-1 (Pichler *et al.*, 2008; Tomita *et al.*, 2012). This up-regulation of miR-146a facilitate the growth of HTLV-1 infected T cells through targeting NFκB. Inhibition of miR-146a by its inhibitor resulted in reduced proliferation of HTLV-1-infected T cell line and overexpression of miR-146a enhanced the growth of this cell line, suggesting miR-146a as a potential therapeutic target for ATL (Tomita *et al.*, 2012). Another two miRNAs miR-93 and miR-130b were also reported to promote the growth of HTLV-1 infected cells or HTLV-1 transformed cells by targeting the tumour protein 53-induced nuclear protein 1 (TP53INP1) (Yeung *et al.*, 2008). On the contrary to miRNAs regulated by HTLV-1, there are also many miRNAs found to target retroviruses (Hakim *et al.*, 2008).

Human papillomavirus (HPV)

Human papillomavirus has a double-stranded circular DNA genome. Some types of HPV infection in human can cause warts, while other rare cases will lead to cancers. No HPV-encoded viral miRNA has been detected so far (Cai *et al.*, 2006a). However, both cellular oncogenic miRNAs and tumour suppressive miRNAs have been shown to be tightly regulated by high-risk HPV infection (Wang *et al.*, 2008) (also see review in Zheng and Wang, 2011). Viral oncogenic protein E6 was explored by HPV to down-regulate tumour repressive miR-34a through destabilizing the cellular TP53 protein, which activates miR-34a expression through direct binding to its promoter (Wang *et al.*, 2009). Besides, E6 protein from high-risk HPV 16 was shown to inhibit expression of cellular miR-218, which can target epithelial cell-specific marker LAMB3 (Martinez *et al.*, 2008). The up-regulation of LAMB3 which was able to enhance cell migration and tumorigenesis may be expected to viral infection and tumorigenesis during HPV infection (Martinez *et al.*, 2008).

Kaposi's sarcoma-associated herpesvirus (KSHV)

Kaposi's sarcoma -associated herpesvirus, also called human herpesvirus 8 (HHV8), is another oncogenic gamma human herpesvirus, which is the aetiological agent of Kaposi's sarcoma (KS), the most common malignance in AIDS patients, primary effusion lymphoma (PEL) and some types of Multicentric Castleman's disease (MCD) (Cesarman *et al.*, 1995; Chang *et al.*, 1994; Soulier *et al.*, 1995). KSHV encodes 12 pre-miRNAs that produce 25 mature miRNAs (Summarized in Table 10.1) through the canonical pathway for miRNA biogenesis (Cai *et al.*, 2005; Grundhoff *et al.*, 2006; Lin *et al.*, 2010; Pfeffer *et al.*, 2005; Samols *et al.*, 2005; Umbach and Cullen, 2010). Ten of these miRNA precursors are organized into clusters within the latent genes locus. The other two pre-miRNAs (miR-K12-10, miR-K12-12) are located within the 3′UTR of ORF K12. All these viral miRNAs share two common promoters with the viral latent transcripts (Cai and Cullen, 2006; Pearce *et al.*, 2005). The function of both viral miRNAs and cellular miRNAs in KSHV associated oncogenesis has been suggested due to that more and more miRNA targets are identified.

KSHV miRNAs and viral life cycle control

In herpes simplex virus type 1 (HSV-1), studies showed that two miRNAs encoded by HSV-1, miR-H2–3p and miR-H6, facilitate the establishment and maintenance of viral latency by targeting viral immediate early transactivators *ICP0* and *ICP4.5*, respectively (Umbach *et al.*, 2008). Similar results in HSV-2, found that miR-III contributes to viral latency control through silencing *ICP0* expression (Tang *et al.*, 2009). Studies in HCMV showed that virus-encoded miR-UL112–1 controls viral latency by inhibiting the viral immediate early gene 72 (IE72) (Murphy *et al.*, 2008). The authors predicted that other herpesviruses might use a similar strategy to control viral latency.

Studies from KSHV-encoded miRNAs confirmed this hypothesis. Several miRNAs have been found to affect the expression level of the viral immediate early gene replication and transcription activator (RTA), either directly (Bellare and Ganem, 2009; Lin *et al.*, 2011; Lu *et al.*, 2010b) or indirectly (Lei *et al.*, 2010; Lu *et al.*, 2010a). Bellare *et al.*, using miRNA mimics or specific inhibitors for KSHV-encoded miRNAs and reporter constructs containing the RTA 3′UTR, found that miR-K12-9-5p targets RTA directly and depends on the canonical 6-mer seed match site. When this miRNA was inhibited by a specific antagomir, a moderate increase in lytic replication was observed (Bellare and Ganem, 2009). A second study by Lu *et al.* (2010b) using constructs expressing KSHV-encoded miRNAs and a reporter containing the RTA 3′UTR demonstrated that miR-K12-5 represses RTA expression, although the RTA 3′ UTR lacks a canonical miR-K12-5 seed sequence. In another study by us, a reporter containing the RTA 3′UTR and constructs expressing all 12 pre-miRNAs, miR-K12-9 and miR-K12-7-5p were found to target RTA directly. MiR-K12-7-5p, targeting RTA, was shown to be mediated by a 7-mer seed match site. Additionally, endogenous RTA expression level was reduced by ectopically overexpressing miR-K12-7 and de-repressed using a miR-K12-7-5p inhibitor. A decrease in viral particles was observed when miR-K12-7 was overexpressed (Lin *et al.*, 2011). Lei *et al.* (2010) used an miR-cluster deletion mutant virus to determine that miR-K12-1 represses IκB, an inhibitor of NFκB. Inhibition of IκB leads to NFκB activation, which suppresses RTA to facilitate viral latency control. Lu *et al.* (2010a) using a lentivirus expressing individual KSHV miRNA, found that miR-K12-3 reduces RTA mRNA levels by targeting NFIB directly. Further studies showed a putative NFIB-binding site is located in the RTA promoter and shRNA knockdown of NFIB resulted in decreased RTA expression.

More recently, a large number of viral and cellular microRNAs (miRNAs), as well as unusual small RNAs (usRNAs), were detected in KSHV virions by using deep sequencing. Both viral and host miRNAs detected in small RNAs extracted from KSHV virions were further shown to colocalize with KSHV virions directly by in situ hybridization (ISH)-electron microscopy (EM) (ISH-EM). The virional miRNAs could be transported into host cells, and they are biologically functional during *de novo* viral infection (Lin *et al.*, 2012).

Taken together, the evidence suggests that multiple KSHV-encoded miRNAs are involved in viral latency maintenance and that herpes viral latency control is an important and complex process. These data support the hypothesis that conservation among herpesviruses allows them to use viral miRNA(s) to target immediate early genes to control viral latency (Murphy *et al.*, 2008).

KSHV miRNA and immune regulation

To establish long-term latent infection in the host cell, KSHV has developed multiple strategies to evade the host innate and adaptive immunity (Areste and Blackbourn, 2009). Some studies suggest that KSHV-encoded miRNAs are involved in immune regulation and favour KSHV infection. For example, several herpesviruses, including KSHV, encode an miRNA that targets the MHC class I-related chain B (MICB), which is a stress-induced ligand recognized by the NKG2D receptor expressed by NK cells and CD8+ T-cells. In 2007, Stern-Ginossar *et al.* (2007) identified that HCMV miRNA miR-UL112–1 targets MICB. They showed that hcmv-miR-UL112 specifically down-regulates MICB expression during viral infection, leading to decreased binding of NKG2D and reduced killing by NK cells. They further showed that both EBV *miR-BART2-5p* and KSHV miR-K12-7 regulate MICB expression (Nachmani *et al.*, 2009). Moreover, Thomas *et al.* (2008) showed that the KSHV immune evasion gene, K5, reduces cell-surface expression of the NKG2D ligands MHC class I-related chain A (MICA) and MICB, probably by K5-mediated ubiquitylation, which signals internalization and causes a potent reduction in NK cell-mediated cytotoxicity. These studies suggested that NKG2D ligands are common targets for both KSHV miRNA and ORF for evading NK cell antiviral function. Another example is that miR-K12-11 directly targets IKK, an important modulator of IFN signalling.

KSHV miRNAs also regulate the host immune response by modulating expression of cytokines. Abend *et al.* (2010) demonstrated that miR-K12–10a robustly down-regulates the expression of tumour necrosis factor (TNF)-like weak inducer of apoptosis (TWEAK) receptor (TWEAKR). The down-regulation of TWEAKR by miR-K10a in primary human endothelial cells results in a decrease in expression of the proinflammatory cytokines IL-8 and monocyte chemo attractant protein 1 in response to TWEAK. This protects cells from apoptosis and suppresses a proinflammatory response (Abend *et al.*, 2010). Qin *et al.* (2009) demonstrated that miR-K12-3 and miR-K12-7 target C/EBPβ p20 (LIP), a negative transcriptional regulator of IL-6 and IL-10, and induces expression of IL-6 and IL-10 in macrophages. These cytokines have broad functions in oncogenesis and immune suppression. The function of miR-K12-3 and miR-K12-7 in immune regulation needs to be further studied (Boss and Renne, 2011).

KSHV miRNAs and viral pathogenesis

Samols *et al.* (2005) identified THBS1 as a target of several KSHV miRNAs. THBS1 is a strong tumour suppressor and anti-angiogenic factor. They proposed that KSHV-encoded miRNAs contribute directly to pathogenesis by down-regulation of THBS1, promoting cell adhesion, migration, and angiogenesis. Gottwein and Cullen (2010) reported that miR-K1 directly targets the cellular cyclin-dependent kinase inhibitor p21, strongly attenuating the cell cycle arrest that normally occurs upon p53 activation. They suggested that this KSHV miRNA likely contributes to the oncogenic potential of KSHV. Hansen *et al.* (2010) demonstrated that multiple KSHV miRNAs contribute to virally induced reprogramming by silencing the cellular transcription factor MAF, which prevents expression of BEC-specific genes, thereby maintaining the differentiation status of LECs. These findings demonstrate that KSHV miRNAs could influence the differentiation status of infected cells, contributing to KSHV-induced oncogenesis (Hansen *et al.*, 2010).

Using developing systems biology methods, the target list of KSHV miRNAs is rapidly increasing (Lieber and Haas, 2011). The functions of KSHV miRNAs include, but are not limited to, viral life cycle control, immune regulation and pathogenesis. KSHV miRNAs are important mediators of viral–host interaction. Using viral miRNA knockout viruses will allow studies that lead to an overall understanding of the functions of unique KSHV miRNAs.

KSHV encodes orthologues of host miRNAs

Early in the identification of KSHV miRNAs, two independent studies reported that miR-K12-11 is the orthologue of host hsa-miR-155 and these miRNAs have identical seed sequence (Gottwein *et al.*, 2007; Skalsky *et al.*, 2007). Hsa-miR-155 is a multifunctional miRNA that is important in immunity, haematopoiesis, inflammation and oncogenesis (McClure and Sullivan, 2008). Not only miR-K12-11, but also miR-M4, which is encoded by the highly oncogenic Marek's disease virus of chickens, is a functional orthologue of hsa-miR-155 (Zhao *et al.*, 2009). EBV does not encode a hsa-miR-155 orthologue. Instead, EBV-induced hsa-miR-155 expression by LMP1 and hsa-miR-155 is important in EBV viral pathogenesis (Gatto *et al.*, 2008; Lu *et al.*, 2008; Motsch *et al.*, 2007). Of note, hsa-miR-155 is down-regulated in KSHV-infected cells (Gottwein *et al.*, 2007; Skalsky *et al.*, 2007); therefore, KSHV might encode miR-K12-11 to replace the function of its orthologue, hsa-miR-155.

Since miR-K12-11 shares an identical seed sequence with hsa-miR-155, these two miRNAs may have similar functions and target the same genes. Gottwein *et al.* (2007) used a range of assays to show that expression of physiological levels of miR-K12-11 or hsa-miR-155 results in the down-regulation of an extensive set of common mRNA targets, including genes with known roles in cell growth regulation, for example BACH1. Moreover, Qin *et al.* (2010) found that KSHV-encoded microRNAs up-regulate xCT expression in macrophages and ECs, largely through miR-K12-11 suppression of BACH-1, a negative regulator of transcription that recognizes antioxidant response elements within gene promoters.

To compare miR-K12-11 and hsa-miR-155 functions *in vivo*, Boss *et al.* (2011) used a foamy virus vector to express the miRNAs in human haematopoietic progenitors and performed immune reconstitutions in NOD/LtSz-scid IL2R (null) mice. They found that ectopic expression of miR-K12-11 or hsa-miR-155 no viral RNase III, leads to a significant expansion of the CD19 (+) B-cell population in the spleen. This *in vivo* study validates miR-K12-11 as a functional orthologue of hsa-miR-155 in the context of haematopoiesis (Boss *et al.*, 2011).

Our study indicated that miR-K12-11 is involved in attenuating interferon signalling and contributing to KSHV latency maintenance through targeting I-kappa-B kinase epsilon (IKKε). We demonstrated that miR-K12-11 attenuated IFN signalling by decreasing IKKε-mediated IRF3/IRF7 phosphorylation. We also demonstrated that IKKε enhances KSHV reactivation synergistically with 12-O-tetradecanoylphorbol 13-acetate treatment. Moreover, inhibition of miR-K12-11 enhances KSHV reactivation induced by vesicular stomatitis virus infection. Taken together, our findings suggest that miR-K12-11 can contribute to maintenance of KSHV latency by targeting IKKε (Liang *et al.*, 2011a).

More recently, we demonstrated that ectopic expression of miR-K12-11 down-regulates TGF-β signalling and facilitates cell proliferation upon TGF-β treatment by directly targeting SMAD5. Our findings highlight a novel mechanism in which miR-K12-11 down-regulates TGF-β signalling, and suggests that viral miRNAs and proteins may exert a dichotomous regulation in virus-induced oncogenesis by targeting the same signalling pathway (Liu *et al.*, 2011b).

KSHV exploits miR-K12-11 to tap into a host miRNA regulatory network (McClure and Sullivan, 2008). Based on the important functions of hsa-miR-155, we believe that miR-K12-11 is of particular importance in KSHV infection and pathogenesis.

Using Target Scan5.0 (Lewis *et al.*, 2005), we found that miR-K12-2 shares the same seed sequence as miR-1183 (Liang *et al.*, unpublished data). The function of miR-1183 is poorly understood. Further study is needed to confirm whether miR-K12-2 is the true orthologue of miR-1183, and to determine the functions of these two miRNAs in KSHV infection.

KSHV modulates the expression profile and/or function of host miRNAs

O'Hara *et al.* (2008) identified 68 miRNAs specifically expressed in PEL cell lines rather than in KSHV non-related lymphoma cell lines and tonsil tissues. However, the question remains whether these PEL-specific miRNAs are regulated directly by KSHV gene products or are the consequence of long-term infection and oncogenesis. To answer this question, in a later study, O'Hara *et al.* (2009) profiled PEL cell lines, KSHV-infected ECs, uninfected ECs, and primary KS biopsies for host miRNAs expression using a qPCR-based array. Eighteen KSHV-regulated miRNAs were identified after adjustment for multiple comparisons. Moreover, they demonstrated that the tumour-suppressor miRNAs in the let-7 family and miR-220/221 are down-regulated in KSHV-associated cancers, including PEL and KS (O'Hara *et al.*, 2009).

To determine the host miRNA profile during early KSHV infection, Lagos *et al.* (2010) detected host miRNA expression at 6 and 72 hours post-KSHV infection of lymphatic endothelial cells (LECs) using a miRNA microarray. They identified two groups of miRNAs induced during primary KSHV infection. The first group consisted of nine miRNAs that reached their peak expression at 6 hours post infection, including hsa-miR-146a, hsa-miR-31 and hsa-miR-132. Expression levels of five miRNAs from the second group, including hsa-miR-193a and hsa-let-7i, steadily increased over the 72 hours post infection (Lagos *et al.*, 2010).

Although the functions of human miRNAs have been extensively studied and KSHV clearly regulates host miRNA expression, the roles of host miRNAs in KSHV infection are less well understood than the functions of the viral miRNAs. Based on a limited number of studies, we infer that KSHV hijacks the host miRNA system to favour infection and pathogenesis.

Lagos *et al.* (2010) showed that KSHV induced hsa-miR-132 expression via a CREB-dependent pathway. The miRNA hsa-miR-132 negatively regulates interferon pathways by targeting the p300 transcriptional co-activator to facilitate viral replication. Interestingly, they show a similar function for hsa-miR-132 during infection of monocytes with herpes simplex virus-1 (HSV-1) and human cytomegalovirus (HCMV). Therefore, induction of hsa-miR-132 might be a common strategy for herpesvirus to control innate immunity at early stage of infection (Lagos *et al.*, 2010). However, more extensive studies are needed to clarify the common signalling pathway that is responsible for CREB-mediated hsa-miR-132 activation.

Another host miRNA up-regulated by KSHV is hsa-miR-146a (Lagos *et al.*, 2010; Punj *et al.*, 2010). Punj *et al.* (2010) showed that vFLIP is responsible for hsa-miR-146a up-regulation. They identified two NFκB sites in the promoter of hsa-miR-146a that are essential for its activation by vFLIP. Up-regulation of hsa-miR-146a suppresses its target CXCR4; down-regulation of CXCR4 might contribute to KS development by promoting premature release of KSHV-infected endothelial progenitors into circulation (Punj *et al.*, 2010). Indeed, hsa-miR-146a is also reported to be up-regulated by EBV LMP1, and hsa-miR-146a might function in a negative feedback loop to modulate the intensity and/or duration of the interferon response (Cameron *et al.*, 2008). Therefore, hsa-miR-146a might be another common host miRNA target of herpesvirus infection.

Tsai *et al.* showed that K15M (the minor form of KSHV K15) induces expression of hsa-miR-21 and hsa-miR-31. Knocking down both of these miRNAs eliminates K15M-induced cell motility. Therefore, K15M might contribute to KSHV-mediated tumour metastasis and angiogenesis via regulation of hsa-miR-21 and hsa-miR-31 (Tsai *et al.*, 2009). In another study, Wu *et al.* (2011) explored the genes and miRNAs involved in KSHV-induced cell motility by combining gene and miRNA profile data. They showed that KSHV induces global changes of miRNA expression in LECs. Specifically, the hsa-miR-221/hsa-miR-222 cluster is down-regulated, whereas hsa-miR-31 is up-regulated. Both LANA and Kaposin B repress the expression of the hsa-miR-221/hsa-miR-222 cluster, which results in up-regulation of their target gene ETS1 or ETS2 and is sufficient to induce EC migration. In contrast, up-regulated hsa-miR-31 stimulates EC migration by reduction of its target gene FAT4 (Wu *et al.*, 2011).

KSHV not only modulates host miRNAs expression, but also modulates the function of some host miRNAs. Kang *et al.* (2010, 2011) showed direct repression of vIL-6 by hsa-miR-1293 and repression of hIL-6 by hsa-miR-608. They also found that KSHV ORF57 and hsa-miR-1293/hsa-miR-608 compete for the same binding site on vIL-6/hIL-6 mRNA. ORF57 binding results in escape of viral and human IL-6 from miRNA-mediated suppression, contributing to KSHV pathogenesis.

The function of host miRNAs in KSHV infection is increasingly a topic of interest in the field. More KSHV-regulated host miRNAs are being identified, and their function is usually clearer than that of viral miRNAs. We can therefore apply knowledge about other host miRNAs to the KSHV field to help understand the functions of KSHV-hijacked host miRNAs in KSHV infection and related pathogenesis.

Merkel cell polyomavirus (MCV or MCPyV)

Merkel cell polyomavirus is recently identified polyomavirus which is supposed to cause the majority of merkel cell carcinoma (MCC) (Feng *et al.*, 2008). One virally encoded miRNA mcv-mir-M1–5p was found in PCV infected cells (Lee *et al.*, 2011; Seo *et al.*, 2009). Mcv-mir-M1–5p was able to target viral large T antigen to and inhibit its expression which may potentially help the virus evade immune surveillance (Seo *et al.*, 2009).

Perspective

MiRNAs have been demonstrated to have profound effects on cellular signalling pathways since discovery (Bartel, 2009). It is not surprising that many viruses have also evolved to encode viral own miRNAs like their cellular counterpart

(Cullen, 2006, 2009, 2010; Plaisance-Bonstaff and Renne, 2011). Both viral miRNAs and cellular miRNAs and protein-coding genes are involved to determine the outcome of virus–host interactions (Liang *et al.*, 2011b). Different human tumour viruses may explore the same strategy for evading host immune surveillance by target viral own gene (such as EBV, KSHV, and MCV). Moreover, miRNAs like miR-146a were also utilized by many viruses for pathogenesis caused by virus. Therefore it is really interesting and critical to investigate the role miRNAs play in the pathogenesis caused by human tumour viruses. This information may shed more light on novel strategy for clinical diagnosis, antiviral treatment, and prognosis of virus-correlated cancers.

Acknowledgements

We thank Karen T. Wilson for critical reading of this manuscript. This work was funded by grants from the Key project of the Natural Science Foundation of China (81230037) and the National Basic Research Program of China (2011CB504805) to KL.

References

Abend, J.R., Uldrick, T., and Ziegelbauer, J.M. (2010). Regulation of tumor necrosis factor-like weak inducer of apoptosis receptor protein (TWEAKR) expression by Kaposi's sarcoma-associated herpesvirus microRNA prevents TWEAK-induced apoptosis and inflammatory cytokine expression. J. Virol. *84*, 12139–12151.

Alajez, N.M., Shi, W., Hui, A.B., Bruce, J., Lenarduzzi, M., Ito, E., Yue, S., O'Sullivan, B., and Liu, F.F. (2010). Enhancer of Zeste homolog 2 (EZH2) is overexpressed in recurrent nasopharyngeal carcinoma and is regulated by miR-26a, miR-*101*, and miR-98. Cell Death Dis. *1*, e85.

Alajez, N.M., Lenarduzzi, M., Ito, E., Hui, A.B., Shi, W., Bruce, J., Yue, S., Huang, S.H., Xu, W., Waldron, J., *et al.* (2011). MiR-218 suppresses nasopharyngeal cancer progression through down-regulation of survivin and the SLIT2-ROBO1 pathway. Cancer Res. *71*, 2381–2391.

Areste, C., and Blackbourn, D.J. (2009). Modulation of the immune system by Kaposi's sarcoma-associated herpesvirus. Trends Microbiol. *17*, 119–129.

Babiarz, J.E., Ruby, J.G., Wang, Y., Bartel, D.P., and Blelloch, R. (2008). Mouse ES cells express endogenous shRNAs, siRNAs, and other Microprocessor-independent, Dicer-dependent small RNAs. Genes Dev. *22*, 2773–2785.

Bartel, D.P. (2009). MicroRNAs: Target Recognition and Regulatory Functions. Cell *136*, 215–233.

Barth, S., Pfuhl, T., Mamiani, A., Ehses, C., Roemer, K., Kremmer, E., Jaker, C., Hock, J., Meister, G., and Grasser, F.A. (2008). Epstein–Barr virus-encoded microRNA miR-BART2 down-regulates the viral DNA polymerase BALF5. Nucleic Acids Res. *36*, 666–675.

Barth, S., Meister, G., and Grasser, F.A. (2011). EBV-encoded miRNAs. Biochim. Biophys. Acta. *1809*, 631–640.

Bellare, P., and Ganem, D. (2009). Regulation of KSHV lytic switch protein expression by a virus-encoded microRNA: an evolutionary adaptation that fine-tunes lytic reactivation. Cell Host Microbe *6*, 570–575.

Bellon, M., Lepelletier, Y., Hermine, O., and Nicot, C. (2009). Deregulation of microRNA involved in hematopoiesis and the immune response in HTLV-I adult T-cell leukaemia. Blood *113*, 4914–4917.

Bennasser, Y., Le, S.Y., Yeung, M.L., and Jeang, K.T. (2004). HIV-1 encoded candidate micro-RNAs and their cellular targets. Retrovirology *1*, 43.

Bogerd, H.P., Karnowski, H.W., Cai, X., Shin, J., Pohlers, M., and Cullen, B.R. (2010). A mammalian herpesvirus uses noncanonical expression and processing mechanisms to generate viral MicroRNAs. Mol. Cell. *37*, 135–142.

Boss, I.W., and Renne, R. (2011). Viral miRNAs and immune evasion. Biochim. Biophys. Acta. *1809*, 708–714.

Boss, I.W., Nadeau, P.E., Abbott, J.R., Yang, Y., Mergia, A., and Renne, R. A. (2011). Kaposi's sarcoma-associated herpesvirus-encoded ortholog of microRNA miR-155 induces human splenic B-cell expansion in NOD/LtSz-scid IL2Rgammanull mice. J. Virol. *85*, 9877–9886.

Cai, X., and Cullen, B.R. (2006). Transcriptional origin of Kaposi's sarcoma-associated herpesvirus microRNAs. J. Virol. *80*, 2234–2242.

Cai, X., Lu, S., Zhang, Z., Gonzalez, C.M., Damania, B., and Cullen, B.R. (2005). Kaposi's sarcoma-associated herpesvirus expresses an array of viral microRNAs in latently infected cells. Proc. Natl. Acad. Sci. U.S.A. *102*, 5570–5575.

Cai, X., Li, G., Laimins, L.A., and Cullen, B.R. (2006a). Human papillomavirus genotype 31 does not express detectable microRNA levels during latent or productive virus replication. J. Virol. *80*, 10890–10893.

Cai, X., Schafer, A., Lu, S., Bilello, J.P., Desrosiers, R.C., Edwards, R., Raab-Traub, N., and Cullen, B.R. (2006b). Epstein–Barr virus microRNAs are evolutionarily conserved and differentially expressed. PLoS Pathogens *2*, e23.

Cameron, J.E., Yin, Q., Fewell, C., Lacey, M., McBride, J., Wang, X., Lin, Z., Schaefer, B.C., and Flemington, E.K. (2008). Epstein–Barr virus latent membrane protein 1 induces cellular MicroRNA miR-146a, a modulator of lymphocyte signaling pathways. J. Virol. *82*, 1946–1958.

Cazalla, D., Xie, M., and Steitz, J.A. (2011). A primate herpesvirus uses the integrator complex to generate viral MicroRNAs. Mol. Cell. *43*, 982–992.

Cesarman, E., Chang, Y., Moore, P.S., Said, J.W., and Knowles, D.M. (1995). Kaposi's sarcoma-associated herpesvirus-like DNA sequences in AIDS-related body-cavity-based lymphomas. N. Engl. J. Med. *332*, 1186–1191.

Chang, Y., Cesarman, E., Pessin, M.S., Lee, F., Culpepper, J., Knowles, D.M., and Moore, P.S. (1994). Identification of herpesvirus-like DNA sequences in AIDS-associated Kaposi's sarcoma. Science *266*, 1865–1869.

Cheloufi, S., Dos Santos, C.O., Chong, M.M., and Hannon, G.J. (2010). A dicer-independent miRNA biogenesis pathway that requires Ago catalysis. Nature *465*, 584–589.

Chen, S.J., Chen, G.H., Chen, Y.H., Liu, C.Y., Chang, K.P., Chang, Y.S., and Chen, H.C. (2010). Characterization of Epstein–Barr virus miRNAome in nasopharyngeal carcinoma by deep sequencing. PloS One *5*, e12745.

Choy, E.Y., Siu, K.L., Kok, K.H., Lung, R.W., Tsang, C.M., To, K.F., Kwong, D.L., Tsao, S.W., and Jin, D.Y. (2008). An Epstein–Barr virus-encoded microRNA targets PUMA to promote host cell survival. J. Exp. Med. *205*, 2551–2560.

Cifuentes, D., Xue, H., Taylor, D.W., Patnode, H., Mishima, Y., Cheloufi, S., Ma, E., Mane, S., Hannon, G.J., Lawson, N.D., *et al.* (2010). A novel miRNA processing pathway independent of Dicer requires Argonaute2 catalytic activity. Science *328*, 1694–1698.

Connolly, E., Melegari, M., Landgraf, P., Tchaikovskaya, T., Tennant, B.C., Slagle, B.L., Rogler, L.E., Zavolan, M., Tuschl, T., and Rogler, C.E. (2008). Elevated expression of the miR-17–92 polycistron and miR-21 in hepadnavirus-associated hepatocellular carcinoma contributes to the malignant phenotype. Am. J. Pathol. *173*, 856–864.

Cosmopoulos, K., Pegtel, M., Hawkins, J., Moffett, H., Novina, C., Middeldorp, J., and Thorley-Lawson, D.A. (2009). Comprehensive profiling of Epstein–Barr virus microRNAs in nasopharyngeal carcinoma. J. Virol. *83*, 2357–2367.

Costinean, S., Zanesi, N., Pekarsky, Y., Tili, E., Volinia, S., Heerema, N., and Croce, C.M. (2006). Pre-B cell proliferation and lymphoblastic leukaemia/high-grade lymphoma in E(mu)-miR155 transgenic mice. Proc. Natl. Acad. Sci. U.S.A. *103*, 7024–7029.

Cullen, B.R. (2006). Viruses and microRNAs. Nat. Genet. *38* (Suppl), S25–30.

Cullen, B.R. (2009). Viral and cellular messenger RNA targets of viral microRNAs. Nature *457*, 421–425.

Cullen, B.R. (2010). Five questions about viruses and microRNAs. PLoS Pathog. *6*, e1000787.

Delecluse, H.J., Feederle, R., O'Sullivan, B., and Taniere, P. (2007). Epstein Barr virus-associated tumours: an update for the attention of the working pathologist. J. Clin. Pathol *60*, 1358–1364.

Diebel, K.W., Smith, A.L., and van Dyk, L.F. (2010). Mature and functional viral miRNAs transcribed from novel RNA polymerase III promoters. RNA *16*, 170–185.

Dolken, L., Malterer, G., Erhard, F., Kothe, S., Friedel, C.C., Suffert, G., Marcinowski, L., Motsch, N., Barth, S., Beitzinger, M., *et al.* (2010). Systematic analysis of viral and cellular microRNA targets in cells latently infected with human gamma-herpesviruses by RISC immunoprecipitation assay. Cell Host Microbe *7*, 324–334.

Kieff, E., and Rickinson, A.B. (2007). Epstein Barr Virus and its Replication (Philadelphia, Lippincott, Williams, and Wilkins).

Feederle, R., Haar, J., Bernhardt, K., Linnstaedt, S.D., Bannert, H., Lips, H., Cullen, B.R., and Delecluse, H.J. (2011). The members of an Epstein–Barr virus microRNA cluster cooperate to transform B lymphocytes. J. Virol. *85*, 9801–9810.

Feng, H., Shuda, M., Chang, Y., and Moore, P.S. (2008). Clonal integration of a polyomavirus in human Merkel cell carcinoma. Science *319*, 1096–1100.

Forte, E., Salinas, R.E., Chang, C., Zhou, T., Linnstaedt, S.D., Gottwein, E., Jacobs, C., Jima, D., Li, Q.J., Dave, S.S., *et al.* (2012). The Epstein–Barr virus (EBV)-induced tumor suppressor microRNA MiR-34a is growth promoting in EBV-infected B cells. J. Virol. *86*, 6889–6898.

Gatto, G., Rossi, A., Rossi, D., Kroening, S., Bonatti, S., and Mallardo, M. (2008). Epstein–Barr virus latent membrane protein 1 trans-activates miR-155 transcription through the NF-kappaB pathway. Nucleic Acids Res. *36*, 6608–6619.

Gottwein, E., and Cullen, B.R. (2010). A human herpesvirus microRNA inhibits p21 expression and attenuates p21-mediated cell cycle arrest. J. Virol. *84*, 5229–5237.

Gottwein, E., Mukherjee, N., Sachse, C., Frenzel, C., Majoros, W.H., Chi, J.T., Braich, R., Manoharan, M., Soutschek, J., Ohler, U., *et al.* (2007). A viral microRNA functions as an orthologue of cellular miR-155. Nature *450*, 1096–1099.

Gourzones, C., Gelin, A., Bombik, I., Klibi, J., Verillaud, B., Guigay, J., Lang, P., Temam, S., Schneider, V., Amiel, C., *et al.* (2010). Extra-cellular release and blood diffusion of BART viral micro-RNAs produced by EBV-infected nasopharyngeal carcinoma cells. Virol. J. *7*, 271.

Grey, F., Hook, L., and Nelson, J. (2008). The functions of herpesvirus-encoded microRNAs. Med. Microbiol. Immunol. *197*, 261–267.

Grundhoff, A., Sullivan, C.S., and Ganem, D. (2006). A combined computational and microarray-based approach identifies novel microRNAs encoded by human gamma-herpesviruses. RNA *12*, 733–750.

Guo, H., Liu, H., Mitchelson, K., Rao, H., Luo, M., Xie, L., Sun, Y., Zhang, L., Lu, Y., Liu, R., *et al.* (2011). MicroRNAs-372/373 promote the expression of hepatitis B virus through the targeting of nuclear factor I/B. Hepatology *54*, 808–819.

Hakim, S.T., Alsayari, M., McLean, D.C., Saleem, S., Addanki, K.C., Aggarwal, M., Mahalingam, K., and Bagasra, O. (2008). A large number of the human microRNAs target lentiviruses, retroviruses, and endogenous retroviruses. Biochem. Biophys. Res. Commun. *369*, 357–362.

Hansen, A., Henderson, S., Lagos, D., Nikitenko, L., Coulter, E., Roberts, S., Gratrix, F., Plaisance, K., Renne, R., Bower, M., *et al.* (2010). KSHV-encoded miRNAs target MAF to induce endothelial cell reprogramming. Genes Dev. 24, 195–205.

He, M.L., Luo, M.X., Lin, M.C., and Kung, H.F. (2012). MicroRNAs: potential diagnostic markers and therapeutic targets for EBV-associated nasopharyngeal carcinoma. Biochim. Biophys. Acta. 1825, 1–10.

Henke, J.I., Goergen, D., Zheng, J., Song, Y., Schuttler, C.G., Fehr, C., Junemann, C., and Niepmann, M. (2008). MicroRNA-122 stimulates translation of hepatitis C virus RNA. EMBO J. 27, 3300–3310.

Hislop, A.D., Taylor, G.S., Sauce, D., and Rickinson, A.B. (2007). Cellular responses to viral infection in humans: lessons from Epstein–Barr virus. Ann. Rev. Immunol. 25, 587–617.

Hou, W., Tian, Q., Zheng, J., and Bonkovsky, H.L. (2010). MicroRNA-196 represses Bach1 protein and hepatitis C virus gene expression in human hepatoma cells expressing hepatitis C viral proteins. Hepatology 51, 1494–1504.

Hussain, M., Torres, S., Schnettler, E., Funk, A., Grundhoff, A., Pijlman, G.P., Khromykh, A.A., and Asgari, S. (2012). West Nile virus encodes a microRNA-like small RNA in the 3′ untranslated region which up-regulates GATA4 mRNA and facilitates virus replication in mosquito cells. Nucleic Acids Res. 40, 2210–2223.

Hutzinger, R., Feederle, R., Mrazek, J., Schiefermeier, N., Balwierz, P.J., Zavolan, M., Polacek, N., Delecluse, H.J., and Huttenhofer, A. (2009). Expression and processing of a small nucleolar RNA from the Epstein–Barr virus genome. PLoS Pathog. 5, e1000547.

Iizasa, H., Wulff, B.E., Alla, N.R., Maragkakis, M., Megraw, M., Hatzigeorgiou, A., Iwakiri, D., Takada, K., Wiedmer, A., Showe, L., *et al.* (2010). Editing of Epstein–Barr virus-encoded BART6 microRNAs controls their dicer targeting and consequently affects viral latency. J. Biol. Chem. 285, 33358–33370.

Jangra, R.K., Yi, M., and Lemon, S.M. (2010). Regulation of hepatitis C virus translation and infectious virus production by the microRNA miR-122. J. Virol. 84, 6615–6625.

Jin, W.B., Wu, F.L., Kong, D., and Guo, A.G. (2007). HBV-encoded microRNA candidate and its target. Comput. Biol. Chem. 31, 124–126.

Jopling, C.L., Yi, M., Lancaster, A.M., Lemon, S.M., and Sarnow, P. (2005). Modulation of hepatitis C virus RNA abundance by a liver-specific MicroRNA. Science 309, 1577–1581.

Kang, J.G., Pripuzova, N., Majerciak, V., Kruhlak, M., Le, S.Y., and Zheng, Z.M. (2010). Kaposi's sarcoma-associated herpesvirus ORF57 promotes escape of viral and human interleukin-6 from microRNA-mediated suppression. J. Virol. 85, 2620–2630.

Kang, J.G., Majerciak, V., Uldrick, T.S., Wang, X., Kruhlak, M., Yarchoan, R., and Zheng, Z.M. (2011). Kaposi's sarcoma-associated herpesviral IL-6 and human IL-6 open reading frames contain miRNA binding sites and are subject to cellular miRNA regulation. J. Pathol. 225, 378–389.

Kincaid, R.P., Burke, J.M., and Sullivan, C.S. (2012). RNA virus microRNA that mimics a B-cell oncomiR. Proc. Natl. Acad. Sci. U.S.A. 109, 3077–3082.

Kiyosawa, K., Sodeyama, T., Tanaka, E., Gibo, Y., Yoshizawa, K., Nakano, Y., Furuta, S., Akahane, Y., Nishioka, K., Purcell, R.H., *et al.* (1990). Interrelationship of blood transfusion, non-A, non-B hepatitis and hepatocellular carcinoma: analysis by detection of antibody to hepatitis C virus. Hepatology 12, 671–675.

Klase, Z., Kale, P., Winograd, R., Gupta, M.V., Heydarian, M., Berro, R., McCaffrey, T., and Kashanchi, F. (2007). HIV-1 TAR element is processed by Dicer to yield a viral micro-RNA involved in chromatin remodeling of the viral LTR. BMC Mol. Biol. 8, 63.

Klase, Z.A., Sampey, G.C., and Kashanchi, F. (2013). Retrovirus infected cells contain viral microRNAs. Retrovirology 10, 15.

Ladewig, E., Okamura, K., Flynt, A.S., Westholm, J.O., and Lai, E.C. (2012). Discovery of hundreds of mirtrons in mouse and human small RNA data. Genome Res. 22, 1634–1645.

Lagos-Quintana, M., Rauhut, R., Lendeckel, W., and Tuschl, T. (2001). Identification of novel genes coding for small expressed RNAs. Science 294, 853–858.

Lagos, D., Pollara, G., Henderson, S., Gratrix, F., Fabani, M., Milne, R.S., Gotch, F., and Boshoff, C. (2010). miR-132 regulates antiviral innate immunity through suppression of the p300 transcriptional co-activator. Nat. Cell Biol. 12, 513–519.

Landgraf, P., Rusu, M., Sheridan, R., Sewer, A., Iovino, N., Aravin, A., Pfeffer, S., Rice, A., Kamphorst, A.O., Landthaler, M., *et al.* (2007). A mammalian microRNA expression atlas based on small RNA library sequencing. Cell 129, 1401–1414.

Lanford, R.E., Hildebrandt-Eriksen, E.S., Petri, A., Persson, R., Lindow, M., Munk, M.E., Kauppinen, S., and Orum, H. (2010). Therapeutic silencing of microRNA-122 in primates with chronic hepatitis C virus infection. Science 327, 198–201.

Lau, N.C., Lim, L.P., Weinstein, E.G. and Bartel, D.P. (2001). An abundant class of tiny RNAs with probable regulatory roles in *Caenorhabditis elegans*. Science 294, 858–862.

Lavanchy, D. (2004). Hepatitis B virus epidemiology, disease burden, treatment, and current and emerging prevention and control measures. J. Viral Hepat. 11, 97–107.

Law, P.T., Qin, H., Ching, A.K., Lai, K.P., Co, N.N., He, M., Lung, R.W., Chan, A.W., Chan, T.F., and Wong, N. (2013). Deep sequencing of small RNA transcriptome reveals novel non-coding RNAs in hepatocellular carcinoma. J. Hepatol. 58, 1165–1173.

Lee, R.C., and Ambros, V. (2001). An extensive class of small RNAs in *Caenorhabditis elegans*. Science 294, 862–864.

Lee, R.C., Feinbaum, R.L., and Ambros, V. (1993). The C. elegans heterochronic gene lin-4 encodes small RNAs with antisense complementarity to lin-14. Cell 75, 843–854.

Lee, S., Paulson, K.G., Murchison, E.P., Afanasiev, O.K., Alkan, C., Leonard, J.H., Byrd, D.R., Hannon, G.J., and Nghiem, P. (2011). Identification and validation of a novel mature microRNA encoded by the Merkel cell polyomavirus in human Merkel cell carcinomas. J Clin. Virol. *52*, 272–275.

Lee, Y., Jeon, K., Lee, J.T., Kim, S., and Kim, V.N. (2002). MicroRNA maturation: stepwise processing and subcellular localization. EMBO J. *21*, 4663–4670.

Lee, Y.S., Shibata, Y., Malhotra, A., and Dutta, A. (2009). A novel class of small RNAs: tRNA-derived RNA fragments (tRFs). Genes Dev. *23*, 2639–2649.

Lei, X., Bai, Z., Ye, F., Xie, J., Kim, C.G., Huang, Y., and Gao, S.J. (2010). Regulation of NF-kappaB inhibitor IkappaBalpha and viral replication by a KSHV microRNA. Nat. Cell Biol. *12*, 193–199.

Lewis, B.P., Burge, C.B., and Bartel, D.P. (2005). Conserved seed pairing, often flanked by adenosines, indicates that thousands of human genes are microRNA targets. Cell *120*, 15–20.

Li, J., Zeng, X.H., Mo, H.Y., Rolen, U., Gao, Y.F., Zhang, X.S., Chen, Q.Y., Zhang, L., Zeng, M.S., Li, M.Z., *et al.* (2007). Functional inactivation of EBV-specific T-lymphocytes in nasopharyngeal carcinoma: implications for tumor immunotherapy. PloS One 2, e1122.

Li, L.M., Hu, Z.B., Zhou, Z.X., Chen, X., Liu, F.Y., Zhang, J.F., Shen, H.B., Zhang, C.Y., and Zen, K. (2010). Serum microRNA profiles serve as novel biomarkers for HBV infection and diagnosis of HBV-positive hepatocarcinoma. Cancer Res. *70*, 9798–9807.

Liang, D., Gao, Y., Lin, X., He, Z., Zhao, Q., Deng, Q., and Lan, K. (2011a). A human herpesvirus miRNA attenuates interferon signaling and contributes to maintenance of viral latency by targeting IKKepsilon. Cell Res. *21*, 793–806.

Liang, D., Lin, X., and Lan, K. (2011b). Looking at Kaposi's Sarcoma-Associated Herpesvirus-Host Interactions from a microRNA Viewpoint. Front. Microbiol. *2*, 271.

Libri, V., Miesen, P., van Rij, R.P., and Buck, A.H. (2013). Regulation of microRNA biogenesis and turnover by animals and their viruses. Cell Mol. Life. Sci. *70*, 3525–3544.

Lieber, D., and Haas, J. (2011). Viruses and microRNAs: a toolbox for systematic analysis. Wiley Interdiscip Rev RNA 2, 787–801.

Lin, J., and Cullen, B.R. (2007). Analysis of the interaction of primate retroviruses with the human RNA interference machinery. J. Virol. *81*, 12218–12226.

Lin, X., Gudgeon, N.H., Hui, E.P., Jia, H., Qun, X., Taylor, G.S., Barnardo, M.C., Lin, C.K., Rickinson, A.B., and Chan, A.T. (2008). CD4 and CD8 T cell responses to tumour-associated Epstein–Barr virus antigens in nasopharyngeal carcinoma patients. Cancer Immunol. Immunother. CII *57*, 963–975.

Lin, X., Liang, D., He, Z., Deng, Q., Robertson, E.S., and Lan, K. (2011). miR-K12-7-5p encoded by Kaposi's sarcoma-associated herpesvirus stabilizes the latent state by targeting viral ORF50/RTA. PloS One *6*, e16224.

Lin, X., Li, X., Liang, D., and Lan, K. (2012). MicroRNAs and unusual small RNAs discovered in Kaposi's sarcoma-associated herpesvirus virions. J. Virol. *86*, 12717–12730.

Lin, Y.T., Kincaid, R.P., Arasappan, D., Dowd, S.E., Hunicke-Smith, S.P., and Sullivan, C.S. (2010). Small RNA profiling reveals antisense transcription throughout the KSHV genome and novel small RNAs. RNA *16*, 1540–1558.

Linnstaedt, S.D., Gottwein, E., Skalsky, R.L., Luftig, M.A., and Cullen, B.R. (2010). Virally induced cellular microRNA miR-155 plays a key role in B-cell immortalization by Epstein–Barr virus. J. Virol. *84*, 11670–11678.

Liu, W.H., Yeh, S.H., and Chen, P.J. (2011a). Role of microRNAs in hepatitis B virus replication and pathogenesis. Biochim. Biophys. Acta *1809*, 678–685.

Liu, X., Lv, X.B., Wang, X.P., Sang, Y., Xu, S., Hu, K., Wu, M., Liang, Y., Liu, P., Tang, J., *et al.* (2012). MiR-138 suppressed nasopharyngeal carcinoma growth and tumorigenesis by targeting the CCND1 oncogene. Cell Cycle *11*, 2495–2506.

Liu, Y., Sun, R., Lin, X., Liang, D., Deng, Q., and Lan, K. (2011b). KSHV-encoded miR-K12-11 attenuates transforming growth factor beta signaling through suppression of SMAD5. J. Virol. *86*, 1372–1381.

Lo, A.K., To, K.F., Lo, K.W., Lung, R.W., Hui, J.W., Liao, G., and Hayward, S.D. (2007). Modulation of LMP1 protein expression by EBV-encoded microRNAs. Proc. Natl. Acad. Sci. U.S.A. *104*, 16164–16169.

Lo, A.K., Dawson, C.W., Jin, D.Y., and Lo, K.W. (2012). The pathological roles of BART miRNAs in nasopharyngeal carcinoma. J. Pathol. *227*, 392–403.

Lu, C.C., Li, Z., Chu, C.Y., Feng, J., Sun, R., and Rana, T.M. (2010a). MicroRNAs encoded by Kaposi's sarcoma-associated herpesvirus regulate viral life cycle. EMBO Rep. *11*, 784–790.

Lu, F., Weidmer, A., Liu, C.G., Volinia, S., Croce, C.M., and Lieberman, P.M. (2008). Epstein–Barr virus-induced miR-155 attenuates NF-kappaB signaling and stabilizes latent virus persistence. J. Virol. *82*, 10436–10443.

Lu, F., Stedman, W., Yousef, M., Renne, R., and Lieberman, P.M. (2010b). Epigenetic regulation of Kaposi's sarcoma-associated herpesvirus latency by virus-encoded microRNAs that target Rta and the cellular Rbl2-DNMT pathway. J. Virol. *84*, 2697–2706.

Lu, J., He, M.L., Wang, L., Chen, Y., Liu, X., Dong, Q., Chen, Y.C., Peng, Y., Yao, K.T., Kung, H.F., *et al.* (2011). MiR-26a inhibits cell growth and tumorigenesis of nasopharyngeal carcinoma through repression of EZH2. Cancer Res. *71*, 225–233.

Lung, R.W., Tong, J.H., Sung, Y.M., Leung, P.S., Ng, D.C., Chau, S.L., Chan, A.W., Ng, E.K., Lo, K.W., and To, K.F. (2009). Modulation of LMP2A expression by a newly identified Epstein–Barr virus-encoded microRNA miR-BART22. Neoplasia *11*, 1174–1184.

McClure, L.V., and Sullivan, C.S. (2008). Kaposi's sarcoma herpes virus taps into a host microRNA regulatory network. Cell Host Microbe 3, 1–3.

Marquitz, A.R., Mathur, A., Nam, C.S., and Raab-Traub, N. (2011). The Epstein–Barr Virus BART microRNAs target the pro-apoptotic protein Bim. Virology *412*, 392–400.

Marquitz, A.R., Mathur, A., Shair, K.H., and Raab-Traub, N. (2012). Infection of Epstein–Barr virus in a gastric carcinoma cell line induces anchorage independence and global changes in gene expression. Proc. Natl. Acad. Sci. U.S.A. *109*, 9593–9598.

Martinez, I., Gardiner, A.S., Board, K.F., Monzon, F.A., Edwards, R.P., and Khan, S.A. (2008). Human papillomavirus type 16 reduces the expression of microRNA-218 in cervical carcinoma cells. Oncogene *27*, 2575–2582.

Matsuoka, M., and Jeang, K.T. (2007). Human T-cell leukaemia virus type 1 (HTLV-1) infectivity and cellular transformation. Nat. Rev. Cancer *7*, 270–280.

Maurin, T., Cazalla, D., Yang, S., Jr., Bortolamiol-Becet, D., and Lai, E.C. (2012). RNase III-independent microRNA biogenesis in mammalian cells. RNA *18*, 2166–2173.

Meckes, D.G., Jr., Shair, K.H., Marquitz, A.R., Kung, C.P., Edwards, R.H., and Raab-Traub, N. (2010). Human tumor virus utilizes exosomes for intercellular communication. Proc. Natl. Acad. Sci. U.S.A. *107*, 20370–20375.

Motsch, N., Pfuhl, T., Mrazek, J., Barth, S., and Grasser, F.A. (2007). Epstein–Barr virus-encoded latent membrane protein 1 (LMP1) induces the expression of the cellular microRNA miR-146a. RNA Biol. *4*, 131–137.

Munoz, N., Castellsague, X., de Gonzalez, A.B., and Gissmann, L. (2006). Chapter 1: HPV in the etiology of human cancer. Vaccine *24* (Suppl) *3*, S3/1–10.

Murakami, Y., Aly, H.H., Tajima, A., Inoue, I., and Shimotohno, K. (2009). Regulation of the hepatitis C virus genome replication by miR-199a. J. Hepatol. *50*, 453–460.

Murphy, E., Vanicek, J., Robins, H., Shenk, T., and Levine, A.J. (2008). Suppression of immediate-early viral gene expression by herpesvirus-coded microRNAs: implications for latency. Proc. Natl. Acad. Sci. U.S.A. *105*, 5453–5458.

Nachmani, D., Stern-Ginossar, N., Sarid, R., and Mandelboim, O. (2009). Diverse herpesvirus microRNAs target the stress-induced immune ligand MICB to escape recognition by natural killer cells. Cell Host Microbe *5*, 376–385.

O'Hara, A.J., Vahrson, W., and Dittmer, D.P. (2008). Gene alteration and precursor and mature microRNA transcription changes contribute to the miRNA signature of primary effusion lymphoma. Blood *111*, 2347–2353.

O'Hara, A.J., Wang, L., Dezube, B.J., Harrington, W.J., Jr., Damania, B., and Dittmer, D.P. (2009). Tumor suppressor microRNAs are underrepresented in primary effusion lymphoma and Kaposi sarcoma. Blood. *113*, 5938–5941.

Omoto, S., and Fujii, Y.R. (2005). Regulation of human immunodeficiency virus 1 transcription by nef microRNA. J. Gen. Virol. *86*, 751–755.

Omoto, S., Ito, M., Tsutsumi, Y., Ichikawa, Y., Okuyama, H., Brisibe, E.A., Saksena, N.K., and Fujii, Y.R. (2004). HIV-1 nef suppression by virally encoded microRNA. Retrovirology *1*, 44.

Ouellet, D.L., Plante, I., Landry, P., Barat, C., Janelle, M.E., Flamand, L., Tremblay, M.J., and Provost, P. (2008). Identification of functional microRNAs released through asymmetrical processing of HIV-1 TAR element. Nucleic Acids Res. *36*, 2353–2365.

Pearce, M., Matsumura, S., and Wilson, A.C. (2005). Transcripts encoding K*12*, v-FLIP, v-cyclin, and the microRNA cluster of Kaposi's sarcoma-associated herpesvirus originate from a common promoter. J. Virol. *79*, 14457–14464.

Persson, H., Kvist, A., Vallon-Christersson, J., Medstrand, P., Borg, A., and Rovira, C. (2009). The non-coding RNA of the multidrug resistance-linked vault particle encodes multiple regulatory small RNAs. Nat. Cell Biol. *11*, 1268–1271.

Pfeffer, S., Zavolan, M., Grasser, F.A., Chien, M., Russo, J.J., Ju, J., John, B., Enright, A.J., Marks, D., Sander, C., *et al.* (2004). Identification of virus-encoded microRNAs. Science *304*, 734–736.

Pfeffer, S., Sewer, A., Lagos-Quintana, M., Sheridan, R., Sander, C., Grasser, F.A., van Dyk, L.F., Ho, C.K., Shuman, S., Chien, M., *et al.* (2005). Identification of microRNAs of the herpesvirus family. Nat. Methods *2*, 269–276.

Pichler, K., Schneider, G., and Grassmann, R. (2008). MicroRNA miR-146a and further oncogenesis-related cellular microRNAs are dysregulated in HTLV-1-transformed T lymphocytes. Retrovirology *5*, 100.

Pineau, P., Volinia, S., McJunkin, K., Marchio, A., Battiston, C., Terris, B., Mazzaferro, V., Lowe, S.W., Croce, C.M., and Dejean, A. (2010). miR-221 overexpression contributes to liver tumorigenesis. Proc. Natl. Acad. Sci. U.S.A. *107*, 264–269.

Plaisance-Bonstaff, K., and Renne, R. (2011). Viral miRNAs. Methods Mol. Biol. *721*, 43–66.

Potenza, N., Papa, U., Mosca, N., Zerbini, F., Nobile, V., and Russo, A. (2011). Human microRNA hsa-miR-125a-5p interferes with expression of hepatitis B virus surface antigen. Nucleic Acids Res. *39*, 5157–5163.

Punj, V., Matta, H., Schamus, S., Tamewitz, A., Anyang, B., and Chaudhary, P.M. (2010). Kaposi's sarcoma-associated herpesvirus-encoded viral FLICE inhibitory protein (vFLIP) K13 suppresses CXCR4 expression by upregulating miR-146a. Oncogene *29*, 1835–1844.

Qin, Z., Kearney, P., Plaisance, K., and Parsons, C.H. (2009). Pivotal Advance: Kaposi's sarcoma-associated herpesvirus (KSHV)-encoded microRNA specifically induce IL-6 and IL-10 secretion by macrophages and monocytes. J. Leukoc. Biol. *87*, 25–34.

Qin, Z., Freitas, E., Sullivan, R., Mohan, S., Bacelieri, R., Branch, D., Romano, M., Kearney, P., Oates, J., Plaisance, K., *et al.* (2010). Up-regulation of xCT by KSHV-encoded microRNAs facilitates KSHV dissemination and persistence in an environment of oxidative stress. PLoS Pathog. *6*, e1000742.

Rahadiani, N., Takakuwa, T., Tresnasari, K., Morii, E., and Aozasa, K. (2008). Latent membrane protein-1 of Epstein–Barr virus induces the expression of B-cell integration cluster, a precursor form of microRNA-*155*, in B lymphoma cell lines. Biochem. Biophys. Res. Commun. *377*, 579–583.

Ramakrishnan, R., Donahue, H., Garcia, D., Tan, J., Shimizu, N., Rice, A.P., and Ling, P.D. (2011). Epstein–Barr virus BART9 miRNA modulates LMP1 levels and affects growth rate of nasal NK T cell lymphomas. PloS One *6*, e27271.

Randall, G., Panis, M., Cooper, J.D., Tellinghuisen, T.L., Sukhodolets, K.E., Pfeffer, S., Landthaler, M., Landgraf, P., Kan, S., Lindenbach, B.D., *et al.* (2007). Cellular cofactors affecting hepatitis C virus infection and replication. Proc. Natl. Acad. Sci. U.S.A. *104*, 12884–12889.

Reinhart, B.J., Slack, F.J., Basson, M., Pasquinelli, A.E., Bettinger, J.C., Rougvie, A.E., Horvitz, H.R., and Ruvkun, G. (2000). The 21-nucleotide let-7 RNA regulates developmental timing in *Caenorhabditis elegans*. Nature *403*, 901–906.

Ruby, J.G., Jan, C.H., and Bartel, D.P. (2007). Intronic microRNA precursors that bypass Drosha processing. Nature *448*, 83–86.

Saito, I., Miyamura, T., Ohbayashi, A., Harada, H., Katayama, T., Kikuchi, S., Watanabe, Y., Koi, S., Onji, M., Ohta, Y., *et al.* (1990). Hepatitis C virus infection is associated with the development of hepatocellular carcinoma. Proc. Natl. Acad. Sci. U.S.A. *87*, 6547–6549.

Samols, M.A., Hu, J., Skalsky, R.L., and Renne, R. (2005). Cloning and identification of a microRNA cluster within the latency-associated region of Kaposi's sarcoma-associated herpesvirus. J. Virol. *79*, 9301–9305.

Saraiya, A.A., and Wang, C.C. (2008). snoRNA, a novel precursor of microRNA in Giardia lamblia. PLoS Pathog. *4*, e1000224.

Schopman, N.C., Willemsen, M., Liu, Y.P., Bradley, T., van Kampen, A., Baas, F., Berkhout, B., and Haasnoot, J. (2012). Deep sequencing of virus-infected cells reveals HIV-encoded small RNAs. Nucleic Acids Res. *40*, 414–427.

Seo, G.J., Chen, C.J., and Sullivan, C.S. (2009). Merkel cell polyomavirus encodes a microRNA with the ability to autoregulate viral gene expression. Virology *383*, 183–187.

Sham, J.S., Wei, W.I., Zong, Y.S., Choy, D., Guo, Y.Q., Luo, Y., Lin, Z.X., and Ng, M.H. (1990). Detection of subclinical nasopharyngeal carcinoma by fibreoptic endoscopy and multiple biopsy. Lancet *335*, 371–374.

Shapiro, J.S., Varble, A., Pham, A.M., and Tenoever, B.R. (2010). Noncanonical cytoplasmic processing of viral microRNAs. RNA *16*, 2068–2074.

Shapiro, J.S., Langlois, R.A., Pham, A.M., and Tenoever, B.R. (2012). Evidence for a cytoplasmic microprocessor of pri-miRNAs. RNA *18*, 1338–1346.

Shi, W., Alajez, N.M., Bastianutto, C., Hui, A.B., Mocanu, J.D., Ito, E., Busson, P., Lo, K.W., Ng, R., Waldron, J., *et al.* (2010). Significance of Plk1 regulation by miR-100 in human nasopharyngeal cancer. Int. J. Cancer *126*, 2036–2048.

Skalsky, R.L., Samols, M.A., Plaisance, K.B., Boss, I.W., Riva, A., Lopez, M.C., Baker, H.V., and Renne, R. (2007). Kaposi's sarcoma-associated herpesvirus encodes an ortholog of miR-155. J. Virol. *81*, 12836–12845.

Soulier, J., Grollet, L., Oksenhendler, E., Cacoub, P., Cazals-Hatem, D., Babinet, P., d'Agay, M.F., Clauvel, J.P., Raphael, M., Degos, L., *et al.* (1995). Kaposi's sarcoma-associated herpesvirus-like DNA sequences in multicentric Castleman's disease. Blood *86*, 1276–1280.

Stern-Ginossar, N., Elefant, N., Zimmermann, A., Wolf, D.G., Saleh, N., Biton, M., Horwitz, E., Prokocimer, Z., Prichard, M., Hahn, G., *et al.* (2007). Host immune system gene targeting by a viral miRNA. Science *317*, 376–381.

Sullivan, C.S., Grundhoff, A.T., Tevethia, S., Pipas, J.M., and Ganem, D. (2005). SV40-encoded microRNAs regulate viral gene expression and reduce susceptibility to cytotoxic T cells. Nature *435*, 682–686.

Taft, R.J., Glazov, E.A., Lassmann, T., Hayashizaki, Y., Carninci, P., and Mattick, J.S. (2009). Small RNAs derived from snoRNAs. RNA *15*, 1233–1240.

Takada, K. (2000). Epstein–Barr virus and gastric carcinoma. Mol. Pathol. *53*, 255–261.

Tang, S., Patel, A., and Krause, P.R. (2009). Novel less-abundant viral microRNAs encoded by herpes simplex virus 2 latency-associated transcript and their roles in regulating ICP34.5 and ICP0 mRNAs. J. Virol. *83*, 1433–1442.

Thomas, M., Boname, J.M., Field, S., Nejentsev, S., Salio, M., Cerundolo, V., Wills, M., and Lehner, P.J. (2008). Down-regulation of NKG2D and NKp80 ligands by Kaposi's sarcoma-associated herpesvirus K5 protects against NK cell cytotoxicity. Proc. Natl. Acad. Sci. U.S.A. *105*, 1656–1661.

Tomita, M., Tanaka, Y., and Mori, N. (2012). MicroRNA miR-146a is induced by HTLV-1 tax and increases the growth of HTLV-1-infected T-cells. Int. J. Cancer *130*, 2300–2309.

Tsai, Y.H., Wu, M.F., Wu, Y.H., Chang, S.J., Lin, S.F., Sharp, T.V., and Wang, H.W. (2009). The M type K15 protein of Kaposi's sarcoma-associated herpesvirus regulates microRNA expression via its SH2-binding motif to induce cell migration and invasion. J. Virol. *83*, 622–632.

Umbach, J.L., and Cullen, B.R. (2010). In-depth analysis of Kaposi's sarcoma-associated herpesvirus microRNA expression provides insights into the mammalian microRNA-processing machinery. J. Virol. *84*, 695–703.

Umbach, J.L., Kramer, M.F., Jurak, I., Karnowski, H.W., Coen, D.M., and Cullen, B.R. (2008). MicroRNAs expressed by herpes simplex virus 1 during latent infection regulate viral mRNAs. Nature *454*, 780–783.

Umbach, J.L., Yen, H.L., Poon, L.L., and Cullen, B.R. (2010). Influenza A virus expresses high levels of an unusual class of small viral leader RNAs in infected cells. MBio. *1*, e00204-10.

Wang, X., Tang, S., Le, S.Y., Lu, R., Rader, J.S., Meyers, C., and Zheng, Z.M. (2008). Aberrant expression of oncogenic and tumor-suppressive microRNAs in cervical cancer is required for cancer cell growth. PLoS One *3*, e2557.

Wang, X., Wang, H.K., McCoy, J.P., Banerjee, N.S., Rader, J.S., Broker, T.R., Meyers, C., Chow, L.T., and Zheng, Z.M. (2009). Oncogenic HPV infection interrupts the expression of tumor-suppressive miR-34a through viral oncoprotein E6. RNA *15*, 637–647.

Wong, A.M., Kong, K.L., Tsang, J.W., Kwong, D.L., and Guan, X.Y. (2012). Profiling of Epstein–Barr virus-encoded microRNAs in nasopharyngeal carcinoma reveals potential biomarkers and oncomirs. Cancer *118*, 698–710.

Wong, Q.W., Lung, R.W., Law, P.T., Lai, P.B., Chan, K.Y., To, K.F., and Wong, N. (2008). MicroRNA-223 is commonly repressed in hepatocellular carcinoma and potentiates expression of Stathmin1. Gastroenterology *135*, 257–269.

Wu, Y.H., Hu, T.F., Chen, Y.C., Tsai, Y.N., Tsai, Y.H., Cheng, C.C., and Wang, H.W. (2011). The manipulation of miRNA-gene regulatory networks by KSHV induces endothelial cell motility. Blood *118*, 2896–2905.

Xia, H., Cheung, W.K., Sze, J., Lu, G., Jiang, S., Yao, H., Bian, X.W., Poon, W.S., Kung, H.F., and Lin, M.C. (2010). miR-200a regulates epithelial-mesenchymal to stem-like transition via ZEB2 and beta-catenin signaling. J. Biol. Chem. *285*, 36995–37004.

Yang, I.V., Wade, C.M., Kang, H.M., Alper, S., Rutledge, H., Lackford, B., Eskin, E., Daly, M.J., and Schwartz, D.A. (2009). Identification of novel genes that mediate innate immunity using inbred mice. Genetics *183*, 1535–1544.

Yang, J.S., and Lai, E.C. (2011). Alternative miRNA Biogenesis Pathways and the Interpretation of Core miRNA Pathway Mutants. Mol. Cell. *43*, 892–903.

Yang, J.S., Maurin, T., Robine, N., Rasmussen, K.D., Jeffrey, K.L., Chandwani, R., Papapetrou, E.P., Sadelain, M., O'Carroll, D., and Lai, E.C. (2010). Conserved vertebrate mir-451 provides a platform for Dicer-independent, Ago2-mediated microRNA biogenesis. Proc. Natl. Acad. Sci. U.S.A. *107*, 15163–15168.

Yeung, M.L., Yasunaga, J., Bennasser, Y., Dusetti, N., Harris, D., Ahmad, N., Matsuoka, M., and Jeang, K.T. (2008). Roles for microRNAs, miR-93 and miR-130b, and tumor protein 53-induced nuclear protein 1 tumor suppressor in cell growth dysregulation by human T-cell lymphotrophic virus 1. Cancer Res. *68*, 8976–8985.

Zhang, G.L., Li, Y.X., Zheng, S.Q., Liu, M., Li, X., and Tang, H. (2010). Suppression of hepatitis B virus replication by microRNA-199a-3p and microRNA-210. Antiviral Res. *88*, 169–175.

Zhang, X., Zhang, E., Ma, Z., Pei, R., Jiang, M., Schlaak, J.F., Roggendorf, M., and Lu, M. (2011). Modulation of hepatitis B virus replication and hepatocyte differentiation by MicroRNA-1. Hepatology *53*, 1476–1485.

Zhao, Y., Yao, Y., Xu, H., Lambeth, L., Smith, L.P., Kgosana, L., Wang, X., and Nair, V. (2009). A functional MicroRNA-155 ortholog encoded by the oncogenic Marek's disease virus. J. Virol. *83*, 489–492.

Zheng, Z.M., and Wang, X. (2011). Regulation of cellular miRNA expression by human papillomaviruses. Biochim. Biophys. Acta *1809*, 668–677.

Zhu, J.Y., Pfuhl, T., Motsch, N., Barth, S., Nicholls, J., Grasser, F., and Meister, G. (2009). Identification of novel Epstein–Barr virus microRNA genes from nasopharyngeal carcinomas. J. Virol. *83*, 3333–3341.

MicroRNAs as Cancer Biomarkers

11

David Otaegui and Charles H. Lawrie

Abstract

Despite having been only formally recognized for just over ten years, microRNAs (miRNAs) have become one of the hottest topics in biology. In this short time these endogenously produced small (19–22 nt) ssRNA post-transcriptional regulators have been found to play crucial roles in many, if not all, physiological and pathological processes including cancer. The importance of miRNAs to carcinogenesis is implied by the fact that many miRNAs are encoded at cancer-associated regions of the genome, and there is now overwhelming evidence that aberrant expression of miRNAs is a ubiquitous characteristic of malignancy. As well as clear therapeutic implications for miRNAs, the remarkable stability of these molecules coupled with their ability to discriminate between different cancers to a degree that surpasses traditional genes, suggests great potential as cancer biomarkers. In this chapter we review the accumulating evidence for this assertion, and provide an review of the major techniques used in miRNA biomarker studies (e.g. microarray, qRT-PCR and next-generation sequencing etc.), discussing their relative strengths and weaknesses.

Introduction

A brief history of small RNAs

The central dogma of molecular biology, first postulated by Francis Crick in 1958 and later refined in 1970, states that biological information flows unidirectionally from DNA to RNA to protein (Crick, 1970). This view implies that non-coding RNA (ncRNA) has little or no intrinsic value, despite accounting for more than 90% of eukaryotic transcriptional output (Mattick, 2001). Consequently, it is perhaps not surprising that microRNAs (miRNAs) were unknown to the scientific community until very recently. Indeed it was only in 1993 when the first miRNA, lin-4, a *C. elegans* developmental timing regulator, was announced simultaneously by the Ambros and Ruvkun laboratories (Lee *et al.*, 1993; Wightman *et al.*, 1993). However, as lin-4 has no clear homologue outside of worms, the biological significance of this discovery was not realized until 7 years later, when let-7, another *C. elegans* developmental regulator, was discovered by the Ruvkun laboratory (Reinhart *et al.*, 2000). Let-7 shared many of the characteristics of lin-4 as it encodes for a small (21 nt) ncRNA transcript that negatively regulates the mRNA of lin family members through complementary RNA–RNA interactions at the 3'UTR of these genes. Unlike lin-4, however, the sequence of let-7 was found to be conserved in most eukaryotic organisms (Pasquinelli *et al.*, 2000; Lagos-Quintana *et al.*, 2001). Together these discoveries instigated the start of the miRNA revolution, a term first coined by Lee and Ambros in 2001 (Lee and Ambros, 2001). Since this time, over twenty-five thousand miRNAs (including more than 2000 human miRNAs) have been identified from a diverse range of more than 190 different species including algae, plants, mycetozoa, arthropods, nematodes, protozoa, vertebrates, plants and viruses (Griffiths-Jones *et al.*, 2006). For a current list of annotated miRNAs the reader should consult the miRBase database (http://www.mirbase.org/).

MiRNAs primarily function as post-transcriptional (negative) regulators of gene expression via binding to complementary sequences located mainly within the UTRs of target genes. Because a single miRNA can target several hundred genes, it is believed ~60% of all human genes are a target for miRNA regulation (Friedman *et al.*, 2009). In addition, a single target gene often contains binding sites for multiple miRNAs that can bind cooperatively (Lewis *et al.*, 2003), allowing miRNAs to form complex regulatory control networks. Perhaps unsurprisingly miRNAs have been shown to play key regulatory roles in virtually every aspect of biology including physiological and pathological processes most notably in cancer, the theme of this chapter.

Biogenesis and function of miRNAs

The majority of human miRNAs are encoded within introns of coding mRNAs whilst others are located exgenically, in non-coding mRNAs or within the 3′UTR sequence of coding mRNA (Rodriguez *et al.*, 2004). MiRNAs are transcribed as 5′-capped large polyadenylated transcripts (pri-miRNA) primarily in a Pol II-dependent manner, although the involvement of Pol III transcription has also been postulated for miRNAs encoded within Alu repeat sequences (Borchert *et al.*, 2006).

Approximately 40% of human miRNAs are co-transcribed as clusters encoding multiple miRNA sequences in a single pri-microRNA transcript (Altuvia *et al.*, 2005; Hertel *et al.*, 2006). Pri-miRNAs are cleaved within the nucleus by Drosha, an RNaseIII-type nuclease, to form 60–110 nt hairpin structures (pre-miRNA) (Fig. 11.1). Drosha by itself possesses little enzymatic activity and requires the cofactor DiGeorge syndrome critical region 8 gene (DGCR8) in humans (Pasha in *Drosophila*) to form the microprocessor complex (Yeom *et al.*, 2006).

Once produced, pre-miRNAs are exported from the nucleus to the cytoplasm by Exportin-5 in a Ran-GTP-dependent manner (Zeng,

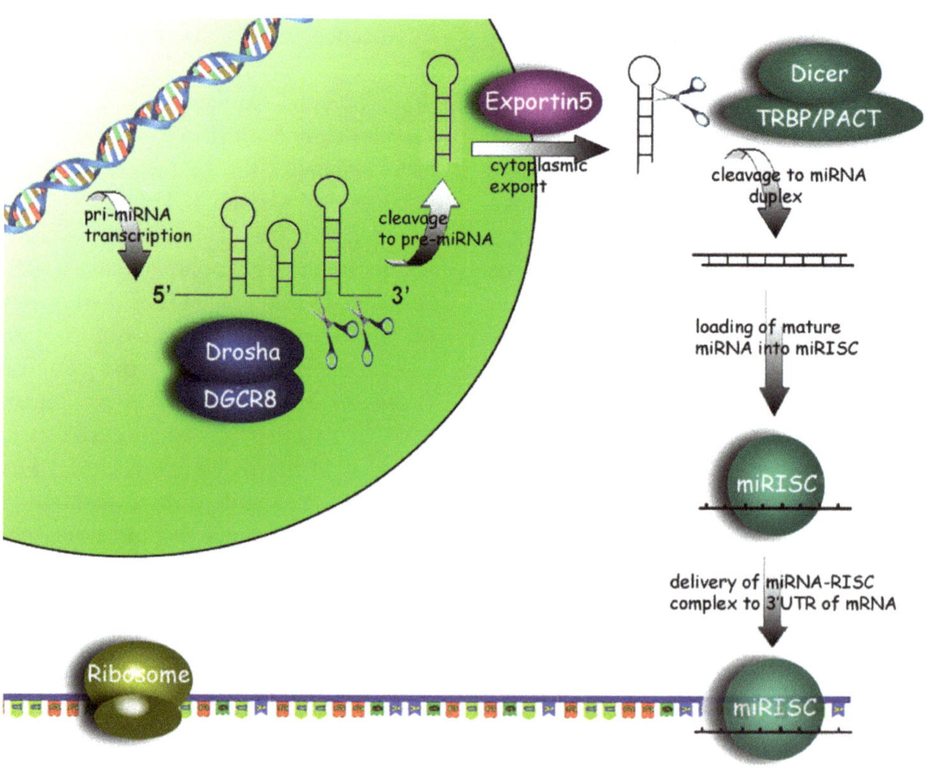

Figure 11.1 Schematic diagram of the canonical miRNA biosynthetic-pathway. Reproduced from Lawrie (2007).

2006). The cytoplasmic pre-miRNA is further cleaved by Dicer, another RNaseIII-type enzyme, to form an asymmetric duplex intermediate (miRNA:miRNA*), consisting of the mature miRNA sequence and the antisense miRNA passenger strand (miRNA*). Similar to Drosha, cofactors such as TRBP and PACT (in humans) are necessary for Dicer activity (Lee *et al.*, 2006). The miRNA:miRNA* duplex is, in turn, loaded into the miRISC complex in which Argonaut (Ago) proteins are the key effector molecules. The strand that becomes the active mature miRNA appears to be dependent upon which has the lowest free energy 5′ end and is retained by the miRISC complex, whilst the passenger strand is generally believed to be degraded by an unknown nuclease (Khvorova *et al.*, 2003; Schwarz *et al.*, 2003). It should be noted, however, that many miRNA passenger strands are also capable of silencing target transcripts and probably play a more important biological role than was previously realized (Okamura *et al.*, 2008; Ghildiyal *et al.*, 2010).

The loaded miRISC is guided by the mature miRNA sequence (19–24 nt) to complementary sequences located primarily within the 3′UTR of the target gene mRNA, although binding sites have additionally been identified in both 5′UTR (Lytle *et al.*, 2007) and coding regions of genes (Tay *et al.*, 2008). In contrast to plant miRNAs that contain extensive regions of complementarity with their target genes, animal miRNAs are only partially complementary and have a propensity to recognize targets via 6–8 nt 'seed' sequences, usually located at nt position 2–8 of the 5′-end of the miRNA (Bartel, 2009), although sometimes also in the centre of the miRNA sequence as well (Shin *et al.*, 2010). There are rare examples of animal miRNAs (e.g. miR-196 and *HOXB8*) that do share near-perfect complementarity resulting in direct cleavage of the mRNA (Yekta *et al.*, 2004).

Whilst the vast majority of animal miRNAs are generated by the canonical miRNA biosynthetic pathway described above (Ghildiyal and Zamore, 2009) (Fig. 11.1), alternative Drosha-independent and Dicer-independent pathways also exist (for detailed review see Yang and Lai, 2011).

Although repression of translation without mRNA degradation was originally believed to be the *modus operandi* of animal miRNAs, the situation appears to be more complex than previously thought, and there is now compelling evidence that miRNAs can also affect transcriptional levels through deadenylation, degradation and/or destabilization of target mRNAs (Giraldez *et al.*, 2006). Indeed, it has been suggested that translational inhibition has only a modest role to play in miRNA function and that mRNA destabilization is the predominant mechanism used to inhibit target genes in mammals (Guo *et al.*, 2010).

MiRNAs as oncogenes and tumour suppressors

The potential importance of miRNAs in cancer is implied by the finding that the majority of human miRNAs are located at cancer-associated genomic regions (Calin *et al.*, 2004b). The first indication that dysregulation of miRNAs was associated with cancer came from the seminal publication by Calin *et al.* (2002), that made the connection between 13q14, a frequently deleted locus in chronic lymphocytic leukaemia (CLL), and down-regulation of the miR-15a/16 cluster that is encoded within this region. There is now overwhelming evidence that dysfunctional expression of miRNAs is a common, if not ubiquitous, hallmark of cancer in general (Lawrie, 2008; Iorio and Croce, 2009).

MiRNAs can function as oncogenes by activating malignant potential, or as tumour suppressors by blocking the cell's malignant potential. Functional evidence for oncogenic potential comes from multiple animal models whereby overexpression of specific miRNAs leads to increased tumour development. For example, overexpression of the miR-17-92 cluster accelerated tumour development in a mouse B-cell *myc* lymphoma model and increased the tumour resistance to apoptosis (He *et al.*, 2005). Similarly, miR-21 and miR-155 overexpression leads to spontaneous development of lymphomas in mice (Medina *et al.*, 2010; Babar *et al.*, 2012), whilst miR-221 overexpression in the liver led to increased hepatocarcinoma levels (Callegari *et al.*, 2012). In contrast, deletion of the region encoding for tumour-suppressor miRNAs miR-15a and miR-16 in mice led to the development of a disease resembling CLL (Klein *et al.*, 2010; Lia *et al.*, 2012). MiRNAs can even function as both oncogene and tumour suppressor depending upon their cellular context. For example miR-222 is overexpressed in hepatocarcinoma where it targets tumour-suppressor PTEN (Garofalo *et al.*, 2009), but down-regulated in erythroblastic leukaemias where it targets the oncogenic KIT protein (Felli *et al.*, 2005).

The reasons for dysfunctional expression of miRNAs in cancers are numerous, and can include chromosomal aberrations, epigenetic deregulation, aberrant expression of transcription factors that regulate promoter regions of miRNAs, as well as factors that change miRNA biosynthesis and/or function (Croce, 2009).

MicroRNAs as cancer biomarkers

MiRNAs show perhaps their greatest, and certainly most immediate clinical potential, as novel biomarkers of cancer diagnosis and prognosis, and as predictors of treatment response. The traditional 'gold standard' of cancer diagnosis, the histological examination of tumour tissue, is often limited by the availability of adequately preserved biopsy material and possibly subjective interpretation by pathologists. The supplementation of traditional morphological classifications with the development of molecular methods, has greatly enhanced the identification of cancer subtypes in recent years. In breast cancer for example, a 97-gene expression profile has been demonstrated to classify histological grade better than lymph node status and tumour size (Sotiriou *et al.*, 2006), whilst in diffuse large B-cell lymphoma (DLBCL) a gene expression signature has led to the identification of two distinct subtypes that are prognostically and mechanistically different (Alizadeh *et al.*, 2000).

MiRNA expression profiling can distinguish cancers according to diagnosis and developmental stage of the tumour to a greater degree of accuracy than traditional gene expression analysis, even discriminating between cancers that are poorly separated histologically (Lu *et al.*, 2005). Additionally, an especially useful characteristic of miRNAs as biomarkers is their remarkable stability which means that they can be robustly measured from routinely prepared formalin-fixed paraffin embedded (FFPE) biopsy material (Lawrie *et al.*, 2007). Consequently retrospective miRNA studies can be carried out on the vast collections of pathology laboratories worldwide, something that is not feasible when investigating traditional genes. Indeed, the vast majority of miRNA studies have been carried out on FFPE material, and this characteristic has greatly facilitated the rapid pace of discovery of miRNA involvement in all but the rarest of malignant pathologies.

The first demonstration that miRNAs could be useful as cancer (diagnostic) biomarkers came in 2002 from Carlo Croce's lab who showed that CLL cases containing the 13q14 deletion could be delineated on the basis of miR-15a/16-1 expression levels (Calin *et al.*, 2002). The earliest study to show a prognostic significance for miRNAs came from Takamizawa and colleagues who in 2004 demonstrated that lung cancer patients with low let-7 levels had significantly shorter post-resection survival times (Takamizawa *et al.*, 2004). In the same year the Croce lab reported the first whole genome miRNA microarray approach to identify diagnostic miRNAs (Calin *et al.*, 2004a), which was expanded in 2005 to include prognostic indicators (Calin *et al.*, 2005). Since this time there have been nearly 5000 publications on the use of miRNAs as cancer biomarkers [source: Pubmed search (11/03/13) string='(microRNA AND cancer) AND (prognosis OR diagnosis OR biomarker)']. In the interests of space we will not attempt to review all of these publications but instead concentrate on those studies that made significant impact on the most common forms of cancer.

MiRNAs as diagnostic markers

The identification of miRNAs that are aberrantly expressed in tumour cells compared with their normal counterpart cells was the initial goal of many early miRNA studies (Lu *et al.*, 2005; Lawrie *et al.*, 2008b). However, whilst of great scientific interest, this approach has limited diagnostic utility as studies are almost exclusively retrospective and based upon the ability to confirm previously made clinical diagnoses. Much more useful is the ability of miRNAs to differentiate between diagnoses of benign and malignant conditions, between confounding pathologies, and to confirm or add to the molecular classification of tumours which is often limited by the prerequisite for fresh or frozen material. For example, a seven miRNA signature that could distinguish between pancreatic ductal adenocarcinomas and benign conditions was identified (Szafranska *et al.*, 2008). Importantly, these findings were independently tested in a multicentre trial and found to be more accurate than traditional cytology (Karim *et al.*, 2008). We reported that blood miR-223 levels could distinguish between patients with Sezary syndrome and those that had the confounding diagnosis of non-erythrodermic mycosis fungoides with >90% accuracy (Ballabio *et al.*, 2010). The miR-17-92 cluster has been demonstrated to be a useful diagnostic differentiator between the potentially conflicting diagnostic classifications of GC-DLBCL and grade 3 FL cases (Fassina *et al.*, 2012), and miRNAs may also be useful for a more accurate classification of the group of B-cell lymphomas with intermediate features between DLBCL and BL (Leucci *et al.*, 2008). MiRNAs can accurately classify breast cancer cases according to the molecular subtype (e.g. luminal A, luminal B, basal-like, etc.) (Blenkiron *et al.*, 2007), and DLBCL cases as either ABC- or GC-molecular subtype (Lawrie *et al.*, 2007, 2009b).

MiRNAs as prognostic indicators

One of the most clinically useful features of miRNAs is their potential as prognostic indicators for cancer (Table 11.1). This can be a result of acting as a surrogate marker of a previously recognized prognostic marker (e.g. clinical, cytogenetic or molecular) or as an independent prognosticator. The majority of studies to date fall within the former category. In one of the first papers to examine the potential of miRNAs in prognosis, Calin *et al.* (2005) identified a 13-miRNA signature on the basis of comparisons between CLL groups with typically good outcome (low ZAP70, mutated IgH) with those with a typically poor outcome (high ZAP70, unmutated IgH). Analysis of 93 breast cancer samples identified 31 miRNAs associated with clinical factors and outcome (Blenkiron *et al.*, 2007); importantly, six of these were related with outcome in an independent study (Rothe *et al.*, 2011). In melanoma, expression of miR-338, miR-565 and miR-193a has been linked with a V660E missense mutation of *BRAF* (Caramuta *et al.*, 2010), and miR-155 expression with *FLT3* mutations in acute myeloid leukaemia (AML) (Garzon *et al.*, 2008a). MiRNAs have also been frequently linked to specific chromosomal abnormalities, particularly in leukaemias. For example in CLL miRNA expression levels have been associated with 17p and 11q deletions, aberrations that frequently indicate poor prognosis (Lawrie *et al.*, 2009a; Visone *et al.*, 2009). In myeloma we identified signatures associated with the most common chromosomal abnormalities including t(4;14), t(11;14), del(13q) and hyperdiploidy (Chi *et al.*, 2010).

Other studies have taken a different approach by retrospectively assigning miRNA signatures to groups of patients on the basis of prognosis in the absence of other pre-defined markers. In lung cancer for example, in a study that examined 70 patient samples, high expression levels of members of the miR-183 family have been linked with poor outcome (Zhu *et al.*, 2011). Low miR-191 and high miR-193a levels were correlated with poor survival in melanoma (Caramuta *et al.*, 2010), and low levels of miR-26 in patients were indicative of poorer survival but a better response to interferon therapy for liver cancer patients (Ji *et al.*, 2009). In gastric cancer a seven miRNA signature was associated with prognostic outcome (Li *et al.*, 2010), whilst in myeloma, a miRNA signature was derived that could predict prognostic outcome independently of other chromosomal or clinical parameters (Zhou *et al.*, 2010).

MiRNAs as indicators of treatment response

There has been a great deal of interest in the use of miRNAs as indicators of response to treatment in various cancers. The NCI-60 panel, a standardized group of cell lines covering many of the most common forms of cancer, and widely used for drug screening, demonstrated a strong correlation between miRNA profile and sensitivity to anticancer compounds (Blower *et al.*, 2007). The observation that miRNAs can have opposing effects with the same anticancer drug depending upon the cell type suggests that relationship between miRNAs and drug resistance is highly complex. MiRNAs that effect drug sensitivity can target many pathways including survival,

Table 11.1 Table of miRNA biomarkers associated with multiple cancer types

miRNA	Cancer	Biomarker	Reference
miR-155	Breast	Serum↑, tumour↑	Sun *et al.* (2012), Zheng *et al.* (2012)
	Oesophageal	Plasma↓	Liu *et al.* (2012a)
	CML	Tumour cells↓	Rokah *et al.* (2012)
	T-ALL	Plasma↑ (poor prognosis)	Ishihara *et al.* (2012)
	AML	FLT3 mutation	Garzon *et al.* (2008b)
miR-21	Lung	Poor prognosis (↓in tumour, ↑in serum)	Voortman *et al.* (2010), Liu *et al.* (2012b)
	Breast	Poor prognosis↑	Yan *et al.* (2008)
	Osteosarcoma	Poor prognosis↑ (serum)	Yuan *et al.* (2012)
	Prostate	Poor prognosis↑ (serum)	Zhang *et al.* (2011)
	Colorectal	Poor prognosis↑ (serum and tumour)	Schetter *et al.* (2008)
	CLL	Poor prognosis↑	Rossi *et al.* (2010)
let-7	Lung	Poor prognosis↓	Takamizawa *et al.* (2004)
	Renal	Prognosis	Heinzelmann *et al.* (2011)
	MDS	Prognosis (serum)	Zuo *et al.* (2011)
miR-15a/16	CLL	Poor prognosis↓	Calin *et al.* (2002)
	Myeloma	Poor prognosis↑	Gao *et al.* (2012)
	ALL	Prognosis	Kaddar *et al.* (2009)
	Melanoma	Poor prognosis↑	Satzger *et al.* (2010)
	MDS	Prognosis (serum)	Zuo *et al.* (2011)
miR-210	Breast	Poor prognosis↑ (tumour and plasma)	Hong *et al.* (2012)
	Osteocarcinoma	Poor prognosis↑	Cai *et al.* (2013)
miR-150	DLBCL	Poor prognosis↓	Alencar *et al.* (2011)
	AML	Poor prognosis↓	Rossi *et al.* (2010)
	MDS	Poor prognosis↓	Hussein *et al.* (2010)
	Oesophageal	Poor prognosis↓	Yokobori *et al.* (2013)
miR-181a	DLBCL	Poor prognosis↓	Alencar *et al.* (2011)
	CLL	Poor prognosis↓	Visone *et al.* (2011)
	AML	Poor prognosis↓	Li *et al.* (2012)
	Gastric	Prognosis	Kim *et al.* (2011)
miR-29	MCL	Prognosis	Zhao *et al.* (2010)
	CLL	Poor prognosis↓	Pekarsky *et al.* (2006)
	Colorectal	Prognosis (serum)	Wang and Gu (2012)
	AML	Poor prognosis↓	Zhu *et al.* (2013)
	Hepatocarcinoma	Poor prognosis↓	Xiong *et al.* (2010)

apoptotic response, DNA repair systems, drug targets and drug transporters and metabolizing enzymes (Fojo, 2007). For a detailed overview of the many studies that have been carried out on miRNAs and drug resistance please see (Giovannetti *et al.*, 2012).

MiRNAs as non-invasive cancer biomarkers

The standard procedure for most cancer diagnoses is the histopathological review of tumour material obtained by invasive biopsy; a procedure that is typically expensive, uncomfortable and sometimes risky for patients. The search for non-invasive tools for the diagnosis and management of cancer has led to great interest in the field of circulating nucleic acids in plasma and serum (Anker *et al.*, 2003; Pathak *et al.*, 2006). Mandel and Metais first reported the existence of cell-free nucleic acids in the plasma of both normal and diseased individuals in 1948 (Mandel and Metais, 1948). It was not until 1975, however, that a relationship between elevated levels of circulating DNA and cancer was hypothesized (Leon *et al.*, 1975), and in 1999 that the potential of circulating RNA was first proposed (Lo *et al.* 1999). However, with a few notable exceptions (e.g. pre-natal diagnosis), the field of circulating nucleic acids has failed to live up to expectations. Unlike other RNA species (including mRNA), the vast majority of which are degraded by high levels of RNases found in the blood (Duttagupta *et al.*, 2011), miRNAs appear stable in the blood and are surprisingly resistant to fragmentation by either chemical or enzymatic agents (Mitchell *et al.*, 2008). We demonstrated for the first time that miRNAs were present in the blood (serum/plasma) of cancer patients at differential levels from healthy controls (Lawrie *et al.*, 2008a). Subsequently, many publications have reported the potential usefulness of miRNAs as non-invasive biomarkers of cancer (and other disease), and miRNAs have now been detected in many biological fluids including plasma, serum, tears, urine, cerebral spinal fluid, breast milk and saliva (Weber *et al.*, 2010).

In the following sections we review some of the most influential studies that have used miRNAs as diagnostic, prognostic and treatment response biomarkers both tumour based and non-tumour based in the most common forms of cancer.

Breast cancer

The first study to address miRNA involvement in breast cancer by Iorio and collegues identified 29 differentially expressed miRNAs that distinguished between normal tissue and tumours (Iorio *et al.*, 2005). A few years later it was demonstrated that miRNA expression could differentiate between samples of known molecular subtype (luminal A, luminal B, basal-like, normal-like and HER+), although the miRNA signature did not predict tumour status (Blenkiron *et al.*, 2007). Members of the miR-200 family are found at high levels in luminal type breast cancer samples, whilst miR-221/222 are overexpressed in triple negative-type tumour samples (Shah and Calin, 2011; Howe *et al.*, 2012). More recently miRNA signatures have been derived from 29 early-stage cases that could distinguish between ER, PR and HER2 positive cases with 100% accuracy compared to immunohistochemical staining (Lowery *et al.*, 2009). In this study miR-210 was identified as a strong prognostic indicator and its sole expression value was proposed to have the same level of predictive power as a 76 gene mRNA signature. In another study, miRNA expression patterns were profiled between normal and tumour breast tissue and visualized by *in situ* hybridization, and although miRNAs were differentially expressed between the two tissue types, expression patterns did not correlate exclusively with ER or HER2 status (Sempere *et al.*, 2007).

The expression levels of miR-21, miR-126, miR-155, miR-199a and miR-335 in sera have been associated with clinicopathologic features of breast cancer, including histological tumour grade and receptor status (Wang *et al.*, 2010). Circulating levels of miR-214 were suggested to have diagnostic potential in breast cancer patients (Schwarzenbach *et al.*, 2012), and levels of circulating miR-21 to have utility in detecting progression of early stage breast cancer (Asaga *et al.*, 2011). In another study, circulating blood levels of miR-122, miR-10b, miR-34a and miR-155 were associated with the presence of overt metastasis (Roth *et al.*, 2010b; Wu *et al.*, 2012). Interestingly serum concentrations of the same miRNAs are also significantly elevated in the sera of patients with ovarian and lung cancer (Roth *et al.*, 2010a; Roth *et al.*, 2011). Heneghan *et al.* found an association between high serum levels of miR-10b and the oestrogen receptor status of breast cancer patients (Heneghan *et al.*, 2010). Additionally, it has been suggested that plasma miR-210 levels could be used for monitoring response of breast cancer patients to trastuzumab (Herceptin™) (Jung *et al.*, 2011).

Prostate cancer

The first study to look at miRNAs in prostate cancer identified 51 aberrantly expressed miRNAs (Porkka *et al.*, 2007). Since then many studies have confirmed and expanded this research (reviewed in Fang and Gao, 2013). MiRNA expression has been linked with many prognostic indicators in prostate cancer including androgen responsiveness and Gleason score [miR-331 (Epis *et al.*, 2009)]. In particular, a link between miR-141, miR-143 and miR-145 progression and Gleason score has been established (Peng *et al.*, 2011). Metastasis in prostate cancer has also been correlated with miRNA expression (Watahiki *et al.*, 2011). A recent study demonstrated that a blood test based upon a combination of the levels of five circulating miRNAs (let-7e, let-7c, miR-30c, miR-622 and miR-1285) could effectively differentiate between prostate cancer and benign prostatic hyperplasia (BPH), as well as healthy controls (Chen *et al.*, 2012). Plasma levels of miR-20a, miR-21, miR-145, and miR-221 were also suggested to be useful in distinguishing between prostate cancer patients of varying aggressiveness of tumour (Shen *et al.*, 2012). MiR-125b and miR-141 have also been shown to be up-regulated in the sera of prostate cancer patients with metastasis in comparison to those of healthy controls (Pang *et al.*, 2010). Patients with hormone-refractory prostate cancer expressed higher serum levels of miR-21 than those with androgen-dependent and localized prostate cancer (Zhang *et al.*, 2010). Finally, miR-107 and miR-574-3p are believed to be particularly promising prognostic candidates as they are represented in high concentrations in the urine of prostate cancer patients compared to controls (Bryant *et al.*, 2012).

Colorectal cancer

There are many studies comparing the miRNA expression of precancerous lesions, adenomas and colorectal tumour tissue in comparison to normal colon tissue. Some of these are consistently dysregulated in multiple studies including miR-143 and miR-145 miR-203, miR-20a and miR-31. The subject is reviewed in detail by (Luo *et al.*, 2011; Vickers *et al.*, 2012). The most intensely studied miRNAs associated with prognosis are miR-21, miR-143, and miR-145. Increased expression of miR-21 in colorectal tumour tissue has been linked to an increased metastatic capacity (Brueckner *et al.*, 2007; Hurst *et al.*, 2009), as well as shortened disease-free survival (Balaguer *et al.*, 2010; Kulda *et al.*, 2010) and overall survival (Schetter *et al.*, 2008). MiR-21 expression has also been associated with the efficacy of 5-fluorouracil (5-FU) chemotherapy (Schetter *et al.*, 2008; Huang *et al.*, 2011). In contrast miR-143 and miR-145 are typically down-regulated in tumour tissue (Slaby *et al.*, 2007; Chen *et al.*, 2009; Motoyama *et al.*, 2009), and expression levels are correlated with increased tumour growth and angiogenesis, and shorter disease-free survival (Akao *et al.*, 2007; Kulda *et al.*, 2010; Xu *et al.*, 2012). Like miR-21, the expression of miR-143 appears to correlate with the sensitivity of colorectal cancer cells to 5-FU treatment (Borralho *et al.*, 2009). In another study, reduced miR-215 levels were associated with patient resistance to methotrexate and tomudex chemotherapy (Song *et al.*, 2010).

Serum levels of miR-29a has been proposed as a potential non-invasive biomarker for early detection of colorectal cancer involving liver metastasis (Wang and Gu, 2012). Plasma miR-141 on the other hand was associated with the presence of distant metastasis and poor prognosis (Cheng *et al.*, 2011). Finally, miR-92 levels in plasma have been shown to be able to differentiate between colorectal cancer and gastric cancer patients, potentially conflicting diagnoses (Ng *et al.*, 2009).

Haematological malignancies

Myeloid disorders

The general importance of miRNA dysregulation to the pathogenesis of myeloid disorders is implied by the fact that more than 70% of all human miRNAs are encoded within regions of recurrent copy-number alterations in myelodysplastic syndrome (MDS) and acute myeloid leukaemia (AML) cell lines (Starczynowski *et al.*, 2011). Many studies have demonstrated that miRNAs are abnormally expressed in myeloid malignancies compared to counterpart controls;

most commonly in AML (reviewed in Marcucci *et al.*, 2011), but also in MDS (reviewed in Rhyasen and Starczynowski, 2012) and chronic myeloid leukaemia (CML). Of particular interest, profiling studies have revealed marked differences in miRNA expression between common cytogenetic subtypes of AML including those harbouring favourable-risk abnormalities such as t(8;21), (Dixon-McIver *et al.*, 2008; Jongen-Lavrencic *et al.*, 2008; Li *et al.*, 2008; Cammarata *et al.*, 2010) inv(16), t(15;17), (Dixon-McIver *et al.*, 2008; Jongen-Lavrencic *et al.*, 2008; Li *et al.*, 2008) and inv(16), and those with less favourable-risk subtypes such as t(11q23)/*MLL* (Dixon-McIver *et al.*, 2008; Garzon *et al.*, 2008b; Jongen-Lavrencic *et al.*, 2008; Li *et al.*, 2008) and trisomy 8 cases (Garzon *et al.*, 2008b). Additionally, karyotype normal AML patients (i.e. CN-AML) also have distinctive miRNA expression patterns associated with recurrent molecular abnormalities including *FLT3-ITD* (Garzon *et al.*, 2008a,b; Jongen-Lavrencic *et al.*, 2008; Cammarata *et al.*, 2010) and *MLL* duplications, mutations in *WT1*, *IDH1*, *IDH2*, *NPM1* (Garzon *et al.*, 2008a; Jongen-Lavrencic *et al.*, 2008) and *CEPBA* (Jongen-Lavrencic *et al.*, 2008), as well as high expression of *BAALC*, (Langer *et al.*, 2008) *ERG* and *MN1* genes (Langer *et al.*, 2009).

The use of miRNA as novel classifiers of myeloid malignancies has drawn a lot of attention recently. In MDS, for example twelve miRNAs were identified as being differentially expressed between high and low risk MDS patients, although numbers in this study were small ($n=25$) (Erdogan *et al.*, 2011). Another study of 52 MDS patients correlated high miR-150 levels with good cytogenetic-risk groups (Hussein *et al.*, 2010). Microarray analyses of samples from 122 AML patients were used to define signatures associated with cytogenetically favourable-risk groups (Garzon *et al.*, 2008b). Importantly, these findings were tested in an independent cohort of 60 AML patients using qRT-PCR and levels of miR-191 and miR-199a were found to be independent predictors of prognosis by multivariate analysis. More recently, high levels of miR-181a were correlated with better prognostic outcome for 187 CN-AML patients (Schwind *et al.*, 2010).

Lymphoid malignancies

Apart from the miR-15a/16-1 cluster mentioned above, other miRNAs have been implicated as biomarkers of CLL. MiR-29 and miR-181b for example, are down-regulated in patients with aggressive CLL, and have been linked to overexpression of the TCL1 oncogene (Pekarsky *et al.*, 2006). Recently high miR-21 levels have been associated with poor prognosis in CLL patients harbouring the 17p deletion (Rossi *et al.*, 2010), and levels of miR-181b were inversely correlated with disease severity over time leading to its proposal as a potential treatment biomarker for CLL progression (Visone *et al.*, 2011). It has also been suggested that a miRNA signature can be used to predict those CLL patients that are refractory to Fludarabine treatment (Ferracin *et al.*, 2010), and that these patients are characterized by low miR-34a levels even in the absence of p53 alterations (Zenz *et al.*, 2009).

Similar to the other leukaemias described above, tumour cells from B-cell acute lymphoblastic leukaemia (ALL) have been shown to have a different miRNA expression profile from their healthy counterpart ($CD33^+$) cells (Schotte *et al.*, 2011). High levels of miR-128 are associated with ALL and can distinguish between AML and ALL with >98% accuracy (Mi *et al.*, 2007). Low levels of miR-16 were linked with superior outcome in paediatric ALL cases (Kaddar *et al.*, 2009). Additionally in this study, various cytogenetic subgroups of B-ALL including patients with 11q23/MLL, TEL-AML1, BCR-ABL and E2A-PBX1 translocations, and hyperdiploid patients were also shown to have distinct miRNA profiles. Let-7b for example, is down-regulated in B-ALL patients with MLL translocations (Schotte *et al.*, 2011), whilst the miR-17-92 cluster is overexpressed (Mi *et al.*, 2010). Differentially expressed miRNAs have also been identified between T-ALL leukaemic cells and healthy thymocytes (Schotte *et al.*, 2011).

The first report to identify aberrantly expressed miRNAs in multiple myeloma (MM), compared the miRNA expression profile of tumour cells ($CD138^+$) from 10 MM patients, 5 monoclonal gammopathy of undetermined significance (MGUS), widely regarded as the progenitor condition of MM, and plasma cells from four healthy individuals (Pichiorri *et al.*, 2008). Subsequently other groups have extended these analyses (Lionetti *et al.*, 2009; Roccaro *et al.*, 2009; Unno *et al.*, 2009; Chi *et al.*, 2010; Gutierrez *et al.*, 2010; Zhou *et al.*, 2010). Recently we identified distinct miRNA patterns associated with differing cytogenetic subgroups, the immunoglobulin isotype (IgG vs. IgA type), and the presence of light-chain disease (Chi *et al.*, 2010). Global overexpression of miRNAs has been linked with poor prognostic outcome for 52 MM patients (Zhou *et al.*, 2010), whilst another study correlated miR-21 and miR-181a expression with resistance to doxorubicin and melphalan treatment (Munker *et al.*, 2010).

DLBCL was one of the first lymphomas to be linked with aberrant miRNA expression, in particular with overexpression of miR-155 (Eis *et al.*, 2005; Kluiver *et al.*, 2005; Lawrie *et al.*, 2007). Early studies linked expression of miRNAs with the molecular subtypes of DLBCL and prognostic outcome (Lawrie *et al.*, 2007, 2009b; Malumbres *et al.*, 2009). A retrospective study of 258 DLBCL cases defined a 9-miRNA signature that was independently indicative of both overall survival (OS) and progression-free survival (PFS) in this cohort (Montes-Moreno *et al.*, 2011). In a recent study of 176 DLBCL patients uniformly treated with R-CHOP, levels of miR-18a were identified as independent variables associated with OS and levels of miR-181a and miR-222 with PFS (Alencar *et al.*, 2011).

In follicular lymphoma (FL), Roehle *et al.* (2008) identified miRNAs that were differentially expressed compared to normal lymph nodes and DLBCL cases. Our group elucidated the miRNA profile of eighteen FL cases in comparison with 80 DLBCL cases and derived a 26-miRNA signature that could differentiate between FL cases and *de novo* cases of DLBCL (Lawrie *et al.*, 2009b). Other studies have compared FL with nodal marginal zone lymphoma (NMZL) (Arribas *et al.*, 2012), and follicular hyperplasia patients (Wang *et al.*, 2012). The latter study also identified miRNAs associated with FL patients responsive to PACE chemotherapy.

A number of miRNA signatures have now been associated with mantle cell lymphoma (MCL) (Navarro *et al.*, 2009; Di Lisio *et al.*, 2010; Zhao *et al.*, 2010; Iqbal *et al.*, 2012). Over-expression of members of the miR-17-92 cluster for example has been associated with high MYC levels in aggressive MCL (Navarro *et al.*, 2009), and inhibition of chemotherapy-induced apoptosis in MCL cell lines (Chaudhuri *et al.*, 2012). In another study miR-29 was identified as a potential prognostic marker for this malignancy (Zhao *et al.*, 2010). In Hodgkin lymphoma (HL) in contrast, low miR-135a levels were associated with significantly poorer prognostic outcome (Navarro *et al.*, 2008; Nie *et al.*, 2008). A number of studies have recently investigated the role of miRNAs in mucosa-associated lymphoid tissue (MALT) lymphoma. For example, a 27-miRNA signature was able to distinguish gastric DLBCL from MALT lymphoma (Craig *et al.*, 2011). In another study a set of five miRNAs (miR-150, miR-550, miR-124a, miR-518b and miR-539) were identified as being differentially expressed in gastritis as opposed to MALT lymphoma (Thorns *et al.*, 2012).

Compared with B-cell lymphoma, there are few studies on miRNAs in T-cell lymphoma. Our group provided the first evidence for a functional role of miRNAs in T-cell lymphomagenesis in a study that identified > 100 aberrantly expressed miRNAs in the cutaneous T-cell lymphoma, Sézary syndrome (SzS) (Ballabio *et al.*, 2010), including miR-223 that was shown to act as a potential diagnostic marker in both training and validation cohorts. The identity of many of these aberrantly expressed miRNAs has now been validated in independent studies (Narducci *et al.*, 2011; Qin *et al.*, 2012). Our group has also recently carried out profiling studies on tumour stage mycosis fungoides (van Kester *et al.*, 2011) and cutaneous anaplastic large cell lymphoma (Benner *et al.*, 2012). A qRT-PCR based classifier (miR-155, miR-203 and miR-205) has recently been proposed that can distinguish between the various forms of cutaneous T-cell lymphomas and related benign disorders (Ralfkiaer *et al.*, 2011). Importantly, both training ($n=90$) and blinded test ($n=58$) sets were used in this study, although further independent studies are still needed to validate these results.

Current state of miRNA biomarker studies

The miRNA biomarker studies described above are by no means an exhaustive list but instead intended to illustrate the rapidly expanding wealth of evidence that points to great potential that miRNAs have as cancer biomarkers of the future. It should be remembered however, that nearly all studies to date originate in the laboratory rather than the clinic, and that currently there are no clinically approved miRNA-based biomarkers. This is a no doubt a reflection of the immaturity of the field and a situation that will surely be remedied in the near future. However, clearly much work remains to be done before these findings can become clinical reality. In addition to the need for further testing in sufficiently powered cohorts, as well as in multi-centre and independent settings, fundamentally there remain many discrepancies between miRNA biomarker studies. These differences are primarily due to biological and technical variation between studies such as the starting material used in experiments (e.g. purification of cells, cell types, control populations used, RNA extraction etc.), technological platforms (e.g. microarray, qRT-PCR, vs. next-generation sequencing (NGS) etc.), and differing statistical methodologies used. Below we discuss some of these points in more detail and in particular describe some of the more common techniques used in miRNA biomarker studies and their relative advantages and disadvantages.

Techniques used in cancer miRNA biomarker studies

MiRNA expression profiling has become a key bioanalytical technique in the miRNA field but due to the relative novelty of these molecules and their specific characteristics (i.e. small size), specialist considerations have to be taken into account before undertaking this type of experiment.

Starting material

Although obvious, the choice of starting material is crucial to initial experimental design particularly when interpreting generated data and trying to repeat previously published observations. As miRNAs are very stable molecules, they can be purified from a wide range of tissue types including serum, plasma, urine and formalin-fixed tissue blocks etc. MiRNAs display a tissue-dependent, even a cell-dependent, expression pattern (Liang *et al.* 2007; Lindberg *et al.* 2010), this fact coupled with the hundreds of potential targets for each miRNA, also expressed in a tissue-dependent way, make the choice of tissue type fundamental. Of particular importance to these comparative designs is the choice of control material that can completely change the results generated in a study. For example the choice of whole blood, PMBCs, or purified cell populations would generate vastly different profiles when compared to the miRNA expression of a leukaemic tumour cell.

miRNA extraction

miRNA isolation requires the same technical cares as other RNA purification procedures. The requirement for sterility at the bench and in the handling procedures should be the same as we are accustomed to using when working with total RNA. Time is also a critical variable in the extraction process, both time of collection of the material from the patient until process initiation, and time taken to process the sample. Delays in the former can lead to changes in cellular state (e.g. induction of apoptosis) and consequent changes in the miRNA profile, whilst delays in processing time could lead to increased risk of (partial) miRNA degradation.

The RNA extraction protocols used in the past did *not* recover the small-size RNA, but these same protocols with some modifications, mainly in the ethanol percentage used in the washes, will also recover small-size RNA (including miRNAs) (Accerbi *et al.*, 2010). When we are working with RNA samples that have already been extracted we should have all the information about which protocol was used in order to decide whether or not the sample is suitable for our current objectives.

Nowadays several manual protocols and commercial kits are available to isolate total RNA (including miRNA), usually based on columns or more recently magnetic beads. TRIZOL (phenol–chloroform) based protocols are widely used for RNA extraction; however, recently it has been shown that its use with samples containing few cells can result in selective loss of low GC-content miRNAs, and hence bias results (Kim *et al.*, 2012). This is particularly important when working with biological fluids, and consequently alternative methods should be sought when using these tissues. Centrifugation steps when working with serum or plasma are also a source of variation as small vesicles and exosomes which are sedimented at high speed, also contain miRNAs. Many kits are available that specifically purify small size RNA from samples as opposed to total RNA purification protocols that retains small RNA. Although popular, most modern detection techniques (except NGS) are sufficiently specific that this enrichment is unnecessary, and may even cause the loss of short RNAs (Redshaw *et al.*, 2013).

RNA quality

The quantity of total RNA can be measured spectrophotometrically (e.g. nanodrop) but this approach does not give us any idea of the quality of the RNA. To measure the quality we can use gel electrophoresis or an automated capillary instrument such as the bioanalyser 2100 (Agilent) or Experion (Bio-Rad) systems. Agilent offers specialized assays (Small RNA kit) for estimating miRNA abundance, based on the proportion of RNA in the 15–40 nt size-window relative to total RNA. It should be noted, however, that both quantity and quality measurements do not allow us to differentiate between miRNA and other RNA species (e.g. mRNA, tRNA, rRNA), particularly when fragmented such is typically the case when studying biological fluids and FFPE biopsy material.

The use of a known amount of synthetic miRNAs (spike-in) at the beginning of the isolation process is a useful tool to monitor and normalize variations in the extraction process (Redshaw *et al.*, 2013).

Choice of profiling platform

The three major approaches used for studying miRNA expression profiling are quantitative reverse transcription PCR (qRT-PCR), hybridization-based methods (e.g. microarray) and increasingly NGS. The decision of which platform to use depends largely upon the question that we want to answer. For example, if the requirement is a complete understanding of the miRnome including the discovery of novel miRNA species, and mutations in existing miRNA sequences, then NGS provides the *de facto* solution. If we are interested only in previously annotated miRNAs then qRT-PCR or microarrays are the most cost-effective, and do not necessarily require the intervention of a bioinformatician in the way that NGS techniques currently do. If the quantity of material is low, then qRT-PCR provides the most sensitive technique. If cost and coverage are the main concerns then microarrays are probably the way to go. As commented above, platform choice is critical when trying to compare results between differing studies and should be an important consideration if trying to expand or compare with previously obtained data sets. As a very fast moving field the techniques listed below are not exhaustive but represent the most commonly used technologies at present. Readers are advised to investigate various commercial websites and technology reviews at the time the experimental design to find the best approach for their objective (Table 11.2).

qRT-PCR

This approach is one of the most attractive because of its ease to incorporate into the current workflow of most modern molecular laboratories. Moreover, most researchers are familiar with qPCR techniques and their analysis. Flexibility of design is also advantageous when compared to microarray platforms for example, and assay availability commercially ranges from whole miRnome, to custom panels, to individual miRNA probes. Some commonly used platforms are Taqman-assays plates (TLDA or open arrays), SABiosciences/Qiagen predispensed plates (miScript miRNA PCR array) and Exiqon LNA based system (miRCURY LNA qPCR).

Due to their small size, the retrotranscription step of miRNA qRT-PCR differs from that of traditional qRT-PCR. This can be achieved using a miRNA-specific stem–loop reverse transcription primers to generate the cDNA (i.e. Taqman assays) (Chen *et al.*, 2005), or with a universal primer that

Table 11.2 Commonly used miRNA profiling techniques

Pros	Cons	Platforms	Company	RNA Required
QPCR				
Sensitive and specific	Only known miRNA	Taqman assays/TLDA/Open Array	Life technologies	10–200 ng
Commonly used in the labs	Medium throughput	miRCURY LNA qPCR	Exiqon	
Absolute quantification	High cost/sample	miSCRIPT miRNA PCR arrays	SABiosciences	
Can study few miRNAs				
Hybridization				
Low cost/sample	Lower specificity	miRCURY LNA microRNA Array	Exiqon	0.5–4 µg
High throughput	Only known miRNAs	miRNA array v3.0	Affymetrix	
Usually more miRNAs than in qPCR profiles		microRNA array	Agilent	
		Sentrix array matrix and bead chips	Illumina	
Next-generation sequencing (NGS)				
Can distinguish isomiRs	Computational support needed	MiSeq	Illumina	1–5 µg
Can detect mutation	Not for absolute quantification	Ion Torrent (PGM)	Life	
Can detect novel miRNAs	Cost	GS Junior (454)	Roche	

relies on the extension of the miRNA by generating a poly(A) tail. However differences between the techniques in efficiency and specificity in the amplification have been reported (Redshaw *et al.*, 2013). Some vendors also propose the use of Locked nucleic acids (LNA) in primers instead of DNA to standardize the hybridization conditions for the hundreds of differing PCR assays that must run simultaneously in a whole miRnome qRT-PCR assay. Unlike microarrays, qRT-PCR techniques can use as little as 10 ng of starting RNA when utilizing a pre-amplification step. Another attractive feature of qRT-PCR is its superior specificity compared to microarray and the fact it can be used for absolute quantification. This technique remains the 'gold standard' of miRNA detection and is still required for the validation of results from microarray or NGS platforms. On the other hand qRT-PCR is an expensive technique, relies on previously annotated miRNA sequences and is limited by the low throughput of samples that can be carried out in a single day.

Hybridization techniques

The most commonly used hybridization system is the classical microarray, a matrix of small RNA probes spotted or lithographically printed upon a glass (or silicon) substrate. Prior to array hybridization, small RNA populations are typically fluorescently labelled. A more classical two-colour approach consisting of separately labelled control and sample populations on the same slide has largely made way for a single channel approach without the need for a control channel. The most widely used platforms are Exiqon (LNA arrays), Affymetrix (GeneChip miRNA array v3.0) or Agilent (MicroRNA microarray). This approach provides a fast and inexpensive tool to obtain lots of data from a large set of small RNAs, not only miRNAs. In most cases more than 100 ng of total RNA are required for a successful experiment.

An alternative hybridization approach is offered by Luminex where instead of using a solid matrix containing specific miRNA probes, labelled beads in solution are used. This technique is based on magnetic microspheres tagged with unique DNA sequences that allows simple identification of the specific chimeric probe that binds to a particular miRNA. It has been proposed that this is better tool to validate array results than the commonly used qRT-PCR which is based upon amplification techniques as the Luminex system depends on a similar technique to arrays (i.e. hybridization) but offers similar levels of specificity to qRT-PCR (Biscontin *et al.*, 2010).

Next-generation sequencing (NGS)

Next-generation sequencing (NGS) systems are increasingly being used for small RNA profiling. The first step in this process is to generate a small RNA library. To do this adaptors are first ligated to each end of the small-RNA, and once the library has been created the RNA is reverse transcribed and PCR amplified. After that millions of amplicons are sequenced in parallel. Bioinformatics analysis of the obtained data allows for identification of both known and novel miRNAs, as well as a relative quantification for each read. Quantification is possible with sufficient coverage (i.e. number of times each sequence is read) based upon the number of reads for a given miRNA relative to the total reads in the sample.

The most obvious advantage of NGS is the ability to identify not only known small RNA species but also unknown and mutated/variant sequences in an unbiased manner. The main disadvantage however is cost which still remains relatively high, although it is falling with the arrival of new technology (e.g. Ion torrent or mySeq instruments) and the use of the DNA-barcodes that allow for multiplexing of samples. Another important limitation of this technique is the requirement for high quantity and high-quality RNA as well as the necessity for complicated bioinformatics analysis. However both factors are becoming less problematic as amplification protocols are being developed (e.g. Ultra-Low kit from Illumina) along with new user-friendly bioinformatics tools such as miREvo (Wen, Shen, Shi, and Tang, 2012) and miRDeep2 (Friedländer, Mackowiak, Li, Chen, and Rajewsky, 2012) that have been developed specifically with small library RNAseq data in mind. Some of the most common platforms used for small RNAseq are the Illumina Genome analyser, Roche 454 sequencing, and Life technologies Solid and Ion Torrent machines.

Normalization

Normalization of the data is imperative in order to standardize data between and within experiments. Normalization in qRT-PCR is usually carried using predefined endogenous or 'housekeeping' genes, although increasingly the mean expression value of the analysed genes in a whole genome or miRnome data set, is being used (Mestdagh *et al.* 2009). Typically snoRNAs such as U6 and U48 are used as miRNA control genes however the subject of which is the best gene to use for miRNA experiments is far from resolved.

The different methods available for normalization of array data are similarly controversial and varied (e.g. specific reference genes, Global geometric mean, quantile, invariant selection, loess or loess M normalization, etc.) (Meyer *et al.*, 2012). The choice of method has a profound effect upon the list of differentially expressed genes obtained from experimental data and should be chosen with great care. To compound the problem there is a lack of consensus in the literature and the choice of method varies between platform as for example Loess or Loess M have been proposed for Agilent arrays while Affymetrix have develop a specific tool for microRNA analysis (miRNAQC software) that uses quantile normalization.

Target analysis

Once you have found your miRNA of interest you naturally want to know more about the function of this miRNA. Whilst great effort has been put into identifying and cataloguing aberrantly expressed miRNAs in cancer, very little is known about their functional role, primarily due to a lack of knowledge about which genes are actually targeted by individual miRNAs. With very few functionally annotated exceptions, current approaches to this problem primarily rely upon the use of the many predictive computational algorithms available. However, these algorithms typically predict hundreds or even thousands of target genes for each miRNA, and in reality perform very poorly (Sethupathy *et al.*, 2006). To compound matters further the function of a particular miRNA is dependent upon cellular context.

That said, predictive algorithms remain the most practical choice for miRNA researchers and many databases are now available- the most common of which are listed in Table 11.3. Other examples include MirWalk (Dweep *et al.*, 2011) and miRDIP (Shirdel *et al.*, 2011) that are comprehensive databases providing information compiling different prediction algorithms. The choice of algorithm to use for this step is not without problems and in reality many researchers chose to use a union approach placing more significance on those target genes that are predicted by multiple algorithms. Another approach to elucidating functional significance *in silico* is to look at the miRNA in the context of gene regulatory networks using tools such as MAGIA (Bisognin *et al.*, 2012) or mirConnX (Huang, Athanassiou, and Benos, 2011). Both tools interact with Cytoscape (Lopes *et al.*, 2010), which is one of the most popular network visualization tools. Under all circumstances such predictions should be functionally proven. Most commonly this is achieved by cloning the 3′UTR sequence of the gene of interest immediately downstream of luciferase reporter vector system and co-transfecting cells with the miRNA in question in comparison with a scrambled control sequence.

Summary and future directions

The miRNA field continues to grow at a phenomenal rate and new biological roles for miRNAs are constantly being uncovered. Nowhere is this pace of discovery more rapid than the cancer field and within this area, the use of miRNAs as novel biomarkers of cancer. As mentioned above, miRNAs are particularly attractive candidates as biomarkers due in large part to their stability compared with other molecular markers. In particular, the ability to purify miRNAs not only from routinely prepared FFPE material, but also from blood and other biological fluids, has meant that the speed of biomarker discovery has been quite astonishing with over 5000 publications in the last 5 years. This 'gold rush', however, should be treated with some caution as the degree of discordancy between seemingly identical studies is worrisome, and in reality very few of the biomarkers studies published will ever make it into clinical practice. As mentioned above there are many reasons for these differences, and consequently there is a clear need for a standardized approach to be taken in future miRNA biomarker studies in order to rationalize these confounding factors. In particular, a systematic approach should be taken with the most common cancer types in a similar fashion to that achieved for other 'omic disciplines (i.e. transcriptomics and genomics).

Table 11.3 Commonly used miRNA target prediction algorithms, including the web addresses and references describing their function

Name	URL	Reference
miRanda	www.microrna.org	Enright *et al.* (2003)
MirTarget2	http://mirdb.org	Wang (2008)
PicTar	http://pictar.mdc-berlin.de/	Grun *et al.* (2005)
rna22	http://cbcsrv.watson.ibm.com/rna22.html	Miranda *et al.* (2006)
RNAhybrid	http://bibiserv.techfak.uni-bielefeld.de/rnahybrid/	Rehmsmeier *et al.* (2004)
TargetScan	http://www.targetscan.org/	Lewis *et al.* (2005)
TargetScan Custom	http://www.targetscan.org/vert_50/seedmatch.html	Lewis *et al.* (2005)
miRecords	http://mirecords.umn.edu/miRecords/index.php	Xiao *et al.* (2009)
PITA	http://genie.weizmann.ac.il/pubs/mir07/mir07_dyn_data.html	Kertesz *et al.* (2007)
miRDIP	http://ophid.utoronto.ca/mirDIP/#	Shirdel *et al.* (2011)
miRWalk	http://diana.cslab.ece.ntua.gr/microT/	Dweep *et al.* (2011)

Another important factor to take into account is the lack of specificity of miRNAs as biomarkers, illustrated by the fact that only a few miRNAs are associated with patient survival in a wide variety of cancer types (Table 11.1). Consequently, miRNAs are more likely to be clinically useful as prognosticators rather than diagnostic molecules. In terms of future developments, asides from the need for more robust biomarker studies, which we might expect will be implemented and driven by commercial entities, an improvement in detection technologies and the development of dedicated biosensors would be the next logical step. Whatever happens, the future for miRNA-based cancer biomarkers is very promising, and we should remember that we are only at the very beginning of our understanding of non-coding (nc)RNA and that in reality miRNAs represent the tip of the ncRNA 'iceberg'. Indeed, although ~75% of the human genome is transcribed (Djebali *et al.*, 2012), the protein-encoding portion of the genome only accounts for 1.5% (Alexander *et al.*, 2010), whilst miRNAs represent another 1.8% (Djebali *et al.*, 2012). There is now emerging evidence that ncRNA species other than miRNAs are essential for both physiological function and development, as well as playing a fundamental role in disease (Mercer *et al.*, 2009; Esteller, 2011). Although relative to miRNAs, the study of other ncRNA molecules is very limited, many classes of ncRNAs are now recognized including short ncRNAs such as miRNAs, piRNAs and tiRNAs; mid-size ncRNAs such as snoRNAs, PASRs, TSSa-RNAs and PROMPTs; and long ncRNAs (lncRNAs) (Esteller, 2011; Harries, 2012). In particular, lncRNAs have been found to be more cell-type and tissue-type specific than protein coding genes and miRNAs (Cabili *et al.*, 2011). It therefore seems highly likely that these molecules will become the next frontier of biomarker discovery.

In summary, although it is clear that miRNAs (and other ncRNAs) have much to offer as cancer biomarkers and as a consequence cancer patients, it is equally obvious that we still have much to learn from these molecules.

Acknowledgements

We apologize to the authors of the many studies that were not included in this review due to space limitations. Charles H. Lawrie and his research are supported by funding from the Starmer-Smith memorial lymphoma fund, and Ikerbasque, the Basque Foundation for Science.

References

Accerbi, M., Schmidt, S.A., De Paoli, E., Park, S., Jeong, D.H., and Green, P.J. (2010). Methods for isolation of total RNA to recover miRNAs and other small RNAs from diverse species. Methods Mol. Biol. *592*, 31–50.

Agirre, X., Jimenez-Velasco, A., San Jose-Eneriz, E., Garate, L., Bandres, E., Cordeu, L., Aparicio, O., Saez, B., Navarro, G., Vilas-Zornoza, A., *et al.* (2008). Down-regulation of hsa-miR-10a in chronic myeloid leukaemia CD34+ cells increases USF2-mediated cell growth. Mol. Cancer Res. *6*, 1830–1840.

Akao, Y., Nakagawa, Y., and Naoe, T. (2007). MicroRNA-143 and -145 in colon cancer. DNA Cell Biol. *26*, 311–320.

Alencar, A.J., Malumbres, R., Kozloski, G.A., Advani, R., Talreja, N., Chinichian, S., Briones, J., Natkunam, Y., Sehn, L.H., Gascoyne, R.D., *et al.* (2011). MicroRNAs are independent predictors of outcome in diffuse large B-cell lymphoma patients treated with R-CHOP. Clin. Cancer Res. *17*, 4125–4135.

Alexander, R.P., Fang, G., Rozowsky, J., Snyder, M. and Gerstein, M.B. (2010). Annotating non-coding regions of the genome. Nat. Rev. Genet. *11*, 559–571.

Alizadeh, A.A., Eisen, M.B., Davis, R.E., Ma, C., Lossos, I.S., Rosenwald, A., Boldrick, J.C., Sabet, H., Tran, T., Yu, X., *et al.* (2000). Distinct types of diffuse large B-cell lymphoma identified by gene expression profiling. Nature *403*, 503–511.

Altuvia, Y., Landgraf, P., Lithwick, G., Elefant, N., Pfeffer, S., Aravin, A., Brownstein, M.J., Tuschl, T., and Margalit, H. (2005). Clustering and conservation patterns of human microRNAs. Nucleic Acids Res. *33*, 2697–2706.

Anker, P., Mulcahy, H., and Stroun, M. (2003). Circulating nucleic acids in plasma and serum as a noninvasive investigation for cancer: time for large-scale clinical studies? Int. J. Cancer. *103*, 149–152.

Arribas, A.J., Campos-Martin, Y., Gomez-Abad, C., Algara, P., Sanchez-Beato, M., Rodriguez-Pinilla, M.S., Montes-Moreno, S., Martinez, N., Alves-Ferreira, J., Piris, M.A. *et al.* (2012). Nodal marginal zone lymphoma: gene expression and miRNA profiling identify diagnostic markers and potential therapeutic targets. Blood *119*, e9-e21.

Asaga, S., Kuo, C., Nguyen, T., Terpenning, M., Giuliano, A.E., and Hoon, D.S. (2011). Direct serum assay for microRNA-21 concentrations in early and advanced breast cancer. Clin. Chem. *57*, 84–91.

Babar, I.A., Cheng, C.J., Booth, C.J., Liang, X., Weidhaas, J.B., Saltzman, W.M., and Slack, F.J. (2012). Nanoparticle-based therapy in an *in vivo* microRNA-155 (miR-155)-dependent mouse model of lymphoma. Proc. Natl. Acad. Sci. U.S.A. *109*, E1695–1704.

Balaguer, F., Link, A., Lozano, J.J., Cuatrecasas, M., Nagasaka, T., Boland, C.R., and Goel, A. (2010). Epigenetic silencing of miR-137 is an early event in colorectal carcinogenesis. Cancer Res. *70*, 6609–6618.

Ballabio, E., Mitchell, T., van Kester, M.S., Taylor, S., Dunlop, H.M., Chi, J., Tosi, I., Vermeer, M.H., Tramonti, D., Saunders, N.J., *et al.* (2010). MicroRNA expression in Sezary syndrome: identification, function, and diagnostic potential. Blood. *116*, 1105–1113.

Bartel, D.P. (2009). MicroRNAs: target recognition and regulatory functions. Cell *136*, 215–233.

Benner, M.F., Ballabio, E., van Kester, M.S., Saunders, N.J., Vermeer, M.H., Willemze, R., Lawrie, C.H., and Tensen, C.P. (2012). Primary cutaneous anaplastic large cell lymphoma shows a distinct miRNA expression profile and reveals differences from tumor-stage mycosis fungoides. Exp. Dermatol. *21*, 632–634.

Bisognin, A., Sales, G., Coppe, A., Bortoluzzi, S., and Romualdi, C. (2012). MAGIA^2: from miRNA and genes expression data integrative analysis to microRNA-transcription factor mixed regulatory circuits (2012 update). Nucleic Acids Res. *40*, W13–21.

Blenkiron, C., Goldstein, L.D., Thorne, N.P., Spiteri, I., Chin, S.F., Dunning, M.J., Barbosa-Morais, N.L., Teschendorff, A.E., Green, A.R., Ellis, I.O., *et al.* (2007). MicroRNA expression profiling of human breast cancer identifies new markers of tumor subtype. Genome Biol. *8*, R214.

Blower, P.E., Verducci, J.S., Lin, S., Zhou, J., Chung, J.H., Dai, Z., Liu, C.G., Reinhold, W., Lorenzi, P.L., Kaldjian, E.P., *et al.* (2007). MicroRNA expression profiles for the NCI-60 cancer cell panel. Mol. Cancer Ther. *6*, 1483–1491.

Borchert, G.M., Lanier, W., and Davidson, B.L. (2006). RNA polymerase III transcribes human microRNAs. Nat. Struct. Mol. Biol. *13*, 1097–1101.

Borralho, P.M., Kren, B.T., Castro, R.E., da Silva, I.B., Steer, C.J., and Rodrigues, C.M. (2009). MicroRNA-143 reduces viability and increases sensitivity to 5-fluorouracil in HCT116 human colorectal cancer cells. FEBS J. *276*, 6689–6700.

Brueckner, B., Stresemann, C., Kuner, R., Mund, C., Musch, T., Meister, M., Sultmann, H., and Lyko, F. (2007). The human let-7a-3 locus contains an epigenetically regulated microRNA gene with oncogenic function. Cancer Res. *67*, 1419–1423.

Bryant, R.J., Pawlowski, T., Catto, J.W., Marsden, G., Vessella, R.L., Rhees, B., Kuslich, C., Visakorpi, T., and Hamdy, F.C. (2012). Changes in circulating microRNA levels associated with prostate cancer. Br. J. Cancer *106*, 768–774.

Cabili, M.N., Trapnell, C., Goff, L., Koziol, M., Tazon-Vega, B., Regev, A., and Rinn, J.L. (2011). Integrative annotation of human large intergenic noncoding RNAs reveals global properties and specific subclasses. Genes Dev. *25*, 1915–1927.

Cai, H., Lin, L., Tang, M., and Wang, Z. (2013). Prognostic evaluation of microRNA-210 expression in pediatric osteosarcoma. Med Oncol. *30*, 499.

Calin, G.A., Dumitru, C.D., Shimizu, M., Bichi, R., Zupo, S., Noch, E., Aldler, H., Rattan, S., Keating, M., Rai, K., *et al.* (2002). Frequent deletions and down-regulation of micro- RNA genes miR15 and miR16 at 13q14 in chronic lymphocytic leukaemia. Proc. Natl. Acad. Sci. U.S.A. *99*, 15524–15529.

Calin, G.A., Liu, C.G., Sevignani, C., Ferracin, M., Felli, N., Dumitru, C.D., Shimizu, M., Cimmino, A., Zupo, S., Dono, M., *et al.* (2004a). MicroRNA profiling reveals distinct signatures in B cell chronic lymphocytic leukaemias. Proc. Natl. Acad. Sci. U.S.A. *101*, 11755–11760.

Calin, G.A., Sevignani, C., Dumitru, C.D., Hyslop, T., Noch, E., Yendamuri, S., Shimizu, M., Rattan, S., Bullrich, F., Negrini, M., *et al.* (2004b). Human microRNA genes are frequently located at fragile sites and genomic regions involved in cancers. Proc. Natl. Acad. Sci. U.S.A. *101*, 2999–3004.

Calin, G.A., Ferracin, M., Cimmino, A., Di Leva, G., Shimizu, M., Wojcik, S.E., Iorio, M.V., Visone, R., Sever, N.I., Fabbri, M., *et al.* (2005). A MicroRNA signature associated with prognosis and progression in chronic lymphocytic leukaemia. N. Engl. J. Med. *353*, 1793–1801.

Callegari, E., Elamin, B.K., Giannone, F., Milazzo, M., Altavilla, G., Fornari, F., Giacomelli, L., D'Abundo, L., Ferracin, M., Bassi, C., *et al.* (2012). Liver tumorigenicity promoted by microRNA-221 in a mouse transgenic model. Hepatology *56*, 1025–1033.

Cammarata, G., Augugliaro, L., Salemi, D., Agueli, C., La Rosa, M., Dagnino, L., Civiletto, G., Messana, F., Marfia, A., Bica, M.G., *et al.* (2010). Differential expression of specific microRNA and their targets in acute myeloid leukaemia. Am J Hematol. *85*, 331–339.

Caramuta, S., Egyhazi, S., Rodolfo, M., Witten, D., Hansson, J., Larsson, C., and Lui, W.O. (2010). MicroRNA expression profiles associated with mutational status and survival in malignant melanoma. J. Invest. Dermatol. *130*, 2062–2070.

Chaudhuri, A.A., So, A.Y., Mehta, A., Minisandram, A., Sinha, N., Jonsson, V.D., Rao, D.S., O'Connell, R.M., and Baltimore, D. (2012). Oncomir miR-125b regulates hematopoiesis by targeting the gene Lin28A. Proc. Natl. Acad. Sci. U.S.A. *109*, 4233–4238.

Chen, C., Ridzon, D.A., Broomer, A.J., Zhou, Z., Lee, D.H., Nguyen, J.T., Barbisin, M., Xu, N.L., Mahuvakar, V.R., Andersen, M.R., *et al.* (2005). Real-time quantification of microRNAs by stem-loop RT-PCR. Nucleic Acid Res. *33*, e179.

Chen, X., Guo, X., Zhang, H., Xiang, Y., Chen, J., Yin, Y., Cai, X., Wang, K., Wang, G., Ba, Y., *et al.* (2009). Role of miR-143 targeting KRAS in colorectal tumorigenesis. Oncogene. *28*, 1385–1392.

Chen, Z.H., Zhang, G.L., Li, H.R., Luo, J.D., Li, Z.X., Chen, G.M., and Yang, J. (2012). A panel of five circulating microRNAs as potential biomarkers for prostate cancer. Prostate. 72, 1443–1452.

Cheng, H., Zhang, L., Cogdell, D.E., Zheng, H., Schetter, A.J., Nykter, M., Harris, C.C., Chen, K., Hamilton, S.R., and Zhang, W. (2011). Circulating plasma MiR-141 is a novel biomarker for metastatic colon cancer and predicts poor prognosis. PLoS One. *6*, e17745.

Chi, J., Ballabio, E., Chen, X.-H., Kušec, R., Taylor, S., Hays, D., Tramonti, D., Ross, F.M., Saunders, N.J., Pezzella, F., *et al.* (2011). MicroRNA expression in multiple myeloma is associated with genetic subtype, isotype and survival. Biol. Direct *18*, e23.

Craig, V.J., Cogliatti, S.B., Imig, J., Renner, C., Neuenschwander, S., Rehrauer, H., Schlapbach, R., Dirnhofer, S., Tzankov, A., and Muller, A. (2011). Myc-mediated repression of microRNA-34a promotes high-grade transformation of B-cell lymphoma by dysregulation of FoxP1. Blood *117*, 6227–6236.

Crick, F. (1970). Central dogma of molecular biology. Nature *227*, 561–563.

Croce, C.M. (2009). Causes and consequences of microRNA dysregulation in cancer. Nat. Rev. Genet. *10*, 704–714.

Di Lisio, L., Gomez-Lopez, G., Sanchez-Beato, M., Gomez-Abad, C., Rodriguez, M.E., Villuendas, R., Ferreira, B.I., Carro, A., Rico, D., Mollejo, M., *et al.* (2010). Mantle cell lymphoma: transcriptional regulation by microRNAs. Leukaemia *24*, 1335–1342.

Dixon-McIver, A., East, P., Mein, C.A., Cazier, J.B., Molloy, G., Chaplin, T., Andrew Lister, T., Young, B.D., and Debernardi, S. (2008). Distinctive patterns of microRNA expression associated with karyotype in acute myeloid leukaemia. PLoS One 3, e2141.

Djebali, S., Davis, C.A., Merkel, A., Dobin, A., Lassmann, T., Mortazavi, A., Tanzer, A., Lagarde, J., Lin, W., Schlesinger, F., *et al.* (2012). Landscape of transcription in human cells. Nature *489*, 101–108.

Duttagupta, R., Jiang, R., Gollub, J., Getts, R.C., and Jones, K.W. (2011). Impact of cellular miRNAs on circulating miRNA biomarker signatures. PLoS ONE *6*, e20769.

Dweep, H., Sticht, C., Pandey, P., and Gretz, N. (2011). miRWalk – database: prediction of possible miRNA binding sites by "walking" the genes of three genomes. J. Biomed. Inform. *44*, 839–847.

Eis, P.S., Tam, W., Sun, L., Chadburn, A., Li, Z., Gomez, M.F., Lund, E., and Dahlberg, J.E. (2005). Accumulation of miR-155 and BIC RNA in human B cell lymphomas. Proc. Natl. Acad. Sci. U.S.A. *102*, 3627–3632.

Enright, A.J., John, B., Gaul, U., Tuschl, T., Sander, C., and Marks, D.S. (2003). MicroRNA targets in *Drosophila*. Genome Biol. *5*, R1.

Epis, M.R., Giles, K.M., Barker, A., Kendrick, T.S., and Leedman, P.J. (2009). miR-331–3p regulates ERBB-2 expression and androgen receptor signaling in prostate cancer. J. Biol. Chem. *284*, 24696–24704.

Erdogan, B., Facey, C., Qualtieri, J., Tedesco, J., Rinker, E., Isett, R.B., Tobias, J., Baldwin, D.A., Thompson, J.E., Carroll, M., *et al.* (2011). Diagnostic microRNAs in myelodysplastic syndrome. Exp. Hematol. *39*, 915–926 e912.

Esteller, M. (2011). Non-coding RNAs in human disease. Nat. Rev. Genet. *12*, 861–874.

Fang, Y.X., and Gao, W.Q. (2014). Roles of microRNAs during prostatic tumorigenesis and tumor progression. Oncogene. *33*, 135–147.

Fassina, A., Marino, F., Siri, M., Zambello, R., Ventura, L., Fassan, M., Simonato, F., and Cappellesso, R. (2012). The miR-17–92 microRNA cluster: a novel diagnostic tool in large B-cell malignancies. Lab. Invest. *92*, 1574–1582.

Felli, N., Fontana, L., Pelosi, E., Botta, R., Bonci, D., Facchiano, F., Liuzzi, F., Lulli, V., Morsilli, O., Santoro, S., *et al.* (2005). MicroRNAs 221 and 222 inhibit normal erythropoiesis and erythroleukaemic cell growth via kit receptor down-modulation. Proc. Natl. Acad. Sci. U.S.A. *102*, 18081–18086.

Ferracin, M., Zagatti, B., Rizzotto, L., Cavazzini, F., Veronese, A., Ciccone, M., Saccenti, E., Lupini, L., Grilli, A., De Angeli, C., *et al.* (2010). MicroRNAs involvement in fludarabine refractory chronic lymphocytic leukaemia. Mol. Cancer. *9*, 123.

Fojo, T. (2007). Multiple paths to a drug resistance phenotype: mutations, translocations, deletions and amplification of coding genes or promoter regions, epigenetic changes and microRNAs. Drug Resist. Updat. *10*, 59–67.

Friedman, R.C., Farh, K.K., Burge, C.B., and Bartel, D.P. (2009). Most mammalian mRNAs are conserved targets of microRNAs. Genome Res. *19*, 92–105.

Gao, X., Zhang, R., Qu, X., Zhao, M., Zhang, S., Wu, H., Jianyong, L., and Chen, L. (2012). MiR-15a, miR-16-1 and miR-17-92 cluster expression are linked to poor prognosis in multiple myeloma. Leuk. Res. *36*, 1505–1509.

Garofalo, M., Di Leva, G., Romano, G., Nuovo, G., Suh, S.S., Ngankeu, A., Taccioli, C., Pichiorri, F., Alder, H., Secchiero, P., *et al.* (2009). miR-221and222 regulate TRAIL resistance and enhance tumorigenicity through PTEN and TIMP3 down-regulation. Cancer Cell. *16*, 498–509.

Garzon, R., Garofalo, M., Martelli, M.P., Briesewitz, R., Wang, L., Fernandez-Cymering, C., Volinia, S., Liu, C.G., Schnittger, S., Haferlach, T., *et al.* (2008a). Distinctive microRNA signature of acute myeloid leukaemia bearing cytoplasmic mutated nucleophosmin. Proc. Natl. Acad. Sci. U.S.A. *105*, 3945–3950.

Garzon, R., Volinia, S., Liu, C.G., Fernandez-Cymering, C., Palumbo, T., Pichiorri, F., Fabbri, M., Coombes, K., Alder, H., Nakamura, T., *et al.* (2008b). MicroRNA signatures associated with cytogenetics and prognosis in acute myeloid leukaemia. Blood *111*, 3183–3189.

Ghildiyal, M., and Zamore, P.D. (2009). Small silencing RNAs: an expanding universe. Nat. Rev. Genet. *10*, 94–108.

Ghildiyal, M., Xu, J., Seitz, H., Weng, Z., and Zamore, P.D. (2010). Sorting of Drosophila small silencing RNAs partitions microRNA* strands into the RNA interference pathway. RNA *16*, 43–56.

Giovannetti, E., Erozenci, A., Smit, J., Danesi, R., and Peters, G.J. (2012). Molecular mechanisms underlying the role of microRNAs (miRNAs) in anticancer drug resistance and implications for clinical practice. Crit. Rev. Oncol. Hematol. *81*, 103–122.

Giraldez, A.J., Mishima, Y., Rihel, J., Grocock, R.J., Van Dongen, S., Inoue, K., Enright, A.J., and Schier, A.F. (2006). Zebrafish MiR-430 promotes deadenylation and clearance of maternal mRNAs. Science *312*, 75–79.

Griffiths-Jones, S., Grocock, R.J., van Dongen, S., Bateman, A., and Enright, A.J. (2006). miRBase: microRNA sequences, targets and gene nomenclature. Nucleic Acids Res. *34*, D140–144.

Grün, D., Wang, Y.L., Langenberger, D., Gunsalus, K.C., and Rajewsky, N. (2005). microRNA target predictions across seven Drosophila species and comparison to mammalian targets. PLoS Comput. Biol. *1*, e13.

Guo, H., Ingolia, N.T., Weissman, J.S., and Bartel, D.P. (2010). Mammalian microRNAs predominantly act to decrease target mRNA levels. Nature *466*, 835–840.

Gutierrez, N.C., Sarasquete, M.E., Misiewicz-Krzeminska, I., Delgado, M., De Las Rivas, J., Ticona, F.V., Ferminan, E., Martin-Jimenez, P., Chillon, C., Risueno, A., *et al.* (2010). Deregulation of microRNA expression in the different genetic subtypes of multiple myeloma and correlation with gene expression profiling. Leukaemia *24*, 629–637.

Harries, L.W. (2012). Long non-coding RNAs and human disease. Biochem. Soc. Trans. *40*, 902–906.

He, L., Thomson, J.M., Hemann, M.T., Hernando-Monge, E., Mu, D., Goodson, S., Powers, S., Cordon-Cardo, C., Lowe, S.W., Hannon, G.J. *et al.* (2005). A microRNA polycistron as a potential human oncogene. Nature *435*, 828–833.

Heinzelmann, J., Henning, B., Sanjmyatav, J., Posorski, N., Steiner, T., Wunderlich, H., Gajda, M.R., and Junker, K. (2011). Specific miRNA signatures are associated with metastasis and poor prognosis in clear cell renal cell carcinoma. World J. Urol. *29*, 367–373.

Heneghan, H.M., Miller, N., Lowery, A.J., Sweeney, K.J., Newell, J., and Kerin, M.J. (2010). Circulating microRNAs as novel minimally invasive biomarkers for breast cancer. Ann. Surg. *251*, 499–505.

Hertel, J., Lindemeyer, M., Missal, K., Fried, C., Tanzer, A., Flamm, C., Hofacker, I.L., and Stadler, P.F. (2006). The expansion of the metazoan microRNA repertoire. BMC Genomics *7*, 25.

Hong, L., Yang, J., Han, Y., Lu, Q., Cao, J., and Syed, L. (2012). High expression of miR-210 predicts poor survival in patients with breast cancer: a meta-analysis. Gene *507*, 135–138.

Howe, E.N., Cochrane, D.R., and Richer, J.K. (2012). The miR-200 and miR-221/222 microRNA families: opposing effects on epithelial identity. J. Mammary Gland Biol. Neoplasia *17*, 65–77.

Huang, Z., Huang, S., Wang, Q., Liang, L., Ni, S., Wang, L., Sheng, W., He, X., and Du, X. (2011). MicroRNA-95 promotes cell proliferation and targets sorting Nexin 1 in human colorectal carcinoma. Cancer Res. *71*, 2582–2589.

Hurst, D.R., Edmonds, M.D., and Welch, D.R. (2009). Metastamir: the field of metastasis-regulatory microRNA is spreading. Cancer Res. *69*, 7495–7498.

Hussein, K., Theophile, K., Busche, G., Schlegelberger, B., Gohring, G., Kreipe, H., and Bock, O. (2010). Significant inverse correlation of microRNA-150/MYB and microRNA-222/p27 in myelodysplastic syndrome. Leuk. Res. *34*, 328–334.

Iorio, M.V., and Croce, C.M. (2009). MicroRNAs in cancer: small molecules with a huge impact. J. Clin. Oncol. *27*, 5848–5856.

Iorio, M.V., Ferracin, M., Liu, C.G., Veronese, A., Spizzo, R., Sabbioni, S., Magri, E., Pedriali, M., Fabbri, M., Campiglio, M., *et al.* (2005). MicroRNA gene expression deregulation in human breast cancer. Cancer Res. *65*, 7065–7070.

Iqbal, J., Shen, Y., Liu, Y., Fu, K., Jaffe, E.S., Liu, C., Liu, Z., Lachel, C.M., Deffenbacher, K., Greiner, T.C., *et al.* (2012). Genome-wide miRNA profiling of mantle cell lymphoma reveals a distinct subgroup with poor prognosis. Blood *119*, 4939–4948.

Ishihara, K., Sasaki, D., Tsuruda, K., Inokuchi, N., Nagai, K., Hasegawa, H., Yanagihara, K., and Kamihira, S. (2012). Impact of miR-155 and miR-126 as novel biomarkers on the assessment of disease progression and prognosis in adult T-cell leukaemia. Cancer Epidemiol. *36*, 560–565.

Ji, J., Shi, J., Budhu, A., Yu, Z., Forgues, M., Roessler, S., Ambs, S., Chen, Y., Meltzer, P.S., Croce, C.M., *et al.* (2009). MicroRNA expression, survival, and response to interferon in liver cancer. N. Engl. J. Med. *361*, 1437–1447.

Jongen-Lavrencic, M., Sun, S.M., Dijkstra, M.K., Valk, P.J., and Lowenberg, B. (2008). MicroRNA expression profiling in relation to the genetic heterogeneity of acute myeloid leukaemia. Blood *111*, 5078–5085.

Jung, E.J., Santarpia, L., Kim, J., Esteva, F.J., Moretti, E., Buzdar, A.U., Di Leo, A., Le, X.F., Bast, R.C. Jr., Park, S.T., *et al.* (2011). Plasma microRNA 210 levels correlate with sensitivity to trastuzumab and tumor presence in breast cancer patients. Cancer *118*, 2603–2614.

Kaddar, T., Chien, W.W., Bertrand, Y., Pages, M.P., Rouault, J.P., Salles, G., Ffrench, M., and Magaud, J.P. (2009). Prognostic value of miR-16 expression in childhood acute lymphoblastic leukaemia relationships to normal and malignant lymphocyte proliferation. Leuk. Res. *33*, 1217–1223.

Karim, B.O., Ali, S.Z., Landolfi, J.A., Mann, J.F., Liu, G., Christian, A., Dicello, J.F., Rosenthal, D.L., and Huso, D.L. (2008). Cytomorphologic differentiation of benign and malignant mammary tumors in fine needle aspirate specimens from irradiated female Sprague-Dawley rats. Vet. Clin. Pathol. *37*, 229–236.

Kertesz, M., Iovino, N., Unnerstall, U., Gaul, U., and Segal, E. (2007). The role of site accessibility in microRNA target recognition. Nat Genet. *39*, 1278–1284.

van Kester, M.S., Ballabio, E., Benner, M.F., Chen, X.H., Saunders, N.J., van der Fits, L., van Doorn, R., Vermeer, M.H., Willemze, R., Tensen, C.P., *et al.* (2011). MiRNA expression profiling of mycosis fungoides. Mol. Oncol. *5*, 273–280.

Khvorova, A., Reynolds, A., and Jayasena, S.D. (2003). Functional siRNAs and miRNAs exhibit strand bias. Cell *115*, 209–216.

Kim, C.H., Kim, H.K., Rettig, R.L., Kim, J., Lee, E.T., Aprelikova, O., Choi, I.J., Munroe, D.J., and Green, J.E.

(2011). MiRNA signature associated with outcome of gastric cancer patients following chemotherapy. BMC Med. Genomics *4*, 79.

Kim, Y.K., Yeo, J., Ha, M., Kim, B., and Kim, V.N. (2012). Retraction notice to: cell adhesion-dependent control of microRNA decay. Mol. Cell *43*, 1005–1014.

Klein, U., Lia, M., Crespo, M., Siegel, R., Shen, Q., Mo, T., Ambesi-Impiombato, A., Califano, A., Migliazza, A., Bhagat, G., *et al.* (2010). The DLEU2/miR-15a/16-1 cluster controls B cell proliferation and its deletion leads to chronic lymphocytic leukaemia. Cancer Cell. *17*, 28–40.

Kluiver, J., Poppema, S., de Jong, D., Blokzijl, T., Harms, G., Jacobs, S., Kroesen, B.J., and van den Berg, A. (2005). BIC and miR-155 are highly expressed in Hodgkin, primary mediastinal and diffuse large B cell lymphomas. J. Pathol. *207*, 243–249.

Kulda, V., Pesta, M., Topolcan, O., Liska, V., Treska, V., Sutnar, A., Rupert, K., Ludvikova, M., Babuska, V., Holubec, L., Jr., *et al.* (2010). Relevance of miR-21 and miR-143 expression in tissue samples of colorectal carcinoma and its liver metastases. Cancer Genet. Cytogenet. *200*, 154–160.

Lagos-Quintana, M., Rauhut, R., Lendeckel, W., and Tuschl, T. (2001). Identification of novel genes coding for small expressed RNAs. Science *294*, 853–858.

Langer, C., Radmacher, M.D., Ruppert, A.S., Whitman, S.P., Paschka, P., Mrozek, K., Baldus, C.D., Vukosavljevic, T., Liu, C.G., Ross, M.E., *et al.* (2008). High BAALC expression associates with other molecular prognostic markers, poor outcome, and a distinct gene-expression signature in cytogenetically normal patients younger than 60 years with acute myeloid leukaemia: a Cancer and Leukaemia Group B (CALGB) study. Blood *111*, 5371–5379.

Langer, C., Marcucci, G., Holland, K.B., Radmacher, M.D., Maharry, K., Paschka, P., Whitman, S.P., Mrozek, K., Baldus, C.D., Vij, R., *et al.* (2009). Prognostic importance of MN1 transcript levels, and biologic insights from MN1-associated gene and microRNA expression signatures in cytogenetically normal acute myeloid leukaemia: a cancer and leukaemia group B study. J Clin Oncol. *27*, 3198–3204.

Lawrie, C.H. (2007). MicroRNAs and haematology: small molecules, big function. Br. J. Haematol. *136* (6), 503–512.

Lawrie, C.H. (2008). MicroRNA expression in lymphoid malignancies: new hope for diagnosis and therapy? J. Cell. Mol. Med. *12*, 1432–1444.

Lawrie, C.H., Soneji, S., Marafioti, T., Cooper, C.D., Palazzo, S., Paterson, J.C., Cattan, H., Enver, T., Mager, R., Boultwood, J., *et al.* (2007). MicroRNA expression distinguishes between germinal center B cell-like and activated B cell-like subtypes of diffuse large B cell lymphoma. Int. J. Cancer. *121*, 1156–1161.

Lawrie, C.H., Gal, S., Dunlop, H.M., Pushkaran, B., Liggins, A.P., Pulford, K., Banham, A.H., Pezzella, F., Boultwood, J., Wainscoat, J.S., *et al.* (2008a). Detection of elevated levels of tumour-associated microRNAs in serum of patients with diffuse large B-cell lymphoma. Br. J. Haematol. *141*, 672–675.

Lawrie, C.H., Saunders, N.J., Soneji, S., Palazzo, S., Dunlop, H.M., Cooper, C.D., Brown, P.J., Troussard, X., Mossafa, H., Enver, T., *et al.* (2008b). MicroRNA expression in lymphocyte development and malignancy. Leukaemia. *22*, 1440–1446.

Lawrie, C.H., Ballabio, E., Dyar, O.J., Jones, M., Ventura, R., Chi, J., Tramonti, D., Gooding, S., Boultwood, J., Wainscoat, J.S., *et al.* (2009a). MicroRNA expression in chronic lymphocytic leukaemia. Br. J. Haematol. *147*, 398–402.

Lawrie, C.H., Chi, J., Taylor, S., Tramonti, D., Ballabio, E., Palazzo, S., Saunders, N.J., Pezzella, F., Boultwood, J., Wainscoat, J.S., *et al.* (2009b). Expression of microRNAs in diffuse large B cell lymphoma is associated with immunophenotype, survival and transformation from follicular lymphoma. J. Cell. Mol. Med. *13*, 1248–1260.

Lee, R.C., and Ambros, V. (2001). An extensive class of small RNAs in *Caenorhabditis elegans*. Science *294*, 862–864.

Lee, R.C., Feinbaum, R.L., and Ambros, V. (1993). The C. elegans heterochronic gene lin-4 encodes small RNAs with antisense complementarity to lin-14. Cell *75*, 843–854.

Lee, Y., Hur, I., Park, S.Y., Kim, Y.K., Suh, M.R., and Kim, V.N. (2006). The role of PACT in the RNA silencing pathway. EMBO J. *25*, 522–532.

Leon, S.A., Green, A., Yaros, M.J., and Shapiro, B. (1975). Radioimmunoassay for nanogram quantities of DNA. J. Immunol. Methods *9*, 157–164.

Leucci, E., Cocco, M., Onnis, A., De Falco, G., van Cleef, P., Bellan, C., van Rijk, A., Nyagol, J., Byakika, B., Lazzi, S., *et al.* (2008). MYC translocation-negative classical Burkitt lymphoma cases: an alternative pathogenetic mechanism involving miRNA deregulation. J Pathol. *216*, 440–450.

Lewis, B.P., Shih, I.H., Jones-Rhoades, M.W., Bartel, D.P., and Burge, C.B. (2003). Prediction of mammalian microRNA targets. Cell *115*, 787–798.

Lewis, B.P., Burge, C.B., and Bartel, D.P. (2005). Conserved seed pairing, often flanked by adenosines, indicates that thousands of human genes are microRNA targets. Cell *120*, 15–20.

Li, X., Zhang, Y., Ding, J., Wu, K., and Fan, D. (2010). Survival prediction of gastric cancer by a seven-microRNA signature. Gut *59*, 579–585.

Li, Z., Lu, J., Sun, M., Mi, S., Zhang, H., Luo, R.T., Chen, P., Wang, Y., Yan, M., Qian, Z., *et al.* (2008). Distinct microRNA expression profiles in acute myeloid leukaemia with common translocations. Proc. Natl. Acad. Sci. U.S.A. *105*, 15535–15540.

Li, Z., Huang, H., Li, Y., Jiang, X., Chen, P., Arnovitz, S., Radmacher, M.D., Maharry, K., Elkahloun, A., Yang, X., *et al.* (2012). Up-regulation of a HOXA-PBX3 homeobox-gene signature following down-regulation of miR-181 is associated with adverse prognosis in patients with cytogenetically abnormal AML. Blood *119*, 2314–2324.

Lia, M., Carette, A., Tang, H., Shen, Q., Mo, T., Bhagat, G., Dalla-Favera, R., and Klein, U. (2012). Functional dissection of the chromosome 13q14 tumor-suppressor locus using transgenic mouse lines. Blood *119*, 2981–2990.

Liang, Y., Ridzon, D., Wong, L., and Chen, C. (2007). Characterization of microRNA expression profiles in normal human tissues. BMC Genomics *8*, 166.

Lindberg, R.L., Hoffmann, F., Mehling, M., Kuhle, J., and Kappos, L. (2010). Altered expression of miR-17-5p in CD4+ lymphocytes of relapsing-remitting multiple sclerosis patients. Eur. J. Immunol. *40*, 888–898.

Lionetti, M., Agnelli, L., Mosca, L., Fabris, S., Andronache, A., Todoerti, K., Ronchetti, D., Deliliers, G.L., and Neri, A. (2009). Integrative high-resolution microarray analysis of human myeloma cell lines reveals deregulated miRNA expression associated with allelic imbalances and gene expression profiles. Genes Chromosomes Cancer *48*, 521–531.

Liu, R., Liao, J., Yang, M., Shi, Y., Peng, Y., Wang, Y., Pan, E., Guo, W., Pu, Y., and Yin, L. (2012a). Circulating miR-155 expression in plasma: a potential biomarker for early diagnosis of esophageal cancer in humans. J. Toxicol. Environ. Health A. *75*, 1154–1162.

Liu, X.G., Zhu, W.Y., Huang, Y.Y., Ma, L.N., Zhou, S.Q., Wang, Y.K., Zeng, F., Zhou, J.H., and Zhang, Y.K. (2012b). High expression of serum miR-21 and tumor miR-200c associated with poor prognosis in patients with lung cancer. Med. Oncol.. *29*, 618–626.

Lo, K.W., Lo, Y.M., Leung, S.F., Tsang, Y.S., Chan, L.Y., Johnson, P.J., Hjelm, N.M., Lee, J.C., and Huang, D.P. (1999). Analysis of cell-free Epstein–Barr virus associated RNA in the plasma of patients with nasopharyngeal carcinoma. Clin. Chem. *45*, 1292–1294.

Lowery, A.J., Miller, N., Devaney, A., McNeill, R.E., Davoren, P.A., Lemetre, C., Benes, V., Schmidt, S., Blake, J., Ball, G., *et al.* (2009). MicroRNA signatures predict oestrogen receptor, progesterone receptor and HER2/neu receptor status in breast cancer. Breast Cancer Res. *11*, R27.

Lu, J., Getz, G., Miska, E.A., Alvarez-Saavedra, E., Lamb, J., Peck, D., Sweet-Cordero, A., Ebert, B.L., Mak, R.H., Ferrando, A.A., *et al.* (2005). MicroRNA expression profiles classify human cancers. Nature *435*, 834–838.

Luo, X., Burwinkel, B., Tao, S., and Brenner, H. (2011). MicroRNA signatures: novel biomarker for colorectal cancer? Cancer Epidemiol. Biomarkers Prev. *20*, 1272–1286.

Lytle, J.R., Yario, T.A., and Steitz, J.A. (2007). Target mRNAs are repressed as efficiently by microRNA-binding sites in the 5′ UTR as in the 3′ UTR. Proc. Natl. Acad. Sci. U.S.A. *104*, 9667–9672.

Malumbres, R., Sarosiek, K.A., Cubedo, E., Ruiz, J.W., Jiang, X., Gascoyne, R.D., Tibshirani, R., and Lossos, I.S. (2009). Differentiation stage-specific expression of microRNAs in B lymphocytes and diffuse large B-cell lymphomas. Blood *113*, 3754–3764.

Mandel, P., and Metais, P. (1948). Les acides nucleiques du plasma sanguine chez l'homme. C. R. Acad. Sci. Paris. *142*, 241–243.

Marcucci, G., Mrozek, K., Radmacher, M.D., Garzon, R., and Bloomfield, C.D. (2011). The prognostic and functional role of microRNAs in acute myeloid leukaemia. Blood *117*, 1121–1129.

Mattick, J.S. (2001). Non-coding RNAs: the architects of eukaryotic complexity. EMBO Rep. *2*, 986–991.

Medina, P.P., Nolde, M., and Slack, F.J. (2010). OncomiR addiction in an *in vivo* model of microRNA-21-induced pre-B-cell lymphoma. Nature *467*, 86–90.

Mercer, T.R., Dinger, M.E., and Mattick, J.S. (2009). Long non-coding RNAs: insights into functions. Nat. Rev. Genet. *10*, 155–159.

Mi, S., Lu, J., Sun, M., Li, Z., Zhang, H., Neilly, M.B., Wang, Y., Qian, Z., Jin, J., Zhang, Y., *et al.* (2007). MicroRNA expression signatures accurately discriminate acute lymphoblastic leukaemia from acute myeloid leukaemia. Proc. Natl. Acad. Sci. U.S.A. *104*, 19971–19976.

Mi, S., Li, Z., Chen, P., He, C., Cao, D., Elkahloun, A., Lu, J., Pelloso, L.A., Wunderlich, M., Huang, H., *et al.* (2010). Aberrant overexpression and function of the miR-17-92 cluster in MLL-rearranged acute leukaemia. Proc. Natl. Acad. Sci. U.S.A. *107*, 3710–3715.

Miranda, K.C., Huynh, T., Tay, Y., Ang, Y.S., Tam, W.L., Thomson, A.M., Lim, B., and Rigoutsos, I. (2006). A pattern-based method for the identification of MicroRNA binding sites and their corresponding heteroduplexes. Cell *126*, 1203–1217.

Mitchell, P.S., Parkin, R.K., Kroh, E.M., Fritz, B.R., Wyman, S.K., Pogosova-Agadjanyan, E.L., Peterson, A., Noteboom, J., O'Briant, K.C., Allen, A., *et al.* (2008). Circulating microRNAs as stable blood-based markers for cancer detection. Proc. Natl. Acad. Sci. U.S.A. *105*, 10513–10518.

Montes-Moreno, S., Martinez, N., Sanchez-Espiridion, B., Diaz Uriarte, R., Rodriguez, M.E., Saez, A., Montalban, C., Gomez, G., Pisano, D.G., Garcia, J.F., *et al.* (2011). MiRNA expression in diffuse large B-cell lymphoma treated with chemoimmunotherapy. Blood *118*, 1034–1040.

Motoyama, K., Inoue, H., Takatsuno, Y., Tanaka, F., Mimori, K., Uetake, H., Sugihara, K., and Mori, M. (2009). Over- and under-expressed microRNAs in human colorectal cancer. Int. J. Oncol. *34*, 1069–1075.

Munker, R., Liu, C.G., Taccioli, C., Alder, H., and Heerema, N. (2010). MicroRNA profiles of drug-resistant myeloma cell lines. Acta Haematol. *123*, 201–204.

Narducci, M.G., Arcelli, D., Picchio, M.C., Lazzeri, C., Pagani, E., Sampogna, F., Scala, E., Fadda, P., Cristofoletti, C., Facchiano, A., *et al.* (2011). MicroRNA profiling reveals that miR-*21*, miR486 and miR-214 are up-regulated and involved in cell survival in Sezary syndrome. Cell Death Dis. *2*, e151.

Navarro, A., Gaya, A., Martinez, A., Urbano-Ispizua, A., Pons, A., Balague, O., Gel, B., Abrisqueta, P., Lopez-Guillermo, A., Artells, R., *et al.* (2008). MicroRNA expression profiling in classic Hodgkin lymphoma. Blood *111*, 2825–2832.

Navarro, A., Bea, S., Fernandez, V., Prieto, M., Salaverria, I., Jares, P., Hartmann, E., Mozos, A., Lopez-Guillermo, A., Villamor, N., *et al.* (2009). MicroRNA expression, chromosomal alterations, and immunoglobulin variable heavy chain hypermutations in Mantle cell lymphomas. Cancer Res. *69*, 7071–7078.

Ng, E.K., Chong, W.W., Jin, H., Lam, E.K., Shin, V.Y., Yu, J., Poon, T.C., Ng, S.S., and Sung, J.J. (2009). Differential expression of microRNAs in plasma of patients with colorectal cancer: a potential marker for colorectal cancer screening. Gut *58*, 1375–1381.

Nie, K., Gomez, M., Landgraf, P., Garcia, J.F., Liu, Y., Tan, L.H., Chadburn, A., Tuschl, T., Knowles, D.M., and Tam, W. (2008). MicroRNA-mediated down-regulation of PRDM1/Blimp-1 in Hodgkin/Reed-Sternberg cells: a potential pathogenetic lesion in Hodgkin lymphomas. Am. J. Pathol. *173*, 242–252.

Okamura, K., Phillips, M.D., Tyler, D.M., Duan, H., Chou, Y.T., and Lai, E.C. (2008). The regulatory activity of microRNA* species has substantial influence on microRNA and 3′ UTR evolution. Nat. Struct. Mol. Biol. *15*, 354–363.

Pang, Y., Young, C.Y., and Yuan, H. (2010). MicroRNAs and prostate cancer. Acta Biochim. Biophys. Sin. (Shanghai). *42*, 363–369.

Pasquinelli, A.E., Reinhart, B.J., Slack, F., Martindale, M.Q., Kuroda, M.I., Maller, B., Hayward, D.C., Ball, E.E., Degnan, B., Muller, P., *et al.* (2000). Conservation of the sequence and temporal expression of let-7 heterochronic regulatory RNA. Nature *408*, 86–89.

Pathak, A.K., Bhutani, M., Kumar, S., Mohan, A., and Guleria, R. (2006). Circulating cell-free DNA in plasma/serum of lung cancer patients as a potential screening and prognostic tool. Clin. Chem. *52*, 1833–1842.

Pekarsky, Y., Santanam, U., Cimmino, A., Palamarchuk, A., Efanov, A., Maximov, V., Volinia, S., Alder, H., Liu, C.G., Rassenti, L., *et al.* (2006). Tcl1 expression in chronic lymphocytic leukaemia is regulated by miR-29 and miR-181. Cancer Res. *66*, 11590–11593.

Peng, X., Guo, W., Liu, T., Wang, X., Tu, X., Xiong, D., Chen, S., Lai, Y., Du, H., Chen, G., *et al.* (2011). Identification of miRs-143 and -145 that is associated with bone metastasis of prostate cancer and involved in the regulation of EMT. PLoS One *6*, e20341.

Pichiorri, F., Suh, S.S., Ladetto, M., Kuehl, M., Palumbo, T., Drandi, D., Taccioli, C., Zanesi, N., Alder, H., Hagan, J.P., *et al.* (2008). MicroRNAs regulate critical genes associated with multiple myeloma pathogenesis. Proc. Natl. Acad. Sci. U.S.A. *105*, 12885–12890.

Porkka, K.P., Pfeiffer, M.J., Waltering, K.K., Vessella, R.L., Tammela, T.L., and Visakorpi, T. (2007). MicroRNA expression profiling in prostate cancer. Cancer Res. *67*, 6130–6135.

Qin, Y., Buermans, H.P., van Kester, M.S., van der Fits, L., Out-Luiting, J.J., Osanto, S., Willemze, R., Vermeer, M.H., and Tensen, C.P. (2012). Deep-sequencing analysis reveals that the miR-199a2/214 cluster within DNM3os represents the vast majority of aberrantly expressed microRNAs in Sezary syndrome. J. Invest. Dermatol. *132*, 1520–1522.

Ralfkiaer, U., Hagedorn, P.H., Bangsgaard, N., Lovendorf, M.B., Ahler, C.B., Svensson, L., Kopp, K.L., Vennegaard, M.T., Lauenborg, B., Zibert, J.R., *et al.* (2011). Diagnostic microRNA profiling in cutaneous T-cell lymphoma (CTCL). Blood *118*, 5891–5900.

Redshaw, N., Wilkes, T., Whale, A., Cowen, S., Huggett, J., and Foy, C.A. (2013). A comparison of miRNA isolation and RT-qPCR technologies and their effects on quantification accuracy and repeatability. Biotechniques *54*, 155–164.

Rehmsmeier, M., Steffen, P., Hochsmann, M., and Giegerich, R. (2004). Fast and effective prediction of microRNA/target duplexes. RNA *10*, 1507–1517.

Reinhart, B.J., Slack, F.J., Basson, M., Pasquinelli, A.E., Bettinger, J.C., Rougvie, A.E., Horvitz, H.R., and Ruvkun, G. (2000). The 21-nucleotide let-7 RNA regulates developmental timing in *Caenorhabditis elegans*. Nature *403*, 901–906.

Rhyasen, G.W., and Starczynowski, D.T. (2012). Deregulation of microRNAs in myelodysplastic syndrome. Leukaemia *26*, 13–22.

Roccaro, A.M., Sacco, A., Thompson, B., Leleu, X., Azab, A.K., Azab, F., Runnels, J., Jia, X., Ngo, H.T., Melhem, M.R., *et al.* (2009). MicroRNAs 15a and 16 regulate tumor proliferation in multiple myeloma. Blood *113*, 6669–6680.

Rodriguez, A., Griffiths-Jones, S., Ashurst, J.L., and Bradley, A. (2004). Identification of mammalian microRNA host genes and transcription units. Genome Res. *14*, 1902–1910.

Roehle, A., Hoefig, K.P., Repsilber, D., Thorns, C., Ziepert, M., Wesche, K.O., Thiere, M., Loeffler, M., Klapper, W., Pfreundschuh, M., *et al.* (2008). MicroRNA signatures characterize diffuse large B-cell lymphomas and follicular lymphomas. Br. J. Haematol. *142*, 732–744.

Rokah, O.H., Granot, G., Ovcharenko, A., Modai, S., Pasmanik-Chor, M., Toren, A., Shomron, N., and Shpilberg, O. (2012). Down-regulation of miR-*31*, miR-*155*, and miR-564 in chronic myeloid leukaemia cells. PLoS One. *7*, e35501.

Rossi, S., Shimizu, M., Barbarotto, E., Nicoloso, M.S., Dimitri, F., Sampath, D., Fabbri, M., Lerner, S., Barron, L.L., Rassenti, L.Z., *et al.* (2010). MicroRNA fingerprinting of CLL patients with chromosome 17p deletion identify a miR-21 score that stratifies early survival. Blood *116*, 945–952.

Roth, C., Kasimir-Bauer, S., Heubner, M., Pantel, K., and Schwarzenbach, H. (2010a). Increase in circulating microRNA levels in blood of ovarian cancer patients. In: Circulating Nucleic Acids in Plasma and Serum, Gahan, P., ed. (Springer, Netherlands), pp. 63–71.

Roth, C., Rack, B., Muller, V., Janni, W., Pantel, K., and Schwarzenbach, H. (2010b). Circulating microRNAs as blood-based markers for patients with primary and metastatic breast cancer. Breast Cancer Res. *12*, R90.

Roth, C., Kasimir-Bauer, S., Pantel, K., and Schwarzenbach, H. (2011). Screening for circulating nucleic acids and caspase activity in the peripheral blood as potential diagnostic tools in lung cancer. Mol. Oncol. *5*, 281–291.

Rothe, F., Ignatiadis, M., Chaboteaux, C., Haibe-Kains, B., Kheddoumi, N., Majjaj, S., Badran, B., Fayyad-Kazan, H., Desmedt, C., Harris, A.L., *et al.* (2011). Global microRNA expression profiling identifies MiR-210 associated with tumor proliferation, invasion and poor clinical outcome in breast cancer. PLoS One *6*, e20980.

San Jose-Eneriz, E., Roman-Gomez, J., Jimenez-Velasco, A., Garate, L., Martin, V., Cordeu, L., Vilas-Zornoza, A., Rodriguez-Otero, P., Calasanz, M.J., Prosper, F. *et al.*

(2009). MicroRNA expression profiling in Imatinib-resistant Chronic Myeloid Leukaemia patients without clinically significant ABL1-mutations. Mol. Cancer *8*, 69.

Satzger, I., Mattern, A., Kuettler, U., Weinspach, D., Voelker, B., Kapp, A., and Gutzmer, R. (2010). MicroRNA-15b represents an independent prognostic parameter and is correlated with tumor cell proliferation and apoptosis in malignant melanoma. Int. J. Cancer *126*, 2553–2562.

Schetter, A.J., Leung, S.Y., Sohn, J.J., Zanetti, K.A., Bowman, E.D., Yanaihara, N., Yuen, S.T., Chan, T.L., Kwong, D.L., Au, G.K., *et al.* (2008). MicroRNA expression profiles associated with prognosis and therapeutic outcome in colon adenocarcinoma. JAMA *299*, 425–436.

Schotte, D., De Menezes, R.X., Akbari Moqadam, F., Khankahdani, L.M., Lange-Turenhout, E., Chen, C., Pieters, R., and Den Boer, M.L. (2011). MicroRNA characterize genetic diversity and drug resistance in pediatric acute lymphoblastic leukaemia. Haematologica *96*, 703–711.

Schwarz, D.S., Hutvagner, G., Du, T., Xu, Z., Aronin, N., and Zamore, P.D. (2003). Asymmetry in the assembly of the RNAi enzyme complex. Cell *115*, 199–208.

Schwarzenbach, H., Milde-Langosch, K., Steinbach, B., Muller, V., and Pantel, K. (2012). Diagnostic potential of PTEN-targeting miR-214 in the blood of breast cancer patients. Breast Cancer Res. Treat. *134*, 933–941.

Schwind, S., Maharry, K., Radmacher, M.D., Mrozek, K., Holland, K.B., Margeson, D., Whitman, S.P., Hickey, C., Becker, H., Metzeler, K.H., *et al.* (2010). Prognostic significance of expression of a single microRNA, miR-181a, in cytogenetically normal acute myeloid leukaemia: a Cancer and Leukaemia Group B study. J. Clin. Oncol. *28*, 5257–5264.

Sempere, L.F., Christensen, M., Silahtaroglu, A., Bak, M., Heath, C.V., Schwartz, G., Wells, W., Kauppinen, S., and Cole, C.N. (2007). Altered MicroRNA expression confined to specific epithelial cell subpopulations in breast cancer. Cancer Res. *67*, 11612–11620.

Sethupathy, P., Megraw, M., and Hatzigeorgiou, A.G. (2006). A guide through present computational approaches for the identification of mammalian microRNA targets. Nat. Methods *3*, 881–886.

Shah, M.Y., and Calin, G.A. (2011). MicroRNAs miR-221 and miR-222, a new level of regulation in aggressive breast cancer. Genome Med. *3*, 56.

Shen, J., Hruby, G.W., McKiernan, J.M., Gurvich, I., Lipsky, M.J., Benson, M.C., and Santella, R.M. (2012). Dysregulation of circulating microRNAs and prediction of aggressive prostate cancer. Prostate *72*, 1469–1477.

Shin, C., Nam, J.W., Farh, K.K., Chiang, H.R., Shkumatava, A., and Bartel, D.P. (2010). Expanding the microRNA targeting code: functional sites with centered pairing. Mol. Cell. *38*, 789–802.

Shirdel, E.A., Xie, W., Mak, T.W., and Jurisica, I. (2011). NAViGaTing the micronome – using multiple microRNA prediction databases to identify signalling pathway-associated microRNAs. PLoS One *6*, e17492.

Slaby, O., Svoboda, M., Fabian, P., Smerdova, T., Knoflickova, D., Bednarikova, M., Nenutil, R., and Vyzula, R. (2007). Altered expression of miR-*21*, miR-*31*, miR-143 and miR-145 is related to clinicopathologic features of colorectal cancer. Oncology *72*, 397–402.

Song, B., Wang, Y., Titmus, M.A., Botchkina, G., Formentini, A., Kornmann, M., and Ju, J. (2010). Molecular mechanism of chemoresistance by miR-215 in osteosarcoma and colon cancer cells. Mol. Cancer *9*, 96.

Sotiriou, C., Wirapati, P., Loi, S., Harris, A., Fox, S., Smeds, J., Nordgren, H., Farmer, P., Praz, V., Haibe-Kains, B., *et al.* (2006). Gene expression profiling in breast cancer: understanding the molecular basis of histologic grade to improve prognosis. J. Natl. Cancer Inst. *98*, 262–272.

Starczynowski, D.T., Morin, R., McPherson, A., Lam, J., Chari, R., Wegrzyn, J., Kuchenbauer, F., Hirst, M., Tohyama, K., Humphries, R.K., *et al.* (2011). Genome-wide identification of human microRNAs located in leukaemia-associated genomic alterations. Blood *117*, 595–607.

Sun, Y., Wang, M., Lin, G., Sun, S., Li, X., Qi, J., and Li, J. (2012). Serum microRNA-155 as a potential biomarker to track disease in breast cancer. PLoS One *7*, e47003.

Szafranska, A.E., Doleshal, M., Edmunds, H.S., Gordon, S., Luttges, J., Munding, J.B., Barth, R.J., Jr., Gutmann, E.J., Suriawinata, A.A., Marc Pipas, J., *et al.* (2008). Analysis of microRNAs in pancreatic fine-needle aspirates can classify benign and malignant tissues. Clin. Chem. *54*, 1716–1724.

Takamizawa, J., Konishi, H., Yanagisawa, K., Tomida, S., Osada, H., Endoh, H., Harano, T., Yatabe, Y., Nagino, M., Nimura, Y., *et al.* (2004). Reduced expression of the let-7 microRNAs in human lung cancers in association with shortened postoperative survival. Cancer Res. *64*, 3753–3756.

Tay, Y., Zhang, J., Thomson, A.M., Lim, B., and Rigoutsos, I. (2008). MicroRNAs to Nanog, Oct4 and Sox2 coding regions modulate embryonic stem cell differentiation. Nature *455*, 1124–1128.

Thorns, C., Kuba, J., Bernard, V., Senft, A., Szymczak, S., Feller, A.C., and Bernd, H.W. (2012). Deregulation of a distinct set of microRNAs is associated with transformation of gastritis into MALT lymphoma. Virchows Arch. *460*, 371–377.

Unno, K., Zhou, Y., Zimmerman, T., Platanias, L.C., and Wickrema, A. (2009). Identification of a novel microRNA cluster miR-193b-365 in multiple myeloma. Leuk. Lymphoma *50*, 1865–1871.

Vickers, M.M., Bar, J., Gorn-Hondermann, I., Yarom, N., Daneshmand, M., Hanson, J.E., Addison, C.L., Asmis, T.R., Jonker, D.J., Maroun, J., *et al.* (2012). Stage-dependent differential expression of microRNAs in colorectal cancer: potential role as markers of metastatic disease. Clin. Exp. Metastasis *29*, 123–132.

Visone, R., Rassenti, L.Z., Veronese, A., Taccioli, C., Costinean, S., Aguda, B.D., Volinia, S., Ferracin, M., Palatini, J., Balatti, V., *et al.* (2009). Karyotype-specific microRNA signature in chronic lymphocytic leukaemia. Blood *114*, 3872–3879.

Visone, R., Veronese, A., Rassenti, L.Z., Balatti, V., Pearl, D.K., Acunzo, M., Volinia, S., Taccioli, C., Kipps, T.J., and Croce, C.M. (2011). miR-181b is a biomarker of disease progression in chronic lymphocytic leukaemia. Blood *118*, 3072–3079.

Voortman, J., Goto, A., Mendiboure, J., Sohn, J.J., Schetter, A.J., Saito, M., Dunant, A., Pham, T.C., Petrini, I., Lee, A., *et al.* (2010). MicroRNA expression and clinical outcomes in patients treated with adjuvant chemotherapy after complete resection of non-small cell lung carcinoma. Cancer Res. *70*, 8288–8298.

Wang, F., Zheng, Z., Guo, J., and Ding, X. (2010). Correlation and quantitation of microRNA aberrant expression in tissues and sera from patients with breast tumor. Gynecol. Oncol. *119*, 586–593.

Wang, L.G., and Gu, J. (2012). Serum microRNA-29a is a promising novel marker for early detection of colorectal liver metastasis. Cancer Epidemiol. *36*, e61–67.

Wang, W., Corrigan-Cummins, M., Hudson, J., Maric, I., Simakova, O., Neelapu, S.S., Kwak, L.W., Janik, J.E., Gause, B., Jaffe, E.S., *et al.* (2012). MicroRNA profiling of follicular lymphoma identifies microRNAs related to cell proliferation and tumor response. Haematologica *97*, 586–594.

Wang, X. (2008). miRDB: a microRNA target prediction and functional annotation database with a wiki interface. RNA *14*, 1012–1017.

Watahiki, A., Wang, Y., Morris, J., Dennis, K., O'Dwyer, H.M., Gleave, M., and Gout, P.W. (2011). MicroRNAs associated with metastatic prostate cancer. PLoS One *6*, e24950.

Weber, J.A., Baxter, D.H., Zhang, S., Huang, D.Y., Huang, K.H., Lee, M.J., Galas, D.J., and Wang, K. (2010). The microRNA spectrum in 12 body fluids. Clin. Chem. *56*, 1733–1741.

Wightman, B., Ha, I., and Ruvkun, G. (1993). Posttranscriptional regulation of the heterochronic gene lin-14 by lin-4 mediates temporal pattern formation in C. elegans. Cell *75*, 855–862.

Wu, X., Somlo, G., Yu, Y., Palomares, M.R., Xuejun Li, A., Zhou, W., Chow, A., Yen, Y., Rossi, J.J., Gao, H., *et al.* (2012). De novo sequencing of circulating miRNAs identifies novel markers predicting clinical outcome of locally advanced breast cancer. J. Transl. Med. *10*, 42.

Xiao, F.F., Zuo, Z.X., Cai, G.S., Kang, S.L., Gao, X.L., and Li, T.B. (2009). miRecords: an integrated resource for microRNA–target interactions. Nucleic Acids Res. *37*, D105-D110.

Xiong, Y., Fang, J.H., Yun, J.P., Yang, J., Zhang, Y., Jia, W.H., and Zhuang, S.M. (2010). Effects of microRNA-29 on apoptosis, tumorigenicity, and prognosis of hepatocellular carcinoma. Hepatology. *51*, 836–845.

Xu, Q., Liu, L.Z., Qian, X., Chen, Q., Jiang, Y., Li, D., Lai, L., and Jiang, B.H. (2012). MiR-145 directly targets p70S6K1 in cancer cells to inhibit tumor growth and angiogenesis. Nucleic Acids Res. *40*, 761–774.

Yan, L.X., Huang, X.F., Shao, Q., Huang, M.Y., Deng, L., Wu, Q.L., Zeng, Y.X., and Shao, J.Y. (2008). MicroRNA miR-21 overexpression in human breast cancer is associated with advanced clinical stage, lymph node metastasis and patient poor prognosis. RNA *14*, 2348–2360.

Yang, J.S., and Lai, E.C. (2011). Alternative miRNA biogenesis pathways and the interpretation of core miRNA pathway mutants. Mol Cell. *43*, 892–903.

Yekta, S., Shih, I.H., and Bartel, D.P. (2004). MicroRNA-directed cleavage of HOXB8 mRNA. Science *304*, 594–596.

Yeom, K.H., Lee, Y., Han, J., Suh, M.R., and Kim, V.N. (2006). Characterization of DGCR8/Pasha, the essential cofactor for Drosha in primary miRNA processing. Nucleic Acids Res. *34*, 4622–4629.

Yokobori, T., Suzuki, S., Tanaka, N., Inose, T., Sohda, M., Sano, A., Sakai, M., Nakajima, M., Miyazaki, T., Kato, H., and Kuwano, H. (2013). MiR-150 is associated with poor prognosis in esophageal squamous cell carcinoma via targeting the EMT inducer ZEB1. Cancer Sci. *104*, 48–54.

Yuan, J., Chen, L., Chen, X., Sun, W., and Zhou, X. (2012). Identification of Serum MicroRNA-21 as a Biomarker for Chemosensitivity and Prognosis in Human Osteosarcoma. J. Int. Med. Res. *40*, 2090–2097.

Zeng, Y. (2006). Principles of micro-RNA production and maturation. Oncogene. *25*, 6156–6162.

Zenz, T., Habe, S., Denzel, T., Mohr, J., Winkler, D., Buhler, A., Sarno, A., Groner, S., Mertens, D., Busch, R., *et al.* (2009). Detailed analysis of p53 pathway defects in fludarabine-refractory chronic lymphocytic leukaemia (CLL), dissecting the contribution of 17p deletion, TP53 mutation, p53-p21 dysfunction, and miR34a in a prospective clinical trial. Blood *114*, 2589–2597.

Zhang, H.L., Yang, L.F., Zhu, Y., Yao, X.D., Zhang, S.L., Dai, B., Zhu, Y.P., Shen, Y.J., Shi, G.H., and Ye, D.W. (2010). Serum miRNA-*21*, Elevated levels in patients with metastatic hormone-refractory prostate cancer and potential predictive factor for the efficacy of docetaxel-based chemotherapy. Prostate. *71*, 326–331.

Zhang, H.L., Yang, L.F., Zhu, Y., Yao, X.D., Zhang, S.L., Dai, B., Zhu, Y.P., Shen, Y.J., Shi, G.H., and Ye, D.W. (2011). Serum miRNA-*21*, elevated levels in patients with metastatic hormone-refractory prostate cancer and potential predictive factor for the efficacy of docetaxel-based chemotherapy. Prostate *71*, 326–331.

Zhao, J.J., Lin, J., Lwin, T., Yang, H., Guo, J., Kong, W., Dessureault, S., Moscinski, L.C., Rezania, D., Dalton, W.S., *et al.* (2010). MicroRNA expression profile and identification of miR-29 as a prognostic marker and pathogenetic factor by targeting CDK6 in mantle cell lymphoma. Blood *115*, 2630–2639.

Zheng, S.R., Guo, G.L., Zhang, W., Huang, G.L., Hu, X.Q., Zhu, J., Huang, Q.D., You, J., and Zhang, X.H. (2012). Clinical significance of miR-155 expression in breast cancer and effects of miR-155 ASO on cell viability and apoptosis. Oncol. Rep. *27*, 1149–1155.

Zhou, Y., Chen, L., Barlogie, B., Stephens, O., Wu, X., Williams, D.R., Cartron, M.A., van Rhee, F., Nair, B., Waheed, S., *et al.* (2010). High-risk myeloma is associated with global elevation of miRNAs and overexpression of EIF2C2/AGO2. Proc. Natl. Acad. Sci. U.S.A. *107*, 7904–7909.

Zhu, C., Wang, Y., Kuai, W., Sun, X., Chen, H., and Hong, Z. (2013). Prognostic value of miR-29a expression in pediatric acute myeloid leukaemia. Clin. Biochem. *46*, 49–53.

Zhu, W., Liu, X., He, J., Chen, D., Hunag, Y., and Zhang, Y.K. (2011). Overexpression of members of the microRNA-183 family is a risk factor for lung cancer: a case control study. BMC Cancer. *11*, 393.

Zuo, Z., Calin, G.A., de Paula, H.M., Medeiros, L.J., Fernandez, M.H., Shimizu, M., Garcia-Manero, G., and Bueso-Ramos, C.E. (2011). Circulating microRNAs let-7a and miR-16 predict progression-free survival and overall survival in patients with myelodysplastic syndrome. Blood *118*, 413–415.

MicroRNA Deregulation in Trinucleotide Repeat Expansion Disorders

12

Edyta Koscianska, Emilia Kozlowska, Edyta Jaworska and Wlodzimierz J. Krzyzosiak

Abstract

Trinucleotide repeat expansion disorders (TREDs) constitute a group of dominantly inherited neurological diseases that are incurable and ultimately fatal. The underlying cause of TREDs is an expansion of trinucleotide repeats that may occur in the coding and non-coding regions of human genes. MicroRNAs (miRNAs) have emerged as potent regulators of gene expression at the posttranscriptional level. They are involved in a variety of physiological and pathological processes in humans, and the alteration of miRNA expression is considered to be a hallmark of many diseases, including TREDs. This review summarizes the current knowledge regarding the involvement of miRNAs in the pathogenesis of TREDs and the experimentally proven associations between specific miRNAs and particular disorders that have been reported to date.

Introduction

MicroRNAs (miRNAs) are endogenous small (21–23 nt in length) non-coding RNAs that control gene expression at the posttranscriptional level. They down-regulate gene expression by imperfect pairing with complementary sites within transcript sequences and suppress their translation, stimulate deadenylation and degradation, or induce target cleavage (reviewed in Chekulaeva and Filipowicz, 2009). The regulatory potential of miRNAs is enormous; more than 2000 mature human miRNAs have been already deposited in miRNA repository (miRBase, Release 19) (Griffiths-Jones *et al.*, 2008), and over 60% of mammalian protein-coding genes may be regulated by miRNAs (Friedman *et al.*, 2009).

The biogenesis of animal miRNAs includes two subsequent RNA cleavage steps, namely, nuclear and cytoplasmic. Two RNase III endonucleases (Drosha and Dicer) process miRNA precursors sequentially to produce mature miRNAs (reviewed in Kim *et al.*, 2009; Winter *et al.*, 2009; Krol *et al.*, 2010). Briefly, the ribonuclease Drosha, acting together with DGCR8 proteins within the complex named Microprocessor, cleaves primary miRNA transcripts (pri-miRNA) into pre-miRNA precursors of approximately 60 nt in length. The pre-miRNAs are exported through the nuclear pores by Exportin-5 to the cytoplasm, where they undergo further cleavage by the ribonuclease Dicer. The final Dicer products are approximately 22-nt-long miRNA duplexes having 2-nt overhangs at their 3′ ends.

The trinucleotide repeat (TNR) expansion diseases (TREDs) constitute a group of neurological disorders caused by an unstable triplet repeat in a specific region of a gene. The group includes disorders known to be triggered by the expansion of a CAG repeat in the translated regions of respective genes (polyglutamine, polyQ, disorders), such as Huntington's disease (HD), six distinct forms of spinocerebellar ataxia (SCA-1, 2, 3, 6, 7, and 17), dentatorubropallidoluysian atrophy (DRPLA), and spinobulbar muscular atrophy (SBMA). The TRED group also comprises fragile X syndrome (FXS) and fragile X-associated tremor/ataxia syndrome (FXTAS), both of which are caused by a CGG expansion in the 5′-untranslated region (5′UTR) of the *FMR1* gene. Moreover, this group includes a neuromuscular disorder, myotonic dystrophy type 1 (DM1), caused by a CUG expansion in the 3′-untranslated region (3′UTR) of the *DMPK* gene. The TRED types and location of the TNRs in gene regions are presented schematically in Fig. 12.1.

The expanded TNRs are proposed to trigger pathogenic effects based on three types of mechanisms: (i) toxic RNA gain-of-function, (ii) toxic protein gain-of-function, or (iii) both transcript and protein loss-of-function (Mirkin, 2007; Orr and Zoghbi, 2007; Lopez Castel *et al.*, 2010). The toxic RNA gain-of-function mechanism has been proposed to explain the pathogenesis caused by the expanded CUG and CGG repeats in non-translated sequences and is best characterized in DM1 (reviewed in Mankodi *et al.*, 2000; Galka-Marciniak *et al.*, 2012). The toxic protein gain-of-function mechanism has been proposed for diseases triggered by the expansions of translated CAG repeats (polyQ disorders). The mechanism explains neurodegeneration as the result of the expression of glutamine-rich proteins that can misfold and form aggregates (Davies *et al.*, 1997; Birman, 2008; Li and Bonini, 2010). More recently, mutant CAG transcripts have been also shown to contribute to the pathogenesis of polyQ diseases (reviewed in Fiszer and Krzyzosiak, 2013).

Genome-wide approaches used in studies of microRNA deregulation in TREDs

Recent studies have demonstrated that interest in miRNA biogenesis and function is growing rapidly and that deep-sequencing technologies in combination with various microarray analyses have become routinely used in the investigation of short RNAs (e.g. Ruby *et al.*, 2006; Landgraf *et al.*, 2007; Morin *et al.*, 2008; Wei *et al.*, 2009; Wyman *et al.*, 2009; Zhang *et al.*, 2009), including studies on miRNA deregulation in TREDs (Lee *et al.*, 2010; Marti *et al.*, 2010; Persengiev *et al.*, 2011; Fernandez-Costa *et al.*, 2012; Reinhardt *et al.*, 2012; Rodriguez-Lebron *et al.*, 2013; Sellier *et al.*, 2013). The identification of global changes in miRNA and mRNA expression using genome-wide approaches has become a vital prerequisite for investigations of the mechanisms that drive the pathogenesis of TREDs.

The technical advancement of deep sequencing used for miRNA profiling has demonstrated an exceptional ability to comprehensively and accurately assess the levels of the miRNA population in a given sample. This ability to assess population levels is particularly important in view of the recent findings on the potential functional role of miRNA sequence and length variants termed isomiRs (isomiRNAs) (Chiang *et al.*, 2010; Lee *et al.*, 2010; Cloonan *et al.*, 2011; Humphreys *et al.*, 2012) (reviewed in Neilsen *et al.*, 2012). Nevertheless, some practical difficulties related to the profiling of miRNAs implicated

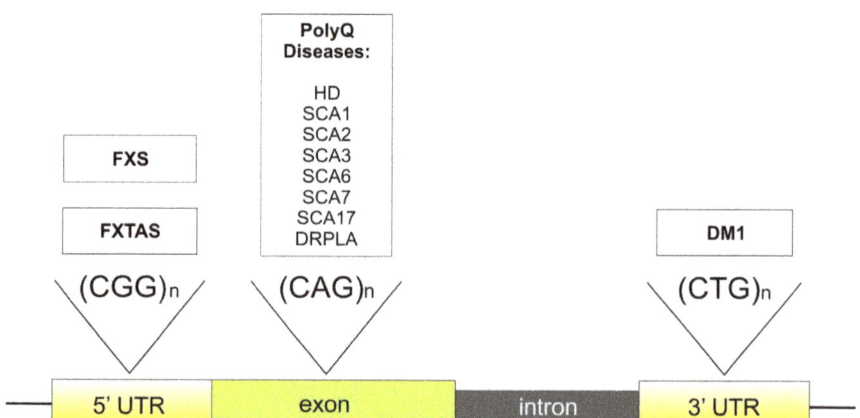

Figure 12.1 Types and location of TNRs in gene regions of the selected TREDs.

in neurodegenerative diseases have been reported (e.g. minimal or no overlap in changes in miRNA expression due to the type and number of samples taken for the analysis, clinical stage of the disease, miRNA isolation method, miRNA detection method, number of available controls, and limitation to the tissue-/cell-specific expression of certain miRNAs), and discussed on the example of studies performed in sporadic Alzheimer's disease (AD) and Parkinson's disease (PD) patients (Hebert and De Strooper, 2009). Similarly, many caveats concerning the use of genome-wide analyses in neurodegenerative disorders, ranging from patient-related parameters to technical issues and data analysis and interpretation, have been discussed elsewhere (Sonntag, 2010).

Generally, both microarrays and next-generation sequencing (NGS) technologies have been used in studies on miRNA expression; however, the utility of various platforms for the detection of differentially expressed miRNAs has not been rigorously evaluated. Three biological samples across six commercially available miRNA microarray platforms were analysed, and their hybridization performance was compared (Git *et al.*, 2010). The following platforms were considered: Agilent Human miRNA Microarray 1.0, Exiqon miRCURY LNA microRNA Array v9.2, Illumina Sentrix Array Matrix 96-well MicroRNA Expression Profiling Assay v1, Ambion mirVana miRNA Bioarrays v2, Combimatrix microRNA 4X2K Microarrays, and Invitrogen NCode Multi-Species miRNA Microarray v2. The differences between these platforms included surface chemistry, printing technology, probe design, labelling techniques, and cost. Their performances also varied considerably, with Ambion, Agilent, and Exiqon microarrays ranking highest in the rate of true differentially expressed (DE) calls.

In terms of miRNA sequencing, 454/Roche, Illumina/Solexa, and SOLiD/ABI are the most frequently used platforms. Although these platforms all have a number of significant advantages, they can generate platform-specific biases that stem from the methods of miRNA library construction (Huse *et al.*, 2007; Dohm *et al.*, 2008; Tian *et al.*, 2010). Therefore, the proper interpretation of deep-sequencing results and the separation of real biological effects from various deep-sequencing artefacts remains ambiguous. Substantial platform-specific biases have been reported for all of these platforms (Huse *et al.*, 2007; Dohm *et al.*, 2008), leading to discordant conclusions regarding miRNA heterogeneity in mammalian cells (Ruby *et al.*, 2006; Wu *et al.*, 2007; Morin *et al.*, 2008; Seitz *et al.*, 2008). Another reported bias was the overabundance of U insertions as compared with U deletions in the case of isomiR sequencing identified in one study (Reid *et al.*, 2008) but not confirmed in another (Cloonan *et al.*, 2011). The latter difference was explained by the different sequencing platforms used; 454 and Illumina sequencing platforms are more susceptible to insertion or deletion artefacts than SOLiD sequencing.

The large-scale miRNA profiling methods have proven highly informative as a whole. They provide a broad picture of the small RNA profile and therefore provide useful information on the pathogenesis of a disease and/or reveal new levels of gene regulation. Nevertheless, despite unquestionable advantages, these techniques have some drawbacks. Some platforms are able to detect stronger changes than others, and their specificity, sensitivity, and accuracy may vary. Thus, many factors should be considered when choosing a platform. Recently, the current knowledge of the relevant methodologies and potential pitfalls in miRNA profiling were extensively reviewed (Chugh and Dittmer, 2012).

The role of microRNAs in the pathogenesis of TREDs

Regulatory miRNAs play a fundamental role in the majority of biological processes. Both the altered miRNA expression and the deregulation of genes controlled by miRNAs have been linked to many disorders, such as cancer and cardiovascular, metabolic, and neurological disorders, including TREDs (e.g. Eisenberg *et al.*, 2007; Johnson *et al.*, 2008; Lee *et al.*, 2008; Croce, 2009; Hebert and De Strooper, 2009; Latronico and Condorelli, 2009; Esteller, 2011; Melo and Esteller, 2011; Chen *et al.*, 2012; Iorio and Croce, 2012).

In terms of neurodegeneration, many studies have directly addressed the role of miRNAs in the development and regulation of the nervous system (reviewed in Schaefer *et al.*, 2007; Bushati and Cohen, 2008; Barbato *et al.*, 2009; Hebert and De Strooper, 2009; Martino *et al.*, 2009; Roshan *et al.*, 2009; Weinberg and Wood, 2009; Lau and de Strooper 2010). It has also been reported that the dysfunction of brain-enriched miRNAs specifically targeting and regulating the expression of disease-associated genes may lead to neurodegeneration and that disease-linked proteins may regulate the miRNA machinery (Savas *et al.*, 2008; Gehrke *et al.*, 2010; Esteller, 2011).

Few reports have been published detailing a close connection between miRNAs and TREDs (see Table 12.1) (earlier reviewed in Witkos *et al.*, 2011); however, there are many publications that strongly suggest such a link. An in-depth *in silico* analysis of the miRNA interactions with mRNAs derived from genes triggering TREDs revealed that these genes may be subject to miRNA regulation (Witkos *et al.*, 2011). Moreover, it has been shown that some miRNAs preferentially repress transcripts with longer CUG repeats (in DM1), which suggests a plausible role for miRNAs in repeat expansion disorders (Hon and Zhang, 2007). Because mammalian miRNAs may decrease target mRNA levels significantly (Guo *et al.*, 2010), targeting the genes bearing TNRs with miRNAs may help to reduce the amount of the mutant protein and/or toxic transcripts.

MiRNA in Huntington's disease

Huntington's disease (HD) is the most common of the nine known polyQ disorders. It is caused by the CAG repeat expansion in exon 1 of the huntingtin (*HTT*) gene. This autosomal dominantly inherited neurodegenerative disease can be characterized by multiple cellular and molecular dysfunctions, especially in striatal and cortical regions, that lead to a massive loss of neurons in these areas. The precise mechanism of neurodegeneration in the case of HD has not been explained; however, both the gain-of-function of the mutant huntingtin (mutHTT) and the loss-of-function of the wild-type HTT protein contribute to the HD pathology (Harper, 2009).

Neurons exhibit considerable sensitivity to mutHTT containing an elongated polyQ chain. Global changes in gene expression induced by the presence of the mutHTT as well as the widespread changes in neuronal microRNAs levels have been described in HD patients (Johnson *et al.*, 2008; Packer *et al.*, 2008; Marti *et al.*, 2010; Sinha *et al.*, 2010, 2011; McCann *et al.*, 2011). Therefore, miRNA deregulation has been recognized as a hallmark of HD and other polyQ diseases.

Several miRNAs whose deregulation are mediated by the repressor element-1 (RE-1) silencing transcription factor (REST), both in the neural mouse ST*Hdh*^{Q7}/*Hdh*^{Q7} cell line and transgenic R6/2 mice, have been identified (Johnson *et al.*, 2008). REST is a transcription repressor that is normally bound by HTT and therefore retained in the neuronal cytoplasm. In the case of HD, the expanded CAG repeats have been shown to prevent the interaction of HTT with REST, which leads to the accumulation of the REST factor in the nucleus (Zuccato *et al.*, 2003). The elevated level of REST in the neuronal nucleus results in the repression of a large cohort of neuron-specific genes, including genes coding miRNAs (Johnson *et al.*, 2008). Conversely, in the same study, the significant up-regulation of several miRNAs was observed when the REST function was lost in the ST*Hdh*^{Q7}/*Hdh*^{Q7} cells (e.g. miR-29a-b, miR-132, miR-135b). Furthermore, the REST factor itself has been shown to be a target for miR-9 and its co-repressor (REST Corepressor 1, CoREST) to be a target for miR-9* (Packer *et al.*, 2008). By being indicative of a double negative feedback loop between the REST silencing complex and miRNAs, the REST-dependent regulation of miRNAs and general miRNA involvement in the regulation of gene expression in the case of HD has been further proven. Moreover, two other miRNAs, namely, miR-137 and miR-153, were recently shown to be direct targets of REST (Soldati *et al.*, 2013). In conclusion, all of the aforementioned papers clearly highlight the importance of REST in the miRNA regulation implicated in HD.

Deregulation of miRNA expression in HD has been systematically studied using animal models of the disease. Specifically, three rodent models of HD (two transgenic mouse lines YAC128 and R6/2 and 3-nitroprioponic acid (3NP)-induced rats) were subjected to miRNA profiling (Lee *et al.*, 2010). In the striatal regions of both mouse models, visible down-regulation of several miRNAs was demonstrated. Additional analyses revealed

Table 12.1 Reported changes in miRNA expression in TREDs

Source of miRNA	Reported Changes	Method	Comments	References
HD				
Human HD, cerebral cortex	↑ miR-29a, ↑ miR-330, ↓ miR-132	RT-qPCR	Information about the number of patients and control groups not provided, inconsistent miR-29a expression data	Johnson *et al*. (2008)
R6/2 mouse, cerebral cortex and hippocampus	↓ miR-29a, ↓ miR-132, ↓ miR-133, ↓ 203			
Human HD, cerebral cortex	↑ miR-132, ↓ miR-9, ↓ miR-9*, ↓ miR-29b, ↓ miR-124a	RT-qPCR	19 patients, 42–78 years old, data for miR-132 inconsistent with other reports, double-negative feedback loop between REST and miRNAs	Packer *et al*. (2008)
R6/2 mouse, striatum, samples from 10-week-old mice	19 miRNAs up-regulated, 61 miRNAs down-regulated	Agilent Mouse miRNA Microarray	Levels of selected miRNAs validated by RT-qPCR, inconsistent rat model data, reduced expression of Drosha and Dicer corresponding with down-regulation of miRNAs in mouse models	Lee *et al*. (2010)
YAC128 mouse, striatum, samples from five-week-old mice	36 miRNAs up-regulated, 10 miRNAs down-regulated			
YAC128 mouse, striatum, samples from 12-week-old mice	7 miRNAs up-regulated, 51 miRNAs down-regulated			
Chemically induced rat model of HD, striatum, 12-week-old rats	4 miRNAs up-regulated, 2 miRNAs down-regulated	Agilent Rat miRNA Microarray		
Human HD, frontal cortex and striatum	19 miRNAs up-regulated, 14 miRNAs down-regulated	Agilent Human miRNA Microarray	Four patients, 28–71 years old, results for 14 miRNAs confirmed by RT-qPCR	Marti *et al*. (2010)
	51 miRNAs up-regulated, 36 miRNAs down-regulated	Illumina Hi-Seq 2000	Two patients, 65 and 59 years old, deep sequencing of miRNA variants (isomiRs)	
ST*Hdh* cells$^{7/7}$ ST*Hdh* cells$^{111/111}$	12 miRNAs up-regulated, 15 miRNAs down-regulated	RT-qPCR	Expression analysis of 90 mature miRNA, miR-146a targets TBP which is recruited by the mutHTT	Sinha *et al*. (2010)
ST*Hdh* cells$^{7/7}$ transfected 83Q; R6/2 mice, striatum	↓ miR-146a, ↓ miR-125b, ↓ miR-150	RT-qPCR	Only three miRNAs investigated, p53-mediated decrease in miR-146a expression by interaction with RelA/NFkB	Ghose *et al*. (2011)
NT2 cells, human HD, plasma	↑ miR-34b, ↑ miR-1285	RT-qPCR miRCURY LNA miRNA Array (Exiqon)	miR-34b is a plasma-stable marker for HD	Gaughwin *et al*. (2011)
ST*Hdh* cells$^{7/7}$ ST*Hdh* cells$^{111/111}$	miR-214, -150, -146a, and -125b down-regulate HTT gene	Luciferase reporter assays	miR-mRNA HTT interaction confirmed in HeLa cells, miRNAs affect formation of mutHTT aggregates	Sinha *et al*. (2011)

Table 12.1 continued

Source of miRNA	Reported Changes	Method	Comments	References
N171-82Q mouse, cerebral cortex, 12-week-old mice	8 miRNAs up-regulated, 7 miRNAs down-regulated	miRNA microarray	miR-200a and miR-200c validated by RT-qPCR (both up-regulated)	Jin *et al.* (2012)
N171-82Q mouse, cerebral cortex, 18-week-old mice	16 miRNAs up-regulated, 8 miRNAs down-regulated			
ST*Hdh* cells$^{7/7}$, ST*Hdh* cells$^{109/109}$	15 miRNAs down-regulated	RT-qPCR	Only miRNAs having REST binding site or down-regulated in HD were investigated	Soldati *et al.* (2012)
SCA1				
Human MCF7, HEK293T, HeLa and mouse NIH3T3 cells	miR-19a, -101, -130a down-regulate *ATXN1* gene	RT-PCR, Western blot	Validation of the selected miRNAs by western blot analysis for ATXN1 only in MCF7 cells	
C57/B6 WT mouse, cerebellar Purkinje cells, 24-week-old mice		Northern blot *In situ* hybridization	miR-101 and miR-130 more abundant in Purkinje cells than miR-19	Lee *et al.* (2008)
HEK293T cells		Cell death assay	Inhibition of endogenous miRNAs enhances cytotoxicity of the polyQ expanded mutant ATXN1	
Human, cerebellum	25 miRNAs up-regulated, ↓ miR-144, ↓ miR-1827	miRCURY LNA human microRNA Array (Exiqon)	65-year-old patients	Persengiev *et al.* (2011)
Human, cortex	77 miRNAs up-regulated		65-year-old patients, only up-regulated miRNAs presented	
Drosophila, heads	↑ miR-33-5p, ↑ miR-34-5p, ↑ miR-92a-5p	Illumina Hi-Seq 2000	Low expression values of most miRNAs, high variability of expression between sample replicates or the set of control samples	Reinhardt *et al.* (2012)
BO5 mouse, Purkinje neurons	34 miRNAs up-regulated, 12 miRNAs down-regulated	miRCURY LNA all species microRNA arrays (Exiqon)	Cerebellar samples; transgenic and control mice analysed at two time points: four weeks (pre-symptomatic stage of pathogenesis), 12 weeks (symptomatic stage); different numbers of changes in miRNA expression in time points	Rodriguez-Lebron *et al.* (2013)
	↑ miR-150	*In situ* hybridization	Other miRNAs not analysed	
BO5 mouse, granule cells	↓ miR-150			
SCA2				
Transgenic recombinant flies, wing imaginal discs cells	↑ *bantam*, ↑ miR-12	Confocal microscopy	Clones of $atx2^{X1}/atx2^{X1}$ with miRNA-dependent translational mir12 reporter were used	McCann *et al.* (2011)

Table 12.1 continued

Source of miRNA	Reported Changes	Method	Comments	References
SCA3				
Drosophila dcr-1 mutants, eyes	*bantam* modulates polyQ degeneration	Phenotype mutants comparison analysis	*bantam* prevents toxicity induced by the SCA3 protein	Bilen *et al.* (2006)
Drosophila, heads	↑ miR-1-3p, ↑ miR-100-5p, ↑ miR-33-5p, ↑ miR-92a-5p	Illumina Hi-Seq 2000	As described for SCA1 with regard to the same analysis	Reinhardt *et al.* (2012)
SCA7				
Drosophila, heads	↑ miR-33-5p, ↑ miR-375-3p, ↑ miR-92a-5p	Illumina Hi-Seq 2000	As described for SCA1 with regard to the same analysis	Reinhardt *et al.* (2012)
DRPLA				
Drosophila miR-8 mutants	dma-miR-8 down-regulates *D. melanogaster* atrophin gene	RT-qPCR	Atrophin overexpression contributes to the *mir-8* mutant phenotype	Karres *et al.* (2007)
DM1				
Human muscle biopsies from the vastus lateralis	↑ miR-206	RT-qPCR	Seven unrelated patients, aged 30–50 years	Gambardella *et al.* (2010)
		Northern blot	Total RNAs were pooled (two samples analysed vs. one control)	
Human muscle biopsies from the biceps brachii	↑ miR-1, ↑ miR-335, ↓ miR-29b, ↓ miR-29c, ↓ miR-33	RT-qPCR	Nine affected males and six females, age = 38±17 years old	Perbellini *et al.* (2011)
Human heart left ventricles samples	↓ miR-1	RT-qPCR	Eight adults	Rau *et al.* (2011)
		Northern blot		
Human muscle biopsies from the biceps brachii	↑ miR-208a, ↑ miR-381, ↓ miR-193b-3p	Gene Chip Human Exon 1.0 ST Array (Affymetrix)	miRNAs that were validated as significantly deregulated in DM2 (11 miRNAs) were also tested in an age- and sex-matched cohort of DM1 patients	Greco *et al.* (2012)
Drosophila i(CTG)480 transgenic line	1 miRNA up-regulated, 19 down-regulated	SOLiD™ 3 sequencing	Deregulation of miR-1, miR-7 (given their conservation in humans) and miR-1003 (given its miRtronic nature) validated by northern blotting	Fernandez-Costa *et al.* (2012)
Human skeletal muscle biopsies (biceps, vastus and deltoid)	↓ miR-1, ↓ miR-7, ↓ miR-10a	RT-qPCR	Five patients, aged 47±5 years	
FXS/FXTAS				
UAS transgenic *Drosophila*, brains and ventral nerve tissues	↑ miR-124a	Immunoprecipitation	Samples derived from the third instar larvae, focused on only one miRNA	Xu *et al.* (2008)

Table 12.1 continued

Source of miRNA	Reported Changes	Method	Comments	References
HEK293T cells	miR-19b, miR-302* and miR-323-3p down-regulate *FMR1* 3′UTR	Luciferase reporter assays	Only interaction with the miRNA targets was shown	Yi *et al.* (2010)
fxr1 knockout E18 mouse embryo, brain tissues and skeletal muscles	FXR1 regulates the level of miR-124, miR-9, and miR-1	Northern blot	Only two brain-specific miRNAs investigated, lower expression of miRNAs in *fxr1* knockout mice	Xu *et al.* (2011)
fmr1/fxr2 double-knockout mouse, brain tissues, 2-month-old mice			Higher expression of brain-specific miRNAs in *fmr1/fxr2* double-knockout individuals	
UAS-CGG_{60} transgenic *Drosophila*, brain samples	↑ *bantam*, ↑ let-7, ↑ miR-1, ↑ miR-13a, ↑ miR-14, ↑ miR-124, ↑ miR-277, ↓ miR-9b, ↓ miR-285	RT-qPCR	Expression profiling of 72 known miRNAs	Tan *et al.* (2012)
Transgenic UAS *Drosophila*, eyes	miR-277 regulates rCGG-mediated neurodegeneration	Scanning electron microscopy (SEM)	Overexpression of miR-277 in flies enhances neurodegeneration compared to controls, blocking the activity of miR-277 by miR-277 sponge suppresses neurodegeneration	
KI mouse, brains, 15-week-old	↓ miR-221	RT-qPCR	Only three miRNAs investigated	Zongaro *et al.* (2013)
Human FXTAS, cerebellum	66 miRNAs up-regulated, 12 down-regulated	miRNA microarray	Results confirmed by RT-qPCR in frontal cortex samples of 13 FXTAS patients for selected miRNAs	Sellier *et al.* (2013)

that the observed decrease in miRNA levels could have been associated with the decreased levels of enzymes pivotal to miRNA biogenesis (i.e. Dicer and/or Drosha). However, the (3NP)-induced rats presented only minor or even no changes in miRNA expression levels. These results suggest that the mutHTT may be directly or indirectly involved in miRNA deregulation or that animals with chemically induced neurodegeneration may not be a proper model for HD.

In addition to animal models, ST*Hdh* cell lines have been used to model HD for miRNA investigation. Indeed, studies performed in such models have revealed important deregulation of miRNA expression. More specifically, miRNA expression in ST*Hdh* cell lines expressing full-length *HTT* gene with either 7 or 111 glutamine residues, STHdh^{Q7}/Hdh^{Q7} and STHdh^{Q111}/Hdh^{Q111} cells, respectively, was compared. Of the 90 miRNAs tested in STHdh^{Q111}/Hdh^{Q111} cells, 15 miRNAs were observed to be down-regulated and 12 were observed to be up-regulated (Sinha *et al.*, 2010). Importantly, these 27 altered miRNAs have been also predicted to target 142 experimentally validated genes, including the TATA binding protein (TBP). This protein is known to be recruited to the mutHTT aggregates (Huang *et al.*, 1998). miR-146a is one of the miRNAs that have been observed to potentially down-regulate TBP. In STHdh^{Q111}/Hdh^{Q111} cells, the increased level of TBP was correlated with the reduced expression of miR-146a. Therefore, TBP has been proposed as a new experimentally determined target of miR-146a (Sinha *et al.*, 2010). The same group subsequently studied miRNA targets in a cell model of HD more extensively and reported that miR-214, -150, -146a, and -125b target the *HTT* gene (Sinha *et al.*, 2011) (Table 12.1). Moreover, they demonstrated that these miRNAs can affect the formation of mutHTT aggregates, toxicity induced by the mutHTT, and the expression of brain-derived neurotrophic factor (BDNF), which collectively contribute to HD pathogenesis.

The significant down-regulation of miR-146a has been further confirmed in other studies performed with the use of the model ST*Hdh* cells. It has been suggested that the observed decrease in miR-146a level might be linked to both the lower expression and activity of p65 subunit of NF-κB (RelA/NF κB) and an increased p53 level (Ghose *et al.*, 2011). The latter finding was also true for the same investigation conducted in the striatum of R6/2 mice, a mouse model of HD.

Very recently, large-scale analyses of miRNA deregulation have been performed. The profiling of miRNA expression in the cerebral cortex of the N171–82Q mouse model of HD revealed that the expression of miRNAs might be dynamically regulated upon disease progression (Jin *et al.*, 2012). Intriguingly, the miR-200 family members displayed a significant up-regulation in the pre-symptomatic stage of HD, suggesting the potential role of those miRNAs in the control of genes involved in neuronal dysfunction caused by the mutHTT. Moreover, several transcripts encoding proteins critical for neuronal plasticity and survival were among the miR-200a and miR-200c targets, and the levels of these proteins were observed to be decreased in the N171-82Q HD model. However, the down-regulation of the miR-200 family has not been reported in other mouse models (Johnson *et al.*, 2008; Lee *et al.*, 2011) or brain samples obtained from HD patients (Marti *et al.*, 2010).

The most comprehensive analysis of miRNA expression in the human brain was performed with the use of the Illumina massively parallel sequencing (Marti *et al.*, 2010). Two forebrain areas, the frontal cortex (FC) and striatum (ST), of both healthy individuals and HD patients were analysed. As expected, the global deregulation of miRNAs in samples taken from the HD patients was reported; however, miRNA variants of length and/or sequence (isomiRs) were also observed for the vast majority of miRNAs. This report confirms the alterations in miRNA profiles in the case of HD but also, and more importantly, highlights the role of isomiRs in gene expression regulation in both physiological and pathological conditions.

MiRNA in spinocerebellar ataxias

There are six spinocerebellar ataxias (SCAs) among the polyQ diseases (Gatchel and Zoghbi, 2005; Huse *et al.*, 2007; Orr, 2012). These SCAs are caused by the expression of CAG repeats of abnormal length in the open reading frame (ORF) in the ataxin (*ATXN*) genes. These ataxias have similar clinical presentations, and molecular studies are necessary to distinguish particular SCA types.

Specific links between miRNA regulation and SCAs have been described in several studies. Specifically, the research on SCA1 has revealed that some miRNAs can regulate the level of ATXN1 protein (see Table 12.1). It has been reported that miR-19a, miR-101, and miR-130a may bind to the 3′UTR of ATXN1 and suppress the translation of ATXN1 (Lee *et al.*, 2008). In a reporter assay, miRs 19a, 101, and 130a exerted their effects when transfected either individually or collectively. Interestingly, miR-101 affected both the mRNA and protein levels, whereas miR-19a and miR-130a were able to decrease the protein levels only. Inhibition of all three of these miRNAs enhanced the cytotoxicity of the mutant ATXN1 protein in human cells and mouse models. Therefore, miRNA-mediated posttranscriptional regulation of ATXN1 may modulate the SCA1-related neuropathology by affecting the protein levels.

Genome-wide microarray analysis has revealed global age-related miRNA down-regulation in human, chimpanzee, and rhesus macaque brains (cerebellum and cortex); however, only a small subset of miRNAs has been found to be selectively up-regulated in these samples (Persengiev *et al.*, 2011). The inhibition of miR-101 and miR-144 resulted in an increased level of ATXN1 in human cells. The latter observation was in accordance with previous reports (Schaefer *et al.*, 2007; Lee *et al.*, 2008). Thus, the activation of specific miRNAs could serve to reduce the cytotoxic effect of the expanded mutATXN1. Recently, global profiling of miRNA expression in the cerebellum of pre-symptomatic and symptomatic BO5 SCA1 transgenic mice revealed changes in the level of 46 miRNAs (Rodriguez-Lebron *et al.*, 2013). Of the miRNAs changed in SCA1 mouse brains, 34 displayed increased expression and 14 of these miRNAs exhibited increased expression for both time points analysed (4- and 12-week-old mice). The miRNA expression level was decreased in 12 miRNAs, and the level of miR-381 was decreased in both the pre- and symptomatic stages.

In SCA1 transgenic mice, the expression of mutant Atxn1 protein is restricted to cerebellar Purkinje cells. Previous reports have suggested that crosstalk between Purkinje and granule cells might play a role in SCA1 (Gatchel *et al.*, 2008). In light of these findings, the differential expression of miR-150 in Purkinje neurons and granule cells detected by *in situ* hybridization appears to be of importance. Specifically, the level of miR-150 was shown to increase in cerebellar Purkinje neurons, whereas its expression in granule cells was slightly decreased when compared with wild-type cells (Rodriguez-Lebron *et al.*, 2013). Therefore, the pathogenic process of SCA1 might be involved in the regulation of miR-150 expression in Purkinje cells.

All of the above-mentioned studies concerned miRNA regulation that occurs mainly in elderly patients or animal models of the disease that are rather advanced in age. However, the miRNA deregulation can be also observed at the beginning of the pathological process. Such deregulation was investigated in *Drosophila* transgenic models of three SCAs (i.e. SCA1, SCA3 and SCA7) with the use of the NGS method. Specifically, two RNA samples from flies and one control were analysed with regard to the specific expanded expression of atxn proteins (atxn1, atxn3, and atxn7). Many of the selected miRNAs displayed low expression or highly variable expression between the tested samples. Nevertheless, a trend towards the over-expression of miR-33 and miR-92a was observed in all three ataxia models. These results suggest that the pathogenesis of SCAs in the early stages of the disease is likely not related to the deregulation of miRNA expression (such a conclusion was reached due to the lack of significance of the observed changes) (Reinhardt *et al.*, 2012). In contrast, *bantam* was identified as a downstream modulator of polyQ toxicity in a *Drosophila* model of SCA3 miRNA. Reduced miRNA processing due to the knockout of Dicer1 enhanced the neurodegeneration induced by ataxin3 in both flies and human cells (Bilen *et al.*, 2006). However, some pathogenic proteins, such as ataxin2, might be required for miRNA functioning (McCann *et al.*, 2011). These findings suggest that miRNA pathways can modulate polyQ-induced neurodegeneration but that the pathogenesis of polyQ diseases can also influence miRNA expression.

MiRNA in dentatorubropallidoluysian atrophy

In dentatorubropallidoluysian atrophy (DRPLA), neurodegeneration is caused by an unstable expansion of a CAG repeat in the atrophin-1 protein (Koide *et al.*, 1994). Little is known about the direct association of miRNAs with the *ATN1* gene in humans, but their potentially significant role has been roughly delineated in studies performed in fruit flies (Karres *et al.*, 2007). More specifically, the transcriptional regulator atrophin was shown to be a direct functional target of the conserved dma-miR-8 (which has the same seed sequence as hsa-miRs-200b and -429). Elevated apoptosis in the brain and behavioural defects were consequences of the reduced atrophin expression. The study demonstrated the involvement of miR-8 in the development and function of the nervous system.

MiRNA in myotonic dystrophy type 1

Myotonic dystrophy type 1 (DM1) is a neuromuscular disorder caused by an expanded CUG repeat occurring in the 3′UTR of the *Dystrophia Myotonica Protein Kinase* (*DMPK*) gene (Mahadevan *et al.*, 1992). The repeat expansion results in the nuclear retention of DMPK mRNA and reduced DMPK protein levels (Davis *et al.*, 1997). Toxic transcripts sequester the Muscleblind-like 1 (MBNL1) splicing factor, leading to the abnormal alternative splicing of multiple transcripts (Mankodi *et al.*, 2002; Lin *et al.*, 2006).

Although spliceopathy is considered to be the major component of the disease, recent findings suggest that changes in the gene expression, translation efficiency, antisense transcripts, and miRNA deregulation may also contribute to DM1 pathogenesis (Sicot *et al.*, 2011). In terms of miRNA deregulation, the altered expression of specific miRNAs has been associated with muscular dystrophies and cardiomyopathies (Eisenberg *et al.*, 2007; van Rooij *et al.*, 2008; Latronico and Condorelli, 2009) and with myotonic dystrophy type 2 (DM2) (Greco *et al.*, 2012).

Several changes in muscle-specific miRNAs (myo-miRs) have been reported for DM1 (Table 12.1). Specifically, miR-1 and miR-335 have been observed to be up-regulated in DM1 skeletal muscle, whereas miRs-29b, -29c, and -33 have been observed to be down-regulated (Perbellini *et al.*, 2011). In addition, miR-1 has been observed to be down-regulated in cardiac muscle (Rau *et al.*, 2011), and miR-206 has been observed to be up-regulated in samples derived from skeletal muscle (Gambardella *et al.*, 2010). Moreover, miR-206 has been observed to be prevalently expressed in the nuclear regions, which may indicate its nuclear-specific function in muscle tissue (Gambardella *et al.*, 2010). The cellular distribution of myo-miRs -1, -133b, and -206 was significantly different in DM1 muscles as compared to control subjects, possibly impairing their function (Perbellini *et al.*, 2011).

The deregulation of DM-associated miRNA has been linked to some alterations in their putative targets' expression, indicating that miRNA misregulation in DM1 is functionally relevant and may contribute to the disease pathology (Perbellini *et al.*, 2011; Rau *et al.*, 2011). Interestingly, the decreased expression of mature miR-1 and increased levels of its targets in the hearts of individuals with myotonic dystrophy is mediated by the functional depletion of MBNL1, a sequestered splicing factor, which affects the processing of pre-miR-1 (Rau *et al.*, 2011). Whether MBNL1 influences the processing of other pre-miRNAs remains unknown (for details, refer to the section 'Biogenesis of miRNAs and pathogenesis of polyglutamine disorders might be related'). Recently, a transgenic fly model of DM1 (i(CTG)480 *Drosophila* line carrying 480 CTG repeats) and SOLiD sequencing were employed to identify miRNA alterations caused directly by CTG expansions (Fernandez-Costa *et al.*, 2012). The analysis revealed that the expression of 20 miRNAs was changed in DM1 flies versus control flies; 19 were down-regulated, and one was up-regulated. These results were validated in skeletal muscles from DM1 patients displaying reduced levels of miR-1, miR-7, and miR-10 as well as the elevated levels of the targets of these miRNAs. Moreover, the involvement of MBNL in pre-miR-1 processing was confirmed, and its similar function for pre-miR-107 has been suggested. Taken together, the above-mentioned reports indicate the pathological potential of miRNA dysregulation in DM1.

MiRNA in fragile X syndrome and fragile X-associated tremor/ataxia syndrome

A disease-causing factor for fragile X syndrome (FXS) is the trinucleotide CGG expansion in the 5′-untranslated region (5′UTR) of the *Fragile X Mental Retardation* (*FMR1*) gene (Verkerk *et al.*, 1991). In contrast to FXS, wherein long expansions result in hypermethylation and silencing of the *FMR1* gene, fragile X-associated tremor/ataxia syndrome (FXTAS) is characterized by the presence of shorter CGG tracts and normal or near-normal FMRP expression (Tassone *et al.*, 2000). FMRP, the protein product of the normal *FMR1* gene, is an RNA-binding protein that interacts with target RNAs and is implicated in mRNA transport and translational control (Eberhart *et al.*, 1996; Bassell and Warren, 2008). FMRP is also involved in RNA-interference (RNAi) machinery (Siomi *et al.*, 2004). FMRP deficiency causes synapse deformity in the neurons essential for cognition and memory activities.

Several studies have revealed links between miRNA pathways and mutant *FMR1* genes in FXS in both *Drosophila* and mammals (Jin *et al.*, 2004a; Cheever and Ceman, 2009a). The FMR protein has been shown to interact with RNAi components, such as AGO1, AGO2, and Dicer, and to associate with endogenous miRNAs (Caudy *et al.*, 2002; Jin *et al.*, 2004a,b; Plante *et al.*, 2006; Yang *et al.*, 2007; Cheever and Ceman, 2009a; Perron and Provost, 2009). In addition, the involvement of FMRP in the processing of pre-miRNAs has been reported (Plante *et al.*, 2006; Xu *et al.*, 2008; Cheever and Ceman, 2009b). Most recently, the RNA-binding protein DGCR8 has been shown to bind to expanded CGG repeats. The resultant partial sequestration of DGCR8 and Drosha within CGG aggregates leads to the reduced activity of these proteins, the inhibition of miRNA processing, and thus to decreased levels of mature miRNAs in the brain tissues obtained from patients with FXTAS (Sellier *et al.*, 2013).

There is considerable evidence linking specific miRNAs with the *Drosophila* homologue of mammalian FMRP (dFMRP) (see Table 12.1). Moreover, growing attention is being paid to the role of miRNA in DNA methylation pathways in the regulation of synaptic plasticity observed in the FXS (Nomura *et al.*, 2008; Smalheiser and Lugli, 2009).

Biogenesis of miRNAs and pathogenesis of polyglutamine disorders might be related

Transcripts containing expanded CUG repeats are toxic and promote DM1 pathogenesis through an RNA-mediated gain-of-function mechanism (Savkur *et al.*, 2001; Kanadia *et al.*, 2003; Ho *et al.*, 2005; Kimura *et al.*, 2005; Wheeler *et al.*, 2007). This mechanism assumes that the mutant transcript is retained in the cell nucleus and forms foci (Taneja *et al.*, 1995; Timchenko *et al.*, 1996; Davies *et al.*, 1997). Specific RNA-binding proteins, such as the MBNL family proteins, are sequestered (Miller *et al.*, 2000; Mankodi *et al.*, 2002). The MBNL protein sequestration results in its compromised function, and an alternative splicing of MBNL-sensitive genes is affected (reviewed in Orr and Zoghbi, 2007; Wojciechowska and Krzyzosiak, 2011; Galka-Marciniak *et al.*, 2012). Recently, similar phenomena have been demonstrated in cells with CAG repeat expansions, in *HTT* and *ATXN3* transcripts, within intranuclear MBNL1-positive RNA foci (de Mezer *et al.*, 2011; Mykowska *et al.*, 2011), suggesting that similar mechanisms may also occur in other polyQ diseases (reviewed in Fiszer and Krzyzosiak 2013).

One of the factors that can increase sequestration of MBNL1 by expanded CUG and CAG repeats is the DEAD-box RNA helicase p68 (DDX5) (Laurent *et al.*, 2012). The DEAD-box helicases are involved in many cellular processes, including the transcription, processing, export, translation, and degradation of RNA (Fuller-Pace, 2006; Janknecht, 2010). Both p68 and its paralogue, p72 (DDX17), have been observed to co-localize with CUG repeats. Furthermore, the binding of p68 to CUG repeats results in a local disruption of the C–G pairs of the CUG hairpin stem, allowing for the recruitment and stabilization of MBNL1 binding to repeats (Laurent *et al.*, 2012). The DEAD-box helicases also play a role in the biogenesis of miRNAs; p68 and p72 are subunits of the larger human Drosha (hDrosha) complex (Gregory *et al.*, 2004), and they are required for the recognition of a subset of pri-miRNAs in the nucleus (Fukuda *et al.*, 2007).

Large-scale analyses conducted recently have shown that in cells having expanded repeats of some types, the expression of miRNAs is altered when compared with cells with repeats of normal length (details in previous subsections). One of the regulators of miRNA biogenesis at the cytoplasmic level is MBNL1 itself. MBNL1 binds to the conserved UGC motif in loops of specific pre-miRNAs, where it competes with LIN28 protein (Rau *et al.*, 2011). LIN28 is an RNA-binding protein regulating the expression of the let-7 miRNA family (Heo *et al.*, 2008; Rybak *et al.*, 2008). Sequestration of MBNL1 by the expanded CUG repeats allows LIN28 to bind to the UGC motif in the terminal loop of pre-miR-1 and trigger its uridylation, which blocks Dicer processing and results in the decreased expression of miR-1. Consequently, the expression of miR-1 targets is increased, which can be linked to DM1 symptoms in both humans and *Drosophila* models (Rau *et al.*, 2011; Fernandez-Costa *et al.*, 2012).

Some proteins, which co-localize with the expanded repeats, may potentially display effects similar to those caused by sequestration. One of these proteins is nucleolin (NCL), a regulator of *rRNA* transcription. It has been shown that CAG-expanded repeats induce nucleolar stress in polyQ diseases, which is related to NCL interaction with the mutant transcript. The expanded repeats prevent NCL from binding to the *upstream control element* (*UCE*) of the *rRNA promoter*. This interaction triggers hypermethylation of the *UCE* and inhibits rRNA transcription. Consequently, the reduced level of rRNA results in the accumulation of ribosomal proteins and the interaction of these proteins with MDM2 E3 ubiquitin ligase, which causes mitochondrial accumulation of the p53 protein. Interaction between p53 and the anti-apoptotic protein Bcl-xL causes oligomerization of the pro-apoptotic protein Bak. This interaction allows for the release of cytochrome *c* to the cytosol and the activation of a caspase cascade, which induces apoptosis (Tsoi *et al.*, 2012; Tsoi and Chan, 2013).

It is also likely that NCL modulates biogenesis of a subset of miRNAs. In line with this possibility, the levels of miR-15a and miR-16 are directly correlated with the NCL expression in cells, which can be explained by the NCL interaction with the Microprocessor complex. NCL is predominantly localized in the nucleolus but can also be detected in the cytoplasm, where it stabilizes the bcl-2 mRNA and thereby inhibits apoptosis. The cellular localization of NCL is critical for the proper functioning of this protein in the apoptosis pathway (Mi *et al.*, 2003; Otake *et al.*, 2007). Furthermore, NCL can bind to pri-miRNA in a specific Microprocessor-independent manner; however, NCL interaction with DGCR8 and Drosha appears to be stronger (Pickering *et al.*, 2011). This mechanism is particularly important in view of recent reports indicating that DGCR8 and Drosha are sequestered by the expanded CGG repeats in brain tissues from patients with FXTAS. In addition, the mechanism is related to miRNA processing alterations that result in the decreased levels of mature miRNAs in neuronal cells (Sellier *et al.*, 2013). Whether similar processes occur in cells with CAG or CUG repeat expansions requires further investigation. In light of recent studies, it appears possible that transcripts containing expanded CAG repeats may also be toxic and promote the pathogenesis of polyQ disorders. Moreover, all of the RNA–protein interactions described above indicate that the biogenesis of miRNAs and the pathogenesis of polyQ disorders may be linked.

Insights into therapeutic approaches using miRNA-mediated repression

Until now, there has been no effective therapy for patients suffering from TREDs. However, RNAi technology and miRNA-based therapeutics have been proposed and have significant medical potential for these types of disorders. There are several miRNA-based therapeutic strategies that could be used (reviewed in Broderick and Zamore, 2011; Junn and Mouradian, 2012). The simplest intervention is to block endogenous miRNAs or deliver exogenous miRNAs. Either synthetic miRNA mimetics or miRNA-encoding expression vectors can be employed for miRNA overexpression. Similarly, to block miRNA activity, synthetic or expressed inhibitors can be used. Antisense oligonucleotides that have been chemically modified (Meister *et al.*, 2004) or expressed sequences corresponding to multiple miRNA seed targets (miRNA sponges) (Ebert *et al.*, 2007) have been successfully used as miRNA inhibitors.

A promising research direction is the construction of artificial miRNAs targeting new sites in relevant 3′UTRs (Liang *et al.*, 2007) and the production of artificial mirtrons mimicking natural pre-miRNA intronic hairpins generated through the splicing of coding mRNAs. The successful knockdown of human DMPK in DM500 murine myoblasts was recently reported; the transfection of the artificial DMPK-Mirt5 corrected splicing abnormalities in Serca-1 mRNA associated with CUG sequestration (Seow *et al.*, 2012). Moreover, synthetic appropriately designed miRNA-like duplexes have been proposed for use instead of standard siRNA duplexes to improve silencing specificity (Hu *et al.*, 2010; Betancur *et al.*, 2012; Fiszer *et al.*, 2012).

The use of miRNAs as potential therapeutics remains challenging with regard to the methods of delivery and their target specificity (unintended off-target effects might appear) (Olejniczak *et al.*, 2010). However, miRNAs can serve as clinically useful disease biomarkers because they are present in human peripheral blood plasma. The usefulness of miRNAs for both the diagnosis and prognosis of a disease (mostly cancer) has been reported (Li *et al.*, 2010; Wang *et al.*, 2011; Witwer *et al.*, 2011). MiRNA has also been used for this purpose for HD (Gaughwin *et al.*, 2011) and other central nervous system diseases (Witwer *et al.*, 2011).

The main drawback in the use of therapeutic miRNAs in TREDs is the fact that miRNA does not discriminate between normal and mutant transcripts. Nevertheless, it has been shown that the partial suppression (45% reduction) of both wild-type and mutant *HTT* alleles in the striatum prevents behavioural and neuropathological abnormalities in rhesus models of HD.

Most importantly, this reduction in the HTT level is safe and well tolerated (McBride *et al.*, 2011)

To date, only a few dozen of miRNA-regulating disease-relevant genes have been identified. It would be advantageous to identify candidate miRNAs that could be functionally linked to a specific pathogenic protein/transcript. With the advances in miRNA detection by sequencing and functional analyses (global miRNA expression studies), increasingly specific miRNA functions are being elucidated. However, isomiRNAs and length and/or sequence miRNA variants are commonly reported in deep-sequencing studies. These variants have been shown to associate actively with the RISC and with translational machinery and have thus been proven to be real physiological miRNA variants (Chiang *et al.*, 2010; Lee *et al.*, 2010; Cloonan *et al.*, 2011; Humphreys *et al.*, 2012). Although their biological significance is yet to be fully resolved, there is no doubt that the existence of isomiRs, and especially their functionality, could have far-reaching implications for therapeutic applications.

Overall, because no conventional therapy exists to cure patients suffering from TREDs, a full understanding of the miRNA machinery and miRNA functions may lead to the development of new treatment strategies for preventing or delaying the neurodegenerative process underlying TREDs.

Conclusions

By targeting multiple transcripts and affecting expression of numerous proteins, miRNAs play a key role in the fine tuning of virtually all cellular functions (i.e. proliferation, differentiation, development, apoptosis, metabolism, and neurodegeneration). In addition, abnormal miRNA expression appears to characterize many diseases. The evidence is growing that a deregulated miRNA/mRNA expression network also contributes to the pathogenesis of TREDs. The exact roles of particular miRNAs as well as the question of whether these miRNAs are causative for or a consequence of these diseases remain unresolved. As there are many unknowns in the pathogenesis of TREDs, defining a possible role for miRNAs could enhance the overall understanding of the disease mechanism and thus provide new angles for therapeutic intervention.

Acknowledgements

This work was supported by the Polish Ministry of Science and Higher Education (N N301 523038) and the National Science Centre (DEC-2012/06/A/NZ1/00094).

References

Barbato, C., Ruberti, F., and Cogoni, C. (2009). Searching for MIND: microRNAs in neurodegenerative diseases. J. Biomed. Biotechnol. *2009*, 871313.

Bassell, G.J., and Warren, S.T. (2008). Fragile X syndrome: loss of local mRNA regulation alters synaptic development and function. Neuron. *60*, 201–214.

Betancur, J.G., Yoda, M., and Tomari, Y. (2012). MiRNA-like duplexes as RNAi triggers with improved specificity. Front. Genet. *3*, 127.

Bilen, J., Liu, N., Burnett, B.G., Pittman, R.N., and Bonini, N.M. (2006). MicroRNA pathways modulate polyglutamine-induced neurodegeneration. Mol. Cell. *24*, 157–163.

Birman, S. (2008). Neurodegeneration: RNA turns number one suspect in polyglutamine diseases. Curr. Biol. *18*, R659–R661.

Broderick, J.A., and Zamore, P.D. (2011). MicroRNA therapeutics. Gene. Ther. *18*, 1104–1110.

Bushati, N., and Cohen, S.M. (2008). MicroRNAs in neurodegeneration. Curr. Opin. Neurobiol. *18*, 292–296.

Caudy, A.A., Myers, M., Hannon, G.J., and Hammond, S.M. (2002). Fragile X-related protein and VIG associate with the RNA interference machinery. Genes Dev. *16*, 2491–2496.

Cheever, A., and Ceman, S. (2009a). Translation regulation of mRNAs by the fragile X family of proteins through the microRNA pathway. RNA Biol. *6*, 175–178.

Cheever, A., and Ceman, S. (2009b). Phosphorylation of FMRP inhibits association with Dicer. RNA *15*, 362–366.

Chekulaeva, M., and Filipowicz, W. (2009). Mechanisms of miRNA-mediated post-transcriptional regulation in animal cells. Curr. Opin. Cell Biol. *21*, 452–460.

Chen, P.S., Su, J.L., and Hung, M.C. (2012). Dysregulation of microRNAs in cancer. J. Biomed. Sci. *19*, 90.

Chiang, H.R., Schoenfeld, L.W., Ruby, J.G., Auyeung, V.C., Spies, N., Baek, D., Johnston, W.K., Russ, C., Luo, S., Babiarz, J.E., *et al.* (2010). Mammalian microRNAs: experimental evaluation of novel and previously annotated genes. Genes Dev. *24*, 992–1009.

Chugh, P., and Dittmer, D.P. (2012). Potential pitfalls in microRNA profiling. Wiley Interdiscip. Rev. RNA *3*, 601–616.

Cloonan, N., Wani, S., Xu, Q., Gu, J., Lea, K., Heater, S., Barbacioru, C., Steptoe, A.L., Martin, H.C., Nourbakhsh, E., *et al.* (2011). MicroRNAs and their isomiRs function cooperatively to target common biological pathways. Genome Biol. *12*, R126.

Croce, C. M. (2009). Causes and consequences of microRNA dysregulation in cancer. Nat. Rev. Genet. *10*, 704–714.

Davies, S.W., Turmaine, M., Cozens, B.A., DiFiglia, M., Sharp, A.H., Ross, C. A., Scherzinger, E., Wanker, E.E., Mangiarini, L., and Bates, G. P. (1997). Formation of neuronal intranuclear inclusions underlies the neurological dysfunction in mice transgenic for the HD mutation. Cell *90*, 537–548.

Davis, B.M., McCurrach, M.E., Taneja, K.L., Singer, R.H., and Housman, D.E. (1997). Expansion of a CUG trinucleotide repeat in the 3′ untranslated region of myotonic dystrophy protein kinase transcripts results in nuclear retention of transcripts. Proc. Natl. Acad. Sci. U.S.A. *94*, 7388–7393.

Dohm, J.C., Lottaz, C., Borodina, T., and Himmelbauer, H. (2008). Substantial biases in ultra-short read data sets from high-throughput DNA sequencing. Nucleic Acids Res. *36*, e105.

Eberhart, D.E., Malter, H.E., Feng, Y., and Warren, S.T. (1996). The fragile X mental retardation protein is a ribonucleoprotein containing both nuclear localization and nuclear export signals. Hum. Mol. Genet. *5*, 1083–1091.

Ebert, M.S., Neilson, J.R., and Sharp, P.A. (2007). MicroRNA sponges: competitive inhibitors of small RNAs in mammalian cells. Nat. Methods *4*, 721–726.

Eisenberg, I., Eran, A., Nishino, I., Moggio, M., Lamperti, C., Amato, A.A., Lidov, H.G., Kang, P.B., North, K.N., Mitrani-Rosenbaum, S., *et al.* (2007). Distinctive patterns of microRNA expression in primary muscular disorders. Proc. Natl. Acad. Sci. U.S.A. *104*, 17016–17021.

Esteller, M. (2011). Non-coding RNAs in human disease. Nat. Rev. Genet. *12*, 861–874.

Fernandez-Costa, J.M., Garcia-Lopez, A., Zuniga, S., Fernandez-Pedrosa, V., Felipo-Benavent, A., Mata, M., Jaka, O., Aiastui, A., Hernandez-Torres, F., Aguado, B., *et al.* (2012). Expanded CTG repeats trigger miRNA alterations in Drosophila that are conserved in myotonic dystrophy type 1 patients. Hum. Mol. Genet. *22*, 704–716.

Fiszer, A., and Krzyzosiak, W.J. (2013). RNA toxicity in polyglutamine disorders: concepts, models, and progress of research. J. Mol. Med. (Berl) *91*, 683–691.

Fiszer, A., Olejniczak, M., Switonski, P.M., Wroblewska, J.P., Wisniewska-Kruk, J., Mykowska, A., and Krzyzosiak, W.J. (2012). An evaluation of oligonucleotide-based therapeutic strategies for polyQ diseases. BMC Mol. Biol. *13*, 6.

Friedman, R.C., Farh, K.K., Burge, C.B., and Bartel, D.P. (2009). Most mammalian mRNAs are conserved targets of microRNAs. Genome Res. *19*, 92–105.

Fukuda, T., Yamagata, K., Fujiyama, S., Matsumoto, T., Koshida, I., Yoshimura, K., Mihara, M., Naitou, M., Endoh, H., Nakamura, T., *et al.* (2007). DEAD-box RNA helicase subunits of the Drosha complex are required for processing of rRNA and a subset of microRNAs. Nat. Cell Biol. *9*, 604–611.

Fuller-Pace, F.V. (2006). DExD/H box RNA helicases: multifunctional proteins with important roles in transcriptional regulation. Nucleic Acids Res. *34*, 4206–4215.

Galka-Marciniak, P., Urbanek, M.O. and Krzyzosiak, W.J. (2012). Triplet repeats in transcripts: structural insights into RNA toxicity. Biol. Chem. *393* 1299–1315.

Gambardella, S., Rinaldi, F., Lepore, S. M., Viola, A., Loro, E., Angelini, C., Vergani, L., Novelli, G., and Botta, A. (2010). Overexpression of microRNA-206 in the skeletal muscle from myotonic dystrophy type 1 patients. J. Transl. Med. *8*, 48.

Gatchel, J.R., and Zoghbi, H.Y. (2005). Diseases of unstable repeat expansion: mechanisms and common principles. Nat. Rev. Genet. *6*, 743–755.

Gatchel, J.R., Watase, K., Thaller, C., Carson, J.P., Jafar-Nejad, P., Shaw, C., Zu, T., Orr, H.T., and Zoghbi, H.Y. (2008). The insulin-like growth factor pathway is altered in spinocerebellar ataxia type 1 and type 7. Proc. Natl. Acad. Sci. U.S.A. *105*, 1291–1296.

Gaughwin, P.M., Ciesla, M., Lahiri, N., Tabrizi, S.J., Brundin, P., and Bjorkqvist, M. (2011). Hsa-miR-34b is a plasma-stable microRNA that is elevated in pre-manifest Huntington's disease. Hum. Mol. Genet. *20*, 2225–2237.

Gehrke, S., Imai, Y., Sokol, N., and Lu, B. (2010). Pathogenic LRRK2 negatively regulates microRNA-mediated translational repression. Nature *466*, 637–641.

Ghose, J., Sinha, M., Das, E., Jana, N.R., and Bhattacharyya, N.P. (2011). Regulation of miR-146a by RelA/NFκB and p53 in STHdh(Q111)/Hdh(Q111) cells, a cell model of Huntington's disease. PLoS One *6*, e23837.

Git, A., Dvinge, H., Salmon-Divon, M., Osborne, M., Kutter, C., Hadfield, J., Bertone, P., and Caldas, C. (2010). Systematic comparison of microarray profiling, real-time PCR, and next-generation sequencing technologies for measuring differential microRNA expression. RNA *16*, 991–1006.

Greco, S., Perfetti, A., Fasanaro, P., Cardani, R., Capogrossi, M.C., Meola, G., and Martelli, F. (2012). Deregulated microRNAs in myotonic dystrophy type 2. PLoS One *7*, e39732.

Gregory, R.I., Yan, K.P., Amuthan, G., Chendrimada, T., Doratotaj, B., Cooch, N., and Shiekhattar, R. (2004). The Microprocessor complex mediates the genesis of microRNAs. Nature *432*, 235–240.

Griffiths-Jones, S., Saini, H.K., van Dongen, S., and Enright, A.J. (2008). miRBase: tools for microRNA genomics. Nucleic Acids Res. *36(Database issue)*, D154–158.

Guo, H., Ingolia, N. T., Weissman, J.S., and Bartel, D.P. (2010). Mammalian microRNAs predominantly act to decrease target mRNA levels. Nature *466*, 835–840.

Harper, S.Q. (2009). Progress and challenges in RNA interference therapy for Huntington disease. Arch. Neurol. *66*, 933–938.

Hebert, S.S., and De Strooper, B. (2009). Alterations of the microRNA network cause neurodegenerative disease. Trends Neurosci. *32*, 199–206.

Heo, I., Joo, C., Cho, J., Ha, M., Han, J., and Kim, V.N. (2008). Lin28 mediates the terminal uridylation of let-7 precursor MicroRNA. Mol. Cell. *32*, 276–284.

Ho, T.H., Bundman, D., Armstrong, D.L., and Cooper, T.A. (2005). Transgenic mice expressing CUG-BP1 reproduce splicing mis-regulation observed in myotonic dystrophy. Hum. Mol. Genet. *14*, 1539–1547.

Hon, L.S., and Zhang, Z. (2007). The roles of binding site arrangement and combinatorial targeting in microRNA repression of gene expression. Genome Biol. *8*, R166.

Hu, J., Liu, J., and Corey, D.R. (2010). Allele-selective inhibition of huntingtin expression by switching to an miRNA-like RNAi mechanism. Chem. Biol. *17*, 1183–1188.

Huang, C.C., Faber, P.W., Persichetti, F., Mittal, V., Vonsattel, J.P., MacDonald, M.E., and Gusella, J.F. (1998). Amyloid formation by mutant huntingtin: threshold, progressivity and recruitment of normal polyglutamine proteins. Somat. Cell Mol. Genet. *24*, 217–233.

Humphreys, D.T., Hynes, C.J., Patel, H.R., Wei, G.H., Cannon, L., Fatkin, D., Suter, C.M., Clancy, J.L., and Preiss, T. (2012). Complexity of murine cardiomyocyte miRNA biogenesis, sequence variant expression and function. PLoS One 7, e30933.

Huse, S.M., Huber, J.A., Morrison, H.G., Sogin, M.L., and Welch, D.M. (2007). Accuracy and quality of massively parallel DNA pyrosequencing. Genome Biol. *8*, R143.

Iorio, M.V., and Croce, C.M. (2012). Causes and consequences of microRNA dysregulation. Cancer J. *18*, 215–222.

Janknecht, R. (2010). Multi-talented DEAD-box proteins and potential tumor promoters: p68 RNA helicase (DDX5) and its paralog, p72 RNA helicase (DDX17). Am. J. Transl. Res. *2*, 223–234.

Jin, J., Cheng, Y., Zhang, Y., Wood, W., Peng, Q., Hutchison, E., Mattson, M.P., Becker, K.G., and Duan, W. (2012). Interrogation of brain miRNA and mRNA expression profiles reveals a molecular regulatory network that is perturbed by mutant huntingtin. J. Neurochem. *123*, 477–490.

Jin, P., Alisch, R.S., and Warren, S.T. (2004a). RNA and microRNAs in fragile X mental retardation. Nat. Cell Biol. *6*, 1048–1053.

Jin, P., Zarnescu, D.C., Ceman, S., Nakamoto, M., Mowrey, J., Jongens, T.A., Nelson, D.L., Moses, K., and Warren, S.T. (2004b). Biochemical and genetic interaction between the fragile X mental retardation protein and the microRNA pathway. Nat. Neurosci. 7, 113–117.

Johnson, R., Zuccato, C., Belyaev, N.D., Guest, D.J., Cattaneo, E., and Buckley, N.J. (2008). A microRNA-based gene dysregulation pathway in Huntington's disease. Neurobiol. Dis. *29*, 438–445.

Junn, E., and Mouradian, M.M. (2012). MicroRNAs in neurodegenerative diseases and their therapeutic potential. Pharmacol. Ther. *133*, 142–150.

Kanadia, R.N., Johnstone, K.A., Mankodi, A., Lungu, C., Thornton, C.A., Esson, D., Timmers, A.M., Hauswirth, W.W., and Swanson, M.S. (2003). A muscleblind knockout model for myotonic dystrophy. Science *302*, 1978–1980.

Karres, J.S., Hilgers, V., Carrera, I., Treisman, J. and Cohen, S.M. (2007). The conserved microRNA miR-8 tunes atrophin levels to prevent neurodegeneration in Drosophila. Cell *131*, 136–145.

Kim, V.N., Han, J., and Siomi, M.C. (2009). Biogenesis of small RNAs in animals. Nat. Rev. Mol. Cell. Biol. *10*, 126–139.

Kimura, T., Nakamori, M., Lueck, J.D., Pouliquin, P., Aoike, F., Fujimura, H., Dirksen, R.T., Takahashi, M.P., Dulhunty, A.F., and Sakoda, S. (2005). Altered mRNA splicing of the skeletal muscle ryanodine receptor and sarcoplasmic/endoplasmic reticulum Ca2+-ATPase in myotonic dystrophy type 1. Hum. Mol. Genet. *14*, 2189–2200.

Koide, R., Ikeuchi, T., Onodera, O., Tanaka, H., Igarashi, S., Endo, K., Takahashi, H., Kondo, R., Ishikawa, A., Hayashi, T., *et al.* (1994). Unstable expansion of CAG repeat in hereditary dentatorubral-pallidoluysian atrophy (DRPLA). Nat. Genet. *6*, 9–13.

Krol, J., Loedige, I., and Filipowicz, W. (2010). The widespread regulation of microRNA biogenesis, function and decay. Nat. Rev. Genet. *11*, 597–610.

Landgraf, P., Rusu, M., Sheridan, R., Sewer, A., Iovino, N., Aravin, A., Pfeffer, S., Rice, A., Kamphorst, A.O., Landthaler, M., *et al.* (2007). A mammalian microRNA expression atlas based on small RNA library sequencing. Cell *129*, 1401–1414.

Latronico, M.V., and Condorelli, G. (2009). MicroRNAs and cardiac pathology. Nat. Rev. Cardiol. *6*, 419–429.

Lau, P., and de Strooper, B. (2010). Dysregulated microRNAs in neurodegenerative disorders. Semin. Cell Dev. Biol. *21*, 768–773.

Laurent, F.X., Sureau, A., Klein, A.F., Trouslard, F., Gasnier, E., Furling, D., and Marie, J. (2012). New function for the RNA helicase p68/DDX5 as a modifier of MBNL1 activity on expanded CUG repeats. Nucleic Acids Res. *40*, 3159–3171.

Lee, L.W., Zhang, S., Etheridge, A., Ma, L., Martin, D., Galas, D., and Wang, K. (2010). Complexity of the microRNA repertoire revealed by next-generation sequencing. RNA *16*, 2170–2180.

Lee, S.T., Chu, K., Im, W.S., Yoon, H.J., Im, J.Y., Park, J.E., Park, K.H., Jung, K.H., Lee, S.K., Kim, M., *et al.* (2010). Altered microRNA regulation in Huntington's disease models. Exp. Neurol. *227*, 172–179.

Lee, Y., Samaco, R.C., Gatchel, J.R., Thaller, C., Orr, H.T., and Zoghbi, H.Y. (2008). miR-*19*, miR-101 and miR-130 co-regulate ATXN1 levels to potentially modulate SCA1 pathogenesis. Nat. Neurosci. *11*, 1137–1139.

Li, L.B., and Bonini, N.M. (2010). Roles of trinucleotide-repeat RNA in neurological disease and degeneration. Trends Neurosci. *33*, 292–298.

Li, L.M., Hu, Z.B., Zhou, Z.X., Chen, X., Liu, F.Y., Zhang, J.F., Shen, H.B., Zhang, C.Y., and Zen, K. (2010). Serum microRNA profiles serve as novel biomarkers for HBV infection and diagnosis of HBV-positive hepatocarcinoma. Cancer Res. *70*, 9798–9807.

Liang, Z., Wu, H., Reddy, S., Zhu, A., Wang, S., Blevins, D., Yoon, Y., Zhang, Y., and Shim, H. (2007). Blockade of invasion and metastasis of breast cancer cells via targeting CXCR4 with an artificial microRNA. Biochem. Biophys. Res. Commun. *363*, 542–546.

Lin, X., Miller, J. W., Mankodi, A., Kanadia, R.N., Yuan, Y., Moxley, R.T., Swanson, M.S., and Thornton, C.A. (2006). Failure of MBNL1-dependent post-natal splicing transitions in myotonic dystrophy. Hum. Mol. Genet. *15*, 2087–2097.

Lopez Castel, A., Cleary, J.D., and Pearson, C.E. (2010). Repeat instability as the basis for human diseases and as a potential target for therapy. Nat. Rev. Mol. Cell. Biol. *11*, 165–170.

McBride, J.L., Pitzer, M.R., Boudreau, R.L., Dufour, B., Hobbs, T., Ojeda, S.R., and Davidson, B.L. (2011). Preclinical safety of RNAi-mediated HTT suppression in the rhesus macaque as a potential therapy for Huntington's disease. Mol. Ther. *19*, 2152–2162.

McCann, C., Holohan, E.E., Das, S., Dervan, A., Larkin, A., Lee, J.A., Rodrigues, V., Parker, R., and Ramaswami, M. (2011). The Ataxin-2 protein is required for microRNA function and synapse-specific long-term olfactory habituation. Proc. Natl. Acad. Sci. U.S.A. *108*, E655–662.

Mahadevan, M., Tsilfidis, C., Sabourin, L., Shutler, G., Amemiya, C., Jansen, G., Neville, C., Narang, M., Barcelo, J., O'Hoy, K., *et al.* (1992). Myotonic dystrophy mutation: an unstable CTG repeat in the 3′ untranslated region of the gene. Science 255, 1253–1255.

Mankodi, A., Logigian, E., Callahan, L., McClain, C., White, R., Henderson, D., Krym, M. and Thornton, C. A. (2000). Myotonic dystrophy in transgenic mice expressing an expanded CUG repeat. Science *289*, 1769–1773.

Mankodi, A., Takahashi, M.P., Jiang, H., Beck, C.L., Bowers, W.J., Moxley, R.T., Cannon, S.C., and Thornton, C.A. (2002). Expanded CUG repeats trigger aberrant splicing of ClC-1 chloride channel pre-mRNA and hyperexcitability of skeletal muscle in myotonic dystrophy. Mol. Cell. *10(1)*, 35–44.

Marti, E., Pantano, L., Banez-Coronel, M., Llorens, F., Minones-Moyano, E., Porta, S., Sumoy, L., Ferrer, I., and Estivill, X. (2010). A myriad of miRNA variants in control and Huntington's disease brain regions detected by massively parallel sequencing. Nucleic Acids Res. *38*, 7219–7235.

Martino, S., di Girolamo, I., Orlacchio, A., and Datti, A. (2009). MicroRNA implications across neurodevelopment and neuropathology. J. Biomed. Biotechnol. *2009*, 654346.

Meister, G., Landthaler, M., Dorsett, Y., and Tuschl, T. (2004). Sequence-specific inhibition of microRNA- and siRNA-induced RNA silencing. RNA *10*, 544–550.

Melo, S.A., and Esteller, M. (2011). Dysregulation of microRNAs in cancer: playing with fire. FEBS Lett. *585*, 2087–2099.

de Mezer, M., Wojciechowska, M., Napierala, M., Sobczak, K., and Krzyzosiak, W.J. (2011). Mutant CAG repeats of Huntingtin transcript fold into hairpins, form nuclear foci and are targets for RNA interference. Nucleic Acids Res. *39*, 3852–3863.

Mi, Y., Thomas, S.D., Xu, X., Casson, L.K., Miller, D.M., and Bates, P.J. (2003). Apoptosis in leukaemia cells is accompanied by alterations in the levels and localization of nucleolin. J. Biol. Chem. *278*, 8572–8579.

Miller, J.W., Urbinati, C.R., Teng-Umnuay, P., Stenberg, M.G., Byrne, B.J., Thornton, C.A., and Swanson, M.S. (2000). Recruitment of human muscleblind proteins to (CUG)(n) expansions associated with myotonic dystrophy. EMBO J. *19*, 4439–4448.

Mirkin, S.M. (2007). Expandable DNA repeats and human disease. Nature *447*, 932–940.

Morin, R.D., O'Connor, M.D., Griffith, M., Kuchenbauer, F., Delaney, A., Prabhu, A.L., Zhao, Y., McDonald, H., Zeng, T., Hirst, M., *et al.* (2008). Application of massively parallel sequencing to microRNA profiling and discovery in human embryonic stem cells. Genome Res. *18*, 610–621.

Mykowska, A., Sobczak, K., Wojciechowska, M., Kozlowski, P. and Krzyzosiak, W.J. (2011). CAG repeats mimic CUG repeats in the misregulation of alternative splicing. Nucleic Acids Res. *39*, 8938–8951.

Neilsen, C.T., Goodall, G.J., and Bracken, C.P. (2012). IsomiRs--the overlooked repertoire in the dynamic microRNAome. Trends Genet. *28*, 544–549.

Nomura, T., Kimura, M., Horii, T., Morita, S., Soejima, H., Kudo, S., and Hatada, I. (2008). MeCP2-dependent repression of an imprinted miR-184 released by depolarization. Hum. Mol. Genet. *17*, 1192–1199.

Olejniczak, M., Galka, P., and Krzyzosiak, W.J. (2010). Sequence-non-specific effects of RNA interference triggers and microRNA regulators. Nucleic Acids Res. *38*, 1–16.

Orr, H.T. (2012). Cell biology of spinocerebellar ataxia. J. Cell. Biol. *197*, 167–177.

Orr, H.T. and Zoghbi, H. Y. (2007). Trinucleotide repeat disorders. Annu. Rev. Neurosci. *30*, 575–621.

Otake, Y., Soundararajan, S., Sengupta, T.K., Kio, E.A., Smith, J.C., Pineda-Roman, M., Stuart, R.K., Spicer, E.K., and Fernandes, D.J. (2007). Overexpression of nucleolin in chronic lymphocytic leukaemia cells induces stabilization of bcl2 mRNA. Blood *109*, 3069–3075.

Packer, A.N., Xing, Y., Harper, S.Q., Jones, L., and Davidson, B.L. (2008). The bifunctional microRNA miR-9/miR-9* regulates REST and CoREST and is downregulated in Huntington's disease. J. Neurosci. *28*, 14341–14346.

Perbellini, R., Greco, S., Sarra-Ferraris, G., Cardani, R., Capogrossi, M.C., Meola, G., and Martelli, F. (2011). Dysregulation and cellular mislocalization of specific miRNAs in myotonic dystrophy type 1. Neuromuscul. Disord. *21*, 81–88.

Perron, M.P., and Provost, P. (2009). Protein components of the microRNA pathway and human diseases. Methods Mol. Biol. *487*, 369–385.

Persengiev, S., Kondova, I., Otting, N., Koeppen, A.H., and Bontrop, R.E. (2011). Genome-wide analysis of miRNA expression reveals a potential role for miR-144 in brain aging and spinocerebellar ataxia pathogenesis. Neurobiol. Aging *32*, 2316 e2317–2327.

Pickering, B.F., Yu, D., and Van Dyke, M.W. (2011). Nucleolin protein interacts with microprocessor complex to affect biogenesis of microRNAs 15a and 16. J. Biol. Chem. *286*, 44095–44103.

Plante, I., Davidovic, L., Ouellet, D. L., Gobeil, L.A., Tremblay, S., Khandjian, E.W., and Provost, P. (2006). Dicer-derived microRNAs are utilized by the fragile X mental retardation protein for assembly on target RNAs. J. Biomed. Biotechnol. *2006*, 64347.

Rau, F., Freyermuth, F., Fugier, C., Villemin, J.P., Fischer, M.C., Jost, B., Dembele, D., Gourdon, G., Nicole, A., Duboc, D., *et al.* (2011). Misregulation of miR-1 processing is associated with heart defects in myotonic dystrophy. Nat. Struct. Mol. Biol. *18*, 840–845.

Reid, J.G., Nagaraja, A.K., Lynn, F.C., Drabek, R.B., Muzny, D.M., Shaw, C.A., Weiss, M.K., Naghavi, A.O., Khan, M., Zhu, H., *et al.* (2008). Mouse let-7 miRNA populations exhibit RNA editing that is constrained in the 5′-seed/cleavage/anchor regions and stabilize predicted mmu-let-7a:mRNA duplexes. Genome Res. *18*, 1571–1581.

Reinhardt, A., Feuillette, S., Cassar, M., Callens, C., Thomassin, H., Birman, S., Lecourtois, M., Antoniewski, C., and Tricoire, H. (2012). Lack of miRNA Misregulation at Early Pathological Stages in Drosophila Neurodegenerative Disease Models. Front. Genet. *3*, 226.

Rodriguez-Lebron, E., Liu, G., Keiser, M., Behlke, M.A., and Davidson, B.L. (2013). Altered Purkinje cell miRNA expression and SCA1 pathogenesis. Neurobiol. Dis. *54*, 456–463.

van Rooij, E., Liu, N., and Olson, E.N. (2008). MicroRNAs flex their muscles. Trends Genet. *24*, 159–166.

Roshan, R., Ghosh, T., Scaria, V., and Pillai, B. (2009). MicroRNAs: novel therapeutic targets in neurodegenerative diseases. Drug Discov. Today *14*, 1123–1129.

Ruby, J.G., Jan, C., Player, C., Axtell, M.J., Lee, W., Nusbaum, C., Ge, H., and Bartel, D.P. (2006). Large-scale sequencing reveals 21U-RNAs and additional microRNAs and endogenous siRNAs in C. elegans. Cell *127*, 1193–1207.

Rybak, A., Fuchs, H., Smirnova, L., Brandt, C., Pohl, E.E., Nitsch, R., and Wulczyn, F.G. (2008). A feedback loop comprising lin-28 and let-7 controls pre-let-7 maturation during neural stem-cell commitment. Nat. Cell Biol. *10*, 987–993.

Savas, J.N., Makusky, A., Ottosen, S., Baillat, D., Then, F., Krainc, D., Shiekhattar, R., Markey, S.P., and Tanese, N. (2008). Huntington's disease protein contributes to RNA-mediated gene silencing through association with Argonaute and P bodies. Proc. Natl. Acad. Sci. U.S.A. *105*, 10820–10825.

Savkur, R.S., Philips, A.V., and Cooper, T.A. (2001). Aberrant regulation of insulin receptor alternative splicing is associated with insulin resistance in myotonic dystrophy. Nat. Genet. *29*, 40–47.

Schaefer, A., O'Carroll, D., Tan, C.L., Hillman, D., Sugimori, M., Llinas, R., and Greengard, P. (2007). Cerebellar neurodegeneration in the absence of microRNAs. J. Exp. Med. *204*, 1553–1558.

Seitz, H., Ghildiyal, M., and Zamore, P.D. (2008). Argonaute loading improves the 5′ precision of both MicroRNAs and their miRNA* strands in flies. Curr. Biol. *18*, 147–151.

Sellier, C., Freyermuth, F., Tabet, R., Tran, T., He, F., Ruffenach, F., Alunni, V., Moine, H., Thibault, C., Page, A., *et al.* (2013). Sequestration of DROSHA and DGCR8 by Expanded CGG RNA Repeats Alters MicroRNA Processing in Fragile X-Associated Tremor/Ataxia Syndrome. Cell Rep. *3*, 869–880.

Seow, Y., Sibley, C.R., and Wood, M.J. (2012). Artificial mirtron-mediated gene knockdown: functional DMPK silencing in mammalian cells. RNA *18*, 1328–1337.

Sicot, G., Gourdon, G., and Gomes-Pereira, M. (2011). Myotonic dystrophy, when simple repeats reveal complex pathogenic entities: new findings and future challenges. Hum. Mol. Genet. *20*, R116–123.

Sinha, M., Ghose, J., Das, E., and Bhattarcharyya, N.P. (2010). Altered microRNAs in STHdh(Q111)/Hdh(Q111) cells: miR-146a targets TBP. Biochem. Biophys. Res. Commun. *396*, 742–747.

Sinha, M., Ghose, J., and Bhattarcharyya, N.P. (2011). MicroRNA -*214*,-*150*,-146a and -125b target Huntingtin gene. RNA Biol. *8*, 1005–1021.

Siomi, H., Ishizuka, A., and Siomi, M.C. (2004). RNA interference: a new mechanism by which FMRP acts in the normal brain? What can Drosophila teach us? Ment. Retard. Dev. Disabil. Res. Rev. *10*, 68–74.

Smalheiser, N.R., and Lugli, G. (2009). MicroRNA regulation of synaptic plasticity. Neuromolecular Med. *11*, 133–140.

Soldati, C., Bithell, A., Johnston, C., Wong, K.Y., Stanton, L.W., and Buckley, N.J. (2013). Dysregulation of REST-regulated coding and non-coding RNAs in a cellular model of Huntington's disease. J. Neurochem. *124*, 418–430.

Sonntag, K.C. (2010). MicroRNAs and deregulated gene expression networks in neurodegeneration. Brain Res. *1338*, 48–57.

Taneja, K.L., McCurrach, M., Schalling, M., Housman, D., and Singer, R.H. (1995). Foci of trinucleotide repeat transcripts in nuclei of myotonic dystrophy cells and tissues. J. Cell Biol. *128*, 995–1002.

Tassone, F., Hagerman, R.J., Loesch, D.Z., Lachiewicz, A., Taylor, A.K., and Hagerman, P.J. (2000). Fragile X males with unmethylated, full mutation trinucleotide repeat expansions have elevated levels of FMR1 messenger RNA. Am. J. Med. Genet. *94*, 232–236.

Tian, G., Yin, X., Luo, H., Xu, X., Bolund, L., Zhang, X., Gan, S.Q., and Li, N. (2010). Sequencing bias: comparison of different protocols of microRNA library construction. BMC Biotechnol. *10*, 64.

Timchenko, L.T., Timchenko, N.A., Caskey, C.T., and Roberts, R. (1996). Novel proteins with binding specificity for DNA CTG repeats and RNA CUG repeats: implications for myotonic dystrophy. Hum. Mol. Genet. *5*, 115–121.

Tsoi, H., and Chan, H.Y. (2013). Expression of Expanded CAG Transcripts Triggers Nucleolar Stress in Huntington's Disease. Cerebellum. *12*, 310–312.

Tsoi, H., Lau, T.C., Tsang, S.Y., Lau, K.F., and Chan, H.Y. (2012). CAG expansion induces nucleolar stress in polyglutamine diseases. Proc. Natl. Acad. Sci. U.S.A. *109*, 13428–13433.

Verkerk, A.J., Pieretti, M., Sutcliffe, J.S., Fu, Y.H., Kuhl, D.P., Pizzuti, A., Reiner, O., Richards, S., Victoria, M.F., Zhang, F.P., *et al.* (1991). Identification of a gene (FMR-1) containing a CGG repeat coincident with a breakpoint cluster region exhibiting length variation in fragile X syndrome. Cell *65*, 905–914.

Wang, B., Komers, R., Carew, R., Winbanks, C.E., Xu, B., Herman-Edelstein, M., Koh, P., Thomas, M., Jandeleit-Dahm, K., Gregorevic, P., *et al.* (2011). Suppression of microRNA-29 expression by TGF-beta1 promotes collagen expression and renal fibrosis. J. Am. Soc. Nephrol. *23*, 252–265.

Wei, B., Cai, T., Zhang, R., Li, A., Huo, N., Li, S., Gu, Y. Q., Vogel, J., Jia, J., Qi, Y., *et al.* (2009). Novel microRNAs uncovered by deep sequencing of small RNA transcriptomes in bread wheat (Triticum aestivum L.) and Brachypodium distachyon (L.) Beauv. Funct. Integr. Genomics *9*, 499–511.

Weinberg, M.S., and Wood, M.J. (2009). Short non-coding RNA Biol. and neurodegenerative disorders: novel disease targets and therapeutics. Hum. Mol. Genet. *18*, R27–39.

Wheeler, T.M., Lueck, J.D., Swanson, M.S., Dirksen, R.T., and Thornton, C.A. (2007). Correction of ClC-1 splicing eliminates chloride channelopathy and myotonia in mouse models of myotonic dystrophy. J. Clin. Invest. *117*, 3952–3957.

Winter, J., Jung, S., Keller, S., Gregory, R.I., and Diederichs, S. (2009). Many roads to maturity: microRNA biogenesis pathways and their regulation. Nat. Cell Biol. *11*, 228–234.

Witkos, T.M., Koscianska, E., and Krzyzosiak, W.J. (2011). Practical Aspects of microRNA Target Prediction. Curr. Mol. Med. *11*, 93–109.

Witwer, K.W., Sarbanes, S.L., Liu, J., and Clements, J.E. (2011). A plasma microRNA signature of acute lentiviral infection: biomarkers of central nervous system disease. AIDS *25*, 2057–2067.

Wojciechowska, M., and Krzyzosiak, W.J. (2011). CAG repeat RNA as an auxiliary toxic agent in polyglutamine disorders. RNA Biol. *8(4)*, 565–571.

Wu, H., Neilson, J.R., Kumar, P., Manocha, M., Shankar, P., Sharp, P.A., and Manjunath, N. (2007). MiRNA profiling of naive, effector and memory CD8 T cells. PLoS One *2*, e1020.

Wyman, S.K., Parkin, R.K., Mitchell, P.S., Fritz, B.R., O'Briant, K., Godwin, A.K., Urban, N., Drescher, C.W., Knudsen, B.S., and Tewari, M. (2009). Repertoire of microRNAs in epithelial ovarian cancer as determined by next generation sequencing of small RNA cDNA libraries. PLoS One *4*, e5311.

Xu, X. L., Li, Y., Wang, F., and Gao, F.B. (2008). The steady-state level of the nervous-system-specific microRNA-124a is regulated by dFMR1 in Drosophila. J. Neurosci. *28*, 11883–11889.

Yang, L., Duan, R., Chen, D., Wang, J., and Jin, P. (2007). Fragile X mental retardation protein modulates the fate of germline stem cells in Drosophila. Hum. Mol. Genet. *16*, 1814–1820.

Zhang, H., Yang, J.H., Zheng, Y.S., Zhang, P., Chen, X., Wu, J., Xu, L., Luo, X. Q., Ke, Z. Y., Zhou, H., *et al.* (2009). Genome-wide analysis of small RNA and novel MicroRNA discovery in human acute lymphoblastic leukaemia based on extensive sequencing approach. PLoS One *4*, e6849.

Zuccato, C., Tartari, M., Crotti, A., Goffredo, D., Valenza, M., Conti, L., Cataudella, T., Leavitt, B.R., Hayden, M.R., Timmusk, T., *et al.* (2003). Huntingtin interacts with REST/NRSF to modulate the transcription of NRSE-controlled neuronal genes. Nat. Genet. *35*, 76–83.

Index

13q14 deletion 204
2′-fluoro (2′-F) ODN 107, 109, 112
2′-O-methoxyethyl (2′-MOE) 10, 107, 109, 112, 114, 152
2′-O-methyl (2′-OMe) 107, 108, 114, 117, 119, 120, 121,163
2′-fluoro 152
2′-O-guanidinopropyl 152
3p-modified siRNAs 159

A

ADAR 6
Adeno-associated vectors (AAV) 53, 72, 128, 132, 158, 163
Adenovirus 28, 53, 128, 133, 158, 163
Adult T-cell leukaemia (ATL) 189
Adult T-cell leukaemia lymphoma (ATLL) 189
AIDS 189
ALN-RSV01 164
α-l-iduronidase (IDUA) 131
Alzheimer's disease (AD) 229
AMPA receptor subunit (GluR-B) 29
Anopheles gambiae 169
Antagomirs 111, 112, 114, 117–119
Antigen-presenting cells (APCs) 128–132
Anti-Mir 105, 106
Antisense 144
AP marker 70
Apolipoprotein AI (Apo AI) 155, 163
Apolipoprotein B (ApoB) 146, 162
Aptamers 152
Argonaute (Ago) 1, 2, 4, 7, 11, 12, 20, 111, 151, 165, 189, 237
- Ago1 34
- Ago2 24–26, 36, 46, 55–56, 111
- MID domain 28
- MILI 22
- MIWI2 23
- PAZ domain 28
- piRNA 22, 23
- PIWI 2, 22, 111

Articifial miR (amiRNA) 140, 144–145
Ataxin (ATXN1) 235 236

B

B cell lymphomas 186
BART miRNAs 185–188, 190
B-cell integration cluster (BIC) 187
BCL2 4
Benign prostatic hyperplasia (BPH) 208
Brain-derived neurotrophic factor (BDNF) 45, 235
Branch and bound analysis (B&B) 94, 95
Breast cancer 207
Burkitt's lymphoma (BL) 184, 186

C

Canonical miRNA biosynthetic pathway 202
cccDNA 154–155, 159
CD34+ haematopoietic stem cells 68
CD81 162–163
CD8T 132
CDC2L6, mRNA 26
CDH1 (E-cadherin) 35
CDR1 7
Cell penetrating peptides (CPPs) 112, 119
Cerebellar degeneration-related protein 1 7
Chip-on-chip analysis 34
Choice of profiling platform 212
Chronic lymphocytic leukaemia (CLL) 203
Circular non-coding RNA (circRNA) 36
ciRS-7 7
Claudin1 162
CLL 204–205
Clustered regularly interspaced short palindromic repeat (CRISPR) 51, 72, 73, 168
c-Myc 5, 53, 56, 60, 66–67, 186
Colorectal cancer 208
Commonly used miRNA profiling techniques 213
Commonly used miRNA target prediction algorithms 215
Competing endogenous RNA (ceRNA) 6, 35
Coxsackie A21 oncolytic virus 134
Coxsackie virus 164
Crigler–Najjar type I (CN-I) 128
Crimean Congo fever virus (CCHFV) 167
CSDC2 (cold-shock domain-containing protein C2) 35
CTCF (CCCTC-motif binding factor) 24

Current state of miRNA biomarker studies 211
Cyclin E2 4
Cytomegalovirus retinitis 146
Cytoplasmic cyclophilins A, B and C 162

D

DEAD-box RNA helicase p68 (DDX5) 2, 5, 238
DEAD-box RNA helicase p72 (DDX17) 2, 5, 238
Decoy PNAs 114
Deep sequencing 26
Dengue virus (DENV) 165–167
Dentatorubropallidoluysian atrophy (DRPLA) 227, 236
DeSeq package 82
DGCR8 19, 25, 238–239
DGCR8 2, 140, 182, 202
Diacylglycerol acyltransferase-1 (DGAT1) 162
Dicer 162
Dicer 2, 19–20, 22, 25 -26, 31, 53–56, 151, 163, 182, 227, 234
Dicer-2 (*Drosophila*) 22
Dicer, knockout 24
Diffuse large B-cell lymphoma (DLBCL) 186, 204
DNA methyltransferase, DNMT3a 45
DNA methyltransferase, DNMT3b 32
DNA oligonucleotides (ODN) 105, 106, 112
Drosha 2, 19, 25, 24, 26, 53, 56, 140, 182, 202, 227, 234, 238
Duchenne muscular dystrophy (DMD) 35
Dystrophia myotonica protein kinase (DMPK) 236

E

EBNA2 186
Ebola (EBOV) 165
EBV 184–187
Egr2 promoter 34
Embroid bodies 55
Embryonic stem cells (ESCs) 28, 51, 54, 67
  embryonic stem cells (ESCs), human 67, 69
  embryonic stem cells (ESCs), mouse 69
Endo-shRNA 26
Endo-siRNA 2, 22, 23
Endosomes 152
Enteroviruses 164
Epstein–Barr nuclear antigens (EBNAs) 184
Epstein–Barr virus (EBV) 181–182
Estrogen receptor-a (ER-α) 5
Examples of RNAi to treat human viral infections 156
Exosome 212
  exosomal miRNA 186
Exploiting miRNAs to accomplish safe and specific transgene expression 129
Exportin-5 karyopherin 2, 5, 143, 202
Expressed RNAi activators 7

F

FastQC 81
FIX 130
Foamy virus vectors 158
Follicular lymphoma (FL) 210
FoxP3 130
Fragile X syndrome (FXS) 227, 237–238
Fragile X-associated tremor/ataxia syndrome (FXTAS) 227, 237–238
FVIII 130
FXR1 (fragile X mental retardation syndrome-related protein 1) 33

G

Ganciclovir (GCV) 133
Gastric carcinoma 187
Gendicinc 146
Giardia lamblia 26
Glial fibrillary acidic protein (GFAP) 11, 134
GP64 131
Granulocyte–macrophage colony stimulating factor (GMCSF) 146
Gunn rats 129
GW bodies 111

H

H1 8, 141–142
H3K27me3 44
Haematological malignancies 208
Haemophilia B 130
HBsAg 188
Hepatitis B virus (HBV) 5, 153–155, 158–159, 181, 188
  HBV genome organization 154
  HBV surface antigen (HBsAg) 131
Hepatitis C virus (HCV) 28, 33, 116, 118, 146, 153, 160–163, 181, 183, 188
  HCV genome organization 160
HEN1 22
Hepatocellular carcinoma 154
Hepcidin 162
HER2 207
Herpes simplex virus (HSV) 167–168, 183
  human herpesvirus 8 (HHV8) 189
Herpes simplex virus thymidine kinase (HSVtk) 133–134
Herpesvirus saimiri (HVS) 182
High-content screening (HCS) 80, 84, 94, 99, 101
High-throughput of RNA isolated by linking immunoprecipitation (HITS-CLIP) 7, 32
High-throughput screening (HTS) 56, 79, 84
Histone deacetylase 4 (HDAC4) 188
HIV 144, 183
Hodgkin's lymphoma (HL) 186
Human papillomavirus (HPV) 31, 167–168, 181, 183, 189
HTLV-1 183, 189
Human heat stress cognate 70 (Hsc70) 158
Human metapneumovirus 164
Human T-cell lymphotropic virus type 1 (HTLV-1) 181
Huntingtin (*HTT*) gene 143
Huntington's disease (HD) 146, 227, 230, 234, 235
Hybridization 212–213
Hydrodynamic injection 158

I

ICP0 190
ICP4.5 190
IFN response 159

IL-10 191
IL-8 191
Image analysis software
CellProfiler 89, 93
FIJI 89, 93
ICY 89, 93
OMERO 93
Immune evasion function of the *BART* miRNAs in NPC 185
Induced pluripotent stem cells (iPSCs) 28, 52, 54, 65, 69–72
Influenza 164–165
Integrin α5 6
Interferon response 154
Interferon-α 154–155, 160
Interleukins 35
IκB 190

J

JANUS liquid handling system 86
JFH1 161, 163
Junín virus (JUNV) 167

K

Kaposi's sarcoma virus (KSHV) 181, 183, 187, 189–193
KSHV encodes orthologues of host miRNAs 191
KSHV miRNAs and viral pathogenesis 191
KSHV miRNAs 189
KSHV modulates the expression profile and/or function of host miRNAs 192
Kaposi's sarcoma 189–192
KIT 204
KLf4 53
Kruppel-associated box (KRAB) 11, 134
K-ras 5
Kupffer cells 130

L

Lagrange multipliers 95
Lamivudine 154
Latent membrane proteins (LMP) 184, 186
LMP1 185
LMP1 191
LMP2A 185
Lentivirus vectors 59, 128, 134, 158, 163, 167
Let-7 5, 6, 10, 32, 65–66, 107, 111, 122, 186, 201, 204
Let-7a 69, 70
Let-7b 115
Let-7c 208
Let-7e 208
Let-7f 34
Long hairpin RNA (lhRNA) 36, 155, 162
Lin 28 5, 6, 53, 60, 238
Lin-4 201
Lipoplex 152, 157
Liver-specific promoter 128
LNA modifications 108, 109, 110, 115, 120, 121
Locked nucleic acids (LNA) 10, 146, 188, 213
LNA/DNA-zyme 106
Long non-coding (lnc) RNAs 6, 43, 44, 45, 46, 47, 48
Low-density lipoprotein receptor (LDLR) 162
Low-density lipoprotein receptor adaptor protein (LDLRAP1) 35
Lymphoid malignancies 209

M

Mammalian miRNA biogenesis pathway 3
MAP kinase interacting serine/threonine kinase 1 (MKNK1) 162
Marburg virus (MARV) 165–166
MBNL1 6
MCC 193
Measles virus (MV) 134, 164
Median absolute deviation (MAD) 96, 97
Membrane-translocating peptides 152
Merkel cell polyomavirus (MCV) 181, 183, 193
Methyl RNA immunoprecipitation 31
Microarray 80, 201
MicroRNA-regulated TetR-KRAB transcriptional control of transgene expression 135
MicroRNAs as cancer biomarkers 204
Mipomersen 146
miPSCs 69, 70
miRs
miR-1 6, 32, 188, 237
miR-10 33, 237
miR-10b 207
miR-100 185–186
miR-101 21, 186, 235
miR-106b 65
miR-106–125 cluster 65
miR-107 208, 237
miR-122 28, 33, 107, 110–111, 115–117, 132–133, 134–135, 144, 163, 188, 207
miR-1225 25
miR-1228 25
miR-123 230
miR-124 32, 132
miR-124 133
miR-124a 67, 210
miR-125a-5p 188
miR-125b 110, 208
miR-126 207
miR-127 134
miR-128 209
miR-1285 208
miR-130a 235
miR-130b 189
miR-132 119, 193
miR-133 135
miR-133b 230, 237
miR-135a 210
miR-137 230
miR-138 7, 36, 186
miR-141 185–186, 208
miR-142-3p 130, 135
miR-143 134, 208
miR-145 5, 208
miR-146a 189, 194
miR-15 239
miR-15a 203
miR-15a/16 cluster 203

miR-15a/16-1 cluster 204, 209
miR-150 210, 236
miR-153 230
miR-155 68–69, 113, 115, 119, 187, 191, 203, 207
miR-16 112, 117, 203, 239
miR-168a 35
miR-17-92 6
miR-17-92 cluster 56, 144, 188, 205, 209
miR-18a 186
miR-181a 210
miR-181b 209
miR-183 family 205
miR-19a 235
miR-191 209
miR-193a 205
miR-196 188
miR-199 188
miR-199a 207, 209
miR-199a-3p 188
miR-2 32
miR-20a 208
miR-200 family 10, 207
miR-200a 185
miR-200b 61
miR-203 122, 208
miR-204 132, 134
miR-206 237
miR-21 67, 109–113, 119, 193, 203, 207–208, 210
miR-210 110, 188, 207
miR-214 207
miR-218 186
miR-22 29
miR-221 186, 203, 208
miR-221/222 207
miR-221/hsa-miR-222 193
miR-222 210
miR-223 68, 188, 205
miR-23 143
miR-233 35
miR-246 33
miR-26a 185–186
miR-29 209–210, 230
miR-291 65
miR-291-5 cluster 60
miR-292 69
miR-294 65
miR-295 65
miR-29a 67, 208
miR-29a/b 186
miR-29b 237
miR-29c 185, 237
miR-30 143–144, 146
miR-30c 208
miR-302 68–69, 121
miR-302-367 cluster 56, 65
miR-302a 67
miR-302b 65
miR-31 134, 144, 193, 208
miR-320 26
miR-33 236–237
miR-331 208
miR-335 207, 237
miR-338 205
miR-34 4, 5, 10, 66–67, 119–120
miR-34a 122, 187, 189, 207
miR-34b 186
miR-34c 186
miR-367-371 cluster 56
miR-369-3 33
miR-371 69
miR-373 35
miR-375 110
miR-376 31
miR-381 236
miR-382 6
miR-427 121
miR-429 236
miR-430 121
miR-451 9, 24
miR-484 26
miR-498 35
miR-518b 210
miR-519a 59
miR-539 210
miR-550 210
miR-565 205
miR-574-3p 208
miR-622 208
miR-671 7, 36
mir-7 7, 237
miR-709 34
miR-9 230
miR-92a 236
miR-93 65, 189
miR-98 186
miR-H2–3p 189
miR-H6 189
miR-III 190
miR-K12-1 190
miR-K12-10 189
miR-K12-11 191
miR-K12-12 189
miR-K12-5 190
miR-K12-9-5p 190
miR-M4 191
miRNA-142 131
miRBase 227
miR-express 82
miRNA arrays 81
miRNA biomarkers associated with multiple cancer types 206
miRNA editing, ADAR 6, 29, 30, 31
miRNA editing, adenylation 28
miRNA editing, uridylation 28
miRNA, extraction 211
miRNA microarrays 229
miRNA off-set RNAs (moRNAs) 27,28
miRNA pathway analysis tools, DIANA miRpath 99
miRNA pathway analysis tools, Ingenuity Pathway Analysis 99
miRNA pathway analysis tools, PathVisio 99
miRNA response elements 6

miRNA star 2
miRNA, naked 64
miRNA, algorithms
- DIANA-microT 98
- Miranda 32, 98
- miRDB, 98
- miRecords 98
- TarBase 98
- Targetscan 32, 98

miRNA, inhibitors 83
miRNA, isomers 28, 29, 31, 235, 228
miRNA, libraries 83, 86, 87
miRNA, miRIDIAN miRNA library 86
miRNA, miRISC 20, 32
miRNA, miRNA degradation 112
miRNA, miRNA mimics 66, 83, 121, 122, 239
miRNA, miRNA sponges 35, 83, 105, 120
miRNA, miRNA-RISC complex 120
miRNA, off target effects 110, 115
miRNA, seed sequence 108
miRNA, target protector (TP) 106, 120, 121
miRNAs as diagnostic markers 204
miRNAs as indicators of treatment response 205
miRNAs as non-invasive cancer biomarkers 206
miRNAs as prognostic indicators 205
miRNome 80, 82, 100
MirTools 82
Mirtrons 2, 24, 25
miR-US25-1 (CMV) 32
Mixmers, 2′-MOE/2′-F mixmers 118
Mixmers, LNA/2′-MOE mixmers 109, 112, 113,
Mixmers, LNA/DNA mixmer 108, 115, 118
Mixmers, LNA/DNA/PS mixmer 118, 119, 120
Mixmers, partial PO/PS mixmers 118
Mixmers, PS/2′-MOE mixmers 118
Monoclonal gammopathy of undetermined significance (MGUS) 209
Mouse bone marrow-derived mononuclear cells (BM-MNC) 65
mRNA decay 21
Multicentric Castleman's disease (MCD) 189
Murine transthyretin (mTTR) promoter 134
Murine γ-herpesvirus 68 (MHV68) 27
Muscleblind-like1 (MBNL1) splicing factor 236, 238
Myeloid disorders 208
Myotonic dystrophy type 1 (DM1) 228
Myotonic dystrophy type 2 (DM2) 237

N

N6-methyladenosine (m6A) 31
N-acetylgalactosamine-conjugated melittin-like peptide (NAG-MLP) 155
Nanog 53
Nanoparticles 66, 72, 101
- nanoparticles, polylactic-co-glycolic acid (PLGA) 113, 119

Nasal NK/T cell lymphomas (NKTCL) 187
Nasopharyngeal carcinoma (NPC) 184, 185–186
Next-generation sequencing (NGS) 80, 81, 211–214, 229
- 454 Sequencing/Roche 81, 229
- Illumina 81, 229
- SOLiD 81, 229

NF kappaB (NFκB) 187, 189–190
NF1-A promoter 35
Niemann–Pick C1 protein (NPC1) 166
NOD/SCID mice 163
NODAL signalling pathway 121
Non-canonical pathways for miRNA biogenesis 182
Non-coding (nc)RNA 122, 216
Normalization 214
NS5A phosphokinase 162
Nucleophosmin-1 35
Nucleoside/nucleotide analogues 154–155

O

Occludin 162
Oct3 53
Oct4 53, 59
Oncolytic viruses 133

P

P bodies 111
p38/MAPK 187
p53 4, 5, 66, 67, 187, 209, 238
PACT 203
Parainfluenza virus 164
Parkinson's disease (PD) 229
PAZ 2
PEL 192
Peptide nucleic acid (PNA) 109, 110, 111, 113, 114, 115, 119
Phosphorodiamidate morpholino oligonucleotide (PMO) 109, 110, 113, 120
PI4-kinase (PI4K) III alpha 162
piRNA 2
Pol II 2, 19, 23, 43, 140, 144, 152, 158, 202
Pol III 8, 141–142, 144, 152, 158, 202
Polo-like kinase 1 (Plk1) 162
Polyethylene glycol (PEG) 157
Polyplex 152
Polypyrimidine track-binding protein (PTB) 162
Population-based incremental learning (PBIL) 94, 95
Post transcriptional gene silencing (PTGS) 34, 45
Precursor microRNA (Pre-miRNA) 25, 227, 237
Primary effusion lymphoma (PEL) 189
Principal component analysis (PCA) 94, 95
Proline-serine-threonine phosphatase-interacting protein 2 (PSTPIP2) 162
Properties of RNAi activators 141
Prostate cancer 208
PS gapmers 112
PS linkages, 108, 109, 117
Pseudogene of PTEN (PTENP1) 36
PTEN 45, 204

Q

Quantitative reverse transcription PCR (qRT-PCR) 212

R

Rabies virus (RABV) 167–168
RDE-1 4

Replication and transcription activator (RTA) 190
Replicon 161
Repressor element-1 (RE-1) silencing transcription factor (REST) 230
Respiratory syncytial virus (RSV) 163–164
Retinal pigment epithelium (RPE) 132
Retinoic acid-inducible gene I protein (RIG-I) 159
Retrovirus 128, 163
Ribavirin 160
Ribozyme 144
Rift Valley fever virus (RVFV) 167
RIP140 (receptor-interacting protein 140) 33
RIP-seq 34
RISC 7, 151–152, 154, 182, 203
RNA demethylase FTO 31
RNA helicase A (RHA) 162
RNA methyltransferase BCDIN3D 31
RNA quality 212
RNA sequencing (MeRIP-seq) 31
RNase P 8 182
RNase Z 182
RNA-seq 28
Robotics 86
Rotaviruses 167–168
RPE 133
RT-PCR 30, 80–82, 97, 110, 114–115, 122

S

Sall4 60
Scavenger receptor class B type 1 (SR-B1) 162
Seed 2, 145, 203
Sendai virus, vectors 53
Severe acute respiratory syndrome (SARS) coronavirus 164
Short hairpin RNAs (shRNA) 7, 9, 145, 151–153, 155, 162
Signalling lymphocyte activation molecule (SLAM) 164
Simtrons 25
SINEs 22
siRNA 9, 140, 145, 151–152, 155, 162
siRNA–SNALP 166
SIRT1 4, 5
Smad 5
Sno-derived RNAs (sdRNAs) 26
snoRNA 26, 182
snoRNA, ACA45 snoRNA 26
snoRNA, C/D box snoRNA (SNORD15A) 27
Spinobulbar muscular atrophy (SBMA) 227
Spinocerebellar ataxia (SCA-1, 2, 3, 6, 7, and 17) 227, 235–236
Splicing 25
SSEA3 marker 70
Support vector machines (SVM) 93, 95

T

TALENs 51, 72
TAR 182
Target analysis 214
Techniques used in cancer miRNA biomarker studies 211
Teratoma 55, 64
Tet operator DNA sequence (*tetO*) 134
Tet repressor protein (TetR) 11, 134
TetR-KRAB 135
TGF-β1 and -β2 147
Trinucleotide repeat (TNR) expansion diseases (TREDs) 227
tiRNAs 22, 24
Tissue-specific promoters 128
TRA-1-60 pluripotency marker 70
Transcriptional gene activation (TGA) 34
Transcriptional gene silencing (TGS) 33, 34, 45
TRBP 2, 20, 28, 182, 203
Trinucleotide-repeat-containing-gene-6 (TNRC6; also known as GW182) 111
tRNA promoter 8
tRNA 2, 182
- tRNA (Lys3) 158
- tRNAase X 27
- tRNA-derived RNA fragments (tRFs) 26
- tRNAse Z 8

Tudor staphylococcal nuclease (Tudor SN) 6, 30
Tumour necrosis factor-alpha (TNF-α) 33

U

U1 interference 159
U1 snRNA 9
U6 8, 10, 141–143
Uridine diphosphate glucoronyltransferase (UGT1A1) 128
USP18 163

V

VA1 RNA 28
Valproic acid 53, 64
Vault RNA (vRNA) 27, 182
Vesicular stomatitis virus 134
Viral haemorrhagic fevers 165
Viral immediate early gene 72 (IE72) 190
Viral latency maintenance 186
Viral miRNAs encoded by human tumour viruses 184
VLDL 162
vmiRNA#2 183

W

West Nile virus (WNV) 165–167, 183
Woodchuck hepatitis virus 155

X

Xrn1 4, 6

Y

Yellow fever virus (YFV) 165–167

www.ingramcontent.com/pod-product-compliance
Lightning Source LLC
LaVergne TN
LVHW071412210926
848486LV00004B/252